Concepts of Nuclear Physics

McGraw-Hill series in fundamentals of physics: an undergraduate textbook program

E. U. CONDON, EDITOR, UNIVERSITY OF COLORADO

INTRODUCTORY TEXTS

Young · Fundamentals of Mechanics and Heat
Kip · Fundamentals of Electricity and Magnetism
Young · Fundamentals of Optics and Modern Physics
Beiser · Concepts of Modern Physics

UPPER-DIVISION TEXTS

Beiser · Perspectives of Modern Physics
Cohen · Concepts of Nuclear Physics
Elmore and Heald · Physics of Waves
Kraut · Fundamentals of Mathematical Physics
Meyerhof · Elements of Nuclear Physics
Reif · Fundamentals of Statistical and Thermal Physics
Tralli and Pomilla · Atomic Theory: An Introduction to Wave Mechanics

Concepts of Nuclear Physics

Bernard L. Cohen
Professor of Physics
Director, Scaife Nuclear Physics Laboratory
University of Pittsburgh

McGraw-Hill Book Company
New York St. Louis San Francisco
Düsseldorf Johannesburg Kuala Lumpur
London Mexico Montreal
New Delhi Panama Rio de Janeiro
Singapore Sydney Toronto

Concepts of Nuclear Physics

Library of Congress Catalog Card Number 70-138856

07-011556-7

1 2 3 4 5 6 7 8 9 0 M A M M 7 9 8 7 6 5 4 3 2 1 0

This book was set in Bodoni Book and News Gothic by The Maple Press Company, and printed on permanent paper and bound by The Maple Press Company. The drawings were done by John Cordes, J. & R. Technical Services, Inc. The editors were Bradford Bayne and Barry Benjamin. Stuart Levine supervised production.

to my parents
Mollie Friedman Cohen
and
Samuel Cohen
on their seventy-fifth and eightieth birthdays
and their fiftieth wedding anniversary

Preface

The structure of nuclei is now about as well-understood as the electronic structure of atoms, but there is a tremendous difference in the extent to which this understanding has been diffused. Atomic structure is taught for the first time in the fourth or fifth grade of elementary school, and more advanced treatments are presented at least twice more before the end of secondary school. Modern-physics courses covering atomic structure are taken by nearly all scientists and engineers on an elementary level and by all physics majors at the senior level. A course on quantum mechanics, required of all physics graduate students, includes more details of atomic structure.

The structure of nuclei, on the other hand, is not taught at all in elementary and secondary schools and is essentially ignored even in the education of physics majors up to the advanced graduate level. Thus while atomic structure is familiar in outline to hundreds of millions of people, nuclear structure is not even familiar to many with Ph.D.s in physics.

Often an elective advanced undergraduate or first-year graduate course in nuclear physics is offered, but it usually devotes a large block of time to experimental aspects of the subject and gives highly phenomenological treatments of decay and reaction processes, heavily influenced by the historical order in which things were discovered. Only near the end does it present a short discussion of nuclear models, in which the impression is given that we are still digging in the dark in our efforts to understand nuclear structure. This would be the equivalent of teaching atomic physics by spending the majority of time on such subjects as the nature of light, optical spectroscopy, and atomic collisions, and only briefly near the end by presenting a brief discussion on models of atomic structure.

Atomic physics is taught in a much more logical way, and this book represents an effort to introduce similar logic into the teaching of an advanced undergraduate or first-year graduate course in nuclear physics. The only absolute prerequisite is an elementary course in modern physics such as the one usually taught as part of the elementary physics sequence. Whenever matters are discussed which would not be understandable to students with that preparation, this is clearly indicated in the text and these discussions can be omitted without loss of continuity. Quantum theory is widely used, but it is reviewed for these students in Chapter 2 with a further extension in Section 10-1. Every physicist has his own way of introducing quantum mechanics, and many may not like my approach. For this I can only apologize and encourage instructors to handle the subject in accordance with their own tastes.

The book originally developed out of courses for students with this minimal preparation given at the University of Pittsburgh during the fall terms of 1967 and 1968, although many sections were omitted or covered only briefly. On the other hand, concepts which can best be understood by the use of more advanced quantum-mechanical techniques are generally treated in that way for the benefit of advanced students. The book was used for a first-year graduate course in the spring of 1969. With all the advanced material included and nearly all the book being covered, there was no indication that the course was too easy.

To my colleagues working in nuclear physics, I would like to offer apologies for weighting the material covered heavily toward areas in which I have had research experience. I find it most difficult to write about subjects I do not thoroughly understand, and most subjects on which I have had no research experience fall into that category. I also want to apologize for so frequently using my own work in examples. This has the advantages that the results are readily available and well-understood, they are usually presented in a manner attuned to my tastes and my style of writing, and the original data are available for replotting or combining in different ways. (A special apology is in order for the use of our old data in Figure 13-4 when so much newer and better data are available, but after hours of searching, I could find nothing that gives coverage to the full mass range.) To avoid giving an unbalanced impression, I have not included authors' names on figures from our data, for which I apologize to my collaborators.

I am greatly indebted to Miss Barbara Ezarik for an outstanding job of typing, to Drs. F. Tabakin, N. Austern, E. Sanderson, R. M. Drisko, D. A. Bromley, and R. A. Sorenson for helpful discussions and suggestions, and to the students who suffered through the developmental stages of this material without a textbook to fall back on.

Bernard L. Cohen

Contents

Chapter 1

Introduction to the Nucleus

An atom consists of a small, massive core called the *nucleus*, surrounded by orbiting electrons. It is the purpose of this book to explain all aspects of the nucleus, its structure, its behavior under various conditions, and its effect on nature and on mankind. In this chapter we introduce some of its most basic characteristics, its mass, size, shape, and other externally observable properties. We also consider some deeper questions such as the force that holds the nucleus together and the mechanical laws that are in effect. We shall introduce more problems than we solve, but our purpose will be to lay out a framework for later discussions.

1-1 Mass, Charge, and Constituents of the Nucleus

Let us begin by reviewing a few fundamental facts that are probably already familiar. The nucleus is made up of neutrons and protons, two particles which are about 1,840 times more massive than electrons. They are spoken of collectively as *nucleons*. The number of protons in a nucleus is just equal to its atomic number Z, and the total number of nucleons A is the integer closest to its atomic weight; hence the number of neutrons is $A - Z$. Thus the nucleus of ${}_{11}\mathrm{Na}^{23}$, a sodium atom which has atomic number 11 and atomic weight very close to 23, contains 11 protons and 12 neutrons. This is a relatively light nucleus; a typical heavy nucleus is ${}_{79}\mathrm{Au}^{197}$, which obviously contains 79 protons and 118 neutrons. The mass of the nucleus is very nearly equal to the mass of the atom; in kilograms it is the atomic weight divided by Avogadro's number, 6.03×10^{26}.

The nucleus was first discovered in 1911 in experiments conducted by Lord Rutherford and his associates on scattering of alpha particles by atoms. He found that the scattering pattern could be explained if atoms consist of a small, massive, positively charged core surrounded by orbiting electrons. While most of his results could be calculated on the basis of an infinitely small nucleus, deviations indicated that the nuclear size is of the order of 10^{-14} m. Since this is 10,000 times smaller than the diameter of atoms, it is small enough to be negligible in practically all atomic problems. For studies of the nucleus itself, however, we must have more accurate size determinations.

1-2 Nuclear Size and the Distribution of Nucleons

The straightforward approach to studying the size and shape of nuclei is to shoot probing particles at them and measure the effects produced. There is, however one well-known limitation in this endeavor: the wavelength of the probing particles must be of the order of the size of the nuclei being studied or smaller. Since ordinary light, for example, has a wavelength of about 10^{-7} m, which is many orders of magnitude larger than the nuclear size, it is not suitable. Light of very short wavelength, i.e., gamma rays, is also unsuitable because nuclei always occur in nature surrounded by electrons and electromagnetic waves interact more strongly with these electrons than with the nucleus. It is therefore better to employ particles such as electrons, protons, and neutrons as probes, all three of which have been used. Neutrons and protons have the advantage that their wavelength is sufficiently short for energies of about 20 MeV,[1] whereas for electrons over 100 MeV of energy is required, which is much more difficult to obtain. However, electrons have the advantage that their interaction with the nucleus is very well known (it is the familiar electromagnetic interaction), so the most accurate results have been obtained with electrons as probes.

The experiments consist of shooting high-energy electrons at a thin target

[1] MeV is million electron volts, the unit of energy we shall generally use; 1 MeV = 10^6 eV = 1.6×10^{-13} joule (J).

FIGURE 1-1 Experimental arrangement for measuring the angular variation of electron scattering from nuclei. The angle θ is varied by moving the detector, and for each θ measurements are made of the ratio between the number of scattered electrons it detects and the number of electrons in the beam as determined by the collector. (Since very few electrons are deflected by large angles, practically all of the beam reaches the collector.) Typical results of these measurements are shown in Fig. 1-2. The detector is actually a very large and complex group of instruments capable of determining the energies of the electrons.

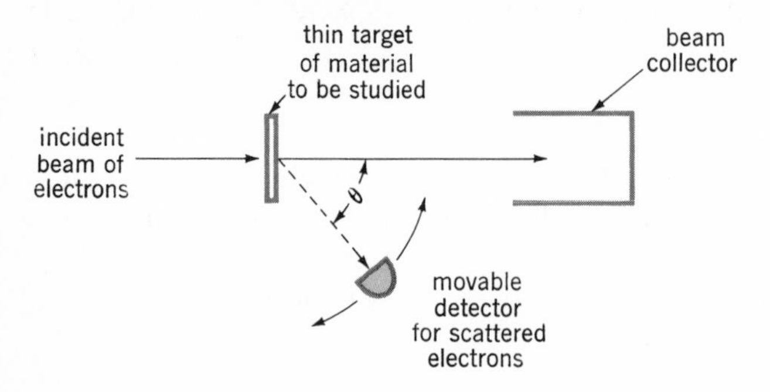

of the material under study and observing the probability of various angular deflections, as shown in Fig. 1-1. In concept, it is very similar to the Rutherford scattering experiments, in which the nucleus was first discovered. Some typical results of these measurements are shown in Fig. 1-2. If one assumes some density distribution $\rho(r)$ for the nucleons in the nucleus and assumes that the neutrons have the same density distribution as the protons, the probability of various

FIGURE 1-2 Angular distributions of 185-MeV electrons scattered from various nuclei. The curves through the data are theoretical fits. [*From B. Hahn, D. G. Ravenhall, and R. Hofstadter, Phys. Rev.,* **101:** 1131 (1956).]

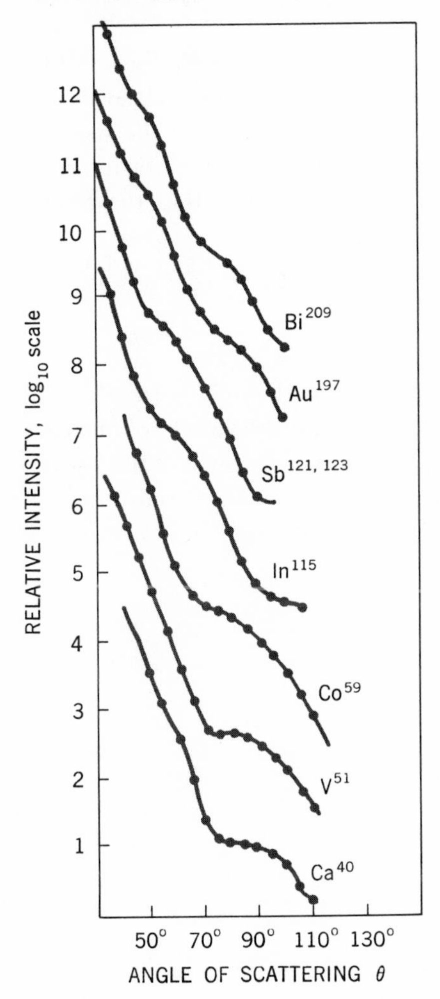

angular deflections can be calculated and compared with the experimental results. If they do not fit, other $\rho(r)$ can be tried until a fit is obtained.

The experiments have been performed and analyzed for a great many nuclei and at several incident electron energies. All the results can be approximately explained by a charge distribution given by

$$\rho(r) = \frac{\rho_0}{1 + \exp\left[(r - R)/a\right]} \tag{1-1}$$

A plot of (1-1) is shown in Fig. 1-3, where the physical significance of the various parameters is illustrated. We see there that ρ_0 is the nucleon density near the center of the nucleus, R is the radius at which the density has decreased by a factor of 2 below its central value, and a is a measure of the surface thickness such that the distance over which the density falls from 90 percent of ρ_0 to 10 percent of ρ_0 is $4.4a$.

The fits to the data obtained with (1-1) are shown by the solid lines in Fig. 1-2, and the density distributions determined by these fits are illustrated by the curves in Fig. 1-4. These give us the answer to our questions about the size and density distributions in nuclei, but it is interesting to see what systematic information about nuclei can be obtained from these fits. It turns out that the results for all nuclei are reasonably well approximated by (1-1) with

$$\begin{aligned} \rho_0 &\simeq 1.65 \times 10^{44} \text{ nucleons/m}^3 = 0.165 \text{ nucleons/F}^3 \\ R &\simeq 1.07 A^{1/3} \text{ F} \\ a &\simeq 0.55 \text{ F} \end{aligned} \tag{1-2}$$

Note that we use the fermi (abbreviated F), 10^{-15} m, as the unit of length.

These results are extremely simple; they indicate that the density of nucleons in the inner regions of all nuclei is about the same and that the surface thicknesses of all nuclei are very similar. These facts are readily discernible from Fig. 1-4. The $A^{1/3}$ variation of the *nuclear radius*, a name frequently used for R, is expected

FIGURE 1-3 Plot of Eq. (1-1) for $\rho(r)$ vs. r. The meaning of ρ_0, R, and a are illustrated.

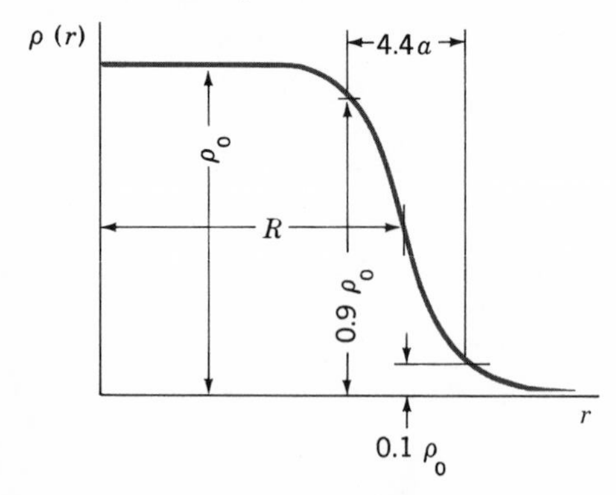

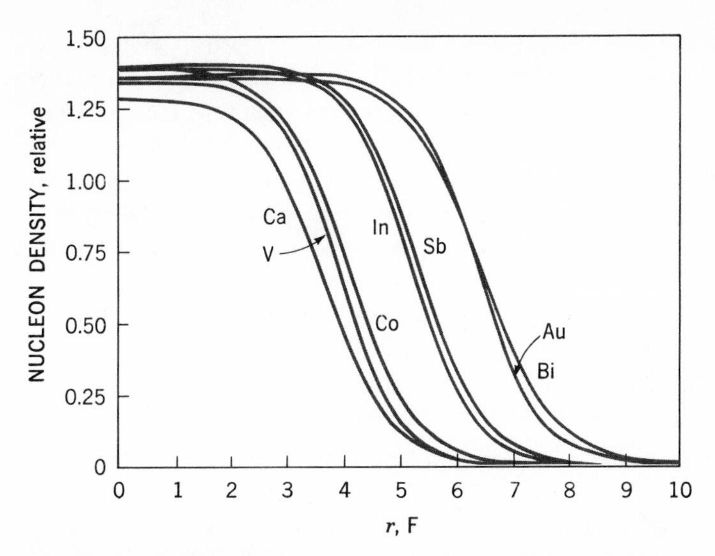

FIGURE 1-4 Nucleon density in various nuclei as obtained from the fits to the data shown in Fig. 1-2. [*From B. Hahn, D. G. Ravenhall, and R. Hofstadter, Phys. Rev.*, **101:** 1131 (1956).]

from the constancy of ρ_0 since this requires that the volume of the nucleus be proportional to A, and the volume is, of course, proportional to R^3.

1-3 Energies of Nucleons in the Nucleus

We shall eventually be treating the kinetic energies of nucleons in the nucleus in great detail, but at this stage we require an order-of-magnitude estimate. The energies of beta rays and gamma rays emitted from nuclei are typically of the order of 1 MeV, but these processes are transitions of nucleons from one state to another so their energies are differences between nucleon energies in two different states; the actual nucleon energy should be much larger.

One approach to the problem is to calculate the electrostatic energy E_c required to insert a proton into a nucleus. This is approximately

$$E_c = \frac{Ze^2}{4\pi\epsilon_0 R} \tag{1-3}$$

which for a medium-weight nucleus ($Z = 50$, $A = 120$) is

$$E_c = \frac{50(1.6 \times 10^{-19})^2 \text{ C}^2}{4\pi(8.9 \times 10^{-12}) \dfrac{\text{C}^2}{\text{N-m}^2} 1.07 \times 120^{1/3} \times 10^{-15} \text{ m}} \frac{1 \text{ J}}{1 \text{ N-m}} \frac{1 \text{ eV}}{1.6 \times 10^{-19} \text{ J}}$$

$$= 13 \times 10^6 \text{ eV} = 13 \text{ MeV} \tag{1-3a}$$

where C is the abbreviation for coulomb and N that for newton. This much coulomb energy would be released if the proton were allowed to come out of the nucleus, but still it does not ordinarily come out. This means that it is *bound* in the nucleus by even more energy.

From these simple arguments, we might guess that energies of nucleons in the nucleus are of the general order of 10 MeV. We shall see later that this is something of an underestimate, but it is the correct order of magnitude. Since the velocity of a 10-MeV nucleon is only about 15 percent of the speed of light, this means that relativistic effects are not important in considering the motion of nucleons in the nucleus. The nonrelativistic relations between mass, velocity, momentum, and kinetic energy may be used freely.

1-4 Is the Nucleus a Classical or a Quantum System?

The next interesting question is whether the wave nature of matter is relevant in a nucleus, as it is in atoms, or whether the nucleus is more like systems encountered in our everyday life, where classical mechanics is a sufficiently good approximation. As a general rule, the wave nature of matter is relevant where the wavelength of the particles is of the order of the size of the system, so let us compare them.

The wavelength of a nucleon with an energy of about 10 MeV is

$$\lambda = \frac{h}{Mv} = \frac{h}{\sqrt{2ME}}$$

$$= \frac{6.6 \times 10^{-34}\ \text{J-s} \times \dfrac{1\ \text{kg m}^2/\text{s}^2}{1\ \text{J}}}{\left(2 \times \dfrac{1\ \text{kg}}{6 \times 10^{26}} \times 10 \times 10^6\ \text{eV} \times 1.6 \times 10^{-19}\ \dfrac{\text{J}}{\text{eV}} \times \dfrac{1\ \text{kg m}^2/\text{s}^2}{1\ \text{J}}\right)^{1/2}}$$

$$= 9.3 \times 10^{-15}\ \text{m} = 9.3\ \text{F}$$

This is clearly of the order of the size of a nucleus as given by (1-2), so the wave nature of matter is indeed relevant. The motions of nucleons in the nucleus are governed by the laws of quantum physics; classical pictures in which nucleons are considered as little balls moving around—applied so successfully in describing gases or liquids—are of limited usefulness. We shall therefore have to use and expand our knowledge of the wave nature of matter. A review of this subject is presented in Chap. 2.

1-5 What Holds the Nucleus Together?

The next question we have to face is perhaps the most difficult if we have only our previous experience to go on: What holds the nucleons together in a nucleus?

Systems are held together by forces, and the only forces we have encountered in classical physics or in atomic physics have been the gravitational and electromagnetic forces. Can these do the job? The electromagnetic force most certainly cannot. Neutrons have no electric charge, so they do not experience the electromagnetic force at all,[1] and the principal electromagnetic force between protons is a strong coulomb repulsion, which tends to tear the nucleus apart. The gravitational force is an attractive one between every pair of nucleons, but it is smaller by a factor of about 10^{39} than the electrical force between two protons. Its effects are completely negligible in all nuclear and atomic phenomena.

Thus, the only two forces we have previously encountered cannot account for the existence of nuclei. The only explanation is to recognize that there is a third force in nature, known as the *nuclear force.* We see immediately that this force must be very strong at distances of the order of the nuclear size, since it must more than compensate the coulomb repulsion between protons. On the other hand, molecular structure can be accurately accounted for by the electromagnetic force alone, so we may conclude that at distances of the order of the spacing between nuclei in molecules ($\sim 10^{-10}$ m) the nuclear force must be negligible. It is therefore a *short-range* force, falling off more rapidly with distance than $1/r^2$.

Before we can proceed very far in studying the structure of the nucleus, we must learn more about the nuclear force. This will form the subject matter of Chap. 3.

1-6 Some Other Properties of Nuclei

We learned in elementary physics that if there are no external torques acting on a system, its angular momentum is conserved. Since an isolated nucleus is such a system, its angular momentum is one of its constant properties. Methods of measuring angular momenta of nuclei by use of atomic beams in Stern-Gerlach experiments and by studying the hyperfine structure of atomic spectral lines with and without applied magnetic fields are generally discussed in modern physics courses. Several other methods will be developed later in this book.

In quantum physics, conserved quantities are represented by quantum numbers. The quantum number for the total angular momentum of a nucleus is I; the two are related by

$$\text{Total angular momentum} = \sqrt{I(I+1)}\,\hbar \tag{1-4}$$

[1] Actually since, as we shall see in the next section, the neutron possesses a magnetic moment, it experiences a force in a nonuniform magnetic field, but this is too small to matter here.

where $\hbar$ is Planck's constant (6.25×10^{-34} J-s) divided by 2π. Values of I will be given and explained in many connections throughout this book. A compilation of directly measured values is given in Table A-2 of the Appendix.

In courses on electromagnetism it is shown that a current loop enclosing an area $\mathcal{A}$ and carrying a current i has a magnetic dipole moment μ given by

$$\mu = ir \tag{1-5}$$

For a circular orbit of radius r, traversed f times per second by a charge e moving with velocity v

$$i = ef = \frac{ev}{2\pi r}$$

$$\mathcal{A} = \pi r^2$$

whence, from (1-5),

$$\mu = \frac{e}{2} vr = \frac{e}{2M} L$$

where L is the angular momentum, Mvr. More generally

$$\mu = \frac{e}{2M} Lg \tag{1-6}$$

where g is a factor called the gyromagnetic ratio. In accordance with the above derivation, $g = 1$ when the charge and mass distributions coincide, as when a particle traverses an orbit. In quantum theory, L is a quantum number times $\hbar$.† For orbital motion with quantum number l, $g_l = 1$, whence, from (1-6),

$$\mu = \frac{e\hbar}{2M} l \tag{1-7}$$

The magnetic moment due to spin is a more complex problem which can be understood only in terms of relativistic quantum theory; the result for an electron, as is well known from atomic physics, is

$$\mu_e = \frac{e\hbar}{2M_e}$$

which, since the spin quantum number is $\frac{1}{2}$, corresponds to $g_s = 2$. Measurements corroborate this result. For nucleons, however, measurements give

$$\mu_p = 2.7925 \frac{e\hbar}{2M_p}$$
$$\mu_n = -1.9128 \frac{e\hbar}{2M_p} \tag{1-8}$$

which corresponds to g_s values equal to 2 times the numerical factors in (1-8). These results, and their contrast with the results for an electron, lead one to

† The statement given here is not quite accurate: it is the maximum component of L in any direction that is equal to $l\hbar$.

believe that an electron is a very simple elementary particle but that neutrons and protons are complex structures including nontrivial electric charge distributions. Efforts have been made to explain this structure in terms of charged *meson clouds* surrounding nucleons (see Sec. 3-9).

Methods of measuring magnetic moments of nuclei are usually discussed in modern-physics courses in connection with Stern-Gerlach atomic-beam experiments and the hyperfine structure of atomic spectral lines in the Zeeman effect. Other methods involve microwave spectroscopy, nuclear magnetic resonance in solids and liquids, molecular band spectroscopy, etc.[1] Measured values of magnetic moments are listed in Table A-2 of the Appendix. In all cases, nuclei with $I = 0$ have $\mu = 0$; this can be shown to be a general quantum-mechanical result.

The magnetic moment of a complex nucleus is the vector sum of contributions from the spins and orbital motion of its component nucleons, each of whose contributions are given by (1-7) and (1-8).[2] An understanding of this vector sum clearly requires an understanding of the detailed structure of nuclei. This problem will be discussed in Sec. 7-4.

In most books on electromagnetism,[3] it is shown that any distribution of electric charge produces an electrical potential which, at a large distance R in the z direction, can be expanded as

$$V = \frac{1}{4\pi\epsilon_0}\left[\frac{1}{R}\int \rho\, dV + \frac{1}{R^2}\int \rho z\, dV + \frac{1}{R^3}\int \rho(3z^2 - r^2)\, dV + \cdots\right] \tag{1-9}$$

where ρ is the charge density and the integral is over the region containing the electric charge. Because of the increasing powers of R in the denominator, this is a rapidly converging series. The integral in the first term is the net charge; for very large R, only this term is important, and the potential is the same as if all the charge were located at a point. The integral in the second term is called the *dipole moment*, the integral in the third term is called the *quadrupole moment*, etc.[4] It will be shown in Sec. 7-4 that a nucleus must have a zero electric dipole moment, so the lowest-order deviation from the field due to a point charge arises from the quadrupole moment.

[1] Brief descriptions of these methods for determining magnetic moments and electric quadrupole moments are given in H. A. Enge, "Introduction to Nuclear Physics," Addison-Wesley, Reading, Mass., 1966, and in *Nucl. Data*, 5: 443 (1969). More complete descriptions are referred to in the Further Reading list at the end of this chapter.

[2] In addition there are small contributions from meson exchange currents, which we ignore here.

[3] For example, W. M. Schwarz, "Intermediate Electromagnetic Theory," Wiley, New York, 1964.

[4] These are not the most general definitions of the dipole and quadrupole moments, as they depend on the choice of the coordinate axes.

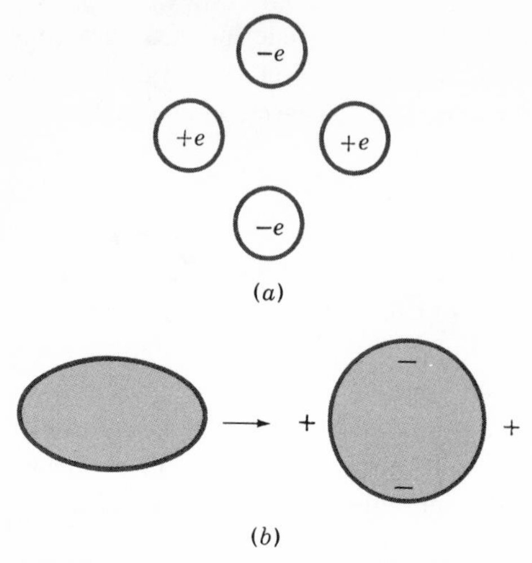

FIGURE 1-5 Two examples of systems having an electric quadrupole moment: *(a)* four point charges located as shown; *(b)* a uniform positively charged distribution of ellipsoidal shape (at the left) which is electrically equivalent to a uniform positively charged sphere plus extra positive and negative charges as shown at the right.

Two examples of charge distributions containing nonzero electric quadrupole moments are shown in Fig. 1-5. The example in part (*a*) is the simplest case, and the fact that it consists of four charges is responsible for the name; note that its net charge and dipole moment are zero, whence the entire field is produced by the quadrupole moment. The example in part (*b*), a uniformly charged ellipsoid, is more like what a nucleus might be; it has both a net charge and a quadrupole moment, but its dipole moment is zero. Its charge distribution can be well approximated, as shown at the right in the figure, by a spherical charge distribution plus an elementary quadrupole.

There is a well-known theorem in electrostatics which states that the electric field due to a uniformly charged shell at all points outside of it is the same as if all the charge were concentrated at the center of the shell. From this theorem it is clear that the field due to any nucleus with a spherically symmetric charge distribution is the same as if all the charge were located at the center of the nucleus. This field can then be represented by the first term only of (1-9), so we may conclude that a nucleus with a spherically symmetric charge distribution has no electric quadrupole moment or higher electric moments. This can, of course, also be easily derived from the expressions in (1-9).

Electric quadrupole moments of nuclei can be determined from hyperfine splitting of atomic spectral lines, microwave absorption spectroscopy, molecular beam resonance methods, etc.[1] A summary of the results is given in Table A-2 of the Appendix. The explanation of electric quadrupole moments from the nuclear structure standpoint is discussed in Sec. 7-5.

Problems

1-1 Calculate the mass of an Fe^{56} nucleus in kilograms.

1-2 From (1-1) and (1-2) calculate and plot the nucleon density distribution in Pb^{208}.

1-3 In the approximation that the nucleus has constant density ρ_1 for $r < R$ and zero density for $r > R$, calculate ρ_1 if R is given by (1-2); compare with ρ_0 from (1-2).

1-4 From the uncertainty principle $\Delta p\, \Delta x \simeq \hbar$ and the fact that a nucleon is confined within the nucleus, what can be concluded about the energies of nucleons in a nucleus?

1-5 Calculate the ratio of the coulomb and gravitational forces between two protons. How does this ratio vary with the distance between them?

1-6 If the total angular-momentum quantum number of a nucleus with $A = 100$ is $I = 1$, and if it is due to a rotation of the nucleus as a rigid body, approximately how many rotations per second would it make and how much energy would be involved in this rotation according to classical mechanics?

1-7 If a proton with $l = 1$ traverses an orbit with radius equal to the radius of a nucleus with $A = 25$, what is the electric current and the magnetic moment due to this motion?

1-8 If the charges shown in Fig. 1-5*a* are due to single electrons and protons and they lie on the surface of a nucleus with $A = 100$, what is the quadrupole moment in units of electron charge–(fermi)^2?

1-9 Show that the electric dipole moment of the charge distribution in Fig. 1-5*b* is zero.

1-10 If the surface of the nucleus shown in Fig. 1-5*b* has the equation

$$x^2 + y^2 + 1.2z^2 = R^2$$

[1] See footnote 1, page 9.

where R is the radius of a nucleus with $A = 200$, calculate its quadrupole moment. Assume that its total charge Ze is uniformly distributed through the volume.

Further Reading

See General References, following the Appendix.

Elton, L. R. B.: "Nuclear Sizes," Oxford University Press, London, 1961.

Hofstadter, R.: "Electron Scattering and Nuclear and Nucleon Structure," Benjamin, New York, 1963.

Kopferman, H.: "Nuclear Moments," Academic, New York, 1958.

Ramsey, N. F.: "Nuclear Moments," Wiley, New York, 1953.

Chapter 2

Quantum Theory of a Particle in a Potential Well

In Sec. 1-4 we found that the wave nature of matter is relevant in considering the motions of neutrons and protons in nuclei, i.e., that the nucleus is a quantum-mechanical system. Therefore we cannot proceed very far in our discussion without using quantum concepts. Most students have already been introduced to these ideas in a modern-physics course. However, since the content of such courses varies considerably, we devote this chapter to a review of these concepts. It includes nearly all the quantum mechanics needed for this book.

We concentrate here on the problem of a particle in a potential well. Since a potential represents a force, this corresponds to the motion of a particle under the influence of a force. The force exerted by one nucleon on another and the average force exerted on one nucleon by all the other nucleons in the nucleus are examples of situations in which this treatment is applicable. These, as we shall see, are among the central problems of nuclear physics.

In general, the forces are complicated functions of coordinates, so the potential well is a three-dimensional one of complex shape. However, the most important concepts can be understood in terms of a one-dimensional well of rectilinear shape, popularly known as a *square well*. The mathematics is greatly simplified in this situation, so we start by using it; in fact, in Sec. 2-1 we even introduce an additional approximation to simplify the treatment further. This simplification and then the rectilinear shape are dropped in Sec. 2-2, the problem is generalized to three dimensions with some simplifying approximations in Sec. 2-3, and finally the results of an accurate treatment are presented in Sec. 2-4. The usefulness of this slow, multistage development transcends the avoidance of mathematical complexity. The results of Sec. 2-1 and of the elementary aspects of Sec. 2-3 will serve the student well in economy of thought. Many sophisticated research physicists think in terms of these elementary solutions and consider everything else to be minor modifications of them. Other simplifications are introduced for economy of thought in Secs. 2-5 and 2-6.

2-1 Particle in a One-dimensional Square Well—Simplified Treatment

When we say that matter has a wave nature, we mean that a particle of matter is associated with a *wave funtion* $\psi(x,y,z)$, which behaves like a wave as a function of position. The physical interpretation of ψ is that $\psi^2\,dx\,dy\,dz$ is the probability that the particle will be found within the interval x to $x + dx$, y to $y + dy$, z to $z + dz$. The wavelength associated with ψ for a particle of mass M and velocity v is

$$\lambda = \frac{h}{Mv} = \frac{h}{\sqrt{2M(E - V)}} \tag{2-1}$$

where we have used the familiar expression for the total energy, $E = \frac{1}{2}Mv^2 + V$, with V representing the potential energy.

These ideas are most easily applied to a particle in a one-dimensional square potential well (Fig. 2-1), and the most interesting cases are when $E < 0$, so that the particle is *bound*, i.e., not free to escape from the well. Based on the above ideas, the simplest assumption would be

$$\psi = \begin{cases} A \cos kx \text{ or } A \sin kx & \dfrac{-L}{2} < x < \dfrac{L}{2} \\ 0 & \text{elsewhere} \end{cases} \tag{2-2}$$

where, in accordance with (2-1),

$$k = \frac{2\pi}{\lambda} = \frac{\sqrt{2M(E + V_0)}}{\hbar} \tag{2-3}$$

FIGURE 2-1 One-dimensional square well.

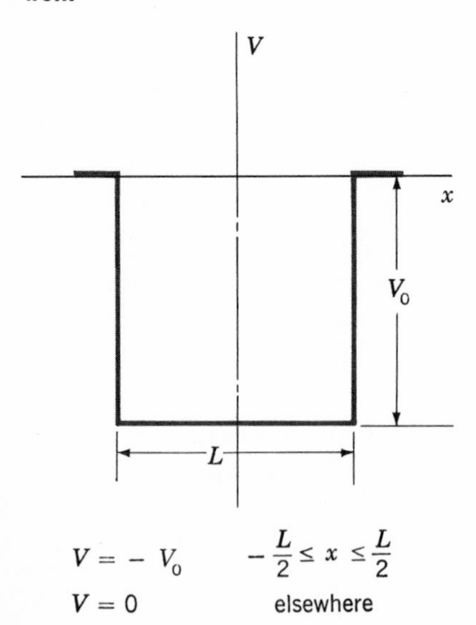

This has the classically expected property that the particle is confined to the inside of the wall. Since ψ is related to the probability, it should be a continuous and single-valued function of x, which leads to the requirement

$$\psi = 0 \qquad \text{at } x = \pm \frac{L}{2}$$

This requirement can be satisfied only if

$$kL = n\pi \tag{2-4}$$

where n is an integer. Equating the values of k from (2-3) and (2-4), we find

$$\frac{2\pi \sqrt{2M(E + V_0)}}{h} = \frac{n\pi}{L}$$

which simplifies to

$$E = V_0 + \frac{n^2h^2}{8ML^2} \qquad n = 1, 2, \ldots \tag{2-5}$$

This shows that the *total energy is quantized* to values obtained from (2-5) for integer values of n; n is therefore known as a *quantum number*. The value of E corresponding to a given n is designated E_n, and the corresponding wave function is designated ψ_n. From (2-2) and (2-4), the ψ_n are

$$\psi_n = \begin{cases} A_n \cos \dfrac{n\pi x}{L} & n = 1, 3, 5 \ldots \\ A_n \sin \dfrac{n\pi x}{L} & n = 2, 4, 6 \ldots \end{cases} \tag{2-2a}$$

These functions are shown as dashed lines in Fig. 2-2.

2-2 Particle in a One-dimensional Potential Well—Accurate Treatment

A more sophisticated and accurate treatment of this problem puts the requirement that ψ be a wave of wavelength given by (2-1) into the form of the differential equation

$$\frac{d^2\psi}{dx^2} + k^2\psi = 0 \tag{2-6}$$

which is easily recognized to have the solutions (2-2) inside the well. Outside the well, however, (2-6) also has solutions, namely,

$$\psi = Be^{\pm\kappa x}$$

where

$$\kappa = \frac{\sqrt{-2ME}}{\hbar} \tag{2-7}$$

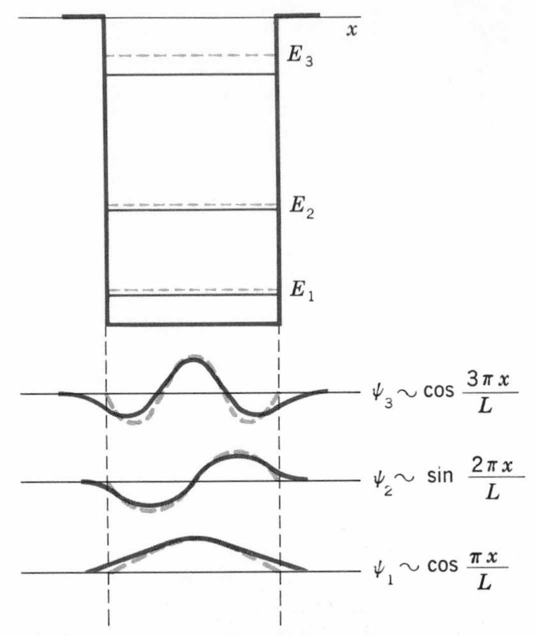

FIGURE 2-2 Energies and wave functions for a particle in the one-dimensional square well shown in the upper figure. Dashed lines are the solutions of the simplified treatment of Sec. 2-1, with energies given by (2-5) and wave functions given by (2-2*a*). Solid lines are the solutions from the accurate treatment of Sec. 2-2, with energies obtained from the solution of the first of (2-10) and wave functions given by (2-9) with the last two equations of (2-10).

(note that E is negative). Acceptance of these solutions goes beyond our simple picture of matter having a wave nature, but it turns out to be correct. Equation (2-6), with the substitution of (2-3) and (2-1), can be written

$$\frac{-\hbar^2}{2M}\frac{d^2\psi}{dx^2} + V\psi = E\psi \tag{2-8}$$

and is known as the time-independent Schrödinger equation in one dimension. In addition, quantum theory imposes the requirements that ψ and $d\psi/dx$ be finite and continuous everywhere. The requirement on ψ is natural in view of the physical interpretation of ψ^2 as a probability, and the requirement on $d\psi/dx$ arises from a relationship between that quantity and the velocity in more advanced treatments.

In our problem, then, some of the solutions are

$$\psi = \begin{cases} A \sin kx & -\dfrac{L}{2} < x < \dfrac{L}{2} \\ Be^{-\kappa x} & x > \dfrac{L}{2} \\ Ce^{\kappa x} & x < -\dfrac{L}{2} \end{cases} \tag{2-9}$$

Note that the solutions $e^{\kappa x}$ for $x > L/2$ and $e^{-\kappa x}$ for $x < L/2$ are physically unacceptable since they become infinite. The requirement on the continuity of ψ at $x = \pm L/2$ leads to

$$A \sin k \frac{L}{2} = Be^{-\kappa L/2} = -Ce^{-\kappa L/2}$$

and the requirement of the continuity of $d\psi/dx$ at $x = \pm L/2$ gives

$$kA \cos k \frac{L}{2} = -\kappa Be^{-\kappa L/2} = \kappa Ce^{-\kappa L/2}$$

Solving these simultaneously, we find

$$\begin{aligned} \cot \frac{kL}{2} &= -\frac{\kappa}{k} \\ C &= -B \\ B &= A \sin \frac{kL}{2} e^{\kappa L/2} \end{aligned} \tag{2-10}$$

The first of (2-10) can be satisfied only by certain values of E, so again energy is quantized.

Another set of solutions based on $\psi = A \cos kx$ inside the well can be obtained by analogous methods. The wave functions and energies for several states are shown by the solid lines in Fig. 2-2. It is readily seen that they correspond closely with the solutions by the approximate method of the last section except that since ψ need not reach quite to zero at $x = \pm L/2$, the "wavelengths" are slightly longer and consequently from (2-1) the energies are slightly lower. Because of this close correspondence, it is often convenient to think of the actual solutions in terms of the approximate ones. For example, the wave function ψ_3 may be referred to as the *one-and-a-half-wavelength solution.*

Once ψ is known, the probability $p(x)\,dx$, that the particle will be between x and $x + dx$ can be calculated as

$$p(x)\,dx = \psi^2\,dx \tag{2-11}$$

The constant A is determined by requiring that the total probability for the particle to be at some value of x is unity, whence

$$\int_{-\infty}^{\infty} \psi^2 \, dx = 1 \tag{2-12}$$

Next we consider a situation where $V(x)$ is not as simple as in Fig. 2-1 but is a more general function of x, as in Fig. 2-3. This greatly complicates the mathematics in solving (2-8), and in general the solutions cannot be given in closed form; but qualitatively, they differ from those for a similar square well in that the wavelength varies with x in accordance with (2-1). The comparison is shown in Fig. 2-3 for the lowest-energy state E_1, where the square-well wave function is shown by dashed lines. The solid curve has a shorter wavelength near $x = 0$, where $E - V$ is larger, and a longer wavelength at larger values of $|x|$, where $E - V$ is smaller. (When we say a portion of a curve has a certain wavelength, we refer crudely to

FIGURE 2-3 A one-dimensional well of complicated shape *(top)* and the wave function for the lowest energy state in it. A similar square well and the wave function for the corresponding state in it are shown by the dashed lines. The difference between the two wave functions is exaggerated.

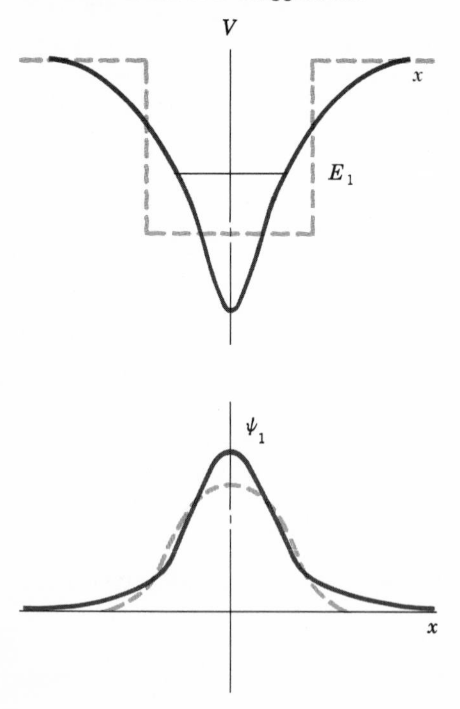

the wavelength of a sine wave approximately fitted to it.) From (2-8), ψ changes from a "wave" (curving toward the x axis) to an "exponential" (curving away from the x axis) at the value of x where E becomes less than V, making $E - V$ negative. The exponential falls off less rapidly for the solid curve since $|E - V|$ is smaller than for the square well.

2-3 Particle in a Three-dimensional Potential Well—Simplified Treatment

While the one-dimensional problem is very instructive, actual physical systems such as an atom or a nucleus are three-dimensional. In a three-dimensional problem the wave function must be a wave in each of the three dimensions. Since systems like atoms and nuclei have basically spherical rather than rectangular symmetry, it is profitable to use spherical coordinates, so we can say crudely that ψ must be a wave in the r direction, a wave in the θ direction, and a wave in the ϕ direction. Let us first concentrate on the θ direction.

The wave in the θ direction must consist of an integral number of wavelengths around a cycle, or else it will cancel itself in successive cycles. This requires that

$$2\pi r = l\lambda_\theta = \frac{lh}{Mv_\theta} = \frac{lh}{M\omega r} \tag{2-13}$$

where l is an integer and $\omega = d\theta/dt$. Equation (2-13) gives an especially interesting result for the orbital angular momentum L, which is familiar in classical mechanics as $L = \mathcal{I}\omega = Mr^2\omega$, where $\mathcal{I}$ is the moment of inertia. Inserting (2-13), we find

$$L = Mr^2\omega = \frac{lh}{2\pi} = l\hbar$$

Thus the requirement that the wave function be a wave of wavelength h/Mv_θ in the θ direction leads to the result that *angular momentum is quantized.* The accurate result obtained from solving the Schrödinger equation is

$$L = \sqrt{l(l+1)}\,\hbar \tag{2-14}$$

The θ dependence of the wave functions in this simplified picture is

$$\psi_\theta \propto \cos l\theta \text{ or } \sin l\theta \tag{2-15}$$

Now let us focus our attention on the fact that the r dependence of the wave function ψ_r must be a wave in the r direction. Actually it turns out that a more convenient function than ψ_r is $u(r)$, defined as

$$u(r) = r\psi_r \tag{2-16}$$

This has the property that $p(r)\,dr$, the probability of the particle's being between r and $r + dr$, is

$$p(r)\,dr \propto \int_{\theta,\phi} \psi^2 r^2 \sin\theta \, dr\, d\theta\, d\phi$$
$$\propto u^2\, dr$$

which is analogous to (2-11). It is $u(r)$, it turns out, that must be a wave.

If the particle is in a potential well $V(r)$, the wavelength of the function $u(r)$ is as given by (2-1), but if we want to concentrate our attention on the r direction alone, we must take into account the fact that the motion in the θ direction gives an effective force in the r direction, namely, the centrifugal force, $F_{cf} = M\omega^2 r$. From the definition of L and (2-14) this is

$$F_{cf} = M\omega^2 r = \frac{L^2}{Mr^3} = \frac{l(l+1)\hbar^2}{Mr^3}$$

which can be expressed in potential form as

$$V_{cf} = \int_\infty^r F_{cf}\, dr = \frac{l(l+1)\hbar^2}{2Mr^2} \qquad \text{(2-17)}$$

The complete potential to be used in calculating the wavelength from (2-1) is the sum of the potential arising from forces, $V(r)$, plus V_{cf}, whence the differential equation for $u(r)$ is, in analogy with (2-8),

$$-\frac{\hbar^2}{2M}\frac{d^2u}{dr^2} + \left[V(r) + \frac{l(l+1)\hbar^2}{2Mr^2}\right]u = Eu \qquad \text{(2-18)}$$

The problem of calculating $u(r)$ is now reduced to a one-dimensional problem like those treated in the previous sections. However, we see that the differential equation must be solved separately for each l. It is conventional to use the spectroscopic notation familiar from atomic physics, in which values of l are designated by letters, as shown in Table 2-1.

The simplest solution of (2-18) is for the s states ($l = 0$), since the centrifugal force term then vanishes. If $V(r)$ is taken to be a square well,

$$V(r) = \begin{cases} -V_0 & r < R \\ 0 & r > R \end{cases}$$

the problem becomes very similar to the one solved in Sec. 2-2. There are two differences, however: (1) r never becomes negative, so we have no region cor-

TABLE 2-1: SPECTROSCOPIC NOTATION FOR l VALUES

l	0	1	2	3	4	5	6	7	8
Spectroscopic notation	s	p	d	f	g	h	i	j	k

responding to the third of (2-9); and (2) we cannot have cosine solutions because they would make ψ infinite at $r = 0$ from (2-16). The only acceptable solutions are therefore

$$u(r) = \begin{cases} A \sin kr & r < R \\ Be^{-\kappa r} & r > R \end{cases}$$

where k and κ are defined in (2-3) and (2-7). From the conditions that the wave function and its derivative must be continuous everywhere, we obtain as boundary conditions at $r = R$

$$A \sin kR = Be^{-\kappa R}$$
$$kA \cos kR = -\kappa Be^{-\kappa R}$$

Solving these simultaneously gives

$$\cot kR = -\frac{\kappa}{k} \qquad B = Ae^{\kappa R} \sin kR \tag{2-19}$$

As was the case in connection with (2-10), the first of (2-19) can be satisfied only for certain values of the energy E, so again we find that energy is quantized. The constant A can again be evaluated by the analog of (2-12) as

$$A^2 \int_0^R \sin^2 kr\, dr + B^2 \int_R^\infty e^{-2\kappa r}\, dr = 1$$

When these integrals are evaluated and the second of (2-19) is inserted, the result is

$$A^2 = \frac{2\kappa}{1 + \kappa R} \qquad B^2 = \frac{2\kappa (\sin^2 kR)e^{2\kappa R}}{1 + \kappa R} \tag{2-19a}$$

The first column of Fig. 2-4 shows the effective potential well for $l = 0$ [just $V(r)$], the energies of the states (labeled E_{1s}, E_{2s}, . . .) and their corresponding wave functions, u_{1s}, u_{2s}, ... It is readily seen that they correspond closely to the solutions for the one-dimensional case shown in Fig. 2-2. We shall refer to the energies and wave functions as E_{nl} and ψ_{nl}, respectively, and call n a quantum number. Its physical significance is that E_{nl} is the nth lowest energy for orbital angular momentum l; we also see from Fig. 2-4 that u_{nl} goes through zero n times counting the one at $r = 0$ (but not counting the one at $r = \infty$).

For $l \neq 0$, the solution of (2-18) is more complicated. The two potential-energy terms, $V(r)$ and V_{cf}, are plotted as dashed curves in the top row of Fig. 2-4, and their sum, the total effective potentials, are shown as solid lines. These potentials are no longer square wells; to find the wave functions we must employ the ideas discussed in the last part of Sec. 2-2. As an aid in drawing them, "equiv-

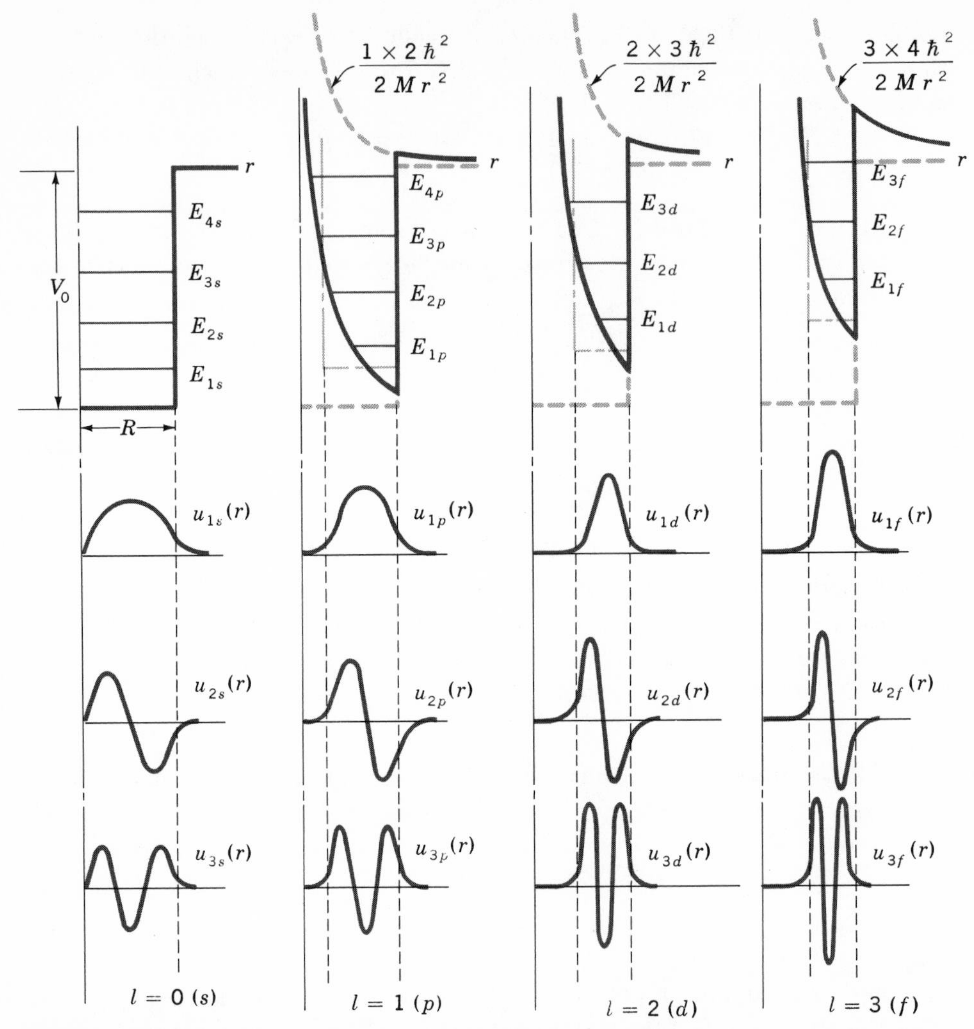

FIGURE 2-4 Energies and wave functions obtained from the solution of (2-18) for a square well. The four columns are the solutions for $l = 0, 1, 2$, and 3, respectively. The top row shows the two contributors to the effective potential, $V(r)$ and V_{cf}, as dashed lines and their sum, the effective potential, as solid lines. Square wells similar to these effective potentials are shown as dot-dash lines; they were used in drawing the wave functions which are plotted below. Note that the wave functions have n half wavelengths in the range of r covered by the potential well with exponential tails going rapidly to zero in the region outside the potential well. The energies of the states, shown in the top row, are moved upward as the well becomes shallower and narrower. The energies shown are not quantitatively correct for a square well.

alent'' square wells have been sketched in as dot-dash lines in the top row of Fig. 2-4. These equivalent square wells are not clearly defined (they are just square wells which roughly approximate the actual potential wells shown by the solid lines), but they will help in our discussion much as the square well in Fig. 2-3 was helpful.

It is immediately evident that as l increases, the potential well becomes both narrower and shallower. The narrowing arises from the fact that a particle with high orbital angular momentum is strongly repelled from small radii by the centrifugal force. As the well becomes narrower, the wavelength of a wave which fits a given number of half wavelengths into the width of the well becomes smaller, and, in accordance with (2-1), a shorter wavelength leads to a higher value of $E - V$. Thus, the distance from the bottom of the equivalent square well to any given energy level, say E_{1l}, increases with l. In addition, the bottom of the well rises with increasing l, thereby pushing the energy levels up even higher. For both these reasons, E_{1l}, E_{2l}, E_{3l}, etc., increase monotonically with increasing l. The wave functions are simultaneously squeezed toward larger radii.

For wells other than a square well, everything happens in a qualitatively similar way. For the $1/r$ well encountered in the hydrogen atom, E_{1p} is pushed up so far that it coincides in energy with E_{2s}. Similarly E_{1d}, E_{2p}, and E_{3s} coincide in energy; etc. This may be expressed as

$$E_{nl} = E_{(n+1)(l-1)} = \cdots = E_{(n+l)0}$$

The n quantum number commonly used in atomic physics is $n + l$ in our notation. This coincidence in energies is valid only for the $1/r$ potential; the potentials applicable in nuclear physics do not deviate from a square well nearly as dramatically as the $1/r$ potential does, so the shifting of E_{nl} with increasing l is much less rapid.

Before closing our discussion of the solutions of (2-18), it is interesting to point out that there are not necessarily *any* bound states of a system even though the forces are attractive. For a state to be bound, E must be less than zero, whence, from (2-3), the maximum value of k is

$$k_{\text{max}} = \frac{\sqrt{2MV_0}}{\hbar}$$

In order to satisfy the first equation of (2-19), the left side must be negative, which requires that $k_{\text{max}}R$ be greater than $\pi/2$. With the above expression for k_{max}, the condition that there be at least one bound state becomes

$$\sqrt{2MV_0}\,\frac{R}{\hbar} > \frac{\pi}{2}$$

which simplifies to

$$V_0 R^2 > \frac{\pi^2 \hbar^2}{8M} \tag{2-20}$$

When this condition is not fulfilled, the system has no bound states even though the forces are attractive.

Up to this point we have made use of the fact that ψ must be a wave in both the θ and r directions, and this has led us to two quantum numbers, l and n, respectively. We have still to consider the consequences of the fact that ψ is a wave in the ϕ direction. Clearly this wave, like that in the θ direction, must consist of an integral number of wavelengths around a cycle to avoid cancellation on successive cycles. The applicable radius for ϕ motion is $r \sin \theta$, so this condition requires

$$2\pi r \sin \theta = m_l \lambda_\phi = m_l \frac{h}{Mv_\phi} = \frac{m_l h}{M\omega_\phi r \sin \theta}$$

where m_l is an integer and ω_ϕ is the ϕ component of angular velocity. In classical mechanics, the z component of orbital angular momentum L_z is

$$L_z = Mr^2 \sin^2 \theta \, \omega_\phi$$

Combining these two equations gives

$$L_z = \frac{m_l h}{2\pi} = m_l \hbar \tag{2-21}$$

We see that L_z is quantized to integer values of $\hbar$, giving rise to a third quantum number, m_l. Since the z component of any vector must not be larger than the vector itself, $L_z < L$; whence from comparing (2-14) with (2-21) we find

$$|m_l| \leq l \tag{2-22}$$

The allowed values of m_l are therefore $-l, -l+1, \ldots, 0, \ldots, +l$, a total of $2l+1$ values. Under ordinary circumstances, there is no reason why the energy of a system should depend on its orientation in space, so the energy is independent of the quantum number m_l. Hence the energy-level diagrams in Fig. 2-4 are still correct and complete if we realize that each level is actually $2l+1$ separate levels corresponding to the various allowed values of m_l.

2-4 Particle in a Three-dimensional Potential Well—Accurate Treatment

We have said that the wave nature of matter has its mathematical manifestation in terms of the differential equation (2-8). In fact, experience has shown that where there is a discrepancy between ideas derived from the wave nature of matter and solutions of (2-8), the latter are to be considered correct. These solutions therefore constitute the accurate treatment of all problems.

The three-dimensional form of (2-8) is clearly

$$-\frac{\hbar^2}{2M}\left(\frac{\partial^2\psi}{\partial x^2}+\frac{\partial^2\psi}{\partial y^2}+\frac{\partial^2\psi}{\partial z^2}\right)+V\psi = E\psi \tag{2-23}$$

which is known as the *time-independent Schrödinger equation.* In polar coordinates, it becomes

$$-\frac{\hbar^2}{2M}\left[\frac{1}{r^2}\frac{\partial}{\partial r}\left(r^2\frac{\partial\psi}{\partial r}\right)+\frac{1}{r^2\sin\theta}\frac{\partial}{\partial\theta}\left(\sin\theta\frac{\partial\psi}{\partial\theta}\right)+\frac{1}{r^2\sin\theta}\frac{\partial^2\psi}{\partial\phi^2}\right]+V\psi = E\psi \tag{2-24}$$

Methods of solving (2-24) are discussed in any textbook on advanced modern physics or quantum mechanics. When, as is often the case, V is a function only of r, the solutions may be written

$$\psi = \frac{1}{r}u_l(r)P_{lm_l}(\theta)e^{im_l\phi} \tag{2-25}$$

where l and m_l are integers with $|m_l| \leq l$ and u_l is a solution of (2-18).

The P_{lm_l}, called *associated Legendre polynomials*, are widely encountered in many areas of physics. Some of their values are listed in Table 2-2. It is seen that the entries in Table 2-2 differ somewhat from the result (2-15) obtained from the simplified picture, although they are qualitatively similar. Using quantum-mechanical operator techniques, it is easily shown that the results (2-14), (2-21), and (2-22) are valid.

TABLE 2-2: SOME ASSOCIATED LEGENDRE POLYNOMIALS P_{lm} Since they have the property $P_{lm} = P_{l\,-m}$, only values for positive m are listed

$P_{00} = 1$
$P_{10} = \cos\theta$
$P_{11} = \sin\theta$
$P_{20} = \frac{1}{4}(3\cos 2\theta + 1)$
$P_{21} = \frac{3}{2}\sin 2\theta$
$P_{22} = \frac{3}{2}(1 - \cos 2\theta)$
$P_{30} = \frac{1}{8}(5\cos 3\theta + 3\cos\theta)$
$P_{31} = \frac{3}{8}(\sin\theta + 5\sin 3\theta)$
$P_{32} = \frac{15}{4}(\cos\theta - \cos 3\theta)$
$P_{33} = \frac{15}{4}(3\sin\theta - \sin 3\theta)$
$P_{40} = \frac{1}{64}(35\cos 4\theta + 20\cos 2\theta + 9)$
$P_{50} = \frac{1}{128}(63\cos 5\theta + 35\cos 3\theta + 30\cos\theta)$

2-5 Orbit Model

It is most helpful in discussing problems physically to have available a model that can be visualized in terms of things familiar in our daily lives. This is not only an artifice for instructing elementary students; some of the most productive scientists employ these models constantly, and a great deal of scientific progress is due to them. These models may be considered as "translations" from the world of quantum physics into the world familiar in our everyday lives. As with language translations, they are useful, but we must always keep in mind that some things are lost in translation.

The orbit model for electrons in atoms is familiar to every student from high school or even elementary school studies. It cannot, of course, be considered accurate; it treats a particle as a very small ball, but no ball in our everyday experience has a wave nature and is subject to uncertainty principles. Nevertheless, the orbit picture is useful for many purposes, and we shall use it widely. It faithfully portrays energies, vector angular momenta, and many collision properties, and on an average it gives correct ideas for positions and momenta of particles.

In the orbit model, the first problem one faces is quantization of energy and angular momentum. No classical orbit problems exhibit quantization. This is therefore taken into account by saying that only very specific orbits are *allowed*, these being specified by quantum numbers. The n and l quantum numbers determine the energy, the latter determines the angular momentum of the orbit, and the m_l quantum number determines its orientation in space. This last statement may be understood from the fact that the ratio of L_z to L gives information on the angle between the plane of the orbit and the z axis. This is all the information on directions in space it is possible to have for a quantum system.

2-6 Vector Model for Addition of Angular Momentum

We have seen that orbital angular momentum is a vector with some very peculiar properties in quantum theory: (1) it is quantized; (2) all we can know about its direction in space is its z component, which is itself quantized; and (3) the maximum value of the z component is $l\hbar$ [from (2-21) and (2-22)], which is less than the total orbital angular momentum $[l(l+1)]^{1/2}\hbar$ from (2-12). Clearly we can expect difficulties in the addition of two such vectors, and this is indeed such an involved matter that it is not usually considered in introductory courses. However, to find the possible sums of two vectors, we can use a simple technique, known as the *vector model*, which can be derived from the correct treatment. In this model, we add angular-momentum quantum numbers just as we ordinarily add vectors, except that we accept only results in which the sum is quantized.

Thus $\mathbf{l}_1 + \mathbf{l}_2$ can[1] give any angular momentum between $l_1 + l_2$ and $|l_1 - l_2|$; $\mathbf{3} + \mathbf{1}$ can give 4, 3, or 2. Note that the maximum and minimum values are the same as in ordinary vector addition.

When half-integer angular momenta are involved (recall that the spin quantum numbers of many particles, including electrons, protons, and neutrons, are ½), there is a corollary to this rule that only results differing from the maximum and minimum by integers are allowed. For example, $\mathbf{5/2} + \mathbf{3/2}$ can add to 4, 3, 2, or 1, and $\mathbf{1} + \mathbf{1/2}$ can add to 3/2 or 1/2. From this, we see that when an even [or odd] number of half-integers is added, the result must be integer [or half-integer].

This last rule played an important part in the early days of nuclear physics, when it was postulated that the nucleus consists of the only two particles known at that time, electrons and protons. On this basis, the nucleus of ${}_7N^{14}$ would contain 14 protons and 7 electrons. This is an odd number of particles, each having spin ½, so in view of our rule, the total spin would be half-integer. According to (2-13), orbital angular momentum can only be integer, so the total angular momentum must be half-integer. But the total angular momentum of ${}_7N^{14}$ was measured by optical spectroscopy and was found to be $I = 1$. Since this is an integer, it was concluded that nuclei cannot be made up of electrons and protons. When the neutron was discovered in 1932, its importance in nuclear structure was therefore immediately appreciated; an ${}_7N^{14}$ nucleus made of seven protons and seven neutrons clearly has an integer total angular momentum.

2-7 Parity

All the potentials we have used in this chapter, including both the one-dimensional and three-dimensional cases, have been symmetric with respect to a reflection of the coordinates about the origin. That is, if we replace x, y, and z by $-x$, $-y$, and $-z$, respectively, nothing is changed. Since this symmetry property is also possessed by the other terms in the Schrödinger equation (2-23), one might expect that the physical results should also be unaffected by such a transformation. One such physical result is the probability for a particle to have certain coordinates, which is represented by ψ^2, whence we expect

$$[\psi(x,y,z)]^2 = [\psi(-x,-y,-z)]^2 \tag{2-26}$$

This requirement is satisfied provided

$$\psi(x,y,z) = \psi(-x,-y,-z) \qquad \text{even parity} \tag{2-27a}$$

or

$$\psi(x,y,z) = -\psi(-x,-y,-z) \qquad \text{odd parity} \tag{2-27b}$$

[1] $\mathbf{l}$ represents the vector whose length is $\sqrt{l(l+1)}\,\hbar$. It is common to speak of *angular momentum* l, meaning $\sqrt{l(l+1)}\,\hbar$.

These are referred to as *even-* and *odd-parity wave functions*, as designated. Even parity is also called *positive parity*, and odd parity is also called *negative*. Examples of even-parity wave functions are the cos kx wave functions in Fig. 2-2, and examples of odd-parity wave functions are the sin kx wave functions in that same figure. The former are symmetric and the latter antisymmetric about the origin.

It is unacceptable to have a wave function that is a sum of functions of opposite parities, for it would not satisfy (2-26). For example, if we take

$$\psi(x) = A \sin kx + B \cos kx$$

we find

$$\begin{aligned} \psi(-x) &= A \sin(-kx) + B \cos(-kx) \\ &= -A \sin kx + B \cos kx \\ [\psi(x)]^2 &= A^2 \sin^2 kx + B^2 \cos^2 kx + 2AB \sin kx \cos kx \\ [\psi(-x)]^2 &= A^2 \sin^2 kx + B^2 \cos^2 kx - 2AB \sin kx \cos kx \end{aligned}$$

which does not satisfy (2-26). On the other hand, it is permissible to have a wave function which is the sum of functions with the same parity. For example, if we take

$$\psi(x) = A \sin kx + B \sin 3kx$$

we find

$$\begin{aligned} \psi(-x) &= A \sin(-kx) + B \sin(-3kx) \\ &= -A \sin kx - B \sin 3kx \\ &= -\psi(x) \end{aligned}$$

whence (2-26) is satisfied.

For our wave functions (2-25) for a particle in a three-dimensional well using (r,θ,ϕ) coordinates, a reflection about the origin is equivalent to the transformation

$$\begin{aligned} \theta &\rightarrow \pi - \theta \\ \phi &\rightarrow \phi + \pi \end{aligned} \tag{2-28}$$

The function $e^{im\phi}$ becomes $e^{im\phi}e^{im\pi} = (-1)^m e^{im\phi}$. The function $P_{lm}(\theta)$ has the property that the transformation (2-28) introduces a multiplying factor $(-1)^{l-m}$; this can easily be checked for the cases listed in Table 2-2. Thus, reflection about the origin of the wave function (2-25) multiplies that wave function by $(-1)^{l-m}(-1)^m = (-1)^l$. We therefore conclude that *wave functions with even l have even parity* and *wave functions with odd l have odd parity*. It is straightforward to show that a product of wave functions with l_1, l_2, l_3, . . . has odd [even] parity if the sum $l_1 + l_2 + l_3 + \cdots$ is odd [even]. This is all we shall have to know about parity in this book.

The concept of parity has played an important part in many areas of physics, and a great deal of spectacular attention was paid when certain interactions were found to lead to wave functions of mixed parity. This does not, however, apply in systems under the influence of the strong nuclear force, so all wave functions used in this book will have a definite parity.

2-8 Measurable Properties of Quantum Systems

In Chap. 1 we discussed such properties of nuclei as their size and magnetic and electric moments. These properties are experimentally measurable, so it is important to be able to calculate them from the quantum description of a system as embodied in the wave function. Let us see how this is done.

As a simple example, we consider the size of a hydrogen atom. The radius of a hydrogen atom is just the average value of r, the distance of the electron from the nucleus. The average value of r, $\bar{r}$, as we know from the usual definition of average, can be found by taking the value of r at each point in space available to the electron, multiplying it by the probability p that the electron is at that point, and summing these products over all space; if all space is divided up into volume elements $d\tau$, this sum becomes an integral over all space and we have

$$\bar{r} = \int rp\, d\tau \tag{2-29}$$

From Sec. 2-1 we know that p is proportional to ψ^2. Normalizing it so that the probability of the electron's being somewhere in space is unity gives

$$p\, d\tau = \frac{\psi^2\, d\tau}{\int \psi^2\, d\tau}$$

Using this in (2-29), we obtain

$$\bar{r} = \frac{\int r\psi^2\, d\tau}{\int \psi^2\, d\tau}$$

In analogy with this expression, if any property of a system q is measured a number of times, the average value obtained is

$$\langle q \rangle = \frac{\int q\psi^2\, d\tau}{\int \psi^2\, d\tau} \tag{2-30}$$

This is called the *expectation value* of q and is designated $\langle q \rangle$.

If the wave function is chosen to satisfy (2-12) in its three-dimensional form, the denominator in (2-30) is unity and need not be written. A more general form for (2-30) applicable when ψ is complex and q is an operator is

$$\langle q \rangle = \int \psi^* q \psi\, d\tau \tag{2-31}$$

Problems

2-1 What energies can a 1-kg ball have in a potential well 1 m deep and 1 m across? Treat the problem one-dimensionally and use (2-5). From the result, what can be concluded about the importance of quantization of energy in macroscopic problems?

2-2 Derive the equations analogous to (2-10) for the case where the first of (2-9) is $A \sin kx$.

2-3 An electron is in a one-dimensional square well of depth 200 eV and width 10^{-10} m. Find the wave functions and energies by the methods of Secs. 2-1 and 2-2. [In the latter, the solution can be obtained by trying different values of k until the first of (2-10) is satisfied.] Compare the energies and plots of the wave functions obtained by the two methods.

2-4 Repeat Prob. 2-3 for the lowest energy state when the depth of the potential well is 1,000 and 40 eV. Discuss the result.

2-5 If a 1-kg ball is swung around on a string 1 m long, how does (2-12) limit the allowable velocities? Consider the same problem if an electron could be swung around on a string of length 10^{-10} m and if a proton could be swung around on a string of length 10^{-14} m. Note that these lengths are roughly the sizes of atoms and nuclei respectively. Calculate the kinetic energies in both cases.

2-6 Find the energies of the 1s and 2s states of a neutron in a three-dimensional square well of depth 50 MeV and radius 5×10^{-15} m. Plot the wave functions. Compare the results obtained by the approximation used in Sec. 2-1 ($\psi = 0$ where $E < V$).

2-7 For the potential in Prob. 2-6, plot the total potential including the centrifugal-force contribution for $l = 1, 2$, and 3. Approximate the results by square wells and with the approximation used in Sec. 2-1 calculate the energies E_{1p}, E_{2p}, E_{1d}, and E_{1f}.

2-8 Calculate the depth of the shallowest square well of $R = 10^{-14}$ m in which a neutron is bound.

2-9 Derive a formula analogous to (2-20) for the condition that the 2s state is bound.

2-10 Find what total angular momenta can be obtained by adding the following individual angular momenta:

(a) $1/2 + 2 + 3/2$
(b) $3/2 + 5/2 + 7/2$
(c) $1 + 4 + 7 + 5/2$

2-11 What are the parities of each of the states shown in Fig. 2-4?

Further Reading

Elementary Treatments

Blanchard, C. H., C. R. Burnett, R. G. Stoner, and R. L. Weber: "Introduction to Modern Physics," Prentice-Hall, Englewood Cliffs, N.J., 1969.

Feynman, R. P., R. B. Leighton, and M. Sands: "The Feynman Lectures on Physics," vol. III, Addison-Wesley, Reading, Mass., 1965.

Sproull, R. L.: "Modern Physics: A Textbook for Engineers," Wiley, New York, 1956.

Wehr, M. R., and J. A. Richards: "Introductory Atomic Physics," Addison-Wesley, Reading, Mass., 1962.

Weidner, R. T., and R. L. Sells: "Elementary Modern Physics," Allyn and Bacon, Boston, 1960.

Intermediate Treatments

Dicke, R. H., and J. P. Wittke: "Introduction to Quantum Mechanics," Addison-Wesley, Reading, Mass., 1960.

Eisberg, R. M.: "Fundamentals of Modern Physics," Wiley, New York, 1961.

Leighton, R. B.: "Principles of Modern Physics," McGraw-Hill, New York, 1959.

Richtmyer, F. K., E. H. Kennard, and J. N. Cooper: "Introduction to Modern Physics," 6th ed., McGraw-Hill, New York, 1969.

Complete Treatments

Any textbook on quantum mechanics, for example:

Landau, L. D., and E. M. Lifshitz: "Quantum Mechanics: Non-relativistic Theory," Pergamon, New York, 1965.

Merzbacher, E.: "Quantum Mechanics," Wiley, New York, 1961.

Messiah, A.: "Quantum Mechanics," North-Holland, Amsterdam, 1961.

Schiff, L. I.: "Quantum Mechanics," 3d ed., McGraw-Hill, New York, 1968.

Stehle, P.: "Quantum Mechanics," Holden-Day, San Francisco, 1966.

Chapter 3
The Nuclear Force

In Sec. 1-5 we concluded that the existence of nuclei can be explained only if a new type of force in nature is assumed, one that is not encountered in atomic physics or in everyday life.

The investigation of this nuclear force has turned out to be a truly monumental task; perhaps more man-hours of work have been devoted to it than to any other scientific question in the history of mankind. In this chapter, we review this development and outline our present understanding of the nuclear force.

3-1 Methods of Approach

It is best to study any phenomenon under the simplest possible conditions, and the simplest case in which the nuclear force is effective is when there are only two nucleons present and interacting. There are two experimentally achievable situations of this type: (1) when a neutron and a proton are bound together, as a deuteron; (2) in collisons between two nucleons, usually referred to as *scattering processes*.

To appreciate how studying these can lead to an understanding of the nuclear force, let us assume that we did not know the coulomb force and tried to learn about it by analogous methods. The analog of the first method would be to study the energy levels of the hydrogen atom; we would find that they obey the well-known relationship $E = -\mathcal{R}/n^2$, where $\mathcal{R}$ is hc times the Rydberg constant and $n = 1, 2, 3, \ldots$. Assuming that we understood quantum theory, we would calculate the energy levels obtained from various potentials using the methods of Chap. 2 and would find that the only potential giving these energy levels is $V = -e^2/r$, where e^2 is a constant that could be evaluated from the measured value of the Rydberg constant. Of course the coulomb force was not found that way; knowing the coulomb force, studies of the energy levels of the hydrogen atom led to the discovery of quantum theory. But that was a historical accident; things could have been the other way. When physicists first began to study the nuclear force in the early 1930s, quantum theory was already well established, while the force was unknown.

The analog of the second method would be to do Rutherford scattering experiments assuming that the structure of atoms was understood. By shooting energetic charged particles at thin foils and observing the scattered particles, it would have been found that the probability of scattering through an angle θ varies as $\csc^4(\theta/2)$ and that as the energy of the incident particle is changed, the scattering probability varies as $1/E^2$. Calculations of scattering probabilities could be made for various forces between the nucleus and the scattered particle, and it would be found that this behavior is to be expected only if the force is of the form $z_1z_2e^2/r^2$. From the measured absolute scattering probability, the value of $z_1z_2e^2$ could be deduced, and by repeating the experiments with various incident particles and various target materials, values of z_1 and z_2 and hence of e^2 could be obtained. Historically things did not happen that way; the coulomb force was already known, and the experiments were used to unravel atomic structure. But if the coulomb force had not been known and somehow it was known that the atom consists of a massive nucleus surrounded by electrons at a large distance, the coulomb force could have been deduced.

It was consequently reasonable to expect that by studying the energy levels of two-nucleon systems and by measuring scattering of nucleons by nucleons, the nature of the nuclear force could be determined. The former studies are discussed in the next four sections, and then after some theoretical developments nucleon-nucleon scattering is considered in Sec. 3-10, leading to a summary of our present knowledge of the nuclear force in the last three sections.

3-2 Bound States of Two Nucleons—Conclusions from the Binding Energy and Size of the Deuteron

The only bound system of two nucleons found in nature is the deuteron, which consists of a neutron and a proton. The other possibilities, two neutrons (the dineutron) and two protons (the diproton, or He^2) do not hold together, but even this information will be useful. Studies of the deuteron reveal that it has no excited states which do not break up very rapidly into a neutron and a proton, so in the three possible systems of two nucleons there is only one bound state. This situation is much less favorable than the analogous case of the hydrogen atom, which has an infinite number of bound states, dozens of which are known experimentally. However let us see what we can learn from the single bound state of the deuteron.

The energy of this state, i.e., the energy by which it is bound, can be found from various experiments. Perhaps the easiest way is to allow slow neutrons to be captured by protons in a material containing hydrogen such as paraffin or a plastic, and measure the energy of the gamma rays emerging; the reaction is

$$n + p \rightarrow d + \gamma \qquad \text{(3-1)}$$

The experimental result is that the binding energy of the deuteron is 2.23 MeV; or, to be consistent with our method of designating the energy in Chap. 2, $E = -2.23$ MeV.

The size of the deuteron has been determined by the experiments described in Sec. 1-2. The root mean square (rms) distance between the neutron and proton r_d was found to be 4.2 F. In the notation of Sec. 2-8, this is the square root of $\langle r^2 \rangle$.

The two-body problem of a neutron and a proton interacting with a force represented by a potential $V(\mathbf{r}_p - \mathbf{r}_n)$ reduces by the well-known center-of-mass transformation of classical mechanics to the problem of a single particle in a potential well $V(r)$, where the coordinates of the effective single particle, r, θ, ϕ, are actually the coordinates of the neutron relative to the proton (or vice versa) and the mass of the "single particle" is the reduced mass $\mu = M_pM_n/(M_p + M_n)$, which is one-half of the nucleon mass.

If we assume that the potential can be represented by a square well with radius R, the problem becomes identical with that discussed in Sec. 2-3. Since the energy level in question is the lowest-energy state of the system, in accordance with Fig. 2-4 it must have $l = 0$ and a wave function like the one designated u_{1s} in that figure. For $r < R$, $u = A \sin kr$, and we see from Fig. 2-4 that for u_{1s}, kR is between $\pi/2$ and π. Moreover the solution leading to (2-19) is valid, whence we have

$$\cot kR = -\frac{\kappa}{k} \tag{3-2}$$

From (2-3) and (2-7) with $M = \mu = \frac{1}{2}M_p$ and $E = -2.23$ MeV, we find

$$\begin{aligned} k &= 0.156\sqrt{V_0(\text{MeV}) - 2.23} \qquad \text{F}^{-1} \\ \kappa &= 0.232\ \text{F}^{-1} \end{aligned} \tag{3-3}$$

A very simple but crude approximation would be to take $R = r_d = 4.2$ F and use the method of Sec. 2-1, which requires

$$kR = \pi$$

This gives immediately, $k = \pi/R = 0.75$. Using this estimate in the right side of (3-2), we see that $\cot kR \approx -0.23/0.75$, whence kR is actually considerably less than π and is, in fact, not much greater than $\pi/2$. The wave function is therefore somewhat like the one shown in Fig. 3-1; the exponential tail represents an important part of the wave function, whence r_d is considerably larger than R.

From (2-31), r_d (which we recall is the rms value of r) can be calculated exactly as

$$r_d^2 = \langle r^2 \rangle = \int r^2\psi^2\, d\tau \tag{3-4}$$

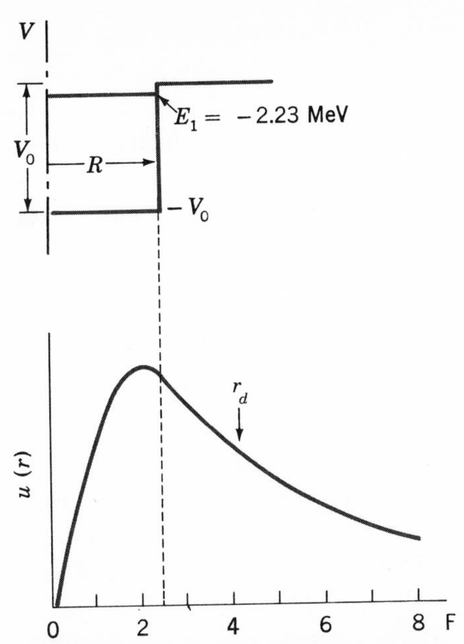

FIGURE 3-1 Square-well solution for the deuteron. The square-well potential is shown at the top with the energy of the state at the experimental position. The lower plot is the wave function $u(r)$.

With the wave function presented before (2-19) and the value of r_d given above, this becomes

$$A^2 \int_0^R r^2 \sin^2 kr\, dr + B^2 \int_R^\infty r^2 e^{-2\kappa r}\, dr = (4.2 \text{ F})^2$$

With A^2 and B^2 taken from (2-19*a*), this equation and the first of (2-19) are two equations in the two unknowns, k and R. Their simultaneous solution gives $R = 2.4$ F, $k = 0.775 \text{ F}^{-1}$ ($kR = 107°$); using this in (3-3) gives $V_0 = 27$ MeV.

The nuclear potential that binds the deuteron is thus, to some approximation, a square well 27 MeV deep and with a width of 2.4 F. The latter dimension gives us our first estimate of the range of the nuclear force, and we see that it does indeed have a short range, as anticipated in Sec. 1-5. Moreover, as also anticipated there, the 27-MeV potential depth is much larger than the coulomb force at these distances; the coulomb force between two protons corresponds to a potential of only 0.6 MeV at 2.4 F.

Finally we note that, in comparison with the depth of the well, the 2.23-MeV binding energy of the deuteron is very small: the deuteron is just barely bound.

3-3 Spin States of the Two-nucleon System

The total angular momentum I for the deuteron can be measured by various techniques including hyperfine structure in optical spectroscopy, which is familiar from atomic physics. The result found is $I = 1$.

In general, the total angular momentum is the vector sum of the orbital and spin angular momenta, or

$$\mathbf{I} = \mathbf{l} + \mathbf{S} \tag{3-5}$$

where S is the total spin quantum number. Since the spins of both the neutron and the proton are 1/2, according to Sec. 2-6,

$$S = |1/2 + 1/2| = 1 \text{ or } 0$$

We may for now think of these two possibilities as parallel or antiparallel spins although a more specific definition will be given in the next section. As we pointed out in Sec. 3-2, the state in question is the lowest-energy state of the system which, from Fig. 2-4, implies that it is the 1s ($l = 0$) state. From (3-5), then, $S = 1$.

The next question one naturally asks is: At what energies are the other states of the neutron-proton system? From Fig. 2-4, the states of $l > 0$ and $n > 1$ are at a considerably higher energy, so it seems not unlikely, in view of the fact that the lowest state is barely bound, that these states should be unbound. But what about the 1s, $S = 0$ state, the state which differs from the ground state only in that the spins of the neutron and proton are antiparallel rather than parallel? Apparently it is also at a high enough energy to be unbound. Actually this state is known from scattering experiments, and it turns out to be unbound by only about 60 keV. But the important point is that here we have a case where the energy of a state is different if the spins are parallel than if they are antiparallel. From Sec. 2-3 we know that the energy of a state depends only on the potential well, which is derived from the force. We can therefore conclude that the force between a neutron and a proton is *spin-dependent;* it depends on whether the spins are parallel or antiparallel, whether their total spin is $S = 1$ or $S = 0$. The nuclear force is not just a function of r like the coulomb and gravitational forces, but it depends on other things; and worse is yet to come.

3-4 Effects of the Pauli Exclusion Principle

We are now at a point where we might begin to wonder about why the two-neutron and two-proton systems do not have stable states. Does this imply that the nuclear force is also different between two neutrons and between a neutron and a proton? The answer to this question is basically *no*, but this takes some explanation. To understand the problem, one must take into account the Pauli exclusion principle. This principle plays a very important part in atomic physics, as we all know, but in most elementary modern-physics courses, it is introduced only in a limited form. For the present application, we must consider it more deeply.[1]

It is commonly stated in elementary school science courses that "no two objects are exactly alike." This gives a good account of the fact that no two snowflakes are exactly alike, but it is just not true in the atomic and nuclear worlds. It is a basic tenet of quantum physics that all the information one can have about a system is its wave function, and certainly the wave functions of all hydrogen atoms are exactly the same. Two hydrogen atoms are therefore indistinguishable *in principle.* It is not only very difficult to tell any difference between them, but it is absolutely impossible. They are absolutely *identical,* as are any two electrons, any two protons, any two neutrons, any two deuterons, etc.

In many cases the fact that two particles or two systems of particles are identical presents no complication since they can be kept separate by their locations. However, let us consider the situation for the two electrons in a helium atom. The probability of electron 1 being at (r_1,θ_1,ϕ_1) and of electron 2 being at (r_2,θ_2,ϕ_2) is proportional to the square of the wave function $\psi(r_1,\theta_1,\phi_1,r_2,\theta_2,\phi_2)$; for short we call this $\psi(1,2)$. For example, if we take the potential as a square well (a poor approximation for a coulomb potential, but it illustrates the point) and consider the region $r < R$, the wave function from Sec. 2-2 might be

$$\psi = A\,\frac{\sin k_1r_1}{r_1}\,\frac{\sin k_2r_2}{r_2} \tag{3-6}$$

where k_1 and k_2 are the solutions of (2-19) corresponding to energies E_1 and E_2. Notice that we have taken the wave function as a product

$$\psi = \psi_1(r_1)\psi_2(r_2) \tag{3-7}$$

This may be understood from the fact that the probability for electron 1 to be at r_1 and for electron 2 to be at r_2 is the *product* of the individual probabilities for 1 to be at r_1 and for 2 to be at r_2, which can be expressed as

$$\psi^2 = \psi_1{}^2(r_1)\psi_2{}^2(r_2)$$

This is just the result we would get from (3-7).

In examining (3-6), we notice something very wrong: this wave function implies that we can distinguish between particles 1 and 2. But this is not possible just from their locations: from (3-6) either particle can easily be at almost any value of r ($<R$). Since we cannot distinguish between particles 1 and 2, the wave function must give the same probability distribution for both particles, not

[1] Less advanced students may omit the remainder of this section, with the exception of the last paragraph, where its results are summarized.

different probability distributions as in (3-6). That is, if we interchange the two particles in ψ, we must get the same ψ^2, or

$$[\psi(1,2)]^2 = [\psi(2,1)]^2$$

This can be satisfied if

$$\psi(1,2) = \pm\psi(2,1) \tag{3-8}$$

For reasons that are not clearly understood, for electrons, protons, neutrons, and all other spin-½ particles, the *minus* sign is chosen in (3-8). This is commonly stated "the wave function must be antisymmetric with respect to the interchange of any two identical spin-½ particles." This is the most fundamental statement of the Pauli exclusion principle.

The Pauli exclusion principle is satisfied if (3-7) is modified to

$$\psi = \psi_1(r_1)\psi_2(r_2) - \psi_2(r_1)\psi_1(r_2) \tag{3-9}$$

which, for our example, is

$$\psi = A\left(\frac{\sin k_1r_1 \sin k_2r_2}{r_1r_2} - \frac{\sin k_2r_1 \sin k_1r_2}{r_1r_2}\right) \tag{3-10}$$

It is readily seen that if r_1 and r_2 are interchanged in either (3-9) or (3-10), the wave function remains the same except for a reversal in sign.

This method breaks down, however, when ψ_1 and ψ_2 are the same functions or, in our example, when $k_1 = k_2$. In these cases, our method gives $\psi = 0$. Thus we find that "no two identical particles can be in the same state." This is the form of the Pauli exclusion principle used in more elementary applications.

Actually, our example (3-10) is really not complete, because a complete wave function must include spin; in fact, it is a product of a space wave function $\psi(r,\theta,\phi)$ and a spin wave function $\chi(\uparrow)$ or $\chi(\downarrow)$ indicating spin *up* and spin *down*, respectively. Equation (3-10) can then be made complete and satisfactory as

$$\psi = A\left(\frac{\sin k_1r_1 \sin k_2r_2}{r_1r_2} - \frac{\sin k_2r_1 \sin k_1r_2}{r_1r_2}\right)\chi_1(\uparrow)\chi_2(\uparrow) \tag{3-11}$$

which indicates that both particles have spin *up*. However, another suitable wave function would be

$$\psi = A\left(\frac{\sin k_1r_1 \sin k_2r_2}{r_1r_2} + \frac{\sin k_2r_1 \sin k_1r_2}{r_1r_2}\right)\left[\chi_1(\uparrow)\chi_2(\downarrow) - \chi_1(\downarrow)\chi_2(\uparrow)\right] \tag{3-12}$$

We see that there are *two* ways of making a wave function antisymmetric with respect to interchange of two particles in order to satisfy the Pauli exclusion principle: it can be symmetric in spin and antisymmetric in space coordinates as in (3-11), or it can be antisymmetric in spin and symmetric in space coordinates as in (3-12). We may note that in (3-12) there is nothing wrong with $k_1 = k_2$, so two identical particles may have the same space wave function provided their spin wave function is antisymmetric; indeed this is the situation for the two electrons in a normal helium atom.

The spin wave function in (3-12) is what we have previously referred to loosely as a state of antiparallel spin, or $S = 0$. The spin wave function in (3-11) is one of the possible states of parallel spin; it corresponds to $S = 1$, $m_s = +1$. The other $S = 1$ wave functions, those with $m_s = 0$ and -1, are also symmetric with respect to interchange of the two particles, so we have the following rule: $S = 1$ states are symmetric, and $S = 0$ states are antisymmetric in the spin coordinates.

Now let us return to systems of two nucleons, such as the deuteron, the dineutron, and the diproton. The space part of their wave functions, in accordance with (2-25), is

$$\psi(\text{space}) \sim \frac{u_l(r)}{r} P_{lm_l}(\theta) e^{im_l\phi}$$

Due to the center-of-mass transformation of Sec. 3-2, we no longer have r_1 and r_2 appearing explicitly: in this case, an interchange of the two particles corresponds to a reflection about the origin, whence the symmetry or antisymmetry of the space part of the wave function depends on whether the wave function changes sign under this transformation. As we found in Sec. 2-7, for all odd values of l, it does indeed reverse its sign, whereas for all even values of l, it does not. We therefore conclude that for even l the wave function is symmetric and for odd l it is antisymmetric in the space coordinates.

Combining the results of the last two paragraphs, we see that only wave functions with even l, $S = 0$ (space symmetric, spin antisymmetric) or odd l, $S = 1$ (space antisymmetric, spin symmetric) satisfy the Pauli exclusion principle for identical nucleons. These states are called $T = 1$ states, where T is the isobaric spin, a concept that will be discussed later. The other states of the two-nucleon system, those with even l, $S = 1$ or odd, l, $S = 0$ are called $T = 0$. We see that $T = 1$ states are available to any of the three two-nucleon systems, proton-proton, proton-neutron, and neutron-neutron, whereas the $T = 0$ states are available only to the proton-neutron system because it does not consist of identical particles and hence need not satisfy the Pauli exclusion principle. The ground state of the deuteron ($l = 0$, $S = 1$) is clearly a $T = 0$ state. We have already mentioned that the lowest-energy $T = 1$ state ($l = 0$, $S = 0$) of the deuteron is unbound, so it is not surprising that the lowest-energy states of the dineutron or diproton, being $T = 1$, are also unbound. The fact that there are no bound states of the dineutron or diproton therefore does not mean that the nuclear force between a neutron and a proton is different from that between two neutrons or between two protons. It is only a manifestation of the Pauli exclusion principle.

3-5 Magnetic Dipole and Electric Quadrupole Moments of the Deuteron—The Tensor Force

In a structure made up of two particles, one expects the total magnetic moment to be the vector sum of the magnetic moments due to spin and the magnetic moments due to orbital motion of charged particles. Since the deuteron is presumably in an $l = 0$ state, no contribution from orbital motion is expected, whence from (1-8) and (1-9) we should find its magnetic moment to be equal to

$$\mu_n + \mu_p = 0.8797 \frac{e\hbar}{2M_p}$$

However, when the magnetic moment of the deuteron was measured, the result was found to be

$$\mu_d = 0.8574 \frac{e\hbar}{2M_p}$$

While the difference between the measured and expected values is not very large and there are some (small) uncertainties in the theory,[1] the discrepancy is difficult to explain. The simplest interpretation is that there is some orbital motion in the deuteron, that our previous assumption that $l = 0$ in the deuteron ground state is not completely correct.

There is even better evidence for this conclusion, derived from a measurement of the electric quadrupole moment of the deuteron. If the deuteron *is* in an $l = 0$ state, we see from (2-25) and Table 2-2 that the wave function has no (θ,ϕ) dependence and hence is a function only of r. It must therefore exhibit spherical symmetry, which was shown in Sec. 1-6 to imply a zero quadrupole moment. However, the quadrupole moment of the deuteron was measured and found to be

$$Q_d = 2.82e \times 10^{-27} \text{ cm}^2$$

This is a relatively small quadrupole moment in comparison with the others listed in Table A-2, but it is not zero. This indicates that the wave function is not a simple $l = 0$ one, in agreement with the conclusion from the magnetic moment.

These results cannot be explained by assuming the state to have some other value of l; in fact they are very much closer to the results for $l = 0$ than for any other l. This suggests that the wave function contains a mixture of l values. For $I = 1$ and a maximum value of $S = 1$, from (3-5) l can only be 0, 1, and 2. However, because of conservation of parity, as shown in Sec. 2-7 even and odd values

[1] See footnote 2, page 9.

of l cannot both be present in the same wave function, so only $l = 2$ can be present with $l = 0$. We therefore take the wave function to be

$$\psi = a_0\psi_{1s} + a_2\psi_{1d}$$

A wave function written as a sum in this way means that the system spends a fraction $|a_0|^2$ of its time in an $l = 0$ state and a fraction $|a_2|^2$ of its time in an $l = 2$ state. It turns out that the results for the electric quadrupole moment can be explained with $|a_0|^2 = 0.96$, $|a_2|^2 = 0.04$; the magnetic moment results, though less certain, are consistent with this. The deuteron is apparently in an $l = 2$ state about 4 percent of the time, while the other 96 percent of the time it is in an $l = 0$ state, as we have been assuming previously. Note that for either $l = 0$ or $l = 2$, (3-5) requires $S = 1$, so that result is unchanged.

In blithely allowing the wave function to include a mixture of l values, we have violated the principle of conservation of orbital (though not of total) angular momentum. Now let us consider the implications of this violation. As we know from elementary mechanics, angular momentum can be changed only by a torque, just as linear momentum can be changed only by a force; and the torque acting on a body is $\mathbf{r} \times \mathbf{F} = rF_\theta = -\partial V/\partial\theta$. A changing orbital angular momentum therefore implies that the potential V is a function of θ and not merely a function of r. Since a *central force* is defined as one for which V is a function only of r, this is a *noncentral force;* it is called the *tensor force.*

The only fixed direction in space that has any relevance in a deuteron is the direction of the spin vector $\mathbf{S}$; the θ upon which the tensor force depends must therefore be measured from this direction. In vector notation, the force is therefore a function of $\mathbf{S} \cdot \mathbf{r}$. For the same separation distance, the tensor force must be different in Fig. 3-2*a* than in Fig. 3-2*b*. It turns out, in fact, that for the deuteron it is repulsive in the former and attractive in the latter. When two nucleons are in an $S = 0$ state, there is no preferred direction in space, so there can be no tensor force.

FIGURE 3-2 The tensor force as it acts in the deuteron. When the spin direction is perpendicular to the line joining the nucleons, as in *(a)*, the force is repulsive. When the two are parallel as in *(b)*, the force is attractive.

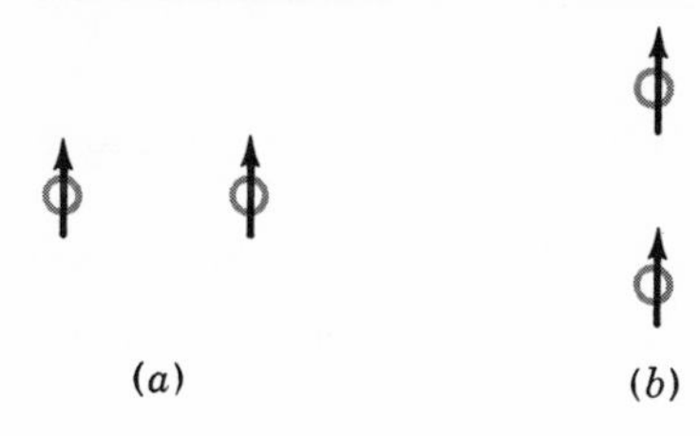

The nuclear potential can be written as the sum of a central and a tensor potential. When done in this way, the tensor potential is some function of r times S_{12} where[1]

$$S_{12} = 2\left[3\,\frac{(\mathbf{S}\cdot\mathbf{r})^2}{r^2} - \mathbf{S}\cdot\mathbf{S}\right] \tag{3-13}$$

It will be shown in Sec. 3-6 that $(\mathbf{S}\cdot\mathbf{r})^2$ is the only acceptable function of $\mathbf{S}\cdot\mathbf{r}$. The r^2 in the denominator is to make the term dimensionless, and the second term in the bracket is added to make the value of S_{12} averaged over all angles equal to zero, thereby eliminating components of the central force from this term.

3-6 General Properties of the Nuclear Force—Static Forces

The previous four sections summarize the information on the nucleon-nucleon force that we can obtain from the deuteron; to learn further details we must use the other approach presented in Sec. 3-1, nucleon-nucleon scattering experiments. However, before getting into that subject it will pay us to consider some of the properties we can reasonably expect of the force we are studying. These will be the topics of this and the following three sections.

First let us list the quantities available in a two-nucleon system for the force to depend on. These are $\mathbf{r}_{12}$, the vector joining the position of particle 1 to that of particle 2; $\mathbf{p}_{12} = \mathbf{p}_2 - \mathbf{p}_1$, their relative momentum; and $\mathbf{S}$ the total spin. To these it is convenient (though not necessary) to add the orbital angular momentum[2] $\mathbf{L} = \mathbf{r}_{12} \times \mathbf{p}_{12}$.

Next let us list the requirements we might expect the potential representing the nuclear force to fulfill:

1. It must be a scalar quantity since it is an energy.
2. Since two neutrons or two protons are indistinguishable, it must remain unchanged if we interchange particles 1 and 2. This would intuitively seem to be a most desirable property even for a force between nonidentical particles. Interchanging particles reverses the sign of $\mathbf{r}_{12}$ and $\mathbf{p}_{12}$ but not of $\mathbf{S}$ and $\mathbf{L}$. (In the latter case, the sign is reversed twice in its definition as $\mathbf{r}_{12} \times \mathbf{p}_{12}$, and two reverses is equivalent to no change.) Hence we cannot have a term like $\mathbf{S}\cdot\mathbf{r}_{12}$ or $\mathbf{S}\cdot\mathbf{p}_{12}$; the tensor force discussed in the last section must therefore be proportional to $(\mathbf{S}\cdot\mathbf{r}_{12})^2$ or to higher even powers of that scalar product. It can be shown that the latter reduce to $(\mathbf{S}\cdot\mathbf{r}_{12})^2$ by mathematical identities.

[1] For students familiar with the σ operators, (3-13) can be written in a more widely used form

$$S_{12} = \frac{3}{r^2}(\boldsymbol{\sigma}_1\cdot\mathbf{r})(\boldsymbol{\sigma}_2\cdot\mathbf{r}) - \boldsymbol{\sigma}_1\cdot\boldsymbol{\sigma}_2$$

[2] Hereafter we shall use $\mathbf{L}$, with quantum number L, to designate orbital angular momentum in the two-nucleon problem.

3. It must remain unchanged if the direction of the flow of time is reversed. This is known as the *principle of time-reversal invariance;* it is valid in all well-understood laws in classical and quantum physics and is widely believed to be true in phenomena in which the nuclear force is effective. If the direction of time were reversed, it is easy to see that $\mathbf{p}_{12}$ and $\mathbf{S}$ would change sign but $\mathbf{r}_{12}$ would not. From its definition, then, the sign of $\mathbf{L}$ would be reversed. We therefore cannot have terms like $\mathbf{r}_{12} \cdot \mathbf{p}_{12}$, $\mathbf{r}_{12} \cdot \mathbf{S}$, or $\mathbf{r}_{12} \cdot \mathbf{L}$.

In order to organize our discussion, we first consider forces that are independent of velocity, the so-called *static forces.* Gravitation is an example of a static force, but electromagnetic forces are not because the force on a charged particle in a magnetic field depends on its velocity. Velocity-dependent forces will be considered in the next section.

If a potential is to be static, it cannot depend on $\mathbf{p}$ (we drop subscripts hereafter) or $\mathbf{L}$ as these are velocity-dependent, whence it must involve only $\mathbf{r}$ and $\mathbf{S}$. The only scalars that can be formed from these are $\mathbf{r} \cdot \mathbf{r}$, $\mathbf{S} \cdot \mathbf{S}$, $\mathbf{r} \cdot \mathbf{S}$, or products of these. All other possible terms, for example, $(\mathbf{r} \times \mathbf{S}) \cdot (\mathbf{r} \times \mathbf{S})$, reduce to linear combinations of these terms. Dependence on $\mathbf{r} \cdot \mathbf{r}$ is ordinary r dependence which we have assumed to be present in all forces; $\mathbf{S} \cdot \mathbf{S}$ is the spin-dependent force discussed in Sec. 3-3; we have already seen that $\mathbf{r} \cdot \mathbf{S}$ is not allowed but $(\mathbf{r} \cdot \mathbf{S})^2$ is allowed and is, in fact, the tensor force discussed in Sec. 3–5. A product of $\mathbf{S} \cdot \mathbf{S}$ and $(\mathbf{r} \cdot \mathbf{S})^2$ would be indistinguishable from the tensor force since that force is applicable only for $S = 1$, in which case the value of $\mathbf{S} \cdot \mathbf{S}$ is determined.

In addition to these vectors, there is one other property that a static nucleon-nucleon force can depend on, namely, the *parity* of the wave function describing the system. While this may seem strange at first, a reasonable explanation will be given in the next section. Since the parity depends on whether L is even or odd, this means that the force is different for even L than for odd L. A complete list of possible static forces therefore includes six members:

$$\begin{array}{ll} (A): & S = 0,\ L \text{ odd, central} \\ (B): & S = 0,\ L \text{ even, central} \\ (C): & S = 1,\ L \text{ odd, central} \\ (D): & S = 1,\ L \text{ even, central} \\ (E): & S = 1,\ L \text{ odd, tensor} \\ (F): & S = 1,\ L \text{ even, tensor} \end{array} \tag{3-14}$$

The most general static potential is the sum of these six terms, each being a function of r, as

$$V = V_A(r) + V_B(r) + V_C(r) + V_D(r) + V_E(r) + V_F(r) \tag{3-15}$$

with the subsidiary condition that terms are zero unless L and S are as in (3-14). For example, in the ground state of the deuteron, all terms are zero except V_D and V_F.

3-7 Exchange Forces

It may seem intuitively repulsive for the nuclear force to depend on whether $S = 0$ or 1 and on whether L is even or odd. This behavior is due to the fact the nuclear force arises from the exchange of mesons, as will be discussed in Sec. 3-9, and meson exchange leads to *exchange forces.* There are three types of exchange forces: the space-exchange, or Majorana force; the spin-exchange, or Bartlett force; and the space-spin exchange, or Heisenberg, force.

The Majorana potential V_M, as it operates on a wave function ψ, may be defined as

$$V_M\psi = v_M(r)P^x\psi$$

where $v_M(r)$ is an ordinary function of r and P^x is an operator which exchanges the positions (but not the spins) of the two particles in the wave function which follows. In a two-nucleon system, exchanging the two particles is a reflection about the origin, and we have already seen in Sec. 2-7 that such a reflection changes the sign of ψ if the parity is odd and leaves ψ unchanged if the parity is even. Since the evenness or oddness or parity was shown in Sec. 2-7 to depend only on the evenness or oddness of L,

$$V_M\psi = \begin{cases} v_M(r)\psi & L \text{ even} \\ -v_M(r)\psi & L \text{ odd} \end{cases} \tag{3-16}$$

In view of the fact that V appears in the Schrödinger equation (2-23) followed by ψ, (3-16) fully takes into account the effects of P^x. We see that the Majorana force provides an explanation of the fact that the nuclear potential depends on whether L is even or odd.

The Bartlett potential V_B is defined as

$$V_B\psi = v_B(r)P^s\psi$$

where $v_B(r)$ is an ordinary function of r and P^s is an operator which interchanges the spins (but not the spatial positions) of the two particles in the wave function which follows. From (3-11) we see that interchanging spins does not affect the wave function if $S = 1$, and from (3-12) we see that if $S = 0$, interchanging spins merely changes the sign of the wave function. Therefore we have

$$V_B\psi = \begin{cases} +v_B(r)\psi & S = 1 \\ -v_B(r)\psi & S = 0 \end{cases} \tag{3-17}$$

This fully takes into account the effects of the P^s operator and provides an explanation for why the nuclear potential is different for $S = 0$ and $S = 1$. Since, from the familiar properties of angular momentum in quantum mechanics,

$$\mathbf{S} \cdot \mathbf{S} = S(S+1) = \begin{cases} 2 & S = 1 \\ 0 & S = 0 \end{cases}$$

(3-17) can be expressed more succinctly as[1]

$$V_B(\psi) = (\mathbf{S} \cdot \mathbf{S} - 1) v_B(r) \psi \tag{3-17a}$$

The Heisenberg potential V_H is defined as

$$V_H \psi = v_H(r) P^x P^s \psi$$

Combining the effects of P^x and P^s, we find

$$V_H \psi = \begin{cases} +v_H(r)\psi & \left.\begin{matrix} S = 1,\ L \text{ even} \\ S = 0,\ L \text{ odd} \end{matrix}\right\} \text{ i.e., } T = 0 \\ -v_H(r)\psi & \left.\begin{matrix} S = 0,\ L \text{ even} \\ S = 1,\ L \text{ odd} \end{matrix}\right\} \text{ i.e., } T = 1 \end{cases} \tag{3-18}$$

where $v_H(r)$ is an ordinary function of r. The isobaric spin $\mathbf{T}$ is a vector which has the same mathematical properties as $\mathbf{S}$, whence (3-18) can be expressed more succinctly as[2]

$$V_H \psi = -(\mathbf{T} \cdot \mathbf{T} - 1) v_H(r) \psi \tag{3-18a}$$

It can also be readily checked that (3-16) can be written[1]

$$V_M \psi = -(\mathbf{S} \cdot \mathbf{S} - 1)(\mathbf{T} \cdot \mathbf{T} - 1) v_M(r) \psi \tag{3-16a}$$

This works out to $v_M(r)\psi$ for L even and $-v_M(r)\psi$ for L odd for either $S = 0$ or $S = 1$, as required by (3-16).

In addition to the three exchange forces, there can also be an ordinary, i.e., no exchange, force called a *Wigner force;* it can be written

$$V_W \psi = v_W(r) \psi$$

[1] For students familiar with σ operators, (3-17*a*) can be written in its more common form

$$V_B(\psi) = \tfrac{1}{2}(1 + \boldsymbol{\sigma}_1 \cdot \boldsymbol{\sigma}_2) v_B(r) \psi$$

[2] In terms of τ operators, which bear the same relationship to σ operators as $\mathbf{T}$ does to $\mathbf{S}$, (3-18*a*) and (3-16*a*) can be written

$$V_H \psi = -\tfrac{1}{2}(1 + \boldsymbol{\tau}_1 \cdot \boldsymbol{\tau}_2) v_H(r) \psi$$
$$V_M \psi = -\tfrac{1}{4}(1 + \boldsymbol{\sigma}_1 \cdot \boldsymbol{\sigma}_2)(1 + \boldsymbol{\tau}_1 \cdot \boldsymbol{\tau}_2) v_M(r) \psi$$

regardless of S or L. In terms of these four basic potentials, the central potentials in (3-14) can be written in accordance with (3-16) to (3-18) as

$$\begin{aligned}
&(A)\ S = 0,\ L \text{ odd:} && V_A = v_W - v_M - v_B + v_H \\
&(B)\ S = 0,\ L \text{ even:} && V_B = v_W + v_M - v_B - v_H \\
&(C)\ S = 1,\ L \text{ odd:} && V_C = v_W - v_M + v_B - v_H \\
&(D)\ S = 1,\ L \text{ even:} && V_D = v_W + v_M + v_B + v_H
\end{aligned} \tag{3-19}$$

Since these are four linearly independent equations in four unknowns, we see that expressing the central part of the nuclear force as a sum of the first four terms in (3-15) is completely equivalent to expressing it as a sum of $V_W + V_M + V_B + V_H$. Since the exchange forces have a physical basis in meson theory, this gives a physical basis to the four central potentials in (3-15). Similarly, the two tensor potentials V_E and V_F of (3-15) can be expressed in terms of tensor components of V_W, V_M, V_B, and V_H, which are proportional to $S_{12}v_{WT}(r)$, $S_{12}v_{MT}(r)$, $S_{12}v_{BT}(r)$, and $S_{12}v_{HT}(r)$, respectively, where v_{WT}, . . . are simple functions of r and S_{12} is from (3-13). Here again the two representations are completely equivalent.

As a result of this equivalence, by using the exchange forces in the form (3-16*a*), (3-17*a*), (3-18*a*), we can write (3-15) as a single expression,

$$\begin{aligned}
V = v_W(r) + S_{12}v_{WT}(r) - (\mathbf{S}\cdot\mathbf{S} - 1)(\mathbf{T}\cdot\mathbf{T} - 1)[v_M(r) + S_{12}v_{MT}(r)] \\
+ (\mathbf{S}\cdot\mathbf{S} - 1)[v_B(r) + S_{12}v_{BT}(r)] - (\mathbf{T}\cdot\mathbf{T} - 1)[v_H(r) + S_{12}v_{HT}(r)]
\end{aligned} \tag{3-20}$$

The form (3-20) is sometimes of value due to its formal simplicity in not requiring subsidiary conditions. However, the experiments directly determine the six potentials in (3-15), so it is more conventional to use that expression.

3-8 Velocity-dependent Forces

While we might have hoped that the nuclear force is static, it turns out not to be. It is therefore useful to consider the simplest velocity-dependent forces. While this name is widely used, what we shall really be discussing is *momentum-dependent potentials.* Why we use potentials instead of forces is obvious from the fact that quantum theory is formulated in terms of potentials (see Chap. 2); one clear reason for using momentum rather than velocity is that the latter behaves peculiarly in the relativistic energy region and there is nothing leading us to believe that the nuclear force behaves that way.

If, as now seems probable, the nuclear force has a very complex velocity dependence, it can be discovered only by continuous studies of two-nucleon systems at ever-increasing energies. However, if our principal interest is in understanding nuclear structure, we need understand only the nuclear force up to the energies of nucleons in nuclei, which is not more than a few hundred MeV. One might hope that it is sufficient for this purpose to consider a low-energy approximation, in which the lowest-order terms in the momentum p are the most important in momentum-dependent potentials.

In view of our discussion of the last section, the simplest terms of lowest order in p would be scalar products of $\mathbf{p}$ and $\mathbf{L}$ with $\mathbf{r}$ and $\mathbf{S}$. Of the four possible combinations, the only one that has not already been eliminated by our rules is $\mathbf{L} \cdot \mathbf{S}$. There is very good evidence for such a term in the potential; it is called the *spin-orbit interaction.* Its magnitude obviously depends on the angle between the spin and orbital motion as well as on the magnitude of L; if $S = 0$, it vanishes.

Many more complicated scalars of the first order in p can be formed from our four vectors, but all turn out to be forbidden by the rules given in the last section or to be equivalent to the spin-orbit term. Let us illustrate this with a few examples. If the velocity dependence is given by a $\mathbf{p}$, rule 2 requires that there also be an odd power of $\mathbf{r}$ and rule 3 requires that there be an odd power of $\mathbf{S}$ (note that there can be no $\mathbf{L}$ if it is to be of first order in p). These requirements would be satisfied by $\mathbf{r} \cdot \mathbf{p} \times \mathbf{S}$ or $\mathbf{p} \cdot \mathbf{r} \times \mathbf{S}$, but in both cases these are identically equal to $\mathbf{r} \times \mathbf{p} \cdot \mathbf{S} = \mathbf{L} \cdot \mathbf{S}$. If the velocity dependence is given by an $\mathbf{L}$, rule 2 requires that there be an even power of $\mathbf{r}$ and rule 3 requires that there be an odd power of $\mathbf{S}$; $(\mathbf{r} \times \mathbf{S}) \cdot (\mathbf{r} \times \mathbf{L})$ would satisfy these requirements, but this reduces by a vector identity to $(\mathbf{r} \cdot \mathbf{r})(\mathbf{L} \cdot \mathbf{S}) - (\mathbf{r} \cdot \mathbf{L})(\mathbf{r} \cdot \mathbf{S})$ and the second term is zero from the definition of $\mathbf{L}$, whence we again have only the $\mathbf{L} \cdot \mathbf{S}$ term. It can be shown quite generally that no other term of first order in velocity dependence is possible.

It would have been very pleasant if the available data on two-nucleon systems could have been fit with no other velocity dependence than an $\mathbf{L} \cdot \mathbf{S}$ term, and several extensive attempts have been made to do just this. However, not even semiquantitative fits to the data could be obtained, so it was concluded that higher-order velocity-dependent terms were needed.

The obvious next step is to include terms of second order in p, which means that they contain two powers of $\mathbf{p}$ or $\mathbf{L}$ or one power of each. If there are two powers of $\mathbf{p}$ or $\mathbf{L}$, the rules of Sec. 3-6 require only that there be even powers of $\mathbf{r}$ and $\mathbf{S}$; satisfactory examples are $\mathbf{p} \cdot \mathbf{p}$, $\mathbf{L} \cdot \mathbf{L}$, or $(\mathbf{L} \cdot \mathbf{S})^2$. If there is one power each of $\mathbf{p}$ and $\mathbf{L}$, the rules require an odd power of $\mathbf{r}$, which in most simple expressions can be converted to $\mathbf{r} \times \mathbf{p} = \mathbf{L}$. In any case, the number of possible terms with second-order velocity dependence is so large that no one has yet tried to determine the importance of all of them. The usual approach is to use only some of the simpler ones, and with these, reasonably satisfactory fits to the experimental data on the two-nucleon system have been obtained.

3-9 Meson Theory of Nuclear Forces

Forces in physics are derived from quantum field theories, but quantum field theory is among the most complex concepts of theoretical physics. It is generally not dealt with before second-year graduate courses, so we cannot go into it here except in a most elementary way.

A somewhat familiar example of a quantum field theory is the one which treats the electromagnetic force. In it, this force is transmitted by the exchange of the *field particle*, namely, the familiar photon. Electromagnetic field theory thus explains the coulomb force without hypothesizing the philosophically distasteful idea of action at a distance. It explains why that force is not transmitted instantaneously but with the velocity of light. It has had great success in predicting probabilities for such electromagnetic interactions as the photoelectric effect, Compton scattering, pair production and annihilation, bremsstrahlung, etc.; none of these probabilities can be calculated in any other way. All in all, it has been a useful and successful theory.

The gravitational field also is presumably derived from a quantum field theory. The force is assumed to be carried by a field particle, the graviton. Like the photon, it has zero mass and travels with the speed of light as it is exchanged between particles of matter.

The nuclear force also derives from a quantum field theory, and in this case the principal field particle is the π meson, which, fortunately, has been observed and studied extensively. It has a mass about 270 times that of the electron ($M_\pi c^2 \simeq 140$ MeV) and occurs in three forms, with positive, negative, and neutral (zero) electric charge.

There is one immediate and simple consequence of the fact that the field particle for the nuclear force, unlike that for the electromagnetic and gravitational forces, has a finite mass; namely, the force is of short range. To see this, we note that when a meson is sent from one nucleon to another (as it must be to transmit the nuclear force), the creation of the meson violates conservation of energy by an amount ΔE of the order of the meson rest mass M_π; or

$$\Delta E \simeq M_\pi c^2$$

Such a violation of energy conservation cannot last longer than a time Δt which, from the uncertainty principle, is

$$\Delta t = \frac{\hbar}{\Delta E} = \frac{\hbar}{M_\pi c^2}$$

Even assuming that the meson travels with a velocity approaching c, the farthest it can go in this time is

$$r = c\,\Delta t = \frac{\hbar}{M_\pi c} = \frac{1}{\mu} \tag{3-21}$$

This is the *range* of the nuclear force. Numerically it works out to be 1.4 F, which is in general agreement with our previous discussion. Note that a similar

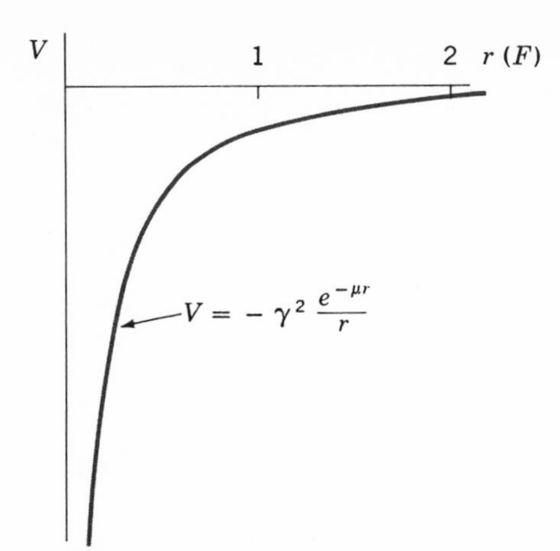

FIGURE 3-3 The Yukawa potential, Eq. (3-22).

calculation for the electromagnetic or gravitational field yields an infinite range since the mass of the field particle is zero.

A very elementary calculation from meson field theory gives the potential energy of interaction between two nucleons as

$$V = -\gamma^2 \frac{e^{-\mu r}}{r} \tag{3-22}$$

where γ is a constant and μ is defined by (3-21). This potential is shown in Fig. 3-3. It was first obtained by H. Yukawa in 1937, and is commonly known as the *Yukawa potential.*

For the more advanced student, the plausibility of (3-22) may be made more evident as follows. The electromagnetic field is well known to be derivable from Maxwell's wave equation

$$\nabla^2\Phi - \frac{1}{c^2}\frac{\partial^2\Phi}{\partial t^2} = 0 \tag{3-23}$$

This may be considered to be derived quantum-mechanically from the energy equation for photons

$$-p^2c^2 + E^2 = 0 \tag{3-24}$$

with the usual operators $-\hbar\nabla$ substituted for $\mathbf{p}$ and $i\hbar\,\partial/\partial t$ substituted for E. The time-independent field equation is obtained from (3-23) by taking

$$\Phi = \phi(r)e^{-i\omega t} \tag{3-25}$$

Using this in (3-23), we find that ϕ obeys

$$\nabla^2\phi = 0 \tag{3-26}$$

which has the well-known solution

$$\phi = \frac{e}{r} \tag{3-27}$$

Proceeding by analogy for the meson field, the energy equation is

$$-p^2c^2 - M_\pi{}^2c^4 + E^2 = 0 \tag{3-24a}$$

which, with the substitution of operators, becomes

$$\nabla^2\Phi - \frac{1}{c^2}\frac{\partial^2\Phi}{\partial t^2} + \mu^2\Phi = 0 \tag{3-23a}$$

where μ is as defined in (3-21). Using (3-25), the field equation becomes

$$\nabla^2\phi - \mu^2\phi = 0 \tag{3-26a}$$

for which the solution is

$$\phi = \gamma\frac{e^{-\mu r}}{r} \tag{3-27a}$$

The constant γ occupies a place in meson theory analogous to that of e in electromagnetic theory, so as the potential energy in the latter case is $V = e\phi$, in the former it is $\gamma\phi$. Using this in (3-27*a*) leads to (3-22).

A more detailed calculation from meson theory gives an elaboration of (3-22), the one-pion exchange potential (OPEP), which may be written as[1]

$$V = \frac{g^2}{12}M_\pi c^2\left(\frac{M_\pi}{M_p}\right)^2(2\mathbf{T}\cdot\mathbf{T} - 3)\left\{(2\mathbf{S}\cdot\mathbf{S} - 3) + S_{12}\left[1 + \frac{3}{\mu r} + \frac{3}{(\mu r)^2}\right]\right\}\frac{e^{-\mu r}}{\mu r} \tag{3-28}$$

where S_{12} is the expression for the tensor force, (3-13). It is important to note that (3-28) is of the form (3-20), with v_W, v_M, v_B, and v_H all nonzero. Hence we see from this example that the tensor force and the four exchange forces of Sec. 3-7 arise in a natural way from meson theory. The constant g^2 in (3-28) has been determined from experiments with mesons (meson-nucleon scattering). Its value is such that the dimensionless quantity $g^2/\hbar c$ is about 0.3. This quantity plays a role in meson field theories analogous to $e^2/\hbar c = 1/137$ in electromagnetic field theory. In treating electromagnetic phenomena, an expansion is frequently made in powers of $e^2/\hbar c$, and we see that this gives a rapidly converging series. In analogous expansions in meson theory, however, the convergence is obviously much more questionable, which leads to many serious difficulties.

Another way to state this problem is to point out that the potential (3-28) corresponds to the exchange of one π meson, whereas, when two nucleons are close

[1] For students familiar with σ operators, $2\mathbf{S}\cdot\mathbf{S} - 3 = \boldsymbol{\sigma}_1\cdot\boldsymbol{\sigma}_2$; $2\mathbf{T}\cdot\mathbf{T} - 3 = \boldsymbol{\tau}_1\cdot\boldsymbol{\tau}_2$.

enough, it is possible for two mesons to be exchanged simultaneously. In comparison with the calculation leading to (3-21), ΔE is twice as large, whence Δt is halved, whence the range of the force is halved. This then contributes terms to the potential proportional to $e^{-2\mu r}$. Similarly, triple meson exchange leads to terms proportional to $e^{-3\mu r}$, etc. Unfortunately, multiple meson exchange is a very complicated process which cannot be treated unambiguously by current field theories, so its effect on the nuclear force cannot be reliably calculated.

In addition to the π meson, there are other more massive mesons which contribute to the nuclear force. These are all *bosons*, which for present purposes means they have integer (or zero) spin, and they have zero *hypercharge*, a classification used in elementary-particle physics. The best known of these are

η meson: $M_\eta c^2 = 549$ MeV

ρ meson: $M_\rho c^2 = 769$ MeV

ω meson: $M_\omega c^2 = 783$ MeV

Each of these leads to terms of the form (3-22) with μ given by (3-21) with the appropriate meson mass.

In accordance with (3-21), the range of the forces due to heavier mesons and multiple meson exchange is considerably less than that due to one-pion exchange, so (3-28) is valid at large distances, say $r > 3$ F.

3-10 Nucleon-Nucleon Scattering

In Sec. 3-1 we pointed out that there are two basic methods for studying the nuclear force, by studying the bound state—the deuteron—and by studying the scattering of nucleons by nucleons. From the properties of the deuteron we learned a great deal about the nuclear potential; we estimated its range and depth and found that it was spin-dependent and had a noncentral component. However, by far the most powerful technique for investigating the nuclear force is nucleon-nucleon scattering.

A simple scattering experiment similar to the one shown in Fig. 1-1 consists of shooting a beam of neutrons or protons at a target of hydrogen and observing the probability for various angles of deflection. A hydrogen target is equivalent to a target of protons, since the electrons in hydrogen cannot deflect an incident neutron or proton any more than a basketball can deflect an automobile; in both cases, the latter is about 2,000 times heavier than the former.

We have here two different experiments, *n-p* and *p-p* scattering. The latter system is restricted by the Pauli exclusion principle to the $T = 1$ states so only potentials B, C, and E of (3-15) are effective, whereas all six potentials are operative in *n-p* scattering. The two experiments therefore give complementary

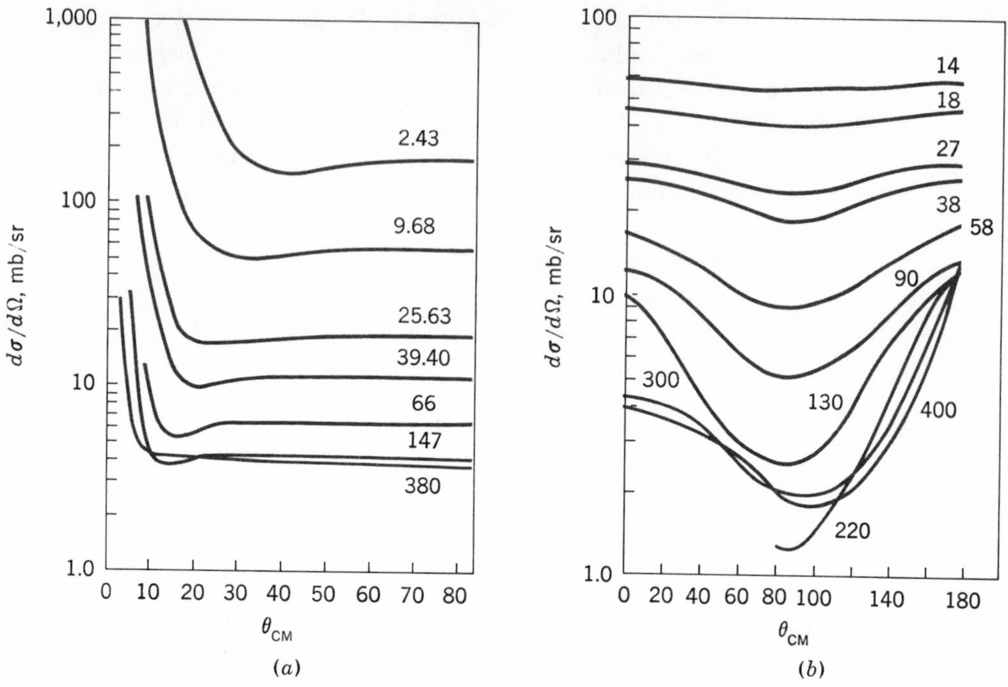

FIGURE 3-4 Results from nucleon-nucleon scattering experiments. Cross sections for *(a)* p-p **and** *(b)* n-p **scattering at various angles for various incident energies (numbers attached to curves, in MeV).** *(From M. J. Moravcsik, "The Two Nucleon Interaction," Clarendon Press, Oxford,* **1963;** *by permission.)*

information. The scattering of neutrons by neutrons, n-n scattering, has not yet been studied experimentally because sufficiently concentrated targets of neutrons are not available.

A summary of some of the results of p-p and n-p scattering experiments at energies above 9 MeV is shown in Fig. 3-4. The data at lower energies would look less spectacular on plots of this type, but they are remarkable for their extreme accuracy.

The analysis of these data makes use of quantum-mechanical scattering theory, which is beyond the scope of this text. It allows the measured cross sections to be transformed into a *phase shift* for each L and $J = |\mathbf{L} + \mathbf{S}|$ in the incident beam. Measurements at various energies determine the energy dependence of these phase shifts. Many shortcuts and tricks have been developed for the analysis. For example, all the low-energy data can be described by just two parameters, a *scattering length* and an *effective range*. Much of the analysis is now computerized.

In addition to simple scattering experiments, there have been double-scattering experiments, in which the scattered particles are deflected again by a second target, and even triple-scattering experiments. In some cases, polarized beams, i.e., beams of particles with their spins predominantly in the same direction, and polarized targets have been used. A complete analysis of these experiments determines 10 independent quantities, each being a function of angle and of energy. The cross section shown in Fig. 3-4 is only one of these; another will be shown in Fig. 3-6.

To illustrate the value of these more complex measurements, let us give an example of how a double scattering experiment can reveal the existence of a spin-orbit force. Consider the experimental arrangement in the upper part of Fig. 3-5. For simplicity let us assume that the only force operating is the spin-

FIGURE 3-5 Double-scattering experiment: ***(a)*** **the experimental arrangement and** ***(b)*** **the two scatterings on a microscopic scale as discussed in the text. The shaded circles are the target nucleons.**

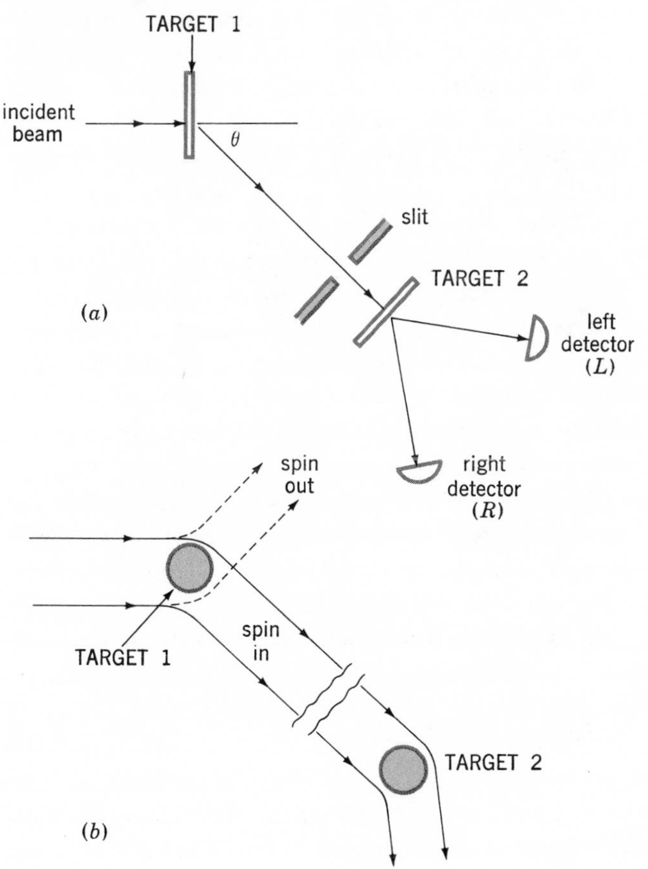

orbit force and that it is attractive if the spin **S** and orbital angular momentum **L** are in the same direction (parallel) and repulsive if they are in opposite directions (antiparallel). Now consider the microscopic description shown in Fig. 3-5*b*, where the circle is a target proton.

The spin of the target proton can be out of the page (*out*) or into the page (*in*). First let us assume that it is *in*. If the incident particle has spin *out*, the total spin is zero, so there is no spin-orbit force and no deflection. If the spin of the incident particle is *in*, the net spin is *in* and we have two possibilities to consider: (1) if it passes to the right of the nucleus, **L** is *out* (a right-hand screw would move *out* if rotated in the way the line joining the two particles is rotated), so **L** and **S** are antiparallel, the motion is therefore repulsive, and the incident particle is deflected to the right; (2) if it passes to the left of the nucleus, the orbital angular momentum is *in*, whence **L** and **S** are parallel, the force is attractive, whence again the particle is deflected to the right. Note that in both cases, the spin is *in* and the particle is deflected to the right.

All the above analysis was under the assumption that the target spin is *in*. If the target spin is *out*, a similar argument shows that incident particles with spin *in* are not deflected and those with spin *out* are deflected to the left regardless of whether they pass to the left or right of the target proton. In summary, spin-*in* particles are either not deflected or are deflected to the right, and spin-*out* particles are either not deflected or are deflected to the left. To look at things another way, the particles deflected to the right are *polarized* in that they all have spin *in* while those deflected to the left are polarized in having spin *out*. In an ordinary single-scattering experiment, this polarization is not observed; from the symmetry of the experimental setup, a detector which is insensitive to polarization must record the same number of particles at a given angle on each side of the beam.

However, if these particles undergo a second scattering in target 2, we again expect that spin-*in* particles will be deflected to the right and spin-*out* particles to the left. But there are no spin-*out* particles striking the target, so no particles are deflected to the left. Hence, if detectors are placed at positions (R) and (L) in Fig. 3-5*a*, the detector at (R) will record a great number of particles while the detector at (L) will record none. In actual cases, of course, there are other forces than the spin-orbit force present, and the situation is more complex, but still among the particles deflected to the right in the first scattering there will be more with spin *in* than with spin *out*, and there will be more particles detected at (R) than at (L). The polarization P is defined by

$$P^2 = \frac{(R) - (L)}{(R) + (L)} \tag{3-29}$$

where (R) and (L) are the number of particles reaching the detectors designated by those symbols. Measured values of P are shown in Fig. 3-6; note that the

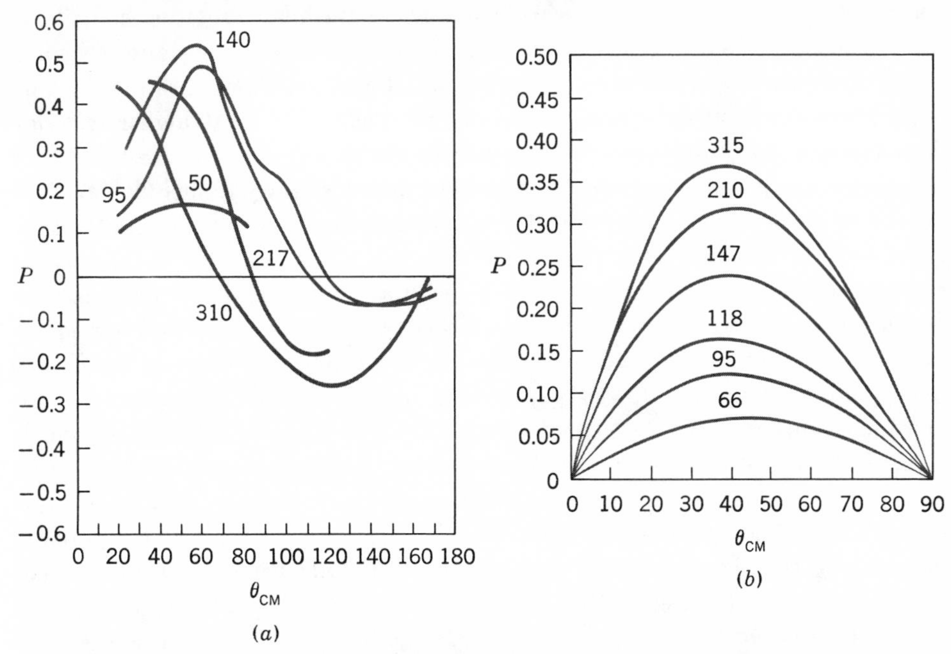

FIGURE 3-6 Polarization in nucleon-nucleon scattering experiments. This is P, as defined by (3-29), as a function of the angle in the two scatterings, for various energies of the incident particle as designated in MeV by the numbers on the curves: *(a)* n-p **and** *(b)* p-p **scattering.** *(From M. J. Moravcsik, "The Two Nucleon Interaction," Clarendon Press, Oxford, 1963; by permission.)*

determination of its sign requires a different type of experiment, which we shall not describe here. From these data, information on the spin-orbit force can be deduced.

3-11 The Nuclear Force as We Now Know It

We are now, at last, in a position to discuss the results of all the experiments and theoretical analyses referred to in the previous sections. These results should not be considered to be *the nuclear force* but an approximation to it. They will explain all our information about the deuteron plus all the information obtained from nucleon-nucleon scattering experiments up to about 350 MeV. In addition, there are some data of importance on the reaction (3-1), on the breakup of deuterons by gamma rays, and on scattering of neutrons by hydrogen molecules. The principal reason for the 350-MeV limitation is that at higher energies the production of π mesons in nucleon-nucleon scattering becomes important (the energy available in the center-of-mass system is sufficient to produce the rest mass of a π meson above 270 MeV), and this fact greatly complicates the analysis. Other contributing reasons are that the velocity dependence becomes more important at higher energies so use of the lowest-order terms is not justified, there are fewer high-quality experimental data available, and one might hope that higher energies do not play a very important role in nuclear structure.

Even to explain the data up to 350 MeV, it is necessary to include velocity-dependent terms of lowest, and next lowest order, so we have a potential with a great many terms, each being an arbitrary function of r. Since there are not nearly enough data to determine all these functions at all radii, meson theory is used as an aid in the analysis. This is done in either of two ways. A purely empirical approach is to take the r dependence as a linear combination of terms like (3-22) with various values of μ and γ adjusted to fit the experimental data. A more ambitious approach is to use a *one-boson exchange potential* (OBEP) in which single exchange of each of the mesons listed in Sec. 3-9 is treated in detail; this ignores multiple meson exchange, but these effects are presumed to be less important except at quite small distances. Unfortunately, the OBEP approach does not fit the experimental data when only the known mesons are included, so it is necessary to assume the existence of one or two additional mesons for which there is no direct experimental evidence. This approach is therefore not much more basic than the purely empirical approach. Since the latter is more accurately developed, it is more widely used in applications to nuclear structure and we shall use it here.

In determining the detailed behavior of the potentials at small radii by scattering experiments, we run into the fundamental diffraction limitations mentioned in Sec. 1-2. The wavelength of 350-MeV nucleons is about 1.4 F, so we cannot hope to determine a great deal of detail about the structure of the potential at radii much smaller than this. Unfortunately, much of the region of greatest interest is blurred by this limitation.

With these apologies and limitations in mind, let us now consider the results. The most important qualitative feature of the results is that the nucleon-nucleon force contains a *repulsive core*; i.e., it becomes very strongly repulsive at distances less than 0.5 F. This is widely believed to be due to forces arising from ω-meson exchange. Because of the diffraction limitation, the detailed nature of this repulsion cannot be discerned, so for simplicity it is often taken as a *hard core*, which means that the potential goes to $+\infty$ at some value of r. This means that two nucleons cannot approach one another closer than this distance under any circumstances.

One of the most widely used nucleon-nucleon potentials is that due to Hamada and Johnston.[1] In addition to the six static potentials (3-14), it includes

[1] T. Hamada and I. D. Johnston, *Nucl. Phys.*, **34:** 382 (1962).

spin-orbit potentials for even L and odd L, referred to as L-S; and second-order velocity-dependent potentials of the form $[\mathbf{L} \cdot \mathbf{L} - (\mathbf{L} \cdot \mathbf{S})^2]$, which are different for even L and odd L and for $S = 1$ and $S = 0$. All the central static potentials are taken to have a hard core at $r \simeq 0.50$ F; that is, the potentials go discontinuously to $+\infty$ at that distance. Since one of these four is applicable in all situations and the addition of anything to ∞ is still ∞, the hard core is effective in the total potential.

The 12 components of the Hamada-Johnston potential are shown in Fig. 3-7. We should remember here that a positive potential is repulsive and a negative

FIGURE 3-7 The components of the Hamada-Johnston potential. [*From T. Hamada and I. D. Johnston, Nucl. Phys.,* **34:** 382 (1962).]

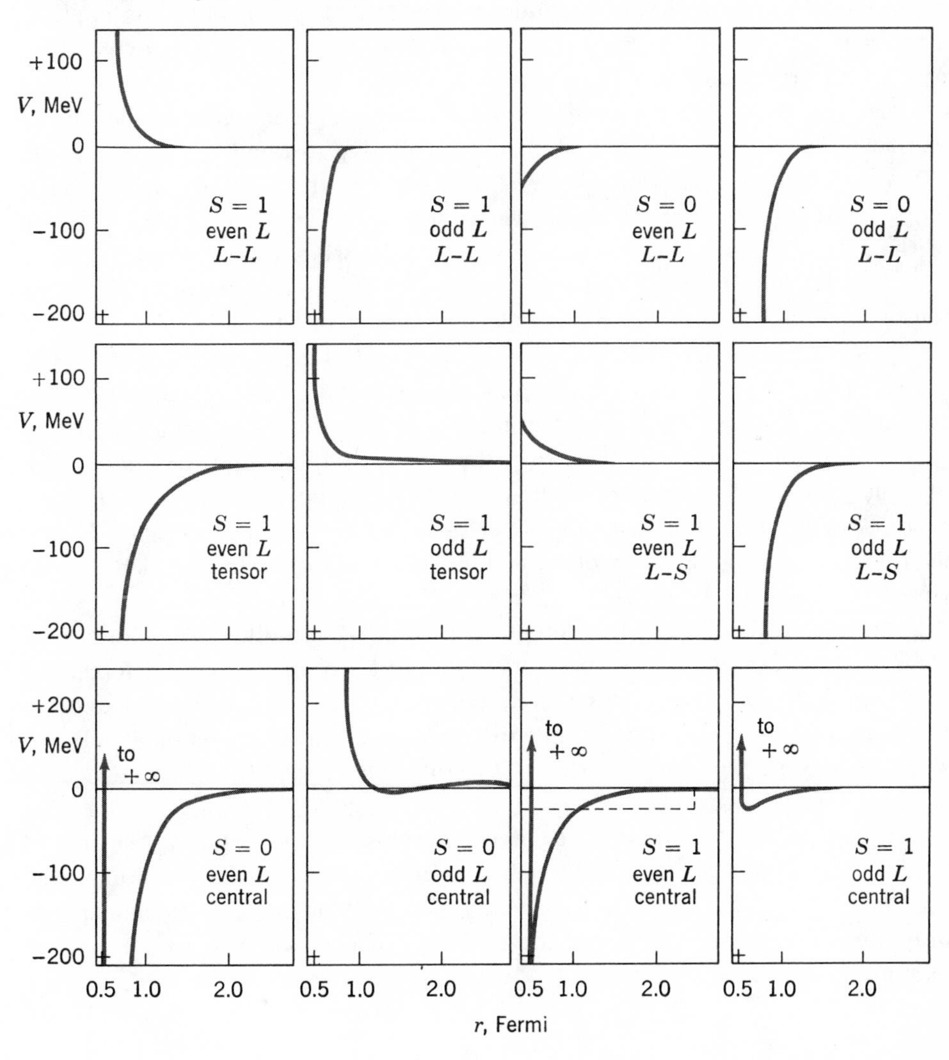

TABLE 3-1: FACTORS MULTIPLYING V_{L-S} AND V_{L-L} IN FIG. 3-7

$\|\mathbf{L} + \mathbf{S}\|$	$\times V_{L\text{-}S}$	$\times V_{L\text{-}L}$
$L + 1$	L	L
$L - 1$	$-(L + 1)$	$-(L + 1)$
$L\ (S = 1)$	-1	$2L^2 + 2L - 1$
$L\ (S = 0)$		$-2L(L + 1)$

potential is attractive. The *L-S* and *L-L* potentials shown are to be multiplied by the factors listed in Table 3-1. These factors depend on the relative orientation of the **L** and **S** vectors, as may be deduced from the first column of that table; they arise from the values of $\mathbf{L} \cdot \mathbf{S}$ and $\mathbf{L} \cdot \mathbf{L}$ themselves (note that we have avoided the problem of products of angular-momentum vectors because of its mathematical complexity). We see that these factors are sometimes much greater than unity and sometimes change the potential from attractive to repulsive or vice versa. (This is obviously true of the $\mathbf{L} \cdot \mathbf{S}$ term.) The tensor potentials, aside from a multiplicative factor close to unity, are as shown in Fig. 3-7 for the orientation of the particles relative to the spin as in Fig. 3-2*b*; when the orientation is as in Fig. 3-2*a*, the multiplicative factor is close to -1. We see that the direction of the tensor force indicated in Fig. 3-2 is valid only in even-*L* states.

It should be understood that the actual potential experienced by two interacting nucleons is the sum of all applicable potentials; e.g., if their spins are parallel ($S = 1$) and they are interacting in an $L = 1$ state, the applicable potentials are the fourth in the bottom row, the second and fourth in the middle row, and the second in the top row. If the two interacting nucleons are both protons, there is, in addition, the coulomb force. Its value is $+1.4$ MeV at $r = 1$ F, and it of course varies as $1/r$; if plotted in Fig. 3-7, it would be too small to be distinguished from the $V = 0$ line.

From the fact that all the potentials in Fig. 3-7 are different we may conclude that exchange forces play a very important role in the nuclear force. We may especially note that corresponding potentials for even *L* and odd *L* tend to be of opposite sign. This implies that space exchange forces are especially strong, since we saw in Sec. 3-6 that this is the force that leads to a dependence on whether *L* is even or odd.

It is of interest to compare the potential that is most important in the ground state of the deuteron, the $S = 1$, even-*L*, central potential, with the square well obtained in Sec. 3-2; the latter is shown by a dashed line in Fig. 3-7. Some degree of qualitative similarity is observed especially if it is noted that our square well extends to $r = 0$.

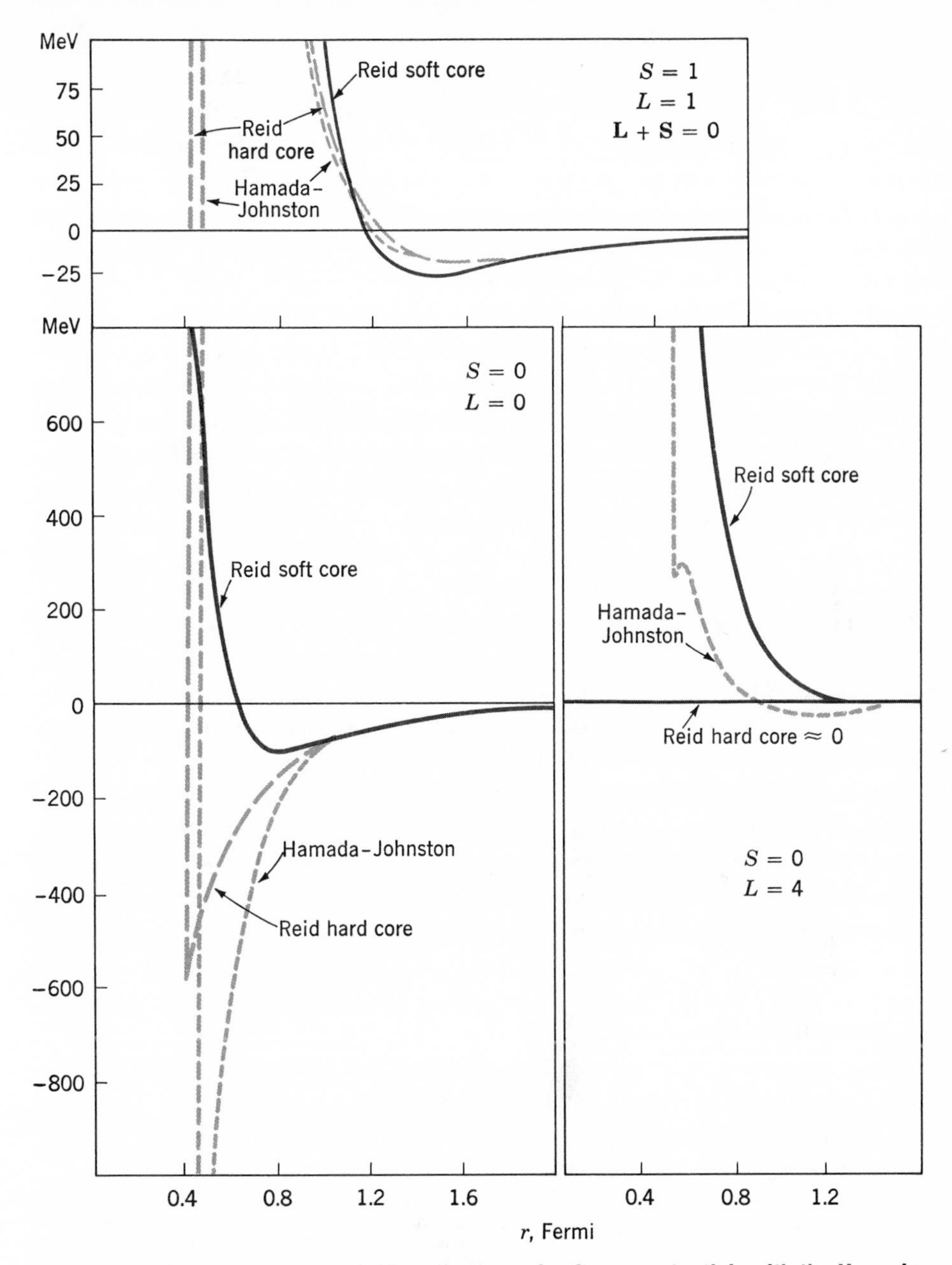

FIGURE 3-8 Comparison of the Reid hard-core and soft-core potentials with the Hamada-Johnston potential for various states of interaction. *(From R. V. Reid, Ph.D. Thesis, Cornell University,* 1968.*)*

In addition to the Hamada-Johnston potential illustrated in Fig. 3-7, several other potentials have been developed which fit all available data on the nucleon-nucleon systems. The best-known of these are those developed by Breit and coworkers at Yale and by Reid at Cornell. The former is quite similar to the Hamada-Johnston potential, but the latter handles the velocity dependence by using different potentials for each L value. In addition, one of the Reid potentials does not use a hard core but employs a repulsive potential at short distances with r dependence of the form (3-22) with $\mu = 4.9\ F^{-1}$; since it does not go abruptly to infinity, it is referred to as a *soft-core potential.* The Reid hard-core and soft-core and Hamada-Johnston potentials for a few typical states of interaction are compared in Fig. 3-8. While they are qualitatively similar, there are clear quantitative differences. These differences may be considered to be a crude measure of the degree of uncertainty in our knowledge of the nuclear force. While they do appear to be substantial, they are not large enough to lead to serious difficulties in applications to nuclear-structure problems.

3-12 Charge Independence of Nuclear Forces

For all the complications that have been encountered with the nuclear force, there is one complication that could very easily have been present but is not; i.e., with the obvious exception of the coulomb force between two protons, the potentials of the last section are the same whether the interaction is between two neutrons, a neutron and a proton, or two protons.[1] This fact, that the force is the same whether the nucleons are charged (protons) or uncharged (neutrons), is referred to as *charge independence of nuclear forces.* It has many interesting and important consequences for nuclear structure (see Sec. 6-3).

3-13 Many-body Forces

Up to this point, we have been tacitly assuming that the nuclear force is a two-body force; i.e., if we have nucleons A, B, and C close together, as in Fig. 3-9, the forces on A are just $F_{AB} + F_{AC}$, where F_{AB} is the force between A and B if C were not present and F_{AC} is the force between A and C if B were not present. This is certainly the way electromagnetic and gravitational forces behave. From our picture of forces arising from meson exchange, this implies that meson ex-

[1] Actually there are small differences that are outside the experimental uncertainty, but it is widely believed that they can be explained away by considerations from meson theory, depending on the (small) differences in mass between charged and neutral mesons.

FIGURE 3-9 Three mutually interacting nucleons.

changes operate only between pairs. But it is immediately clear that other exchange patterns are present. For example, when only two nucleons are present and one nucleon emits two mesons, both must be absorbed by the other nucleon, but if two other nucleons are present, one can be absorbed by each. This would give rise to a *three-body force*, one in which the above definition of a two-body force is not fulfilled. It is easy to see that the meson-exchange picture also predicts four-body forces, five-body forces, etc., which are collectively referred to as *many-body forces*.

Since many meson masses must be created simultaneously in these processes, the discussion leading up to (3-21) requires that the range of the forces decrease as the number of "bodies" increases; crudely we may estimate that the range of n-body forces is $1/(n-1)$ times the range of the two-body force. If we take the range to be $1/\mu$, this is about 1.4, 0.7, 0.47, 0.35, . . . F for 2-, 3-, 4-, 5-, . . . body forces, respectively.

To estimate the distances between nucleons in a nucleus, we note that the volume per nucleon is $1/A$ times the nuclear volume, which, from (1-2), is $\frac{4}{3}\pi(1.07\ \text{F})^3$; this implies that, on an average, each nucleon occupies the volume of a sphere with radius 1.07 F, whence the average distance between nucleons in the nucleus is about 2.1 F. Moreover, as was pointed out in Sec. 3-10, nucleons almost never approach each other closer than about 0.5 F due to the very strong repulsion experienced at these distances. We may therefore expect that the two-body force would be by far the most important one in determining nuclear structure and that four- or higher-body forces would be unimportant. The principal many-body force which need concern us is therefore the three-body force. This is most easily studied in systems containing three nucleons.

One method of approach to the problem is then to study three-nucleon systems and see if their properties can be calculated from two-body forces. If they cannot, this will be evidence for three-body forces. Three-nucleon systems can be studied as bound states, the H^3 and He^3 nuclei, or by scattering of neutrons or protons from deuterons, and a great deal of information is available.

The analysis is very difficult since the three-body problem cannot be solved exactly in classical mechanics and these same difficulties extend to quantum mechanics. However, a great effort has been expended on calculating the binding energy of H^3 (the triton), which is experimentally known to be 8.48 MeV. The results seem to indicate that two-body forces give a binding of only about 7 MeV, which implies that about 1.5 MeV of binding derives from three-body forces. This is in agreement with estimates from other approaches to the problem. From this we infer that three-body forces are about 20 percent as important as two-body forces in nuclei; in more complex nuclei the estimates are about 15 percent.

Problems

3-1 If the binding energy of the deuteron had been found to be 10 MeV, what would be the approximate depth of the potential in Fig. 3-1? Plot the wave function and estimate r_d.

3-2 Solve (3-2) and (3-4) simultaneously and accurately without introducing the approximations used in the text.

3-3 From the fact that the $S = 0$ state of the deuteron is just barely unbound ($E \simeq 0$) obtain a relationship between V_0 and R for the square-well approximation to the $S = 0$, even-l potential. [*Hint:* Note the development used to obtain (2-20).] Compare the result with the corresponding one for the $S = 1$, even-l potential.

3-4 Since both the neutron and proton have magnetic moments, there is a difference in their energy in the two arrangements shown in Fig. 3-2*a* and *b*. Estimate this when they are separated by r_d and when the r vector joining them is parallel to their spins and perpendicular to them. Compare this with the tensor force from Fig. 3-7.

3-5 Explain why the following dependences for potentials are not acceptable for use in this chapter:

(*a*) $[(\mathbf{r} \times \mathbf{S}) \cdot (\mathbf{r} \times \mathbf{S})](\mathbf{S} \cdot \mathbf{S})$
(*b*) $(\mathbf{r} \times \mathbf{L}) \cdot \mathbf{p}$
(*c*) $(\mathbf{L} \cdot \mathbf{S})(\mathbf{L} \cdot \mathbf{L})$
(*d*) $(\mathbf{r} \cdot \mathbf{p})(\mathbf{r} \cdot \mathbf{S})$
(*e*) $(\mathbf{r} \cdot \mathbf{p})(\mathbf{L} \cdot \mathbf{S})$

3-6 When a 10-MeV proton interacts with an electron which is initially at rest, what is the maximum angle by which the proton can be deflected? Treat both particles nonrelativistically.

3-7 Which of the potentials in Fig. 3-7 are applicable when a neutron and a proton interact in an $L = 1$, $S = 0$ state? In an $L = 2$, $S = 0$ state? Which are applicable when two neutrons interact in an $L = 3$, $S = 0$ state? In an $L = 3$, $S = 1$ state?

3-8 Solve the deuteron problem in Prob. 3-2 if the potential is a square well with a hard core at $r = 0.5$ F.

3-9 Look up the Yale nucleon-nucleon potential[1] and compare it with the Hamada-Johnston potential.

Further Reading

See General References, following the Appendix.

Brink, D. M.: "Nuclear Forces," Oxford University Press, New York, 1965.
Marshak, R. E.: "Meson Physics," McGraw-Hill, New York, 1952.
Moravcsik, M. J.: "The Two Nucleon Interaction," Clarendon Press, Oxford, 1963.
Rosenfeld, L.: "Nuclear Forces," North-Holland, Amsterdam, 1948.
Schweber, S. S., H. A. Bethe, and F. deHoffman: "Mesons and Fields," Row, Peterson, Evanston, Ill., 1955.
Wilson, E.: "The Nucleon-Nucleon Interaction," Interscience, New York, 1963.

[1] E. Segre, "Nuclei and Particles," Benjamin, New York, 1965.

Chapter 4
Complex Nuclei: Shell Theory

If an accurate solution for the three-nucleon system is as difficult as described in Sec. 3-11, it is evident that an accurate solution for a nucleus with dozens or hundreds of nucleons is all but impossible. In order to understand the structure of complex nuclei, we must therefore resort to approximations. In the search for appropriate approximations, nuclear physicists have been extremely lucky: the simplest and easiest to use approach has turned out to be surprisingly accurate. This chapter deals with the choice of this approximation and its development.

4-1 Choice of an Appropriate Approximation

The approximation used in the study of complex nuclei is to assume that *from the standpoint of any one nucleon, the forces exerted on it by all the other nucleons in the nucleus can be represented, to a first approximation, by a potential well.* For reasons that will become evident later, this potential well is known as the *shell-theory potential.* The power of this approximation is evident; the complex many-body problem is immediately converted to the simplest problem in all of quantum physics—a single particle in a potential well, which is just the subject we reviewed in Chap. 2. Moreover, the problems of nuclear structure then become closely related to those of atomic structure, whence nuclear-structure theory inherits a rich legacy of calculational techniques and a wide variety of other experience.

But while the convenience of this approach is obvious, its usefulness is not immediately clear. We know that the solution of the problem in this way will lead to an orbit picture, much like that for electrons in atoms. This method can be useful only if the nucleons in the nucleus make one or more turns around their orbits before undergoing an orbit-changing disturbance, somewhat as is illustrated in Fig. 4-1*a*. On the other hand, if the motion is as in Fig. 4-1*b*, where the direction is radically changed after a small fraction of the orbit has been traversed, the notion of orbits is almost meaningless and our approximation is of little value. In the early years of nuclear physics research it was believed that the true situation resembled case (*b*), so other approximations were tried. Models based on the motion of atoms in liquids or gases were developed, but these approaches turned

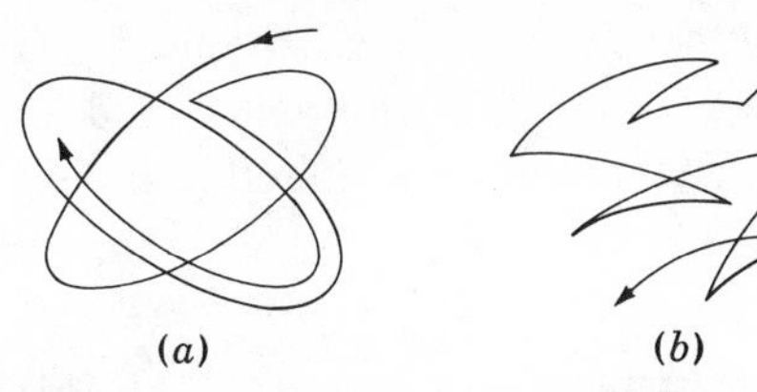

FIGURE 4-1 Situations in which an orbit approximation is *(a)* valid and *(b)* not valid. In *(a)* nucleons typically go around an orbit one or more times before having a collision. In *(b)* nucleons traverse only a small fraction of an orbit between collisions.

out to be largely unfruitful; the motions of nucleons in the nucleus are more like case (*a*). When this was first realized from an analysis of experimental data in the late 1940s (M. G. Mayer and J. H. D. Jensen later received a Nobel prize for this work), it seemed astounding. With so many other nucleons to collide with and with such strong forces acting between them, how can a nucleon go around complete orbits without a collision?

In spite of extensive efforts, the answer to this question was not forthcoming for nearly a decade, until K. A. Brueckner and collaborators succeeded in developing approximate solutions to the many-body problem. Starting with the two-nucleon forces we have discussed in Chap. 3, they were able to show that case (*a*) is actually closer to the correct one.

The explanation relies heavily on the Pauli exclusion principle. There are endless numbers of collisions that could take place if a nucleus were a classical system, an assemblage of little balls moving with high velocity and confined to a small volume. But in reality the nucleus is a quantum system in which nucleons are restricted to a very limited number of allowed orbits, with the further restriction from the Pauli exclusion principle that there can never be more than one nucleon of a given type (neutron or proton) in any orbit. This severely limits the possibilities for collisions; in fact we shall see in Chap. 5 that in some nuclei collisions are all but impossible.

4-2 The Shell-theory Potential

Now let us proceed under the assumption that our approximation is correct, that nucleons move as though they are in a potential well. The first problem is to determine the shape of this shell-theory potential. One might hope that the calculations of the Brueckner group would answer this question, but unfortunately they are not sufficiently accurate. We must rely on less basic approaches.

Since the potential well is caused by the forces exerted by all the nucleons in the nucleus, it seems reasonable to assume that the depth of the well should be roughly proportional to the density of nucleons. The potential is therefore taken, in accordance with (1-1), as

$$V = -\frac{V_0}{1 + \exp[(r - R)/a]} \tag{4-1}$$

The constants in (4-1) have been determined by methods to be described in Sec. 13-3. They are

$$\begin{aligned} V_0 &\simeq 57 \text{ MeV} + \text{corrections} \\ R &\simeq 1.25A \; A^{1/3} \text{ F} \\ a &\simeq 0.65 \text{ F} \end{aligned} \tag{4-2}$$

It may be noted that both R and a in (4-2) are larger than the corresponding quantities in (1-2). This is due to the fact that the potential extends beyond the positions of the outermost nucleons by the range of the nuclear force.

The most important correction to the value of V_0 given in (4-2) is due to what is known as the *symmetry energy*, which arises from unequal numbers of neutrons and protons in the nucleus. As we have seen in Sec. 3-4, a neutron and a proton can interact in more ways than two neutrons or two protons because in the latter cases, many of the interactions are forbidden by the Pauli exclusion principle. The effective force between a neutron and a proton is thus stronger than the others. Therefore, if a nucleus has more neutrons than protons, V_0 is stronger than (4-2) for a proton, since its interaction in the nucleus is mostly with neutrons, and weaker than (4-2) for a neutron, since its interaction is mostly with other neutrons. The shift in V_0 due to this symmetry effect ΔV_s has also been determined by the methods of Sec. 13-3. It is approximately

$$\Delta V_s = \pm 27 \text{ MeV} \times \frac{N - Z}{A} \qquad \begin{cases} - \text{ neutrons} \\ + \text{ protons} \end{cases} \tag{4-3}$$

The shell-theory potential for a proton must include, in addition to (4-1) and (4-3), a repulsive coulomb potential. In order to estimate it, let us for simplicity assume that the nucleus has a sharp edge at $r = R_c$ and a constant electric charge density ρ inside. There is a well-known theorem in electrostatics which states that for a spherically symmetric charge distribution, the electric field at radius r' is the same as if all the charge at $r < r'$ were concentrated at $r = 0$. This field $E(r)$ is then

$$E(r) = \frac{\frac{4}{3}\pi r^3 \rho}{4\pi\epsilon_0 r^2} = \frac{r}{R_c} E(R_c) \qquad r < R_c$$

Since all the charge Ze is inside the radius R_c,

$$E(r) = \frac{Ze^2}{4\pi\epsilon_0 r^2} \qquad r > R_c$$

whence

$$E(R_c) = \frac{Ze^2}{4\pi\epsilon_0 R_c^2}$$

The potential energy $V(r)$ is defined in electrostatics as

$$\begin{aligned} V(r) &= \int_\infty^r E(r)\, dr \\ &= \int_\infty^{R_c} \frac{Ze^2}{4\pi\epsilon_0 r^2}\, dr + \int_{R_c}^r \frac{Ze^2}{4\pi\epsilon_0 R_c^2} \frac{r}{R_c}\, dr \\ &= \begin{cases} \dfrac{Ze^2}{4\pi\epsilon_0 R_c} \left\{1 + \dfrac{1}{2}\left[1 - \left(\dfrac{r}{R_c}\right)^2\right]\right\} & r < R_c \\ \dfrac{Ze^2}{4\pi\epsilon_0 r} & r > R_c \end{cases} \end{aligned} \tag{4-4}$$

This potential is therefore 1½ times larger at the center of the nucleus than at the edge. Examples of the coulomb potential (4-4), the shell-theory potential for a neutron (4-1), and their sum, which is the shell-theory potential for a proton, are shown in Fig. 4-2.

FIGURE 4-2 Comparison of shell-theory potentials for neutrons and protons. The dashed curve is the potential for neutrons from (4-1) and (4-2). The proton potential, shown by the solid curve, is the sum of this and the coulomb potential. The latter, from (4-4), is shown by the dot-dash curve.

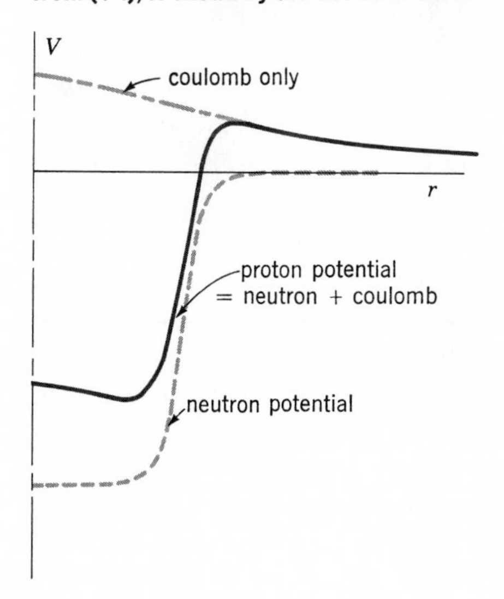

Now let us consider the effect on the shell-theory potential of all the complications in the nucleon-nucleon force—the spin and parity dependence arising from the exchange character, the noncentral component, the velocity dependence, etc. The only vectors that have meaning for a nucleon in the nucleus are its spin **s**, its position relative to the center of the nucleus **r**, and its linear and angular momentum relative to the center of the nucleus, **p** and **l**, respectively. As was pointed out in Sec. 1-6 and as is evident from Table A-2, the total angular momentum of nuclei is near zero, so there are approximately equal numbers of nucleons with spin *up* and spin *down*. Consequently, a nucleon with spin *up* would encounter the same spin-dependent forces on an average as a nucleon with spin *down*, so the shell-theory potential cannot depend explicitly on **s**. Since we are interested only in the average force experienced by a nucleon, we need not consider a possible $\mathbf{r} \cdot \mathbf{s}$ force arising from the nucleon-nucleon tensor force; due to the symmetry of the situation, this would give a constant average force around an orbit and would therefore effectively add to the central forces. The nucleon-nucleon force depends on whether the orbital angular momentum L relative to some nucleon with which it interacts is even or odd, but there is no correlation between the evenness or oddness of l and of L, so the shell-theory potential cannot depend on whether l is even or odd. This exhausts the static forces of Chap. 3.

The velocity-dependent forces of Chap. 3 cause much more difficulty. First let us consider the spin-orbit interaction. From Fig. 3-7 we see that it is strongest in odd-L states, where it causes a potential-energy decrease when **L** and **S** are parallel and a potential-energy increase when they are antiparallel. It is zero when $S = 0$, that is, when the spins of the two nucleons are antiparallel.[1] In order to understand the effects of this force, let us consider the shaded nucleon in Fig. 4-3, which clearly has an **l** vector out of the page, as it interacts with two other nucleons in the nucleus, labeled 1 and 2. In its encounter with nucleon 1, **L** is out of the page and hence is parallel to **l**, while in its encounter with nucleon 2, **L** is into the page and hence antiparallel with **l**. If **s** is parallel to **l**, there can be a spin-orbit interaction with the other nucleons only if their spins are also parallel to **l** (so as to make $S = 1$), in which case there is a potential-energy decrease in the interaction with 1 (**L** and **S** are parallel) and a potential-energy increase in the interaction with 2. If, on the other hand, **s** is antiparallel with **l**, similar reasoning reveals that if there is a spin-orbit interaction, i.e., if $S = 1$ rather than 0, the potential-energy changes will be positive with nucleon 1 and negative with nucleon 2. As our shaded nucleon passes through the interior of the nucleus, it encounters nucleons on both sides equally often, whence there is no average

[1] The true meanings of $S = 0$ and $S = 1$ are explained in Sec. 3-4. We use the crude meanings here to simplify the discussion.

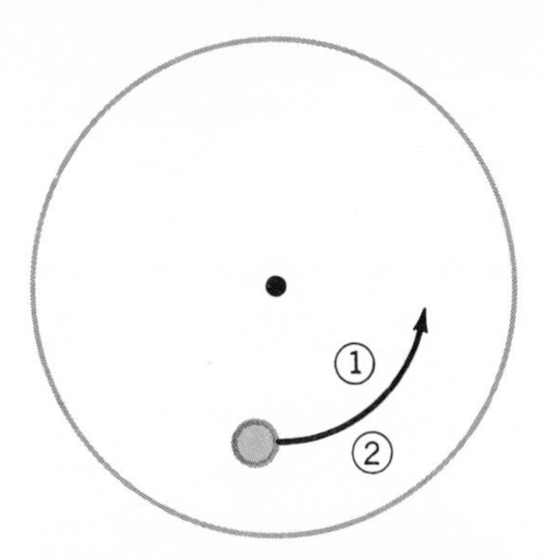

FIGURE 4-3 A nucleon moving through the nucleus interacting with other nucleons 1 and 2. This is used to explain the spin-orbit term in the shell-theory potential.

change in potential energy, regardless of whether **s** is parallel or antiparallel with **l**. However, when it is near the surface of the nucleus, where the density of nucleons is decreasing, it will pass more nucleons on the inside of its orbit, like nucleon 1, than on the outside of its orbit, like nucleon 2, because the density of the former is larger. In this situation, therefore, there is a net potential-energy decrease if **s** is parallel with **l** and a net potential-energy increase if **s** is antiparallel with **l**. The shell-theory potential therefore does depend on $\mathbf{l} \cdot \mathbf{s}$ in the surface region.[1] Moreover, it is evident that the magnitude of this interaction is, on an average, proportional to l since both l and $\boldsymbol{L}$ are proportional to the velocity. The effect of this *spin-orbit interaction* on the shell-theory potential is shown schematically in Fig. 4-4.

4-3 Effective Mass

Finally let us consider the effect of the second-order velocity dependence of the nucleon-nucleon force on the shell-theory potential. If the nucleon-nucleon force becomes more attractive (or repulsive) as the relative velocity of the two nucleons increases, it is clear that the average interaction experienced by a nucleon becomes more attractive (or repulsive) toward the center of the nucleus as its velocity in the nucleus increases, whence the shell-theory potential becomes deeper (or

[1] A small additional dependence on $\mathbf{l} \cdot \mathbf{s}$ is introduced by the tensor force.

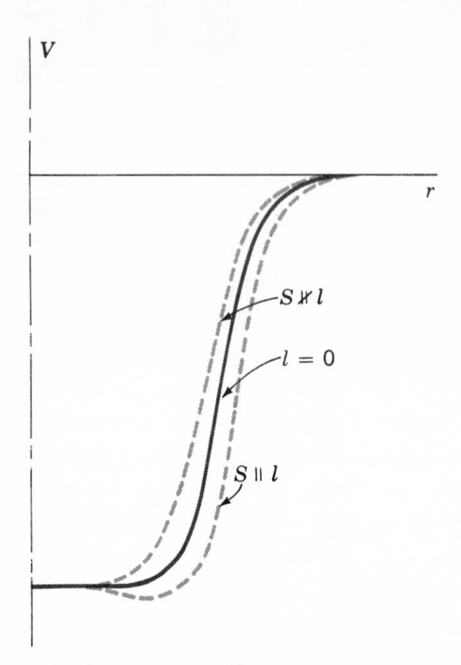

FIGURE 4-4 The effect of the spin-orbit interaction on the shell-theory potential. It has no effect for $l = 0$, so that curve is from (4-1) and (4-2). The spin-orbit interaction introduces added attraction if l and s are parallel and added repulsion if l and s are antiparallel in the surface region.

shallower). This can be expressed as a power series in the momentum dependence of V_0; since V_0 is a scalar, this must be a power series in $\mathbf{p} \cdot \mathbf{p}$, or p^2, as

$$V_0 = V_{00} + \alpha p^2 + \beta p^4 + \cdots \tag{4-5}$$

First we consider only the lowest-order term. The expression for energy

$$\frac{p^2}{2M} - V_0 = E$$

which serves as the basis for much of classical mechanics and essentially all of quantum mechanics then becomes

$$\frac{p^2}{2M} - V_{00} - \alpha p^2 = E$$

This can be rewritten

$$\frac{p^2}{2M^*} - V_{00} = E$$

where we have introduced the *effective mass* M^* defined by

$$\frac{1}{2M^*} = \frac{1}{2M} - \alpha \tag{4-6}$$

We see that the lowest-order velocity dependence can be taken into account in practically all problems merely by redefining the mass. In this way, no new formal difficulties are introduced into the solution of problems; e.g., all methods and results given in Chap. 2 are still valid if we only make the trivial replacement of M by M^*. Very roughly, higher-order terms in (4-5) can be considered to be variations in M^* with energy.

Much of our knowledge of nuclear structure comes from a rather limited energy region, and from (2-3) we see that in a limited energy region any change in M can be compensated for by a change in V_0 without altering the values of k and E. Since energies and wave functions depend only on k and E, we cannot determine M and V_0 separately from studies in a limited energy region. It is therefore conventional to use $M^* = M$ and determine V_0 from these studies. This method was used in obtaining the value ($\simeq$ 57 MeV) given in (4-2), and that value will be used here. Our present knowledge is insufficient to justify any other procedure.

On the other hand, it should be made clear that this procedure is at best an approximation valid over a limited energy region. Energies of proton orbits have been reliably measured to be as low as -60 MeV, which would be impossible in the potential well (4-1), (4-2), which, including the coulomb term, is less than 50 MeV deep. Calculations of V_0 by methods to be discussed in Sec. 7-3 indicate that it is closer to 100 MeV for very low velocity nucleons. Measurements of V_0 at higher energies, to be described in Sec. 13-3, indicate that it does change with energy. A rather direct determination of M^*/M will be discussed in connection with (4-19), and the ratio is not unity.

One problem introduced by the velocity dependence of V_0 arises because, from Fig. 4-2, the total potential well is shallower for protons than for neutrons due to coulomb forces. For the same total energy E, the kinetic energy is therefore lower for protons, whence, in accordance with (4-5), V_0 is different for neutrons and protons. This difference is a function of ΔV_c and hence of $Z/A^{1/3}$. One analysis[1] of measurements of V_0 for protons in the energy range between 9 and 22 MeV gives

$$V_0\,(\text{MeV}) = 53.3 - 0.55E\,(\text{MeV}) + 0.4\frac{Z}{A^{1/3}} + 27\frac{N-Z}{A} \tag{4-7}$$

[1] F. G. Perey, *Phys. Rev.*, **131**, 745 (1963).

The second term is the velocity dependence (4-5), the third term represents the difference between protons and neutrons discussed above, and the last term is just (4-3). Note that the minus sign in the second term of (4-7) implies that α in (4-5) is negative, whence from (4-6), M^* is smaller than M in this region. Unfortunately, (4-7) is valid only over a limited energy region, and this does not include the region of greatest interest in studies of nuclear structure. We therefore have no reasonable alternative to the use of (4-2) in conjunction with (4-3).

4-4 Allowed Orbits in the Shell-theory Potential

Once we have our shell-theory potential, the problem of solving for the energies and wave functions available to nucleons in the nucleus (we shall use our orbit-model language and refer to them as the *allowed orbits*) is just the problem considered in Chap. 2. In fact, the energy-level scale in Fig. 2-4 was distorted so as to make the results correspond roughly to those in the shell-theory potential (4-1) rather than to those in a square well. Hence, the solutions are essentially the ones from Fig. 2-4 in conjunction with the other terms in (2-25). The energy levels from Fig. 2-4 (extended) are plotted on the left side of Fig. 4-5. However, before using them we must consider the effects of the spin-orbit interaction.

We have had some experience in atomic physics with spin-orbit interactions. For example, the yellow line in the optical spectrum from sodium is actually a doublet, with the two components being split by the atomic spin-orbit interaction. We found in atomic physics that when there is a spin-orbit interaction, it is profitable to introduce the total angular momentum of a particle $\mathbf{j} = \mathbf{l} + \mathbf{s}$. Associated with it is a quantum number j such that, in analogy with other angular momenta,

$$|\mathbf{j}| = \sqrt{j(j+1)}\,\hbar \qquad \textbf{(4-8a)}$$

Also, the z component of $\mathbf{j}$ is quantized as

$$j_z = m\hbar \qquad \textbf{(4-8b)}$$

where m may be $-j, -j+1, \ldots +j$, a total of $2j+1$ allowed values. Using the methods of Sec. 2-7 and recognizing that $s = \frac{1}{2}$ for nucleons, we find

$$j = |\mathbf{l} + \mathbf{s}| = \begin{cases} l + \frac{1}{2} & l \parallel s \\ l - \frac{1}{2} & l \parallel s \end{cases} \qquad \textbf{(4-8c)}$$

For a given l, there are these two possible values of j. For example, nucleons in $l = 2$ orbits can have $j = \frac{5}{2}$ or $\frac{3}{2}$. In using the spectroscopic notation introduced in Table 2-1, the value of j appears as a subscript, so the cases in our example are designated $d_{5/2}$ or $d_{3/2}$. We shall think of these as different orbits even though their (θ,ϕ) dependences are the same.

Now let us consider the effects of the spin-orbit interaction on the energies of these states. From Fig. 4-4, we see that the potential well is narrower and shallower when **l** and **s** are antiparallel ($j = l - \frac{1}{2}$) than when they are parallel ($j = l + \frac{1}{2}$). In Sec. 2-3 we showed that when a well is made narrower and shallower, the energy level is pushed up. The $1d_{3/2}$ level is therefore at a higher energy than the $1d$ level in Fig. 2-4, and, conversely, the $1d_{5/2}$ level is at a lower energy. The $1d$ level is said to be *split* by the spin-orbit interaction, and the energy difference between the two is referred to as the *spin-orbit splitting.* Since the strength of the spin-orbit interaction is proportional to l, this energy difference increases approximately linearly with l; for example, the energy difference between $p_{3/2}$ and $p_{1/2}$ orbits is about one-third of the energy difference between $f_{7/2}$ and $f_{5/2}$. As we see from Fig. 4-4, the spin-orbit interaction changes the potential only near the outer edge of the nucleus; since the relative importance of the outer edge relative to the entire volume of the nucleus decreases as the well gets wider, the magnitude of the spin-orbit splitting decreases with increasing A.

When the locations of the states from Fig. 2-4 are modified by this spin-orbit splitting, the various energy levels fall approximately as in Fig. 4-5. Since this figure will be extremely important in many of our future discussions, the student should be certain to understand its origin thoroughly.

The most striking observation from Fig. 4-5 is that the energies occur in groups: several orbits are close together in energy, and then there is a large gap, and this pattern repeats itself many times. A grouping of energy levels with large gaps between the groups also occur in atomic physics, as we all know very well. In the atomic-physics problem, the groups of orbits with the same energy also have the property that their average radius is about the same, whence these electrons lie in spherical *shells.* The groups are therefore referred to as shells. This name has carried over to the nuclear case, where the energy groupings are also referred to as shells. As we can see from the wave functions in Fig. 2-4, there is no very strong tendency for nucleons in the same shell to have the same average radius, so this term is something of a misnomer.

We have assigned numbers $\mathfrak{N}$ to the shells in Fig. 4-5 as an aid in our discussion. It is well worthwhile for any student of nuclear physics to remember the states in each shell. This requires very little memorizing if one recognizes that, as a first approximation, the shells contain the very orderly sequence listed in Table 4-1. This is the sequence obtained from Fig. 2-4, and more advanced students may recognize that it is exactly the one derived from a harmonic oscillator potential,

$$V = -V_1 + \tfrac{1}{2}M\omega^2r^2 \qquad \textbf{(4-9)}$$

where V_1 and ω are constants. The harmonic oscillator potential is a rather good approximation to the potential (4-1), and because of its mathematical simplicity

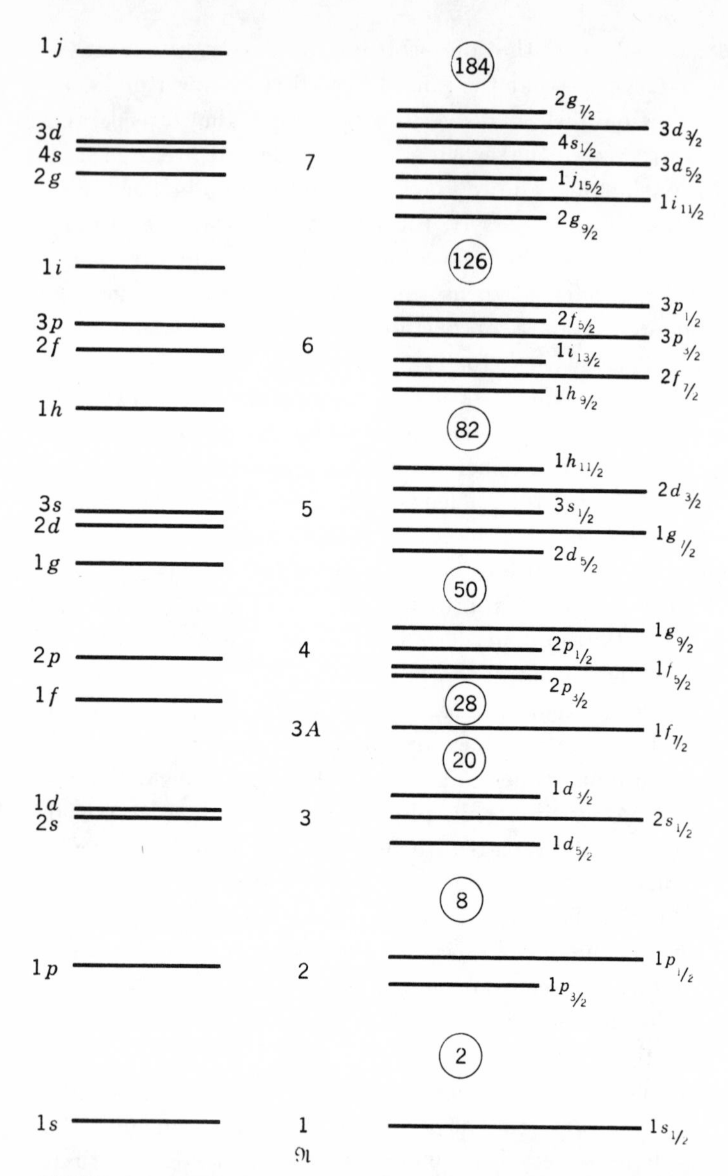

FIGURE 4-5 Energy levels in the shell-theory potential. The energy levels at the left are roughly what is expected without spin-orbit interaction. They are just the levels from Fig. 2-4 extended to larger potential wells. The right side shows the effect of the spin-orbit interaction in splitting each level (except $l = 0$) into two with $j = l + \frac{1}{2}$ and $j = l - \frac{1}{2}$. The amount of this splitting is roughly proportional to l.

TABLE 4-1: ORBITS IN EACH SHELL IN FIRST APPROXIMATION
These are also the orbits in each grouping for a harmonic oscillator potential

$\mathfrak{N}$ \ l	0	1	2	3	4	5	6	7
1	1*s*							
2		1*p*						
3	2*s*		1*d*					
4		2*p*		1*f*				
5	3*s*		2*d*		1*g*			
6		3*p*		2*f*		1*h*		
7	4*s*		3*d*		2*g*		1*i*	
8		4*p*		3*f*		2*h*		1*j*

it is often used in calculations. The wave functions then become associated Hermite polynomials. It should be clearly realized, however, that (4-9) is merely a potential-well shape; nothing is oscillating in the classical sense.

In order to go from Table 4-1 to Fig. 4-5, we must recognize that, due to spin-orbit splitting, the highest j state in each shell above the fourth is moved down to the next shell. The highest j state in the $\mathfrak{N} = 4$ shell, the $1f_{7/2}$, is also moved down but not far enough to become a member of the $\mathfrak{N} = 3$ shell. It is therefore between the $\mathfrak{N} = 3$ and 4 shells, and is hence often treated as a separate shell, which we designate $\mathfrak{N} = 3A$. Aside from these changes, the orbits in each shell are as listed in Table 4-1.

The next point of interest in connection with Fig. 4-5 is the number of allowed orbits in each shell. Each state in Fig. 4-5 represents $2j + 1$ allowed orbits corresponding to the various values of m, so the total number of allowed orbits in $\mathfrak{N} = 1$ is 2, in $\mathfrak{N} = 2$ is $4 + 2 = 6$, in $\mathfrak{N} = 3$ is $6 + 2 = 12$, in $\mathfrak{N} = 3A$ is 8, in $\mathfrak{N} = 4$ is $4 + 6 + 2 + 10 = 22$, etc. The cumulative number of orbits up to the top of each shell is then 2, 8, 20, 28, 50, etc. These numbers are circled in Fig. 4-5.

4-5 Filling of Allowed Orbits in the Shell-theory Potential

We now turn our attention to the question of how the allowed orbits shown in Fig. 4-5 fill as nucleons are put together to form nuclei. Due to the Pauli exclusion principle, there can be no more than one neutron and one proton in each orbit, i.e., with the same n, l, j, m quantum numbers, so if we confine our attention to nuclei in their lowest-energy states, we expect the orbits to fill in order of increasing energy as in the familiar case of electrons in atoms. For example, the nucleus with two protons and two neutrons, ${}_2\mathrm{He}^4$, would have all $\mathfrak{N} = 1$ orbits full and all others empty. A nucleus of this type, with all shells either completely full or completely empty, is called a *closed-shell* nucleus. Other closed-shell nuclei found in nature are ${}_8\mathrm{O}^{16}$, ${}_{20}\mathrm{Ca}^{40}$, ${}_{20}\mathrm{Ca}^{48}$, and ${}_{82}\mathrm{Pb}^{208}$. They play a role in nuclear structure analogous to that of the noble gases, He, Ne, Ar, Kr, Xe, and Rn, in atomic structure.

Another interesting type of nucleus is the so-called *single particle* nucleus, which has all shells completely full or completely empty except for a single particle in the lowest-energy otherwise empty shell. Examples of these are ${}_2\mathrm{He}^5$, ${}_3\mathrm{Li}^5$, ${}_8\mathrm{O}^{17}$, ${}_9\mathrm{F}^{17}$, ${}_{20}\mathrm{Ca}^{41}$, and ${}_{21}\mathrm{Sc}^{41}$. Their role is analogous to that of the alkali metals, Li, Na, K, etc., in atomic structure. A closely related group are the single-hole nuclei, ones in which all shells are completely full or completely empty except for a single missing nucleon in the highest-energy filled shell. Some examples of single-hole nuclei are ${}_8\mathrm{O}^{15}$, ${}^7\mathrm{N}^{15}$, ${}_{19}\mathrm{K}^{39}$, and ${}_{20}\mathrm{Ca}^{39}$. Their role in nuclear structure is similar to that of the halogens, F, Cl, Br, etc., in atomic structure.

4-6 Separation Energies of Nucleons

In order to get a feel for the properties of the shell-theory potential, in this section and the next we do some simple calculations with it. If time is short, they may be omitted without loss of continuity provided the principal results, (4-11) and (4-15), are accepted. In this section, we do a rough calculation to see how far up in the well the levels are filled in nuclei we normally encounter.

The easiest calculation we can do is for $s_{1/2}$ states in a square well. Inside the well, the wave functions are of the form sin kr and if we use the simplified approach of Sec. 2-1, there are $n/2$ wavelengths fitting into the radius R, or

$$\frac{n}{2}\frac{h}{\sqrt{2M(E-V)}} = R$$

This can be solved to give

$$E = -V + \frac{n^2h^2}{8MR^2} \tag{4-10}$$

The calculations are outlined in Table 4-2; let us go through them. In accordance with the ordering of orbit energies in Fig. 4-5, the $2s_{1/2}$ orbit should fill after the $1s_{1/2}$, $1p_{3/2}$, $1p_{1/2}$, and $2d_{5/2}$ orbits; as these accommodate respectively $2 + 4 + 2 + 6 = 14$ nucleons, the fifteenth neutron and proton go into the $2s_{1/2}$ orbits. Similarly we can easily calculate that the sixty-fifth neutron and proton go into the $3s_{1/2}$ orbits and the one hundred seventy-first go into $4s_{1/2}$ orbits. We shall do our calculation for nuclei with these numbers of neutrons and protons.

TABLE 4.2: CALCULATION OF THE ENERGY OF THE LEAST BOUND NUCLEON, $E = -S_n$ OR $-S_p$ IN VARIOUS NUCLEI IN WHICH IT IS IN AN $s_{1/2}$ ORBIT
Energies in Mev; the last line is the experimentally determined separation energy S_n or S_p averaged between the two $s_{1/2}$ nucleons in each case

	Neutrons			*Protons*	
	$2s$	$3s$	$4s$	$2s$	$3s$
N	15	65	171	16	94
Z	14	48	~104	15	65
A	29	113	~275	31	159
$A^{1/3}$	3.07	4.84	6.51	3.14	5.43
R,F	3.84	6.04	8.13	3.93	6.78
n	2	3	4	2	3
$\frac{2n - \frac{1}{2}}{0.9}$	3.9	6.1	8.3	3.9	6.3
$E - V$	55	50	50	53	41
V_0	57	57	57	57	57
ΔV_s	−1.0	−4.1	−6.6	+1.0	+4.9
ΔV_c	0	0	0	−7.2	−17.9
V	56.0	52.9	50.4	50.8	44.0
$-E$	1.0	2.9	0.4	−2.2	3.0
S_n, S_p	9.5	7.8	~5	8.1	6.8

The number of protons and neutrons, respectively, in these nuclei can be found from atomic numbers and atomic weights of elements found in nature; a short extrapolation is needed for the nucleus with 171 neutrons. The next few lines of Table 4-2 are devoted to determining R from (4-2), and then $E - V$ from (4-10) (the seventh line is for a later use).

The following four lines of Table 4-2 show a calculation of the depth of the potential well V as the sum of V_0 (from (4-2), ΔV_s from (4-3), and the coulomb potential for protons ΔV_c. For the sake of simplicity we take ΔV_c to be a constant equal to the average of the coulomb potentials (4-4) at $r = 0$ and $r = R$. By adding V to $E - V$, we obtain E.

This calculation is not very accurate because of the square-well assumption and the use of the simplification of Sec. 2-1. However, it is correct in indicating that E is close to zero; the shell-theory potential is filled with nucleons "nearly to the top." Moreover, as is also indicated by the calculation in Table 4-2, E is nearly the same for stable nuclei of all masses, and it is the same for both neutrons and protons.

This approximate equality of the highest neutron and proton energies in a nucleus is, as we shall see in Sec. 8-2, a condition for stability against beta decay and must therefore be a property of all stable nuclei. It is this condition that determines the relative number of neutrons and protons in a stable nucleus. The potential well for protons is shallower than the one for neutrons (by $\Delta V_c - 2\Delta V_s$ in Table 4-2 or, more qualitatively, from Fig. 4-2), whence it takes fewer protons than neutrons to fill all orbits up to a given energy. This explains why heavy nuclei have more neutrons than protons. It is basically a coulomb effect ΔV_c; the symmetry effect ΔV_s works against this inequality but is much weaker.

The energy required to remove the least bound neutron or proton from a nucleus is called the *nucleon separation energy* S_n or S_p. For the nuclei considered in Table 4-2, it corresponds to $-E$ because the $s_{1/2}$ nucleons are the least bound in those nuclei, and if a nucleon is to be removed from a nucleus, its energy must be raised to $E = 0$. Separation energies can be determined experimentally in various ways including the one mentioned in connection with (3-1) and another to be described in Sec. 7-1. The pertinent values are listed in the last row of Table 4-2. We see there that separation energies are typically

$$S_p \approx S_n \approx 8 \text{ MeV} \tag{4-11}$$

The accuracy of the calculation in Table 4-2 may be judged by comparing the last two lines; it is in error by a few MeV, which is not too bad considering that the total effects being calculated are of the order of 50 MeV.

4-7 Energy Spacings between Shells

For future applications we shall find it useful to know the energy difference between successive shells, $dE/d\mathfrak{N}$. Since the $1s_{1/2}$, $2s_{1/2}$, $3s_{1/2}$, . . . orbits are in the $\mathfrak{N} = 1, 3, 5, \ldots$ shells of Fig. 4-5, if we restrict our considerations to the $s_{1/2}$ orbits, we can write

$$\mathfrak{N} = 2n - 1 \tag{4-12}$$

$dE/d\mathfrak{N}$ is then $\frac{1}{2}dE/dn$, whence, from (4-10),

$$\frac{dE}{d\mathfrak{N}} \propto n \propto \mathfrak{N} + 1$$

However, this result, implying that the energy spacing between shells increases rapidly with increasing $\mathfrak{N}$, is incorrect. Once again the difficulty arises from the use of a square well and the simplification of Sec. 2-1. Calculations with the potential (4-1), (4-2) indicate that $dE/d\mathfrak{N}$ is roughly constant for all shells in a given nucleus. More advanced students may recognize this to be the same as the situation in a harmonic oscillator potential, for which[1]

$$E = -V_1 + (\mathfrak{N} + \tfrac{1}{2})\hbar\omega \tag{4-13}$$

whence

$$\frac{dE}{d\mathfrak{N}} = \hbar\omega = \text{const} \tag{4-14}$$

Complete calculations for either the potential (4-1), (4-2) or for the harmonic oscillator potential (4-9) give more specifically

$$\frac{dE}{d\mathfrak{N}} \simeq \frac{41 \text{ MeV}}{A^{1/3}} \tag{4-15}$$

For the harmonic oscillator this may be understood crudely if we realize that the energy of the least bound nucleon $-S_n$ is related to the radius R' of the potential at that energy by (4-9). Equating this energy as obtained from (4-13) and (4-9) gives

$$(\mathfrak{N} + \tfrac{1}{2})\hbar\omega = \tfrac{1}{2}M\omega^2 R'^2 \tag{4-16}$$

From (4-12) and line 7 of Table 4-2 we see that, to a rather good approximation,

$$\mathfrak{N} + \tfrac{1}{2} = 2n - \tfrac{1}{2} = 0.9R \qquad \text{F}^{-1}$$

If we take R' as the radius of the potential (4-1), (4-2) at $E = -S_n$, for nuclei of medium mass it is approximately

$$R' \approx 1.15R$$

Using these in (4-16) reduces that equation to

$$\hbar\omega = \frac{1.36\hbar^2}{MR} = \frac{41 \text{ MeV}}{A^{1/3}} \tag{4-17}$$

which substituted into (4-14) gives (4-15).

If we had used the effective mass M^* in this calculation, we see from (4-17) that the result would have been

$$\frac{dE}{d\mathfrak{N}} = \hbar\omega = \frac{41 \text{ MeV}}{A^{1/3}} \frac{M}{M^*} \tag{4-18}$$

[1] The second term in this formula is usually given as $(\mathfrak{N} + \tfrac{3}{2})\hbar\omega$, but this assumed $\mathfrak{N} = 0$ for the lowest energy state whereas we call this state $\mathfrak{N} = 1$.

In the energy region where orbits are filling in stable nuclei, $E \approx -8$ MeV, experimental determinations by methods to be discussed in connection with Fig. 14-7 give[1]

$$\frac{dE}{d\mathfrak{N}} \simeq \frac{33 \text{ MeV}}{A^{1/3}} \tag{4-19}$$

Comparing this with (4-18), we find $M^* \approx 1.3M$ in this energy region. The strange variations in the ratio M^*/M are considered in Prob. 4-3.

4-8 Nonspherical Nuclei

The fact that the shell-theory potential represents the average interactions between a given nucleon and all the other nucleons in the nucleus has a very important consequence: it causes many nuclei to be nonspherical in shape. This can be understood with the help of the following greatly oversimplified model.

In a completely filled shell, the fact that all orbits of each j are occupied means that all possible values of the spatial orientation quantum number m are equally represented among occupied orbits, so closed-shell nuclei must be spherically symmetric, as shown by the circle in Fig. 4-6. Now let us assume that the allowed orbits in the next shell, when viewed edge on, can have six orienta-

[1] *Phys. Rev.*, **130:** 227 (1963); *Am. J. Phys.*, **33:** 1011 (1965).

FIGURE 4-6 Schematic representation of a closed-shell nucleus (spherically symmetric and hence represented by a circle) and the orbits in the next shell viewed edge on.

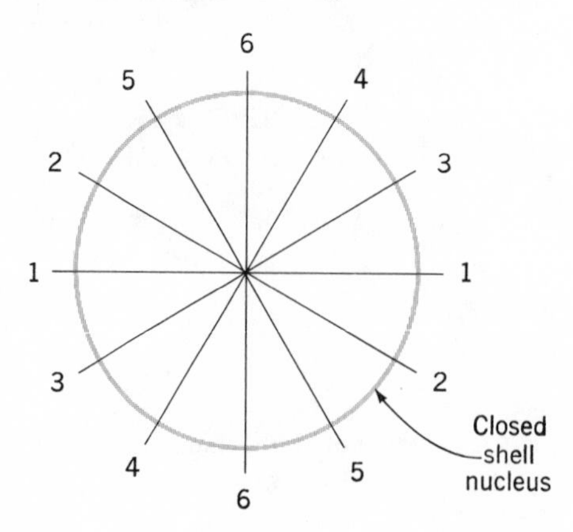

tions in space, as shown by the six numbered lines in the figure. Let us see what happens as these orbits are filled with nucleons.

Let us say the first nucleon in the new shell goes into a 1 orbit. The lowest energy situation for a second nucleon is in another 1 orbit, since in this orbit it feels the added attraction of the first nucleon. Further added nucleons will go into 1 orbits for the same reason, until all 1 orbits are filled. The next added nucleons will go into 2 or 3 orbits, as these bring them as close as possible to the nucleons already there. By the time all the 1, 2, and 3 orbits are filled, there is a considerably greater number of nucleons in horizontal than in vertical orbits.

What does this do to the shell-theory potential? That potential represents the average force exerted on a nucleon by all the other nucleons in the nucleus, so it is roughly proportional to the density distribution of nucleons. The fact that most nucleons are in horizontal orbits then means that the density distribution is not spherically symmetric. As a result, the shell-theory potential is deformed into an ellipsoidal shape, elongated in the horizontal direction. Since the potential is deformed, even the orbits in the closed shells are deformed, and the whole nucleus takes on an ellipsoidal shape.

Now let us return to our filling procedure based on Fig. 4-6. Once the 1, 2, and 3 orbits are filled, the only orbits in that shell left to fill are the vertical ones, 4, 5, and 6. Of course there are many more horizontal orbits in the next higher shell, but to put nucleons into them requires a sizable additional energy given by (4-15). It is therefore more economical in energy for the added nucleons to go into the vertical orbits. As these fill, the excess of nucleons in horizontal orbits decreases, and consequently the density of nucleons (and therefore also the shell-theory potential) becomes less ellipsoidal. Finally, when all orbits of Fig. 4-6 are filled (the shell is closed), all directions are equally represented and the nucleus is spherically symmetric.

This simple model is, of course, far from realistic. The space orientation of orbits is a much more complex matter than the two-dimensional representation in Fig. 4-6. Moreover, no account was taken of the fact that the average distance between nucleons is a minimum for a given nucleon density when the nucleus is spherical, so the strongest average force is obtained in a spherical shape. This leads to an especially strong preference for a spherical shape when the forces are of very short range. Nevertheless, the basic content of the model is correct. Nuclei whose nucleon numbers are far from closed shells take on a rather stable ellipsoidal shape.

The principal regions of ellipsoidal nuclei are where the $\mathfrak{N} = 2$ shells are about half full (Li^7, Be, B, C), where the $\mathfrak{N} = 3$ shells are about half full (Mg, Al, Si), where the $\mathfrak{N} = 5$ proton and $\mathfrak{N} = 6$ neutron shells are simultaneously about half full (heavy rare earths, Ta, W) and where the $\mathfrak{N} = 6$ proton and $\mathfrak{N} = 7$ neutron shells are about half full (Th, U, transuranics). Note that there is a large region through the middle of the periodic table ($A \simeq 35$ to $A \simeq 150$) where either the neutrons or protons are always near enough to closed shells to prevent the nucleus from assuming a stable ellipsoidal deformation. Such nuclei will be referred to as spherical although, as we shall see in Sec. 6-7, this is not a very accurate description.

It turns out that in almost all cases the ellipsoidal shape assumed by non-spherical nuclei is an *ellipsoid of revolution*, otherwise called a *spheroid;* i.e., two of the three principal axes are equal. In cases where the third (unequal) axis is longer than the others, the nucleus has something of a football shape; it is called *prolate*. In the other case, where the unequal axis is shorter than the others so that the nucleus has something of a pumpkin shape, it is called *oblate*. More extreme prolate shapes would be those of a cigar or a hot dog; more extreme oblate shapes would be those of a pancake or a hamburger. However, deviations from the spherical shape are not that large in nuclei.

Among the heavy nonspherical nuclei, the longest and shortest axes differ in length by about 20 percent. In lighter nuclei, however, the fraction of the nucleons in unfilled shells is larger, so larger deformations are obtained, and there is one nucleus at least, Mg^{24}, in which all three axes are believed to be unequal. Where deformations are very large, the ellipsoid approximation breaks down and more complex shapes are encountered. An extreme example is the C^{12} nucleus; one calculation of its shape is shown in Fig. 4-7.

FIGURE 4-7 Calculated density distribution of nucleons in a C^{12} nucleus. Numbers attached to curves are relative nucleon densities. [*From G. Ripka, Advan. Nucl. Phys.*, **1:** 254 (1967); *by permission.*]

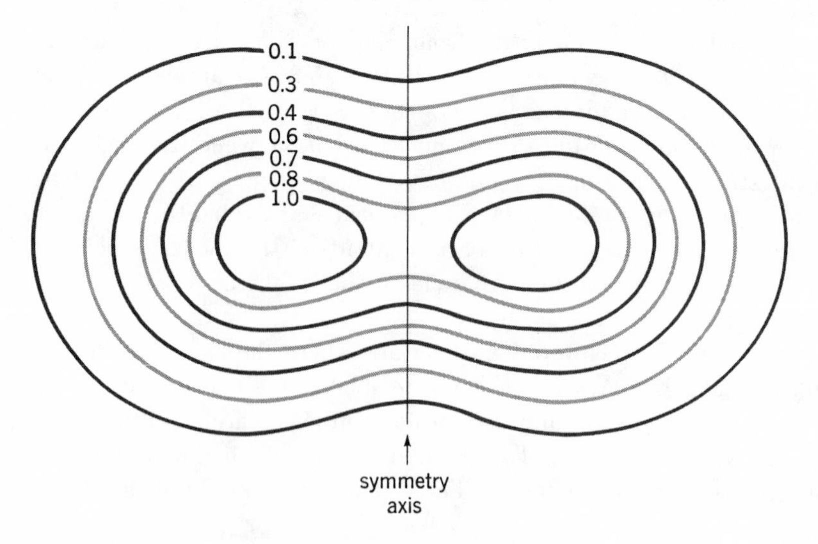

Problems

4-1 Calculate the depth of the shell-theory potential for neutrons and for protons in Pb^{208}; in Ca^{40}.

4-2 If V_0 increases by 1 MeV in a 5-MeV nucleon energy interval, what is M^*?

4-3 (*a*) If V_{00} is 110 MeV and V_0 is 57 MeV near $E = 0$, calculate M^*. This represents the value obtained from the calculations described in Sec. 7-3. (*b*) Estimate M^* from (4-7) under the assumption that $V_{00} = 100$ MeV. (*c*) Compare these with the value given at the end of Sec. 4-7.

4-4 If the spin-orbit interaction were much smaller, i.e., close to zero, and Table 4-1 were accurate, what total numbers of nucleons would constitute closed shells?

4-5 List at least 10 nuclei found in nature which have closed neutron shells and 10 others which have closed proton shells. List at least three single-particle and three single-hole nuclei that are found in nature but not mentioned in the text.

4-6 Plot the potentials for $l = 0$ neutrons and protons in Pb^{208}.

4-7 Assuming that a nucleus with 171 protons contains 400 neutrons, carry out the calculations of Table 4-2.

4-8 Give some justification for taking the square well used in the calculation of Table 4-2 to have a radius $R + a$. Repeat the calculation for this radius.

4-9 Explain qualitatively why $dE/d\mathfrak{N}$ increases more rapidly with increasing $\mathfrak{N}$ for a square well than for the potential (4-1).

Further Reading

See General References, following the Appendix.

Mayer, M. G., and J. H. D. Jensen: "Elementary Theory of Nuclear Shell Structure," Wiley, New York, 1955.

deShalit, A., and I. Talmi: "Nuclear Shell Theory," Academic, New York, 1963.

Chapter 5

The Structure of Complex Nuclei: Spherical Even-Even Nuclei

We have said from the beginning that the shell theory is an approximation. It is clearly impossible to take into account accurately all the complex interactions in a nucleus with a simple potential well. In particular, the frequency and strength with which one nucleon interacts with other nucleons in close encounters depends on what orbit it is in and on what other orbits are occupied, properties that a simple potential well does not portray. The forces resulting from these close encounters, known as *residual interactions*, produce energy changes and other effects which must then be considered as corrections to the results obtained from solutions to the potential-well problem. In this chapter we study these effects and thereby develop a description of the structure of nuclei in its simpler aspects; the more complicated aspects are left for Chap. 6.

5-1 Collisions

In order to keep the development on a simple level, we shall treat residual interactions as *collisions* between nucleons. From this point of view, the effect of a particle's being in the shell-theory potential well is to cause it to move in a stable orbit. However, the motion is actually like that in Fig. 4-1*a*, where every once in a while the particle undergoes a collision, causing it to change its orbit. We now consider these collisions.

For the orbit model to work here we must impose on the collisions a few rather reasonable-sounding conditions, as follows (we label the two participating nucleons 1 and 2 and use primed and unprimed symbols to designate properties after and before the collision respectively):

1. Both nucleons involved in a collision must be in *allowed* and otherwise unoccupied orbits after, as well as before, the collision. By an allowed orbit we mean one specified by a definite set of quantum numbers n, l, j, m satisfying (4-8).

2. Energy must be approximately conserved. This requirement, as we shall see, is equivalent to

$$\mathfrak{N}_1 + \mathfrak{N}_2 = \mathfrak{N}_1' + \mathfrak{N}_2' \tag{5-1}$$

3. Angular momentum must be conserved; that is,

$$\begin{aligned} \mathbf{j}_1 + \mathbf{j}_2 &= \mathbf{j}_1' + \mathbf{j}_2' \\ m_1 + m_2 &= m_1' + m_2' \end{aligned} \tag{5-2}$$

4. Parity must be conserved; that is,

$$\text{If } l_1 + l_2 \text{ is even, } l_1' + l_2' \text{ must be even} \tag{5-3a}$$

or

$$\text{If } l_1 + l_2 \text{ is odd, } l_1' + l_2' \text{ must be odd} \tag{5-3b}$$

These rules follow in a straightforward way from the rules for classical collisions and previously stated quantum restrictions on the orbit model; the only exception is rule 2, where we require that energy be only approximately, rather than exactly, conserved. This difference arises from the well-known uncertainty principle (3-20), which, as we have already seen in Sec. 3-9, says that energy conservation may be violated by an amount ΔE provided the violation does not last longer than a time Δt given by

$$\Delta t \simeq \frac{\hbar}{\Delta E} = \frac{6 \times 10^{-22} \text{ MeV-s}}{\Delta E} \tag{5-4}$$

But in our situation this violation of energy conservation must last long enough for the nucleons to have another collision, since each orbit has a definite energy (as shown in Fig. 4-5) and the only way for a nucleon to change its energy is to change its orbit by a collision. Hence Δt is of the order of the average time between collisions. But we have previously said that the time between collisions is at least of the order of the time required for a nucleon to go around its orbit, or

$$\Delta t \gtrsim \frac{2\pi r}{v} \approx \frac{2\pi \times 6 \times 10^{-15} \text{ m}}{8 \times 10^{7} \text{ m/s}} \approx 5 \times 10^{-22} \text{ s} \tag{5-5}$$

In evaluating (5-5) we have taken the radius of a medium-mass nucleus from (1-2) and the velocity corresponding to a nucleon of about 40 MeV kinetic energy for reasons discussed in Sec. 4-5.

Solving (5-4) for ΔE and inserting (5-5), we find

$$\Delta E \lesssim 1 \text{ MeV} \tag{5-6}$$

This is of the order of the energy difference between orbits in the same shell but is much less than the energy difference between different shells as given by (4-19).

Therefore a collision in which one nucleon stays in the same shell while the other changes shells is in violation of (5-6). However, due to the fact that the energy spacing between shells is constant in a given nucleus (Sec. 4-7), energy conservation is still satisfied if the $\mathfrak{N}$ of one nucleon increases by the same number as the $\mathfrak{N}$ of the other nucleon decreases. Hence we obtain (5-1).

5-2 Cases where Collisions Are Forbidden

Let us now apply the rules of Sec. 5-1 to a few special cases. First we consider a nucleus with all shells up to and including $\mathfrak{N}_0$ filled and with a few nucleons in the shell $\mathfrak{N}_0 + 1$. Can the nucleons in the filled shells have collisions?

If two nucleons in the shell $\mathfrak{N}_0$ were to have a collision, according to rule 2 either both would have to end up in other $\mathfrak{N} = \mathfrak{N}_0$ orbits or one would have to go to a higher- and the other to a lower-$\mathfrak{N}$ orbit. But all other $\mathfrak{N}_0$ orbits are filled, and all orbits with lower $\mathfrak{N}$ are filled, so neither of these is possible. If a nucleon in an $\mathfrak{N}_0$ orbit were to have a collision with one in an $\mathfrak{N}_0 + 1$ orbit, rules 1 and 2 require that one end up in the original $\mathfrak{N}_0$ orbit, whence the occupancy of $\mathfrak{N}_0$ orbits is not changed. We might think that the other nucleon could go into a different $\mathfrak{N}_0 + 1$ orbit, but this is forbidden by rule 3 since, from Fig. 4-5, each orbit in a given shell has a different angular momentum. Thus no collisions of this type can occur.

We may therefore conclude that in the nucleus described above, the *nucleons in filled shells cannot have collisions.* A special case of this type is a closed-shell nucleus, so we may conclude that in closed-shell nuclei like He^4, O^{16}, Ca^{40}, etc., there can be no collisions at all.

Another special case is the single-particle nucleus. If the nucleons in the closed shell cannot have collisions, there is no nucleon with which the extra particle can have collisions, so it cannot have collisions either. There can therefore be no collisions in a single-particle nucleus like O^{17}, Ca^{41}, etc. A third case of interest is the single-hole nucleus. From energy conservation, there are no possibilities for changing shells, and collisions between nucleons in the same shell can only change the orbit of one nucleon into the unoccupied orbit. But since all orbits in the same shell have different angular momenta, this would violate angular-momentum conservation. We therefore find that there can be no collisions in single-hole nuclei.

The fact that collisions are forbidden in closed-shell, single-particle, and single-hole nuclei[1] makes their wave functions trivially simple. Since, as ex-

[1] This is not completely true, as will be explained in Sec. 5-12.

plained in connection with (3-7), the total wave function is the product of the wave functions of each particle,[1] we can write for He^4, for example,

$$\psi(\mathrm{He}^4) = (\psi_{1s_{1/2}}\psi_{1s_{1/2}})_{\text{protons}}(\psi_{1s_{1/2}}\psi_{1s_{1/2}})_{\text{neutrons}} \tag{5-7}$$

where $\psi_{1s_{1/2}}$ is the wave function (2-25) with $l = m = 0$ and $u(r)$ given by the curve labeled $1s$ in Fig. 2-4. However, a much simpler notation is generally used, with the following changes: (1) there is no point in writing ψ's everywhere: they are understood; (2) there is no need for writing the n quantum number since one would not ordinarily consider the $2s_{1/2}$ orbit, which is two shells removed, to be occupied in He^4; and (3) the subscripts protons and neutrons can be replaced by symbols, π and ν, respectively. With these simplifications, (5-7) can be written

$$\psi(\mathrm{He}^4) = \pi(s_{1/2})^2\nu(s_{1/2})^2 \tag{5-7a}$$

Similarly we can write, from Fig. 4-5,

$$\begin{aligned} \psi(\mathrm{O}^{16}) &= \pi(s_{1/2}{}^2p_{3/2}{}^4p_{1/2}{}^2)\,\nu(s_{1/2}{}^2p_{3/2}{}^4p_{1/2}{}^2) \\ \psi(\mathrm{O}^{17}) &= \pi(s_{1/2}{}^2p_{3/2}{}^4p_{1/2}{}^2)\,\nu(s_{1/2}{}^2p_{3/2}{}^4p_{1/2}{}^2d_{5/2}) \\ \psi(\mathrm{O}^{15}) &= \pi(s_{1/2}{}^2p_{3/2}{}^4p_{1/2}{}^2)\,\nu(s_{1/2}{}^2p_{3/2}{}^4p_{1/2}) \end{aligned} \tag{5-8}$$

Since nucleons in closed shells do not undergo collisions, their orbits are not usually of immediate concern and in any case they can be inferred, so they need not be written out explicitly. Hence only the deviations from the closed shell are ordinarily written, for example,

$$\begin{aligned} \psi(\mathrm{O}^{17}) &= d_{5/2} \\ \psi(\mathrm{O}^{15}) &= (p_{1/2})^{-1} \end{aligned} \tag{5-8a}$$

Note that where it is clear from the nature of the nucleus whether it is protons or neutrons which deviate from closed shells, the π or ν are not ordinarily written. From the nature of the nuclei—O^{17} with eight protons and nine neutrons, and O^{15} with eight protons and seven neutrons—it is evident that in both these cases it is the neutrons which deviate from the closed shell of eight. Similarly we could write

$$\begin{aligned} \psi({}_7\mathrm{N}^{15}) &= (p_{1/2})^{-1} \\ \psi({}_9\mathrm{F}^{17}) &= d_{5/2} \end{aligned}$$

where it is clear that the terms on the right represent deviations of the protons from closed shells. In cases where there is a chance for confusion between protons and neutrons, the π and ν are written explicitly.

[1] It may be noted that a simple product does not have the antisymmetry under exchange of any two nucleons required by the Pauli exclusion principle, but we ignore that complication here. It can be readily taken care of without changing the results we shall obtain.

5-3 An Important Example of Collisions—The Pairing Interaction

In nuclei with two nucleons in a shell, many collisions are possible. Let us study one particular example. Consider two nucleons with the same n, l, and j quantum numbers and with m quantum numbers equal but of opposite sign. In our orbit model, this corresponds to two nucleons moving in the same orbit but in opposite directions. Let them have a collision in which they go into another orbit, again moving in opposite directions. An example is shown in Fig. 5-1. Now let us check to see if this type of collision satisfies the rules of Sec. 5-1. The first and second rules are easily satisfied if the new orbits are in the same shell as the old; if there are only two nucleons in a shell, there are certainly many unoccupied orbits in that shell. The angular momenta of the two particles are equal in magnitude and opposite in sign, whence the total angular momentum is zero, both before and after the collision; hence the third rule is satisfied. Since the two nucleons have the same l value, $l_1 + l_2 = 2l_1$ and $l_1' + l_2' = 2l_1'$; both sums are *even* regardless of the values of l_1 and l_1', so parity is conserved.

We see that all the rules are satisfied, so this type of collision can occur. In fact it turns out to be the most common and important type of collision in nuclei. In O^{18}, which has two neutrons outside of closed shells, for example, there are six pairs of orbits that can be reached from one another by this type of collision. They are listed in Table 5-1 and consist of all 12 allowed orbits in the $\mathfrak{N} = 3$ shell arranged in pairs with equal but opposite m values.

FIGURE 5-1 A simple type of allowed collision between two nucleons. They are moving in the same orbit in opposite directions before the collision; after the collision they are in a different orbit than the original one but again in the same orbit as each other and moving in opposite directions.

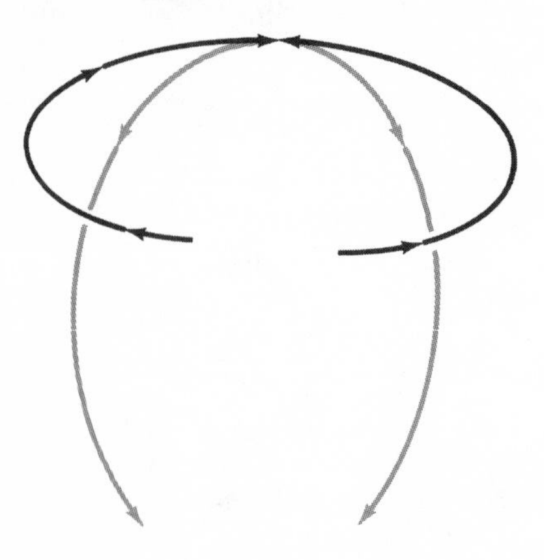

TABLE 5-1: ORBIT PAIRS IN O^{18} WITH $I = 0$

Designation	A	B	C	D	E	F
j_1	$5/2$	$5/2$	$5/2$	$1/2$	$3/2$	$3/2$
m_1	$+5/2$	$+3/2$	$+1/2$	$+1/2$	$+3/2$	$+1/2$
j_2	$5/2$	$5/2$	$5/2$	$1/2$	$3/2$	$3/2$
m_2	$-5/2$	$-3/2$	$-1/2$	$-1/2$	$-3/2$	$-1/2$
Spectroscopic notation		$(d_{5/2})^2$		$(s_{1/2})^2$	$(d_{3/2})^2$	

Since any one of these arrangements can be reached from any other by collisions, we may assume that in some states of O^{18} the two neutrons spend a fraction of their time in each. We may therefore write the wave function for one of these states as

$$\psi(O^{18}) = c_1A + c_2B + c_3C + c_4D + c_5E + c_6F \tag{5-9}$$

Since probabilities are proportional to squares of wave functions, (5-9) implies that c_1^2 is[1] the fraction of the time the nucleus spends in arrangement A, c_2^2 is the fraction of the time it spends in arrangement B, etc. As the total probability for the two nucleons to be in one of the six arrangements is unity (no other arrangement can be reached by collisions), we have

$$c_1^2 + c_2^2 + c_3^2 + c_4^2 + c_5^2 + c_6^2 = 1 \tag{5-10}$$

Since A, B, and C differ only in the orientation of the orbits in space, there is no reason why the nucleus should spend different amounts of time in them. Even if there were some spatial anisotropy such as a magnetic field, its effect would be equal and opposite for the two nucleons and would therefore cause no energy differences or other distinctions between A, B, and C. Hence we expect

$$c_1^2 = c_2^2 = c_3^2 = \frac{a_1^2}{3} \tag{5-11}$$

where a_1 is a new coefficient. The total fraction of the time the nucleus spends in A, B, and C is now a_1^2. These arrangements can be labeled $(d_{5/2})^2$ in view of our notation developed in the last section. Since $(d_{5/2})^2$ can have several different

[1] More advanced students may recognize that c_1 can sometimes be a complex number. If this is so, c_1^2 should be interpreted as $|c_1|^2$: we shall leave off the absolute-value signs in this book.

angular momenta I, we shall add a subscript 0 to indicate that I is zero, so the designation becomes $(d_{5/2})_0{}^2$, commonly read "$d_{5/2}$ squared coupled to zero."

With these considerations in mind, we can now write (5-9) in its final form

$$\psi(\mathrm{O}^{18}) = a_1(d_{5/2})_0{}^2 + b_1(s_{1/2})_0{}^2 + e_1(d_{3/2})_0{}^2 \tag{5-12}$$

where, in analogy with the definition of a_1,

$$\begin{aligned} c_4{}^2 &= b_1{}^2 \\ c_5{}^2 &= c_6{}^2 = \frac{e_1{}^2}{2} \end{aligned} \tag{5-13}$$

In view of (5-10), (5-11), and (5-13),

$$a_1{}^2 + b_1{}^2 + e_1{}^2 = 1 \tag{5-14}$$

If all six c coefficients were equal, we would have $a_1{}^2/3 = b_1{}^2 = e_1{}^2/2$, where the denominators are the number of arrangements in Table 5-1 they represent. These numbers are $(2j + 1)/2$, so we can say that the coefficients in a wave function like (5-12) contain an inherent proportionality to $\sqrt{2j+1}$. However, while there is no reason for differences between $c_1{}^2$, $c_2{}^2$, and $c_3{}^2$ or between $c_5{}^2$ and $c_6{}^2$, there is every reason to expect differences between $c_1{}^2$, $c_4{}^2$, and $c_5{}^2$ since these represent orbital arrangements with different energies. If the energy available for each nucleon is close to the energy of the $1d_{5/2}$ orbit in Fig. 4-5, they can only reach states D, E, and F by collisions which violate energy conservation; such violations, according to the uncertainty principle, can last for only a very short time. The system therefore spends more time in A, B, and C than in D, and still less time in E and F since, from Fig. 4-5, the latter two represent a larger violation of energy conservation. Hence we have

$$\frac{a_1{}^2}{3} > b_1{}^2 > \frac{e_1{}^2}{2} \tag{5-15}$$

On the other hand, if the energy of the system were close to that required to put the two neutrons into the $1d_{3/2}$ orbit of Fig. 4-5, situations A, B, and C could be reached only by violation of energy conservation (of opposite sign, but this does not matter), whence the nucleus spends less time in the latter so that

$$\frac{e_1{}^2}{2} > b_1{}^2 > \frac{a_1{}^2}{3}$$

It is easy to conceive of an energy intermediate between these where

$$b_1{}^2 > \frac{e_1{}^2}{2} \qquad b_1{}^2 > \frac{a_1{}^2}{3}$$

However, there is no choice of energy that could lead to

$$\frac{a_1^2}{3} > \frac{e_1^2}{2} > b_1^2$$

Before proceeding further with our discussion it is necessary to pause here to establish a few basic definitions.

A *configuration* is a listing of what orbits are occupied without regard to m values. For example, $(d_{5/2})^2$ is called a configuration. Note that we do not specify the total angular momentum, so it is not necessarily zero. Other configurations we might encounter in O^{18} are $d_{5/2}s_{1/2}$, $d_{5/2}d_{3/2}$, etc. The energy of a configuration is the sum of the energies of all occupied orbits as taken from Fig. 4-5.

A *term* is a configuration with angular-momentum coupling specified. For example, $(d_{5/2})_0{}^2$ is a term. A different term would be $(d_{5/2})_2{}^2$ although these have the same configuration. Other terms in O^{18} would be $(d_{5/2}s_{1/2})_2$, $(d_{5/2}s_{1/2})_3$, $(d_{5/2}d_{3/2})_3$, etc. Note that in all cases, I is one of the values that can be obtained from

$$\mathbf{I} = \mathbf{j}_1 + \mathbf{j}_2$$

in accordance with the rules of Sec. 2-7. The energy of a term is just the energy of the configuration from which it is derived.

A *state* is a very definite quantum-mechanical concept which is not to be confused with the meaning of this word in other contexts. A state is the solution of the Schrödinger equation for the system; it has a definite energy, a definite wave function, a definite total angular momentum, a definite parity, and several other definite properties, such as electric and magnetic moments. For example, (5-12) is the wave function for a state provided the values of a_1, b_1, and c_1 are chosen to satisfy the Schrödinger equation for O^{18}. The total angular momentum of this state is 0 and its parity is positive since for each component, $l_1 + l_2$ is even. The total angular momentum and parity of a state are conventionally written together as I^π, which in this case is 0^+.

In order to clarify the difference between terms and states, we recall that the orbit picture including Fig. 4-5 was derived by solving the Schrödinger equation (2-24) with V replaced by the shell-theory potential. If all the forces in the problem could be accurately represented by the shell-theory potential, the states would therefore be identical with the terms. However, we know that the shell-theory potential does not take into account short-range interactions that depend on what orbits particles are in, which we refer to as collisions. The inclusion of these collisions in the V of (2-24) is what distinguishes states from terms. The effect of including collisions, as we have seen in (5-12) for example, is to convert states into linear combinations of terms. However, there is a fundamental theorem in quantum theory which requires that in this process the number of states be unchanged. Therefore, since there are three terms in (5-12), there must be three states of that form. We can write their wave functions as

$$\begin{aligned}\psi_1(O^{18}) &= a_1(d_{5/2})_0{}^2 + b_1(s_{1/2})_0{}^2 + e_1(d_{3/2})_0{}^2 \\ \psi_2(O^{18}) &= a_2(d_{5/2})_0{}^2 + b_2(s_{1/2})_0{}^2 + e_2(d_{3/2})_0{}^2 \\ \psi_3(O^{18}) &= a_3(d_{5/2})_0{}^2 + b_3(s_{1/2})_0{}^2 + e_3(d_{3/2})_0{}^2\end{aligned} \tag{5-16}$$

where in each case, the coefficients satisfy (5-14). All three of these states are, of course, 0^+. Quantum theory requires that certain additional relationships, known as *orthogonality conditions*, be satisfied by the coefficients in (5-16). These are

$$\begin{aligned}a_1a_2 + b_1b_2 + e_1e_2 &= 0 \\ a_2a_3 + b_2b_3 + e_2e_3 &= 0 \\ a_3a_1 + b_3b_1 + e_3e_1 &= 0\end{aligned}$$

It may be noted from these equations that some of the coefficients in (5-16) must be negative. In fact it is readily seen that no more than one of the three wave functions in (5-16) can have all positive coefficients.

Let us now turn our attention to energies. Up to this point, all energy considerations have been derived from Fig. 4-5, which gives the energy of each orbit. An expanded view of the $\mathfrak{N} = 3$ shell is shown in Fig. 5-2. Orbit energies are defined at the left; since we are interested only in relative energies, we have set the energy of the lowest state equal to zero.

The energy of a configuration, or of the terms derived from it, is just the sum of the energies of the occupied orbits. For example, the energies of the configura-

FIGURE 5-2 Energies of orbits in the $\mathfrak{N} = 3$ shell. This is an enlargement of a section of Fig. 4-5 with definitions of ϵ_s and ϵ_d added.

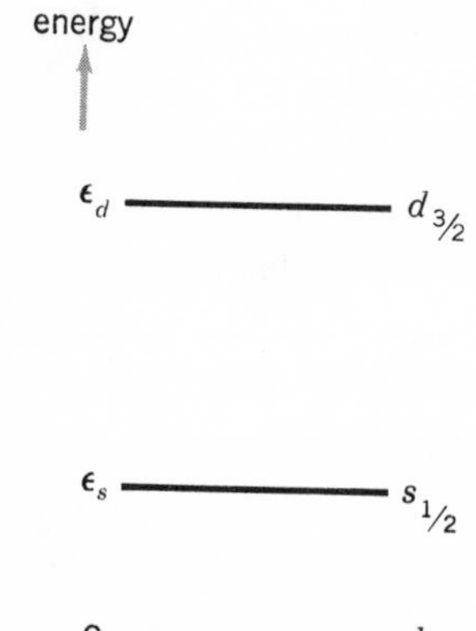

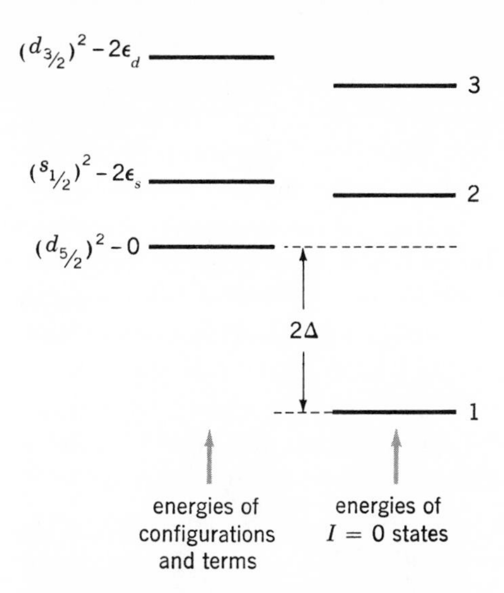

FIGURE 5-3 Engeries of $I = 0$ terms and states in O^{18} (two neutrons in the $\mathfrak{N} = 3$ shell). The energies of the terms are the sum of the energies of the occupied orbits; 2Δ is the energy gap.

tions in (5-16) are 0, $2\epsilon_s$, and $2\epsilon_d$ for $(d_{5/2})^2$, $(s_{1/2})^2$, and $(d_{3/2})^2$ respectively. These are shown at the left in Fig. 5-3.

The energy of a *state*, however, is a much more subtle and complicated matter. The orbit energies of Fig. 4-5 were determined from the shell-theory potential, which represents the average force of all the other nucleons in the nucleus. However, such an average force cannot possibly take into account the short-range forces that come into effect when two nucleons are colliding, i.e., the residual interactions. These forces cause small changes in the energy of a state, making it deviate from the average energy of the configurations it includes by an amount depending on the pattern of collisions. In most cases, the deviation is not large, and we might expect the three states (5-16) to have energies in the region between 0 and $2\epsilon_d$. This is indeed true for two of the three states, but something very strange happens to the third: it gets pushed far down in energy, as shown for state 1 in Fig. 5-3! This behavior is easily derived using the mathematical techniques of quantum theory, as will be shown in Sec. 5-4. In our orbit-model picture, the effect may seem plausible if we say that in state 1 a very orderly and rhythmic collision pattern develops such that all the residual interactions combine in a coherent way to give the maximum possible net attractive force. In the other two states, 2 and 3, the collision pattern is not so regular and coherent, so the residual interactions are sometimes attractive and sometimes repulsive, whence the total energy is not much affected.

The energy 2Δ by which the lowest-energy state is below the energy of the lowest configuration[1] is known as the *energy gap*. It plays an important part in the energy-level structure of all even-even nuclei, i.e., nuclei with even numbers of neutrons and protons, and has an interesting counterpart in the theory of superconductivity, an important phenomenon in low-temperature and solid-state physics.

The type of collision we have been discussing in this section is known as *pairing*. Recall that it is an interaction between a *pair* of particles moving in the same orbit but in opposite directions; they undergo frequent collisions of the type shown in Fig. 5-1; these collisions change their orbits, but they always remain paired with both particles in the same orbit going in opposite directions. As a result of the pairing interaction, the nucleus is taken from one 0^+ term to another, so the wave functions for 0^+ states are linear combinations of these terms. The energies of these states behave as shown in Fig. 5-3, with one state pushed far down in energy so as to give an energy gap.

5-4 Quantum-mechanical Treatment of the Energy Gap

In the last section we discussed the orderly and rhythmic collision pattern which causes one state to be substantially lowered in energy, thereby creating an energy gap. While this may give some impression of the phenomenon, a true understanding of it can only be obtained from a quantum-mechanical treatment. We give elementary treatments in this section, which should be understandable to students who have had an advanced course in modern physics or the equivalent. For others, the remainder of this section may be omitted without serious loss of continuity.

We give two treatments here, one based on perturbation theory, which is more likely to be familiar to less advanced students, and a more complete and general treatment which should appeal more to students who have had a regular course in quantum mechanics. In the first treatment, we must assume that the energies of all terms are equal, that is, $\epsilon_s = \epsilon_d = 0$ in the example of Sec. 5-3, and use *degenerate-perturbation theory*. If the energies of several states are the same before perturbations V' (in our case, the residual interactions) are considered, the effect of the perturbations is to give states whose energies differ from the unperturbed energy by the eigenvalues of the matrix.

$$\begin{bmatrix} v_{11} & v_{12} & v_{13} & \cdots \\ v_{21} & v_{22} & v_{23} & \cdots \\ v_{31} & v_{32} & v_{33} & \cdots \\ \cdots & \cdots & \cdots & \cdots \end{bmatrix} \tag{5-17}$$

[1] A more quantitative definition of Δ will be given in Sec. 7-2.

where $$v_{mn} = \int \phi_m^* V' \phi_n \, d\tau \tag{5-18}$$

and ϕ_m, ϕ_n are the unperturbed wave functions. In the case discussed in Sec. 5-3, these are $\phi_1 = (d_{5/2})^2$, $\phi_2 = (s_{1/2})^2$, and $\phi_3 = (d_{3/2})^2$. In order to illustrate our effect, let us say all v_{mn} are equal. The matrix then becomes

$$v_{11} \begin{bmatrix} 1 & 1 & 1 & \cdots \\ 1 & 1 & 1 & \cdots \\ 1 & 1 & 1 & \cdots \\ \cdots & \cdots & \cdots & \cdots \end{bmatrix} \tag{5-17a}$$

If the matrix is 3×3, as in our example, the eigenvalues are $3v_{11}$, 0, and 0. Thus the energies of two of the states are unchanged by the residual interactions, while the third is pushed down[1] in energy to $3v_{11}$. This is very crudely the situation shown in Fig. 5-3.

From degenerate-perturbation theory, the wave functions of the states are the eigenfunctions of (5-17) with each component multiplied by the corresponding wave function. In our example, this gives[2]

$$\begin{aligned} \psi_1 &= 0.557(d_{5/2})^2 + 0.577(s_{1/2})^2 + 0.577(d_{3/2})^2 \\ \psi_2 &= -0.707(s_{1/2})^2 + 0.707(d_{3/2})^2 \\ \psi_3 &= 0.816(d_{5/2})^2 - 0.408(s_{1/2})^2 - 0.408(d_{3/2})^2 \end{aligned}$$

Note that for the wave function of the lowest-energy state ψ_1 all terms carry the same sign whereas this is not true for any of the other states (as can easily be understood from the orthogonality conditions). The property of having all terms with the same sign is spoken of as *coherence;* it corresponds to the "orderly and rhythmic collision pattern" described in the last section.

If the matrix (5-17*a*) includes p rows and columns, i.e., if there are p states which have the same energy before residual interactions are considered, the eigenvalues are

$$\begin{aligned} E_1 &= pv_{11} \\ E_2 &= E_3 = \cdots = E_p = 0 \end{aligned} \tag{5-19}$$

Here again one state is pushed far down in energy while the others remain unchanged. Once again the wave function for the state with lowered energy has all terms of the same sign, i.e., it is coherent, while the wave functions corresponding to the zero eigenvalues do not have this property.

[1] If the residual interactions are attractive, V' and hence v_{mn} are negative.

[2] The coefficients in these wave functions are not proportional to $\sqrt{2j + 1}$, as suggested by the discussion following (5-14), because the assumption that all v_{mn} are equal is an oversimplification.

It is interesting to point out from (5-19) that the amount by which the energy of the coherent state is lowered is proportional to p, which is the number of terms of which it is a linear combination. It is also proportional to v_{11}, which is in turn proportional to V', the strength of the residual interaction.

The use of degenerate-perturbation theory in this discussion is subject to criticism on two basic accounts: (1) the unperturbed energies in Fig. 5-3 are 0, $2\epsilon_s$, $2\epsilon_d$, which are not equal to one another, and (2) the residual interactions might be too strong to be considered as a small perturbation. These difficulties can be eliminated by the following treatment which, however, requires more experience with quantum theory.

Let the configuration wave functions, $(d_{5/2})_0{}^2$, $(s_{1/2})_0{}^2$, and $(d_{3/2})_0{}^2$ in our example, be designated ϕ_i. They are solutions of the Schrödinger equation with V taken as the shell-theory potential V_s or

$$H\phi_i = \epsilon_i\phi_i \tag{5-20}$$

where H is the operator

$$H = -\frac{\hbar^2}{2M}\nabla^2 + V_s$$

Since they are solutions of a Schrödinger equation, the ϕ_i are orthonormal, or

$$\int \phi_j^* \phi_i \, d\tau = \delta_{ij} \tag{5-21}$$

If we represent the potentials introduced by the residual interactions as V', the actual energies E are given by solutions of the complete Schrödinger equation

$$(H + V')\psi = E\psi \tag{5-22}$$

where ψ are the wave functions for the actual states. Expanding ψ as

$$\psi = \sum_i c_i\phi_i$$

and inserting this into (5-22) gives

$$H\Sigma c_i\phi_i + V'\Sigma c_i\phi_i = E\Sigma c_i\phi_i$$

Using (5-20) in the first term, multiplying from the left with $\phi_j^* \, d\tau$, and integrating then gives, by use of (5-21),

$$c_j\epsilon_j + \sum_i c_i \int \phi_j^* V' \phi_i \, d\tau = Ec_j$$

Making use of (5-18) and writing this out explicitly gives for the case where there are three terms:

for $j = 1$: $\quad c_1(v_{11} + \epsilon_1 - E) + c_2v_{12} + c_3v_{13} = 0$
for $j = 2$: $\quad c_1v_{21} + c_2(v_{22} + \epsilon_2 - E) + c_3v_{23} = 0$
for $j = 3$: $\quad c_1v_{31} + c_2v_{32} + c_3(v_{33} + \epsilon_3 - E) = 0$

which is equivalent to

$$\begin{bmatrix} \epsilon_1 + v_{11} & v_{12} & v_{13} \\ v_{21} & \epsilon_2 + v_{22} & v_{23} \\ v_{31} & v_{32} & \epsilon_3 + v_{33} \end{bmatrix} \begin{bmatrix} c_1 \\ c_2 \\ c_3 \end{bmatrix} = E \begin{bmatrix} c_1 \\ c_2 \\ c_3 \end{bmatrix} \tag{5-23}$$

The energies and wave functions are therefore the eigenvalues and eigenfunctions of the matrix in (5-23). We see that it differs from (5-17) only in that ϵ_1, ϵ_2, and ϵ_3 appear in the diagonal terms. Moreover, there has been no assumption that V' is very small. So long as ϵ_1, ϵ_2, and ϵ_3 are not too different, the results will be similar to those obtained from (5-17).

Unfortunately, the method we have described is not applicable when the residual interactions between two nucleons have $V' = \infty$ over a finite range of r, for in such cases v_{mn} in (5-18) is infinite. Since the basic interaction between two nucleons as given in Fig. 3-7 has a hard core ($V = \infty$ for $r \gtrsim 0.5$ F) and this short-range repulsion is clearly not taken into account in the shell-theory potential, it must be part of the residual interactions. A great deal of effort has therefore been devoted to treating residual interactions with hard cores, and very elaborate techniques have been developed for this purpose.

5-5 The Ground States of Even-Even Nuclei

Since nucleons in the same orbit (even though moving in opposite directions) spend more time near each other than in any other combination of orbits, interactions in this situation occur more frequently than in any other, so the pairing interaction is the strongest of all residual interactions. Since the energy shifts caused by an interaction are proportional to its strength [note the proportionality to v_{11} and hence to V' of the energy shifts in (5-19)], the energy lowering of the coherent state by the pairing interaction is larger than that of any other state. As a result, state 1 in Fig. 5-3 is the lowest-energy state of O^{18}. The lowest-energy state is called the *ground state* or *normal state* since it is the state in which the nucleus is normally found. All other states are called *excited states;* they ordinarily make rapid transitions to ground states by decay mechanisms to be discussed in Chap. 8.

Now that we have an understanding of the ground states of nuclei with two neutrons outside of closed shells, it is very easy to discuss two-hole nuclei, i.e., nuclei which lack only two nucleons from having all closed shells. As an example, consider Pb^{206}, in which the proton shells are closed (it has 82 protons) and there are 124 neutrons, which is 2 less than the closed shell at 126. As in Sec. 5-2, we need only keep track of the holes, which, according to Fig. 4-5 can be in the $p_{1/2}$, $f_{5/2}$, $p_{3/2}$, $i_{13/2}$, $f_{7/2}$, or $h_{9/2}$ orbits. In analogy with our discussion of O^{18}, the ground-state wave function for Pb^{206} is

$$\psi(\mathrm{Pb}^{206}) = a_1(p_{1/2})^{-2} + b_1(f_{5/2})^{-2} + c_1(p_{3/2})^{-2} + d_1(i_{13/2})^{-2} + e_1(f_{7/2})^{-2} + f_1(h_{9/2})^{-2} \tag{5-24}$$

In general, all further discussion about nuclei with a given number of nucleons outside a closed shell applies equally well to nuclei with the same number of holes inside a closed shell. When a shell is less than half full, it is easier to talk about particles, and when it is more than half full, it is easier to talk about holes.

Where there are more than two nucleons in a shell, the pairing interaction is even more spectacular than in the nuclei we have been discussing. Consider for example O^{20}, which has four neutrons outside of the O^{16} closed shells. There are now five 0^+ terms in which all neutrons are paired. These are shown at the left in Fig. 5-4; their energies can be readily calculated from Fig. 5-2. As a result

FIGURE 5-4 Energies of some $I = 0$ terms and states in O^{20} (four neutrons in the $\mathfrak{N} = 3$ shell). These are the terms in which all neutrons are coupled in pairs to zero angular momentum. The term energies shown at the left are just the sum of the energies of occupied orbits with the energy definitions of Fig. 5-2. The energy gap 2Δ is also shown.

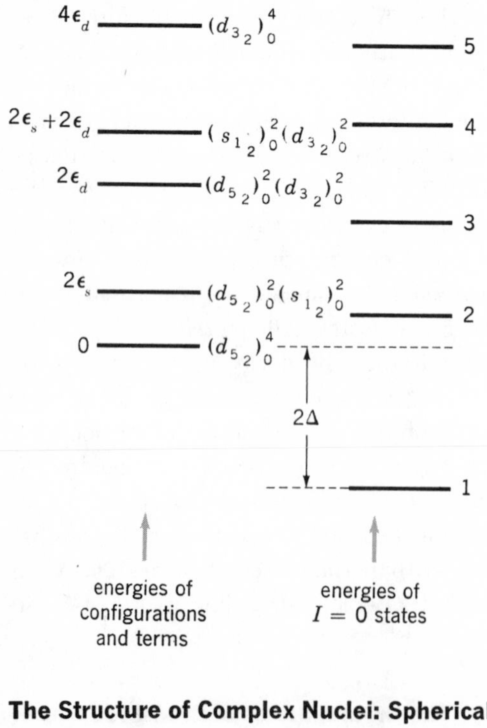

$4\epsilon_d$ — $(d_{3/2})^4_0$ — 5

$2\epsilon_s + 2\epsilon_d$ — $(s_{1/2})^2_0(d_{3/2})^2_0$ — 4

$2\epsilon_d$ — $(d_{5/2})^2_0(d_{3/2})^2_0$ — 3

$2\epsilon_s$ — $(d_{5/2})^2_0(s_{1/2})^2_0$ — 2

0 — $(d_{5/2})^4_0$

2Δ

1

energies of configurations and terms

energies of $I = 0$ states

of residual interactions, these lead to five states whose wave functions are linear combinations of these terms, as

$$\psi(\mathrm{O}^{20}) = \alpha_i(d_{5/2})^4 + \beta_i(d_{5/2})_0{}^2(s_{1/2})_0{}^2 + \gamma_i(d_{5/2})_0{}^2(d_{3/2})_0{}^2 + \delta_i(s_{1/2})_0{}^2(d_{3/2})_0{}^2 + \epsilon_i(d_{3/2})_0{}^4 \tag{5-25}$$

with $i = 1, 2, 3, 4$, and 5. The energies of these states are as shown at the right of Fig. 5-4; once again there is one state with a very regular and rhythmic collision pattern which lowers its energy far below the others, leaving an energy gap 2Δ. Once again, the effect is so strong that this is the ground state of O^{20}.

This problem can, of course, be solved by the methods used in connection with (5-23). Here the five terms in (5-25) are the ϕ_i, and their configuration energies are the ϵ_i. The energies and wave functions are the eigenvalues and eigenfunctions of a 5×5 matrix analogous to (5-23). From (5-19) we see that the energy lowering of the ground state is proportional to the dimensionality of the matrix p, so it is even larger for O^{20} than for O^{18}.

The pairing interaction is so strong that it is the dominant effect in the ground states of all non-closed-shell nuclei with even numbers of neutrons and protons. In the ground state of all of these even-even nuclei, all neutrons not in closed shells are paired with another neutron, and all protons are paired with another proton. Each pair consists of two partners in the same orbit but moving in opposite directions; i.e., they have the same n, l, and j quantum numbers and equal but opposite m quantum numbers. They undergo frequent orbit-changing collisions, but the two partners always remain paired. The total angular momentum of each pair is zero since it consists of two equal and oppositely directed angular momenta. The parity of each pair is even since they both have the same l quantum number, whence $\Sigma l = 2l$, which is always *even* regardless of the value of l. Hence the total angular momentum and parity of the state are 0^+. As in O^{18} and O^{20}, there are a large number of states of the type just described, but in the ground state the collision pattern is the most orderly and rhythmic one. It is therefore lower in energy than any of the others by the energy gap 2Δ.

The wave function of a typical even-even nucleus has a great many terms; a relatively simple example is shown in Table 5-2, where we see that the ground-state wave function for ${}_{44}\mathrm{Ru}^{98}$, which has six proton holes in the $\mathfrak{N} = 4$ shell and four neutron particles in the $\mathfrak{N} = 5$ shell, can have terms consisting of any of the 15 proton configurations and any of 14 neutron configurations; since any combination of a proton configuration and a neutron configuration is an acceptable term in the wave function, the wave function has $15 \times 14 = 210$ terms. There are 210 wave functions of 210 states made up of linear combinations of these terms, but one of these 210 is a coherent combination which causes its energy to be much lower than that of any of the others; it is therefore the ground state of Ru^{98}.

TABLE 5-2: TERMS IN THE GROUND-STATE WAVE FUNCTION OF $_{44}Ru^{98}$ (SIX PROTON HOLES IN $p_{3/2}$, $f_{5/2}$, $p_{1/2}$, $g_{9/2}$, AND FOUR NEUTRON PARTICLES IN $d_{5/2}$, $g_{7/2}$, $s_{1/2}$, $d_{3/2}$, $h_{11/2}$). The product of any entry in the left column with any entry in the right column is a term in that wave function.

Protons	*Neutrons*
$(p_{3/2})_0^{-4}(f_{5/2})_0^{-2}$	$(d_{5/2})_0^4$
$(p_{3/2})_0^{-4}(p_{1/2})_0^{-2}$	$(d_{5/2})_0^2(g_{7/2})_0^2$
$(p_{3/2})_0^{-4}(g_{9/2})_0^{-2}$	$(d_{5/2})_0^2(s_{1/2})_0^2$
$(p_{3/2})_0^{-2}(f_{5/2})_0^{-4}$	$(d_{5/2})_0^2(d_{3/2})_0^2$
$(p_{3/2})_0^{-2}(f_{5/2})_0^{-2}(p_{1/2})_0^{-2}$	$(d_{5/2})_0^2(h_{11/2})_0^2$
$(p_{3/2})_0^{-2}(f_{5/2})_0^{-2}(g_{9/2})_0^{-2}$	$(g_{7/2})_0^4$
$(p_{3/2})_0^{-2}(p_{1/2})_0^{-2}(g_{9/2})_0^{-2}$	$(g_{7/2})_0^2(s_{1/2})_0^2$
$(p_{3/2})_0^{-2}(g_{9/2})_0^{-4}$	$(g_{7/2})_0^2(d_{3/2})_0^2$
$(f_{5/2})_0^{-6}$	$(g_{7/2})_0^2(h_{11/2})_0^2$
$(f_{5/2})_0^{-4}(p_{1/2})_0^{-2}$	$(s_{1/2})_0^2(d_{3/2})_0^2$
$(f_{5/2})_0^{-4}(g_{9/2})_0^{-2}$	$(s_{1/2})_0^2(h_{11/2})_0^2$
$(f_{5/2})_0^{-2}(p_{1/2})_0^{-2}(g_{9/2})_0^{-2}$	$(d_{3/2})_0^4$
$(f_{5/2})_0^{-2}(g_{9/2})_0^{-4}$	$(d_{3/2})_0^2(h_{11/2})_0^2$
$(p_{1/2})_0^{-2}(g_{9/2})_0^{-4}$	$(h_{11/2})_0^4$
$(g_{9/2})_0^{-6}$	

The example of Ru^{98} was chosen because of its relative simplicity! The largest number of configurations occurs when a shell is about half full, which in the $\mathfrak{N} = 4$ and 5 shells is 11 and 16 particles, respectively; these are many more than the four and six used in our example. In heavier nuclei, shells have increasingly larger numbers of different orbits, which also increases the number of configurations, as we see from Table 5-2, where four particles filling five different orbits give almost as many configurations as six holes filling four different orbits. When these effects are considered, it is easy to see that in some nuclei the number of terms in the ground-state wave function runs into the tens of thousands.

In our description of the ground states of even-even nuclei, we said that each neutron is paired with another neutron and each proton is paired with another proton, but none of the arguments used would preclude pairing between a neutron and a proton. In our example of Ru^{98}, it is clear that this is not possible because the neutrons and protons are filling different shells. The shell being filled by protons is already full for neutrons, and the shell being filled by neutrons is not the lowest one available for protons, so it would not be occupied in the ground state. In light nuclei, however, neutrons and protons are filling the same shell and are even filling principally the same orbits. For example, in Ne^{20}, the two protons and two neutrons outside the O^{16} closed shell all spend most of their time in $d_{5/2}$ orbits. In this type of situation pairing between neutrons and protons does occur.

An intermediate situation occurs in a nucleus like Zn^{70}, which has 30 protons and 40 neutrons. Both neutrons and protons are filling the $\mathfrak{N} = 4$ shell of Fig. 4-5, but the protons are filling principally the lower orbits of that shell ($p_{3/2}, f_{5/2}$) while the neutrons are mainly filling the upper orbits ($p_{1/2}, g_{9/2}$). Relatively little pairing between neutrons and protons will therefore occur in such a nucleus.

5-6 Broken Pairs and Quasi-particle Number

An interesting aspect of Table 5-2 is that it does not by any means contain *all* the 0^+ terms in Ru^{98}. For example, it does not contain neutron terms like

$$[(d_{5/2})_2{}^2(g_{7/2})_2{}^2]_0 \tag{5-26a}$$

$$(d_{5/2}g_{7/2}s_{1/2}d_{3/2})_0 \tag{5-26b}$$

These terms have two *broken pairs;* i.e., two pairs of nucleons, a total of four, which do not have partners with the same n, l, j and equal but opposite m quantum numbers. Such terms can be reached by a collision between two members of different pairs, but once in a configuration like (5-26), the probability of getting back to a completely paired situation is small; nucleons in these situations do not often come close enough to have a collision, and when they do, it is much more likely to lead to another two-broken-pair term. We do not therefore expect terms like (5-26) to participate in the pattern of very frequent and rhythmic collisions characteristic of ground states of even-even nuclei.

More advanced students may understand this in terms of (5-23) as follows. The fact that collisions going between the fully paired and the two-broken-pair terms occur infrequently means that v_{mn} is much smaller between these than between two fully paired or between two two-broken-pair terms; let us say they are zero. We can take the fully paired terms to be the ϕ_i with $i = 1$ to q and the two-broken-pair terms to be the ϕ_i with $i = q + 1$ to N. The matrix analogous to (5-23) is then

$$\left[\begin{array}{ccc|ccc} v_{11} + \epsilon_1 & \cdots & v_{1q} & & & \\ \cdots & \cdots & \cdots & & 0 & \\ v_{q1} & \cdots & v_{qq} + \epsilon_q & & & \\ \hline & & & v_{q+1,q+1} + \epsilon_{q+1} & \cdots & v_{qN} \\ & 0 & & \cdots & \cdots & \cdots \\ & & & v_{N,q+1} & \cdots & v_{NN} + \epsilon_N \end{array}\right]$$

This is equivalent to two independent matrices, and the ground state, which is an eigenfunction of the upper left one, does not contain any ϕ_i for $i > q$.

There are, of course, terms in which there is only one broken pair, but a broken pair cannot couple to angular momentum zero (note that no two levels in the same shell in Fig. 4-5 have the same j), so one-broken-pair terms are of no interest in connection with ground states. As we shall see, they are important in excited states.

A concept closely related to the number of broken pairs is the quasi-particle number of a state. This is defined roughly as the number of particles plus holes that must be introduced into the ground state of an even-even nucleus to obtain the state in question. Under this definition, a change in orbit by one nucleon from its orbit in the ground state leads to a two-quasi-particle (abbreviated 2QP) state, since it differs from the ground state in having an additional particle-occupied orbit (the new orbit) and an additional hole (in the old orbit). The term (5-26*a*) could be a part of a 4QP state since two nucleons, a $d_{5/2}$ and a $g_{7/2}$, must have changed from their orbits in the ground state to orbits of different m, that is, different orientation in space. The term (5-26*b*) could also be part of a 4QP state.

The quasi-particle number of a state must clearly be at least as large as twice the number of broken pairs, but it can easily be larger. For example, from the definition there can only be one zero-quasi-particle state, namely, the ground state. All the other 0^+ states in which the collision pattern is not the coherent one must be at least 2QP states, even though they have no broken pairs. In many situations to be discussed, an important property of a state is how it differs from the ground state of an even-even nucleus, so the quasi-particle number will be most useful.

5-7 Occupation Numbers

Since the wave functions of the ground states of even-even nuclei are so complicated, often consisting of thousands or even tens of thousands of terms, it is highly desirable to develop a simpler, even if less accurate, way of describing them. Such a method is available in the use of occupation numbers V_j^2, which indicate how fully occupied the orbits are with given values of n, l, j. (In any given shell, there is only one type of orbit with a given j; a specification of j therefore implies the n and l quantum numbers, which are not needed as subscripts.) First let us give some examples of V_j^2 for cases where we have already discussed the wave functions. In the O^{18} ground-state wave function (5-12) the two neutrons are in $d_{5/2}$ orbits a fraction of the time a_1^2, and when they are in $d_{5/2}$ orbits, those orbits are one-third full since $d_{5/2}$ orbits can accommodate $2j + 1 = 6$ neutrons; hence

$$V_{5/2}^2 = \tfrac{1}{3}a_1^2 \tag{5-27a}$$

The same method gives

$$V_{1/2}^2 = b_1^2 \tag{5-27b}$$

$$V_{3/2}^2 = \tfrac{1}{2}e_1^2 \tag{5-27c}$$

The fraction is $1/(2j + 1)$ times the number of orbits occupied. In the O^{20} wave function (5-25) with $i = 1$ similar reasoning can easily be seen to lead to

$$\begin{aligned} V_{5/2}^2 &= \tfrac{4}{6}\alpha_1^2 + \tfrac{2}{6}(\beta_1^2 + \gamma_1^2) \\ V_{1/2}^2 &= \beta_1^2 + \delta_1^2 \\ V_{3/2}^2 &= \tfrac{2}{4}(\gamma_1^2 + \delta_1^2) + \epsilon_1^2 \end{aligned} \tag{5-28}$$

One great advantage in using the occupation numbers V_j^2 is that we can estimate their values quite easily; let us see how this is done. The V_j^2 would have a very simple behavior if there were no collisions in nuclei. As neutrons are added to the $\mathfrak{N} = 3$ shell, for example, the first six would go into $d_{5/2}$ orbits, whence

$$V_{5/2}^2 = \begin{cases} \tfrac{1}{6}N & N < 6 \\ 1.0 & N \geq 6 \end{cases}$$

where N is the number of neutrons. The seventh and eighth neutrons would go into $s_{1/2}$ orbits, and the next four neutrons would go into $d_{3/2}$ orbits, so the V_j^2 would behave as illustrated in Fig. 5-5.

However we know that there are collisions in nuclei, and as a result the simple system of filling described by Fig. 5-5 is not valid. In fact, as we see from (5-12), even in O^{18} with $N = 2$ the $s_{1/2}$ and $d_{3/2}$ orbits are partly full, so the curves of Fig. 5-5 must be modified. However, this modification occurs in a smooth and non-drastic way, as illustrated in Fig. 5-6. The features of Fig. 5-5 are retained, but the sharp corners are rounded off.

FIGURE 5-5 The occupation numbers V_j^2 vs. N, the number of nucleons in the $\mathfrak{N} = 3$ shell, in the approximation that there are no residual interactions.

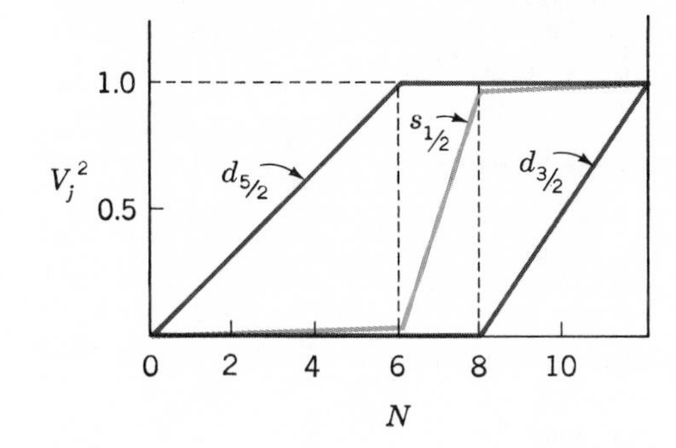

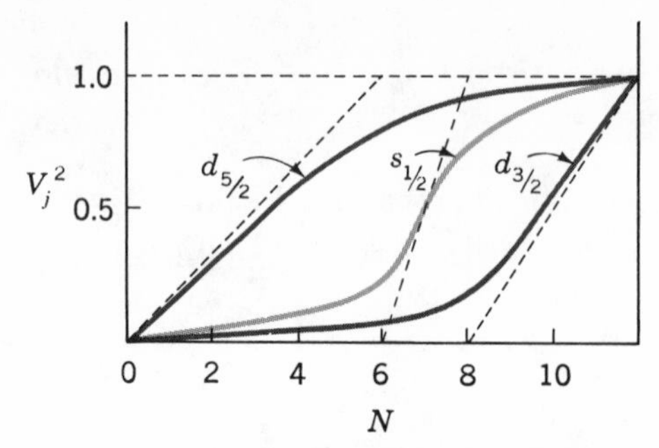

FIGURE 5-6 V_j^2 **vs.** N **corrected for the existence of residual interactions. The dashed lines are a reproduction of Fig. 5-5, and the curves, which show the effects of residual interactions, are essentially a rounded-off version of them.**

From (5-27) and Fig. 5-6 we see that for O^{18}

$$\tfrac{1}{3}a_1^2 \gg b_1^2 \gg \tfrac{1}{2}e_1^2$$

This is in agreement with experimental determinations of the O^{18} ground-state wave function through determinations of the V_j^2 by methods to be described in Sec. 14-3 and application of (5-27).[1] These give $a_1^2 = 0.81$, $b_1^2 = 0.155$, $e_1^2 = 0.035$.

When considering holes, the number of holes in a state j is proportional to $1 - V_j^2$. For two-hole nuclei, as for $N = 10$ in Fig. 5-6, it can be seen that the holes will most probably occupy the highest-energy orbit, which in that case is $d_{3/2}$. This can be illustrated for Pb^{206}, whose wave function is given in (5-24). In this case, the equivalent of Fig. 5-6 for the $N = 82$ to 126 shell leads to

$$a_1^2 > \tfrac{1}{3}b_1^2 > \tfrac{1}{2}c_1^2 > \tfrac{1}{7}d_1^2 > \tfrac{1}{4}e_1^2 > \tfrac{1}{5}f_1^2$$

Experimental determinations of the coefficients give[2]

$$\begin{aligned} a_1^2 &= 0.54 & b_1^2 &= 0.20 \\ c_1^2 &= 0.12 & d_1^2 &= 0.12 \\ e_1^2 &= 0.03 & f_1^2 &\lessapprox 0.01 \end{aligned}$$

These clearly satisfy the above conditions.

[1] J. C. Armstrong and K. S. Quisenberry, *Phys. Rev.*, **122:** 150 (1961).

[2] *Nucl. Phys.*, **20:** 370 (1960); *Phys. Rev.*, **127:** 1284 (1962).

The fact that V_j^2 must behave as shown in Fig. 5-6 puts severe restrictions on the very complex wave functions previously discussed, e.g., in Table 5-2, but it does not completely determine them except in cases where there are only two nucleons outside of closed shells. Even in the relatively simple case of O^{20} the fact that the three V_j^2 are known from Fig. 5-6 is not sufficient to determine the five constants α_1, β_1, γ_1, δ_1, ϵ_1 with (5-28). However the V_j^2 give all the important information on wave functions needed for many practical purposes. Methods are available for calculating them accurately, i.e., for putting Fig. 5-6 on a quantitative basis, and have proved to be very useful.

We give here one of these methods,[1] known as *pairing theory*, without derivation. In it

$$V_j^2 = \frac{1}{2}\left[1 - \frac{\epsilon_j - \lambda}{\sqrt{(\epsilon_j - \lambda)^2 + \Delta^2)}}\right] \tag{5-29}$$

where λ is determined from the subsidiary requirement that the number of nucleons in the shell have its correct value, N. This is just the sum over j of the number of nucleons with each j, or

$$N = \sum_j (2j + 1) V_j^2$$

For the shell illustrated in Fig. 5-6 there are three values of j: $j = \frac{5}{2}(d_{5/2})$ with $\epsilon_j = 0$, $j = \frac{1}{2}(s_{1/2})$ with $\epsilon_j = \epsilon_s$, and $j = \frac{3}{2}\ (d_{3/2})$ with $\epsilon_j = \epsilon_d$. The Δ in the denominator of (5-28) is half of the energy gap, as shown in Figs. 5-3 and 5-4. A method for determining it from measurements of separation energies is given in Sec. 7-2.

5-8 Low-energy Excited States—An Introduction

Now that we have an understanding of the ground states of even-even nuclei, let us turn our attention to their excited states. As an introduction to this very complex problem, let us again begin with a study of the O^{18} nucleus and consider all states that can be formed without exciting a neutron into a higher shell. On the left side of Fig. 5-7 are shown the energies of the three orbits in the $\mathfrak{N} = 3$ shell from Fig. 4-5, and the second column shows the six possible configurations that can be reached by placing two nucleons in these orbits; as stated previously, the configuration energy is just the sum of the energies of the occupied orbits. The next part of the figure shows the terms classified according to $\boldsymbol{I}$ values. For example, a simple application of the methods of Sec. 2-6 shows that angular momenta $\frac{5}{2} + \frac{3}{2}$ can combine to give $I = 1$, 2, 3, or 4. There are therefore

[1] L. S. Kisslinger and R. A. Sorenson, *Kgl. Danske Videnskab. Selskab, Mat. Fys. Medd.*, 32(9) (1960).

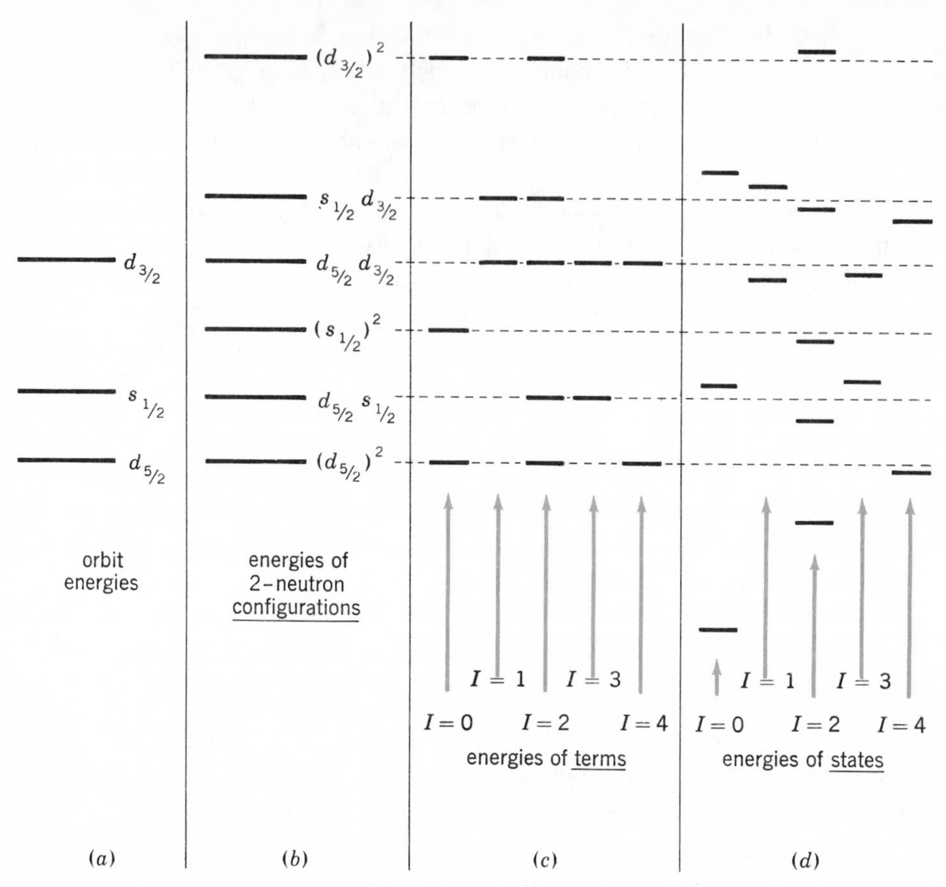

FIGURE 5-7 Development of the states of O^{18} formed from all possible configurations of the two neutrons in the $\mathfrak{N} = 3$ shell. The orbit energies (*a*) are from Fig. 4-5. The energies of the various two-neutron configurations are shown in (*b*); they are just the sum of the energies of the occupied orbits taking the lowest as the zero of energy. (*c*) shows the terms of each angular momentum I that can be derived from the configurations; for example, a $d_{5/2}d_{3/2}$ configuration gives terms with $I = 1, 2, 3$, and 4; their energy is the same as the configuration energy. (d) shows the states arising from the terms of each I.

$d_{5/2}d_{3/2}$ terms with these I values in Fig. 5-7. It is shown in Sec. 5-13 that, due to the Pauli exclusion principle, $(d_{5/2})^2$ can give only $I = 0$, 2, and 4 terms and $(d_{3/2})^2$ can give only $I = 0$ and 2 terms. All the terms in Fig. 5-7 have positive parity since all three orbits have even l and the sum of two even numbers is always even. The energies of the terms, as we know, are just the energies of the configurations from which they arise.

We are now ready to consider the effect of collisions. From the rules given in Sec. 5-1, collisions must conserve the total angular momentum I, so the changing of configurations through collisions can occur only among terms of the same I. We have already seen in Fig. 5-3 how collisions affect the energies of the $I = 0$ states. Their effect on $I = 2$ states is somewhat similar although not as strong. Since there are five terms with $I = 2$ in Fig. 5-7, there must be five $I = 2$ states, each with wave functions of the form

$$\psi_i = a_i(d_{5/2})_2{}^2 + b_i(d_{5/2}s_{1/2})_2 + c_i(d_{5/2}d_{3/2})_2 + d_i(s_{1/2}d_{3/2})_2 + e_i(d_{3/2})_2{}^2$$

In one of these there is a very systematic and rhythmic collision pattern, which causes its energy to be pushed down well below the lowest configuration energy. The quantum-mechanical explanation is similar to that discussed in Sec. 5-4. The effect is so strong that this state becomes the first, i.e., lowest-energy, excited state of the nucleus, as we see in Fig. 5-7.

As in the case of pairing and the ground state, this effect occurs in all spherical non-closed-shell even-even nuclei. If there are a number q of 2^+ terms with one broken pair, there are q states which are linear combinations of these terms. In one of these q states, the collision pattern is a very rhythmic one, which results in its energy being so lowered that it becomes the first excited state of the nucleus.

The reason for this large energy lowering has a simple physical interpretation. The collision pattern here is such a rhythmic one that it causes a deformation of the orbits of all the other nucleons in the nucleus in unison with it for reasons similar to those given in Sec. 4-8. As a result, the shape of the entire nucleus changes with time in a manner very much like the vibration of a spherical liquid drop. Before continuing our discussion, it will pay us to pause for a discussion of this type of vibration in some detail.

5-9 Shape Oscillations of a Liquid Drop

Let us consider the classical physics problem of the shape oscillations of a macroscopic drop of liquid. Aside from a few complicating effects, they can readily be observed in nature as the oscillations of drops of water on a sizzling hot greasy surface or of drops of liquid nitrogen from spillings on a floor or table. The theory was developed by Lord Rayleigh in the nineteenth century; the vibrations may be considered to be composed of standing waves on the surface of the drop analogous to the familiar standing waves on a vibrating string (e.g., a violin string) or in an air column (e.g., an organ pipe). In order to be standing waves, the waves on the surface of the drop must consist of an integral number of wavelengths going around the surface of the sphere lest there be cancellation from successive cycles as the waves travel around. The problem is therefore similar to that considered in Secs. 2-3 and 2-4 for the θ, ϕ dependence of the wave function for a particle in a potential well, whence the results are rather similar: the θ, ϕ dependence of the displacement from the spherical equilibrium shape in these waves is given by the associated Legendre polynomials listed in Table 2-1.

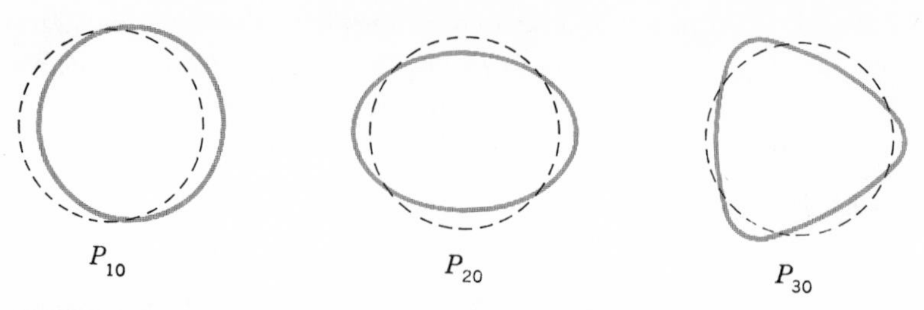

FIGURE 5-8 Displacements from spherical symmetry in standing waves on the surface of a liquid drop are proportional to the associated Legendre polynomials $P_{\lambda\mu}$. Shown here are waves for $\mu = 0$ and $\lambda = 1, 2,$ and 3. All figures have rotational symmetry around the horizontal axis through the center. Dashed line is zero displacement, i.e., spherical symmetry.

We refer to them as $P_{\lambda\mu}$. (Note that λ here is not to be confused with the symbol for wavelength.)

The displacement in a $\lambda = 0$ wave, P_{00}, has no θ dependence, so it is uninteresting in this connection. The displacements in $\lambda = 1$, $\lambda = 2$, and $\lambda = 3$ waves for $\mu = 0$ are shown in Fig. 5-8. We see immediately that the $\lambda = 1$ wave is unphysical since it corresponds to a movement of the center of mass of the drop. The $\lambda = 2, 3, \ldots$ waves do not suffer from this shortcoming and are therefore fundamental vibrational shapes in the same sense that an integral number of half wavelengths is a fundamental vibrational shape of a string. These fundamental shapes are called *normal modes*, and in courses in classical mechanics techniques are given for describing any possible vibrational pattern as a linear combination of normal modes. Problems in vibration are thereby reduced to problems of understanding the vibration of each individual normal mode.

The treatment of an individual normal mode is analogous to the familiar treatment of a wave on a string. The velocity of the wave v is roughly determined by

$$v \approx \sqrt{\frac{T'}{M'}} \tag{5-30}$$

where M' is the mass density and T' is the *surface tension*, the force per unit displacement tending to resist deviations from a spherical shape (this will be discussed in Sec. 7-2). The wavelength, which we designate W_λ here to avoid confusion, is determined by the geometry. For example, when the shape is as shown in Fig. 5-8 for P_{30}, the standing wave goes through three wavelengths in going around, whence the wavelength is one-third of $2\pi R$. In general, the number of wavelengths in going around is equal to λ, whence

$$W_\lambda \approx \frac{2\pi R}{\lambda} \tag{5-31}$$

The frequency ν_λ is then determined by the familiar relation between frequency, velocity, and wavelength for a wave,

$$\nu_\lambda = \frac{v}{W_\lambda} \tag{5-32}$$

We know from Sec. 1-2 that the mass densities and the surface structure are quite similar for all nuclei, so from (5-31) and (5-32) we see that to some approximation

$$\nu_\lambda \approx \text{const} \times \frac{\lambda}{R} \tag{5-33}$$

In a system small enough for quantum effects to be important, such as a molecule or a nucleus, vibrations are quantized into energy units of size $h\nu_\lambda$, known as *phonons*. A vibrational state must contain an integral number of phonons of each λ. Some of the lowest-energy vibrational states therefore have energy $h\nu_2$, $2h\nu_2$, $3h\nu_2$, $h\nu_3$, $h\nu_3 + h\nu_2$, $h\nu_4$, etc. With a more accurate calculation than that leading to (5-33), the energies of these states are found to be approximately as exhibited in Fig. 5-9. It can be shown that each phonon of vibration carries an angular momentum λ and parity $(-1)^\lambda$. Thus the vibrational states of energy $h\nu_2$, $h\nu_3$,

FIGURE 5-9 Energy spectrum of vibrational states expected if the nucleus could be considered to be a liquid drop. Groupings are according to λ with states for increasing numbers of phonons in the same vertical columns. States shown close together actually are at the same energy. The energy gap 2Δ is shown at the relative energy at which it occurs in typical nuclei. The energies of the one-phonon vibrations have been adjusted to agree with the actual situation in nuclei.

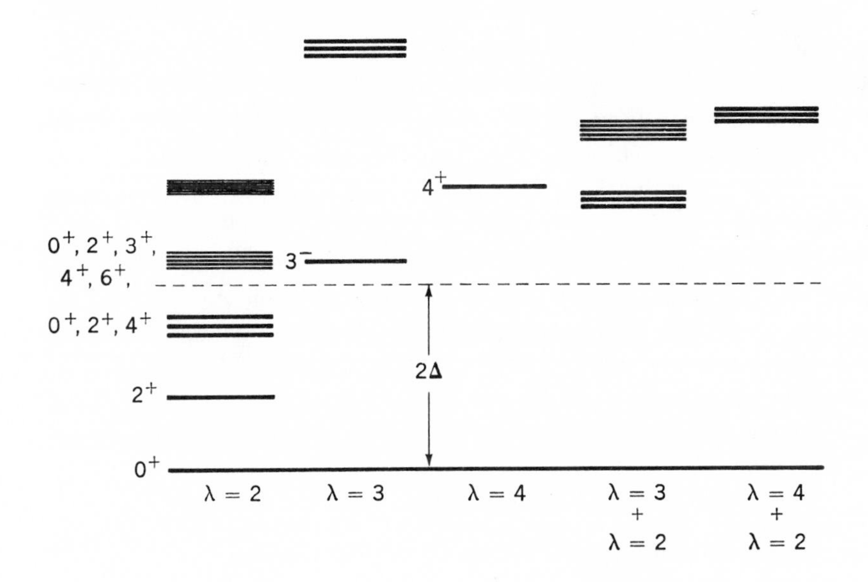

and $h\nu_4$ have I^π of 2^+, 3^-, and 4^+, respectively. In the two-phonon vibration of energy $2h\nu_2$, it is shown in Sec. 5-13 that the total angular momentum can be 0, 2, or 4, so we have three states at an energy $2h\nu_2$ with I^π of 0^+, 2^+, and 4^+. These states are so labeled in Fig. 5-9.

It is interesting to point out that in our everyday world vibrations are generally combinations of many normal modes and hence of many frequencies. In musical instruments these are known as *overtones* or *harmonics*. In quantum language, we would say that these vibrations contain large numbers of phonons for each λ. In nuclei (and also in molecular vibrations), on the other hand, the most readily observable states are those of lowest energy, and, as we see from Fig. 5-9, these are states with only one or two phonons. Quantum vibrations are therefore generally much simpler than those encountered in everyday life.

5-10 Collective Vibrations of Spherical Even-Even Nuclei

A nucleus, of course, does not have all the properties of a liquid drop; the number of nucleons in a nucleus is rather small for a theory designed for use on a continuous medium like a liquid, and the orbits of the nucleons are subject to many quantum restrictions (such as quantized energies and angular momenta and the Pauli exclusion principle) which are not considered in treatments like that of Sec. 5-9. However, there is similarity enough for us to expect states in nuclei which correspond roughly to those in Fig. 5-9. States of this type are referred to as *collective* since they are made up of the motions of many nucleons acting collectively to produce effects that could not be produced by one nucleon alone.

As mentioned at the end of Sec. 5-8, the first excited state of all spherical non-closed-shell even-even nuclei is such a state, and from a comparison of the discussion there with Fig. 5-9 we see that it is the one-phonon $\lambda = 2$ vibration. From the standpoint of our orbital picture, it is a two-quasi-particle (2QP) state[1] arising from a linear combination of one-broken-pair 2^+ terms in a rhythmic collision pattern which leads to an oscillating deformation of all nucleon orbits in unison with it. In this state, the surface of the nucleus vibrates between the two shapes shown in Fig. 5-10*a*. As a result of the rhythmic collision pattern—or more accurately, as a result of the considerations of Sec. 5-4—its energy is lowered well below that of the lowest-energy 2^+ term, which puts it into the lower middle part of the energy gap discussed at the end of Sec. 5-3.

Since a phonon arises from a combination of one-broken-pair terms, a two-

[1] The fact that it is a 2QP state is not obvious, but it can be demonstrated. The same is true of other states whose quasi-particle number is given in this section.

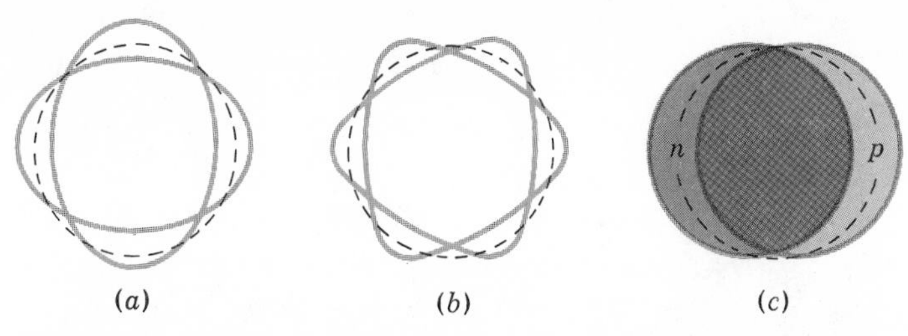

FIGURE 5-10 Different types of nuclear vibrations. The two extremes of the vibration are shown by the solid curves, and the midpoint of the vibration, a spherical shape, is shown by the dashed line. All figures have rotational symmetry about a horizontal axis through the center. *(a)* **$\lambda = 2$ vibration (2^+ state);** *(b)* **$\lambda = 3$ vibration (3^- state); in (*a*) and (*b*) the neutrons and protons move together;** *(c)* **$\lambda = 1$ (1^- state), in which the neutrons and protons move in opposite directions. At one extreme of the vibration, the neutrons are displaced to the left and the protons are displaced to the right as shown, while at the other extreme of the vibration their positions are reversed.**

phonon vibration must arise principally from two-broken-pair terms and must be a 4QP state. A nucleus with only two nucleons outside of closed shells, such as O^{18}, cannot, of course, have such a state (without exciting nucleons into higher shells, which takes a lot of energy), but in a nucleus having four or more nucleons outside of closed shells, there will generally be many 4QP states with $I = 0^+, 2^+$, and 4^+. In one of these 0^+ states there is a very rhythmic collision pattern which corresponds roughly to the two-phonon 0^+ vibration; its energy is therefore pushed down below that of the other 4QP 0^+ states to an energy near that of the corresponding vibrational state in Fig. 5-9. A similar situation prevails among the 4QP 2^+ states and again among the 4QP 4^+ states.

The vibrational description of these states is generally oversimplified for reasons already mentioned and for further reasons to be discussed in Sec. 6-7. As a result, the simple energy relationships of Fig. 5-9, where the two-phonon states are at twice the energy of the one-phonon state, are not accurately fulfilled; these energy ratios vary typically between about 1.8 and 2.6. Nevertheless the simple picture of Fig. 5-9 is at least crudely fulfilled for the $\lambda = 2$ vibrations. In essentially all non-closed-shell spherical even-even nuclei the first excited state is 2^+, and the next three excited states are 0^+, 2^+, and 4^+ (not necessarily in that order) and are located at about twice the energy of the first 2^+ state. The energy levels of some of these nuclei are shown in Fig. 5-11. There is some evidence for three-phonon $\lambda = 2$ vibrations, but things become so complicated at the rather high excitation energy where they occur that it is very difficult to draw detailed correspondences.

The one-phonon 4^+ vibration is a 2QP state arising from a linear combina-

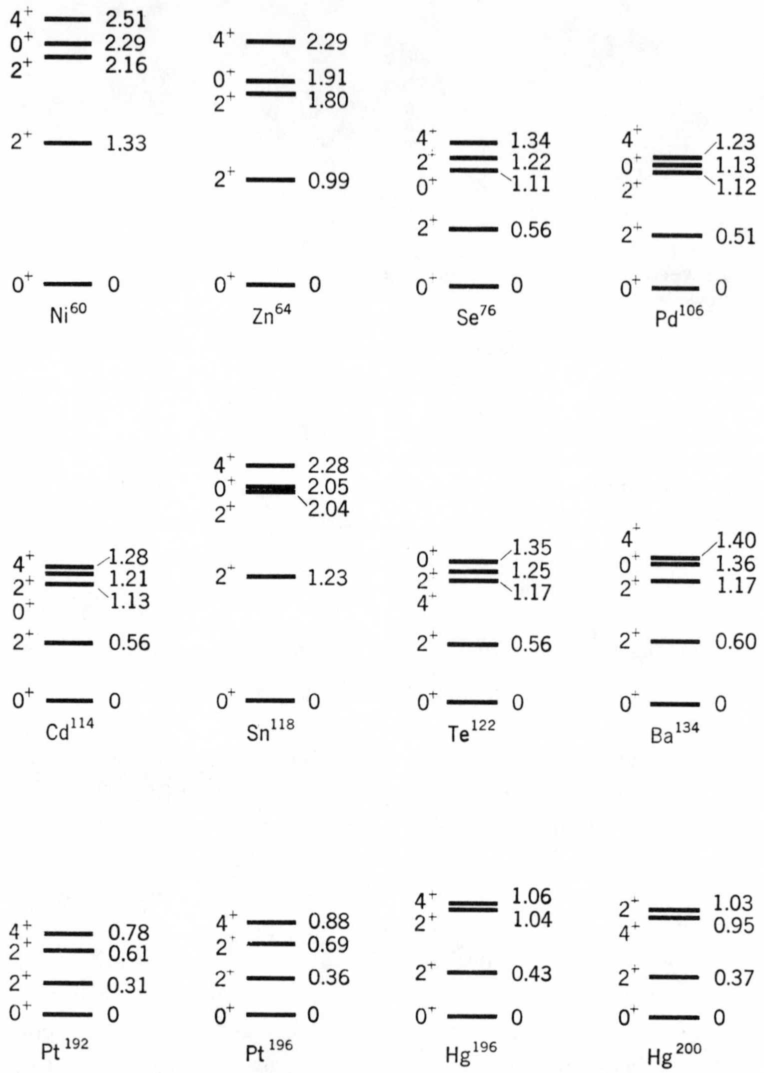

Ni^{60}: 4^+ 2.51; 0^+ 2.29; 2^+ 2.16; 2^+ 1.33; 0^+ 0

Zn^{64}: 4^+ 2.29; 0^+ 1.91; 2^+ 1.80; 2^+ 0.99; 0^+ 0

Se^{76}: 4^+ 1.34; 2^+ 1.22; 0^+ 1.11; 2^+ 0.56; 0^+ 0

Pd^{106}: 4^+ 1.23; 0^+ 1.13; 2^+ 1.12; 2^+ 0.51; 0^+ 0

Cd^{114}: 4^+ 1.28; 2^+ 1.21; 0^+ 1.13; 2^+ 0.56; 0^+ 0

Sn^{118}: 4^+ 2.28; 0^+ 2.05; 2^+ 2.04; 2^+ 1.23; 0^+ 0

Te^{122}: 0^+ 1.35; 2^+ 1.25; 4^+ 1.17; 2^+ 0.56; 0^+ 0

Ba^{134}: 4^+ 1.40; 0^+ 1.36; 2^+ 1.17; 2^+ 0.60; 0^+ 0

Pt^{192}: 4^+ 0.78; 2^+ 0.61; 2^+ 0.31; 0^+ 0

Pt^{196}: 4^+ 0.88; 2^+ 0.69; 2^+ 0.36; 0^+ 0

Hg^{196}: 4^+ 1.06; 2^+ 1.04; 2^+ 0.43; 0^+ 0

Hg^{200}: 2^+ 1.03; 4^+ 0.95; 2^+ 0.37; 0^+ 0

FIGURE 5-11 The lowest-energy states of various spherical even-even nuclei. Excitation energies (in MeV) are given at the right, and I^π are shown at the left. Note that in all cases the ground state is 0^+, the first excited state is 2^+, and there are three states at about twice its excitation energy with $I^\pi = 0^+$, 2^+, and 4^+, not necessarily in that order. In some of these nuclei, all three of the latter group are not yet known experimentally.

tion of one-broken-pair 4^+ terms, e.g., the two 4^+ terms in Fig. 5-7. As in the several cases already discussed, one linear combination of these contains a rhythmic collision pattern which leads to oscillations in the nuclear shape resembling the one-phonon $\lambda = 4$ vibration.

In the lighter even-even nuclei at least, all configurations obtained without exciting nucleons to another shell have even parity. For example all $\mathfrak{N} = 2$ orbits have odd l, so an even number of these must have sums of l values which are even and hence have even (+) parity. In the higher-$\mathfrak{N}$ shells, all orbits but one have the same parity, so again the great majority of configurations formed from an even number of these orbits are of even parity. It is clear, therefore, that the 3^- collective vibration cannot easily arise from combinations of these configurations. Odd-parity states can arise principally from exciting nucleons to the next higher shell. For example, negative-parity states of O^{18} can arise from exciting nucleons normally in $\mathfrak{N} = 2$ orbits to $\mathfrak{N} = 3$ orbits or from exciting nucleons normally in $\mathfrak{N} = 3$ orbits into $\mathfrak{N} = 3A$ or 4 orbits. Configurations of the first type are $(1p_{3/2})^{-1}(1d_{5/2})^3$, $(1p_{1/2})^{-1}(1d_{5/2})^2 2s_{1/2}$, etc. Configurations of the second type are $(1d_{5/2})(1f_{7/2})$, $(1d_{3/2})(2p_{3/2})$, etc. Any of these can lead to 3^- terms, and there is little difficulty in seeing that there are many others of these types.

These configurations, of course, have very high excitation energies, of the order of the energy spacing between shells. On the other hand, there is a very large number of 3^- terms (see Prob. 5-16). One linear combination of these has a rhythmic collision pattern corresponding to the one-phonon $\lambda = 3$ vibration, which causes the shape of the nucleus to undergo oscillations like those shown in Fig. 5-10*b*. Since there are so many of these terms, the energy lowering of the collective state is especially large here. [From (5-19) the energy shift is proportional to the number of terms involved.] It is large enough to bring this state approximately into the position shown in Fig. 5-9.[1]

Since the $\lambda = 2$ vibrational states arise from orbital configurations of nucleons in the shell that is filling (as in Fig. 5-7), there can be no such states in closed-shell nuclei. Excited states of closed-shell nuclei can arise only from exciting nucleons into higher shells. As this requires an energy of the order of the spacing between shells, there are *no* low-energy excited states in closed-shell nuclei. The lowest-energy state formed by exciting nucleons into the next shell may be the $\lambda = 3$ vibration. The first excited state of Pb^{208} is this 3^- state, occurring at the exceptionally high energy (for so heavy a nucleus) of 2.6 MeV. In O^{16} and Ca^{40} this state is among the lowest-energy excited states, but the first

[1] Actually, it is not quite large enough to bring the energy down to that expected for a liquid drop, which is at less than twice the energy of the first 2^+ state, as we see from (5-33). However, we have set the energy of the 3^- state in Fig. 5-9 higher than the liquid-drop value to agree with the nuclear situation.

excited state is a nonspherical one. As mentioned in Sec. 4-7, nonspherical nuclei usually arise when there are several nucleons outside of closed shells; in O^{16} this can be realized by having nucleons excited from the $\mathfrak{N} = 2$ to the $\mathfrak{N} = 3$ shell. This, of course, requires a great deal of energy, but now both the $\mathfrak{N} = 2$ and $\mathfrak{N} = 3$ shell can assume ellipsoidal shapes, thereby lowering the energy considerably. Still, this first excited state is at 6.05 MeV excitation energy, which is much higher than in non-closed-shell nuclei. The 3^- state is just above it at 6.13 MeV.

The energies of the lowest 2^+ and 3^- states in various nuclei are plotted vs. A in Fig. 5-12. It is seen that in nuclei with both neutron and proton shells closed (O^{16}, Ca^{40}, Pb^{208}), the lowest 2^+ state is at a very high energy, indicating that it is not collective. In all other cases, except where nuclei are nonspherical, the states shown are the one-phonon vibrations. The curves labeled "liquid drop" are calculated from a liquid-drop model as described in Sec. 5-9; they have been multiplied by an adjustable constant to best fit the data. It is seen that they reproduce the trends in the data reasonably well except near closed shells and in regions where nuclei are nonspherical. The increase in these curves with decreasing A may be qualitatively understood from the fact that the energy is $h\nu_\lambda$, which, from (5-33), is inversely proportional to R and hence to $A^{1/3}$.

The principal deviations in the data occur for nuclei in which either the neutrons or the protons have closed shells. This is because these nuclei have a smaller number of low-energy 2^+ terms. For example, in the $_{50}Sn$ isotopes, these terms can arise only from neutron configurations, whereas in neighboring nuclei they can come from proton configurations as well. Since from (5-19) the lowering of the collective states is proportional to the number of contributing configurations, the energy lowering is less in Sn than in its neighbors. The 3^- states are much less affected by such considerations since they arise largely from excitations from one shell to the next, and these can also occur in closed-shell nuclei. In heavy nuclei, however, the effect of the opposite-parity orbits in the shell that is filling is important enough to give the 3^- state a lower energy in non-closed-shell nuclei.

The dashed line in Fig. 5-9 is the energy gap 2Δ, which we have defined as the energy by which the ground state is lowered by the pairing interaction. If excitations are measured from the ground state, we expect to find the great number of states that are not lowered by the effect discussed in Sec. 5-4 at excitation energies of 2Δ and higher. This region is therefore a very complicated one with large numbers of states of various types. In the region within the energy gap (excitation energy $< 2\Delta$), there are very few states, so the one- and two-phonon $\lambda = 2$ vibrational states are easily distinguished. The other collective states shown in Fig. 5-9 are much more difficult to distinguish unless they are strongly excited in some special processes that do not excite the other states. There are such processes for the one-phonon 3^- and 4^+ states, as we shall see in Sec. 14-6, but there are not any good processes for exciting the two- and higher-phonon states. These are consequently not generally known for any but the $\lambda = 2$ vibra-

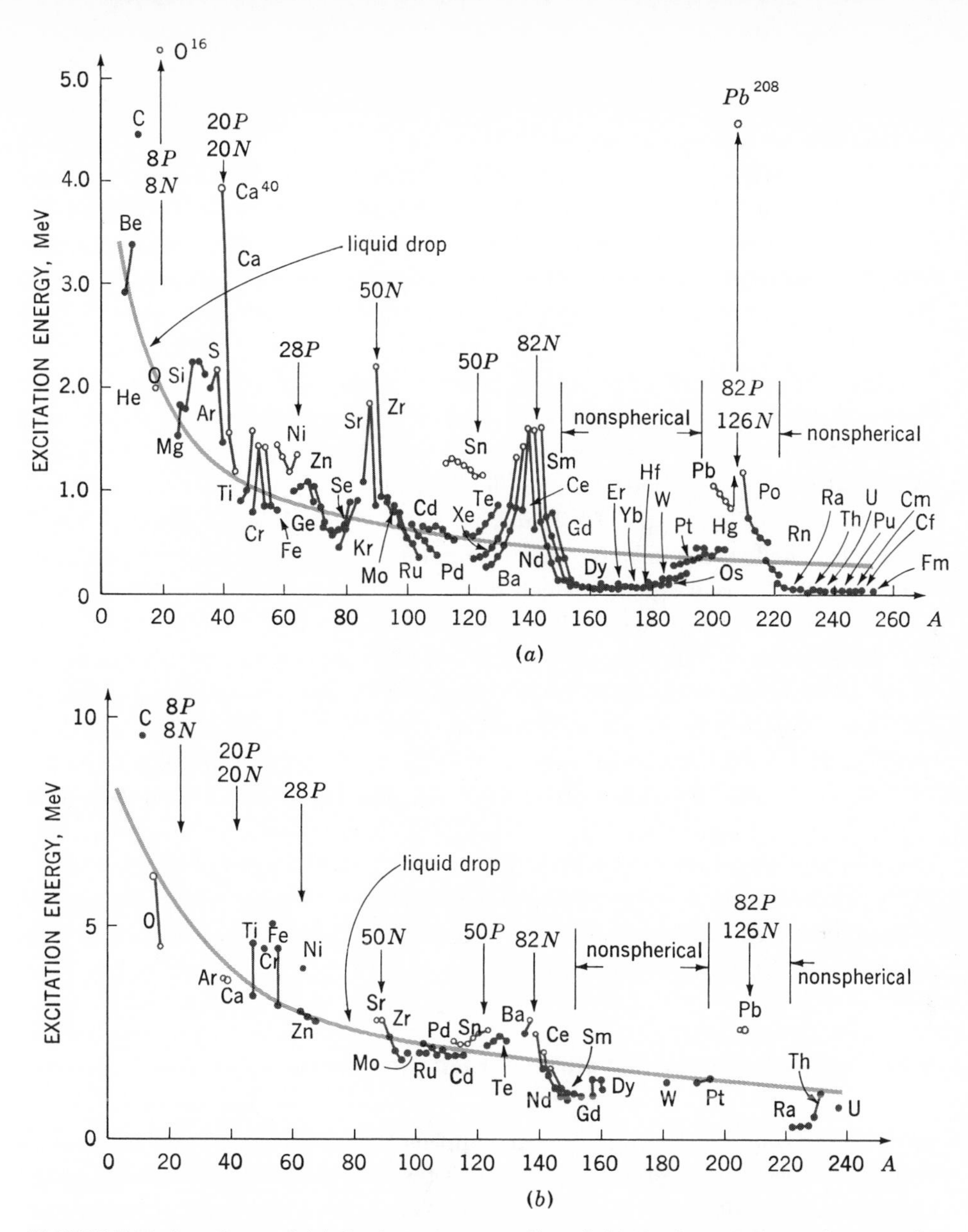

FIGURE 5-12 **Locations of** *(a)* **the lowest-energy 2^+ and** *(b)* **the lowest-energy 3^- states in various even-even nuclei. With the exception of the 2^+ states in the closed-shell nuclei (O^{16}, Ca^{40}, and Pb^{208}) and the nonspherical nuclei, these are the one-phonon $\lambda = 2$ and $\lambda = 3$ vibrations. Lines labeled "liquid drop" are energies expected if the nucleus vibrated like a classical liquid drop; they have been multiplied by an arbitrary constant (twice as large for the 3^- as for the 2^+) to obtain the best fit to the data. Locations of closed shells are shown by vertical lines with arrows. States of different isotopes of a given element are connected by straight lines and marked by the chemical symbol for the element.** [*Adapted from O. Nathan and S. G. Nilsson in K. Siegbahn (ed.), "Alpha, Beta, and Gamma Ray Spectroscopy," North-Holland Publishing Company, Amsterdam,* **1966**; *by permission.*]

tions. The states that have not been observed experimentally are those for which I^π designations are omitted in Fig. 5-9.

Before leaving the subject of collective vibrations, we should mention a class of vibrations other than the ones discussed here. In the vibrations described above, the neutrons and protons move in concert to give the overall vibration of the nuclear shape. In the other type, the neutrons and protons move in opposition to one another. In this type of vibration, the $\lambda = 1$ deformation of Fig. 5-8 is possible, and the deformation is shown in Fig. 5-10*c*. When the neutrons move one way, the protons move the other, and vice versa, so there is no displacement of the center of mass. Because of the strong attraction between neutrons and protons, the restoring forces are very large in this type of deformation, whence the vibration frequencies are quite high. Typically, $h\nu_1$ is of the order of 14 MeV in heavy nuclei. This type of vibration is easily excited when a gamma ray of energy $h\nu_1$ strikes a nucleus, since a gamma ray incorporates an oscillating electric field of this frequency which drives the protons back and forth relative to the neutrons. This process will be discussed in Sec. 13-10.

Other types of vibrations are known at high excitation energies. In one of them, all nucleons with spin *up* are displaced in one direction while all those with spin *down* are displaced in the other. These then oscillate back and forth, much as the neutrons and protons do in Fig. 5-10*c*. In still another vibration, neutrons with spin *up* and protons with spin *down* go one way, while protons with spin *up* and neutrons with spin *down* go the other. Both these oscillations are easily excited by gamma rays of the proper energy, since gamma rays are associated with a varying magnetic field, which pushes nucleons in different directions depending on the orientation of their magnetic moments, which are in turn determined by their spin orientation.

5-11 Noncollective Excited States of Even-Even Nuclei

At this stage one might wonder what happens to terms that are not involved in collective vibrations, such as the 1^+ and 3^+ terms of Fig. 5-7. The answer is that nothing very spectacular happens. Where two terms of the same I^π are close together in energy, they mix (in the two states that arise from them the nucleus goes frequently from one term to the other by collisions), but the energies of these states are not much different from the energy of the terms. In more complex nuclei with several neutrons and protons outside of closed shells, there are often a number of terms of the same I^π in the same energy region, so nucleons spend only a small fraction of their time in any given configuration; i.e., the wave functions contain a large number of terms. However, as long as no collective vibrations are involved, nothing spectacular happens to the energies of these states; they are close to the average energy of the configurations of which the states are composed.

The same might be said of the 0^+, 2^+, etc., states in which the collision pattern is not the rhythmic one leading to the ground state or to one of the collective vibrations. In Sec. 5-5 we pointed out that even if we restrict the nucleons to the shells that are filling in the ground state, in complex nuclei there may be tens of thousands of 0^+ states and there are many more of each larger I. At higher excitation energies, states in which nucleons are excited from one shell to another are encountered; at still higher energies, there are states in which two or more nucleons are excited to the next shells or even to higher shells. At excitation energies where this is possible, the number of terms and hence the number of states increases very rapidly, and since there are many states of each I^π, mixing of configurations becomes very widespread.[1] In a typical heavy nucleus, there may be millions of states up to 10 MeV excitation. Clearly we can never hope to understand many of these in any detail. The ones nuclear physicists usually study are the collective states and a few other low-energy states with simple configurations.

On the other hand, the multitude of higher-energy and more complicated states cannot be completely ignored because they play important roles in nuclear reactions and decay processes. For these purposes, we can often treat them statistically by introducing a level density $\omega(\epsilon)$ defined such that the number of states between excitation energies ϵ and $\epsilon + d\epsilon$ is $\omega(\epsilon)\, d\epsilon$. Expressions for $\omega(\epsilon)$ can be derived from various statistical arguments. A common result is

$$\omega(\epsilon) = C \exp\,[2(\mathcal{a}\epsilon)^{1/2}] \qquad \textbf{(5-34)}$$

where C and $\mathcal{a}$ are constants depending on the mass number A.

Students familiar with statistical mechanics will recognize that one can derive an entropy $S(\epsilon)$ and a temperature $T(\epsilon)$ from this expression, as

$$S(\epsilon) = \ln \omega(\epsilon) = 2(\mathcal{a}\epsilon)^{1/2}$$
$$kT(\epsilon) = \left[\frac{\partial S(\epsilon)}{\partial \epsilon}\right]^{-1} = \left(\frac{\epsilon}{\mathcal{a}}\right)^{1/2} \qquad \textbf{(5-35)}$$

where k is the Boltzmann constant. The last equation can be rearranged to give

$$\epsilon = \mathcal{a}k^2T^2 \qquad \textbf{(5-36)}$$

which is the well-known relationship for a Fermi gas, i.e., a gas in which Fermi-Dirac statistics is applicable. It differs from the familiar relationship for a gas in Maxwell-Boltzmann (classical) statistics,

$$\epsilon = \tfrac{3}{2}N_0kT$$

[1] An additional reason for this will be given in Sec. 13-4.

where N_0 is the number of particles, because only a small fraction of the nucleons in the nucleus are excited. Even in rather highly excited states of a heavy nucleus, the nucleons in the $\mathfrak{N} = 1$ and $\mathfrak{N} = 2$ shells are not disturbed. A more explicit example of this is given in Sec. 13-1.

This so-called *degeneracy effect* leads inevitably to an expression of the form (5-36), except that the power to which T is raised can deviate slightly from 2 if the system is not a perfect gas. Once we have (5-36), a simple integration gives $S(\epsilon)$ as in (5-35), and thence (5-34).

Many elaborate treatments of the level-density problem have been given, incorporating shell effects, empirical information, etc. In one of them,[1] the level density at energies above about 4 MeV is estimated for states of each I and either parity as

$$\omega(\epsilon, I) = \frac{1.1\ \text{MeV}^{-1}}{A\mathfrak{a}\epsilon^2}(2I + 1)\{\exp\,[2(\mathfrak{a}\epsilon)^{1/2}]\}\left[\exp\left(-\frac{I + \frac{1}{2}}{2\sigma^2}\right)\right] \tag{5-37}$$

where

$$\mathfrak{a} = 0.14A + \text{small shell corrections}$$
$$\sigma^2 = 0.089A^{2/3}(\mathfrak{a}\epsilon)^{1/2}$$

and ϵ is in MeV. When this expression is integrated over I, the result which elaborates on (5-34) is

$$\omega(\epsilon) = \frac{0.20\ \text{MeV}^{-1}}{A^{1/3}\mathfrak{a}^{1/2}\epsilon^{3/2}}\exp\,[2(\mathfrak{a}\epsilon)^{1/2}] \tag{5-34a}$$

The excitation energy ϵ to be used in these formulas is not the excitation above the ground state. As we have seen, the ground-state energy is lowered very much in even-even nuclei by pairing, whereas this does not affect any of the other states of the nucleus. In order to correct for this, ϵ is taken as the excitation above the ground state minus 2Δ and Δ for even-even and odd-A nuclei, respectively. The parameter Δ has been introduced in Fig. 5-3, and methods for obtaining it will be given in Sec. 7-2.

5-12 Limitations of the Shell Approximation

In all our discussions we have tacitly assumed that the energy spacing between shells is extremely large while the energy spacings between different orbits in the same shell are very small. This assumption, illustrated in Fig. 5-13*a*, is oversimplified. The actual situation is more like that in Fig. 5-13*b*. The energy region spanned by a shell (A to B in Fig. 5-13*b*) is about as large as the energy spacing between shells (B to C).

[1] A. Gilbert and A. G. C. Cameron, *Can. J. Phys.*, **43**: 1446 (1965).

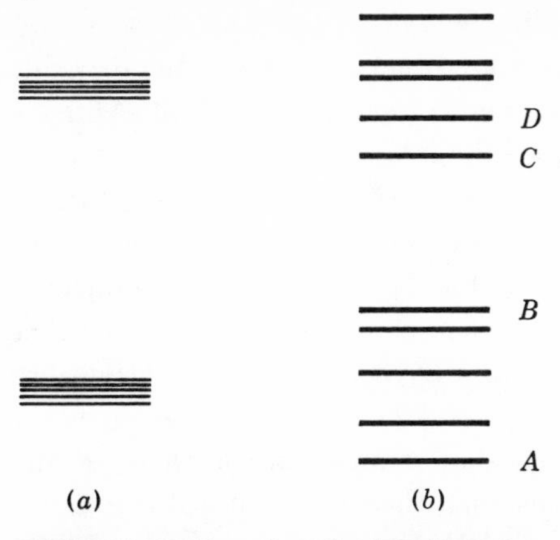

FIGURE 5-13 Energies of orbits from two successive shells. *(a)* **The approximation we have been using in our discussion, assuming that the energy difference between the various orbits in a shell is very much less than the energy difference between adjacent shells.** *(b)* **More typical of the actual situation; the maximum energy difference between orbits in the same shell, for example, A and B, is about the same as the minimum energy difference between orbits in adjacent shells, for example, B and C.**

Another oversimplification we have used widely is the estimate (5-6) of the allowable violation of energy conservation in collisions. The uncertainty principle does not say that energy-conservation violations absolutely cannot last for a longer time than that given by (5-4); it says only that (5-4) gives an *estimate* of the length of time for which such energy-conservation violations can *easily* persist. Larger violations are possible, and for longer times, but with diminished probability.

These two oversimplifications were the basis for obtaining (5-1), which has the effect of preventing collisions of nucleons in closed shells while allowing a large variety of collisions among nucleons in nonclosed shells. Both of these effects have therefore been exaggerated. For example, in a typical closed-shell nucleus, where the lower of the two shells shown in Fig. 5-13*b* is supposedly filled while the upper is empty, a pair of nucleons is in orbit C rather than in orbit B for a few percent of the time, and occasionally nucleons may be found in orbit D.

For many purposes, this deviation from ideal behavior is small enough to be ignored. But similarly, separations of orbit energies within a shell are often sufficiently large for some of the orbits to be considered as belonging to separate shells. For example, we can see from the results given in Sec. 5-7 that there would be little error in calculating the ground-state wave function of O^{18} if the $d_{3/2}$ orbit were ignored; the two neutrons spend only 3.5 percent of their time in it. Approximations of this type are often made to reduce the labor in computation.

A different and unrelated oversimplification we have been using is in assuming that the relative orbit energies of Fig. 4-5 are the same in all nuclei. Since the shell-theory potential represents the average force exerted on a given nucleon by all other nucleons in the nucleus, it depends to some extent on what other orbits are occupied. For example, we see from Fig. 2-4 that high-l orbits are largely confined to the outer edge of the nucleus. Therefore two nucleons in high-l orbits spend more time near each other than near a low-l nucleon, so the filling of a high-l orbit has more effect on the shell-theory potential for a high-l nucleon than for a low-l nucleon and vice versa. However, there are many more nucleons in a high-l orbits than in low-l orbits, so as orbits fill, the potential experienced by high-l nucleons becomes a little stronger relative to that experienced by low-l nucleons. When the potential is stronger, the well is deeper, so the orbit energies are lowered. We therefore expect high-l orbits to move down in energy relative to low-l orbits as a shell fills. An example of this can be seen in Fig. 5-14, which shows experimental determinations of orbit energies in the $\mathcal{N} = 5$ and $\mathcal{N} = 6$ shells for various A. In that figure the $g_{7/2}$ and $h_{11/2}$ orbits in the $\mathcal{N} = 5$ shell and the $h_{9/2}$ and $i_{13/2}$ orbits in the $\mathcal{N} = 6$ shell move down in energy relative to the others as A increases.

While the oversimplifications discussed in this section must be recognized by the research worker, we shall generally ignore them in our future discussions

FIGURE 5-14 Experimentally determined energies of orbits in the $\mathcal{N} = 5$ and $\mathcal{N} = 6$ shells for various A. [*From Phys. Lett.*, **27B:** 271 (1968).]

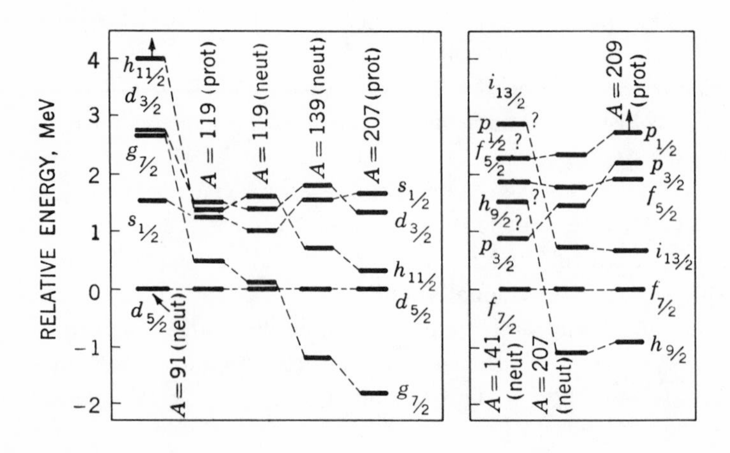

here. To do otherwise would greatly complicate our task of understanding the simpler aspects of nuclear structure.

5-13 Some Results from Angular-momentum Coupling

In this chapter we have used some results from angular-momentum coupling, referring to this section for their derivation. These results were for two rather different situations: (1) the angular momentum I of terms formed from several nucleons of the same type with the same n, l, j quantum numbers and (2) the values of I obtained from the coupling of two phonons. To demonstrate the derivations of the first type, let us consider the terms that can be formed from $(d_{5/2})^2$.

There are six $d_{5/2}$ orbits, with $m = 5/2, 3/2, \ldots, -5/2$. The various possible ways of putting two particles into these orbits consistent with the requirement from the Pauli exclusion principle that no two identical particles can be in the same orbit, i.e., have all the same quantum numbers, are shown in Table 5-3. Each vertical column represents a different possibility; e.g., the first column is the situation where the two nucleons are in the $m = +5/2$ and $m = +3/2$ orbits. We refer to the different arrangements, i.e., different columns in the table, as m configurations; each has a z component of total angular momentum m_I equal to the sum of the m values for occupied orbits (Σm). This is given in the last row of Table 5-3. From this last row we compile Table 5-4, which is the number of m configurations with each value of m_I.

We know from the general properties of angular momentum in quantum mechanics that a term with total angular momentum I has $m_I = I, I-1, \ldots, -I$. From Table 5-4, then, the largest value of I is 4. This accounts for m configurations with $m_I = 4, 3, 2, 1, 0, -1, -2, -3$, and -4. The largest remaining

TABLE 5-3: VARIOUS WAYS OF PUTTING TWO IDENTICAL NUCLEONS IN $d_{5/2}$ ORBITS

m	*Occupied orbits*														
$+5/2$	x	x	x	x	x										
$+3/2$	x					x	x	x	x						
$+1/2$		x				x				x	x	x			
$-1/2$			x				x			x			x	x	
$-3/2$				x				x			x		x		x
$-5/2$					x				x			x		x	x
Σm	+4	+3	+2	+1	0	+2	+1	0	−1	0	−1	−2	−2	−3	−4

TABLE 5-4: NUMBER OF m CONFIGURATIONS WITH EACH m_I IN TABLE 5-3

m_I	+4	+3	+2	+1	0	−1	−2	−3	−4
Number of configurations	1	1	2	2	3	2	2	1	1

m_I is 2, so the term with next highest I must have $I = 2$. This accounts for m configurations with $m_I = 2, 1, 0, -1, -2$, leaving only a single m configuration with $m_I = 0$. This can only be accounted for by an $I = 0$ term. In summary, there are three terms derived from $(d_{5/2})^2$, and they have $I = 4$, 2, and 0. A more detailed quantum-mechanical treatment is required to show such things as what combination of the $m_I = 0$ configurations of Table 4-3 are in the $I = 4$, $m_I = 0$ term, but these matters need not concern us here.

With a treatment similar to that given above, it can be shown that for two identical particles in orbits of the same n, l, j, terms result with $I = 0, 2, 4, \ldots, 2j - 1$. For example, $(d_{3/2})^2$ gives states with $I = 0$ and 2, and $(g_{7/2})^2$ gives states with $I = 0$, 2, 4, and 6. By a completely analogous treatment it is readily shown that the same results are obtained for two holes with the same n, l, j quantum numbers.

Now let us turn our attention to the second problem of this section, the coupling of two phonons. Each $\lambda = 2$ phonon carries angular momentum 2 with z component $m_\lambda = 2, 1, 0, -1$, or -2. The Pauli exclusion principle does not apply for phonons since they are not spin-½ particles, so the various possibilities are those listed in Table 5-5. Note that we do retain the principle that the two phonons are indistinguishable, so, for example, it makes no sense to consider as separate situations the case where phonon 1 has $m_\lambda = +2$ and phonon 2 has $m_\lambda = +1$ and the case where these are interchanged; both are encompassed by the single m_λ configuration given by the second column of Table 5-5. More

TABLE 5-5: VARIOUS m_λ CONFIGURATIONS FOR TWO-PHONON STATES

m_λ	*m_λ of the two phonons*														
+2	xx	x	x	x	x										
+1		x				xx	x	x	x						
0			x				x			xx	x	x			
−1				x				x			x		xx	x	
−2					x				x			x		x	xx
Σm_λ	+4	+3	+2	+1	0	+2	+1	0	−1	0	−1	−2	−2	−3	−4

TABLE 5-6: NUMBER OF STATES WITH VARIOUS m_I IN TABLE 5-5

m_I	+4	+3	+2	+1	0	−1	−2	−3	−4
No. of config.	1	1	2	2	3	2	2	1	1

advanced students may recognize that the treatment used here is appropriate for particles obeying Einstein-Bose statistics.

The sum of the m_λ values, Σm_λ, is shown in the last row of Table 5-5. For a state of total angular momentum I, the values of m_I must be $I, I-1, \ldots, -I$. But m_I is just Σm_λ, so here again we can determine what I values are possible by seeing how many m_λ configurations are available for each m_I. These numbers, as compiled from the last line of Table 5-5, are listed in Table 5-6. It is clear from Table 5-6 that the largest value of I is 4; this accounts for $m_I = +4, +3, +2, +1, 0, -1, -2, -3, -4$. The largest remaining m_I is 2, so this must be derived from an $I = 2$ state which then accounts for $m_I = +2, +1, 0, -1, -2$. The only remaining m_I is zero which must therefore be from an $I = 0$ state. In summary, there are three states derived from the coupling of two $\lambda = 2$ phonons, and these have $I = 0$, 2, and 4. This result is used in Fig. 5-9.

Problems

5-1 Give a reason why the following collisions between two nucleons are not possible:

	Initial orbits								*Final orbits*							
	$\mathfrak{N}_1$	l_1	j_1	m_1	$\mathfrak{N}_2$	l_2	j_2	m_2	$\mathfrak{N}_1'$	l_1'	j_1'	m_1'	$\mathfrak{N}_2'$	l_2'	j_2'	m_2'
(*a*)	2	1	3/2	3/2	2	1	3/2	−1/2	3	2	3/2	3/2	1	0	1/2	+1/2
(*b*)	4	3	5/2	3/2	4	4	9/2	−3/2	4	3	7/2	5/2	4	3	7/2	−5/2
(*c*)	5	0	1/2	1/2	5	0	1/2	−1/2	5	4	7/2	3/2	5	2	3/2	−3/2
(*d*)	3	2	5/2	3/2	4	3	5/2	−1/2	4	3	5/2	5/2	4	4	9/2	−3/2

5-2 Write the wave functions for the ground states of Bi^{209}, Pb^{207}, Ca^{49}, Pb^{209}, Ca^{41}, Ca^{39}, K^{39}, and Sc^{41}.

5-3 Write the wave function for the ground states of O^{14}, Ne^{18}, and Pb^{210}.

5-4 If a nucleus has closed shells except for two neutrons in $\mathfrak{N} = 4$ orbits, write the wave function for its ground state and give a relationship between the coefficients.

5-5 In Fig. 5-2, assume $\epsilon_s = 0.9$ MeV, $\epsilon_d = 5.1$ MeV (these are estimates of the actual values in O^{18}) and take $v_{mn} = -G[(2j_m + 1)(2j_n + 1)]^{1/2}$ with $G = 1.5$ MeV. Find the energies and wave functions of the 0^+ states in O^{18} by finding the eigenvalues and eigenfunctions of the matrix in (5-23).

5-6 List the terms in the wave function for the ground state of Pd^{102}.

5-7 Under the assumption that the energy difference between the $d_{5/2}$ and $d_{3/2}$ orbits is 3 times that between the $d_{5/2}$ and $s_{1/2}$ orbits, construct the left half of Fig. 5-4.

5-8 From Fig. 5-6 estimate how many neutrons, on an average, are in various orbits in Si^{28}.

5-9 Construct the equivalent of Fig. 5-6 for the $\mathfrak{N} = 6$ shell. As part of the construction, plot the values of V_j^2 obtained from the Pb^{206} wave function given in the text and be sure the curves pass through them.

5-10 Construct the equivalent of Fig. 5-7 for He^6.

5-11 By use of (5-28), calculate the V_j^2 for O^{18} and compare with the experimental results given in the text. Use $\Delta = 2$ MeV and ϵ_j from Prob. 5-5.

5-12 Do Prob. 5-11 for O^{20}, Si^{28}, and Ca^{38}.

5-13 Construct the equivalent of Fig. 5-7 for Pb^{206}. Make separate plots for terms of positive and negative parity. Why is it expedient to do this?

5-14 A circular wire loop carrying a current experiences electromagnetic forces tending to increase the diameter of the circle; this puts the wire under tension, and we can use it as a two-dimensional model of a vibrating nucleus. If the radius of the loop is 0.1 m, the tension is 1 N, and the mass per unit length of the wire is 0.01 kg/m, what are the values of the fundamental frequencies ν_λ? What are the energies of the lowest-energy 2^+, 3^-, and 4^+ vibrational states? Ignore the curvature of the wire.

5-15 Look up the energy levels[1] of Ni^{62}, Zn^{66}, and Se^{78}. Explain the character of the low-energy states.

5-16 List all 3^- terms in O^{18} that can be formed by exciting neutrons from $\mathfrak{N} = 2$ to $\mathfrak{N} = 3$ and from $\mathfrak{N} = 3$ to $\mathfrak{N} = 3A$ or 4 orbits. Compare this with the number

[1] Good compilations of energy levels can be found in *Nucl. Data, Sec. B*. Several other compilations are available.

of 2^+ terms that can be formed by exciting nucleons by no more than one shell. Use the result to explain the energy of the 3^- collective state.

5-17 Calculate the number of 0^+ states per MeV and the total number of states per MeV in Sn^{116}, Sn^{117}, Fe^{57}, and Co^{58} at an excitation energy of 8 MeV. Calculate these for Sn^{116} at 5, 12, and 16 MeV.

5-18 What states (i.e., what are their I^π) can be formed from two $g_{7/2}$ neutrons?

5-19 What states can be formed from three $d_{5/2}$ neutrons?

5-20 What are the angular momenta of the two-phonon $\lambda = 3$ vibrational states?

Further Reading

See General References, following the Appendix.

deShalit, A., and I. Talmi: "Nuclear Shell Theory," Academic, New York, 1963.

Eisenbud, L., and E. P. Wigner: "Nuclear Structure," Princeton University Press, Princeton, N.J., 1961.

Feenberg, E.: "Shell Theory of the Nucleus," Interscience, New York, 1959.

Kisslinger, L. S., and R. A. Sorenson, *Kgl. Danske Videnskab. Selskab, Mat. Fys. Medd.* **27**(16) (1953).

Lane, A. M.: "Nuclear Theory," Benjamin, New York, 1964.

Mayer, M. G., and J. H. D. Jensen: "Elementary Theory of Nuclear Shell Structure," Wiley, New York, 1955.

Chapter 6

The Structure of Complex Nuclei: Other Nuclei

In Chap. 5 we described the structure of spherical even-even nuclei. The other types—those with odd numbers of neutrons, of protons, or of both, and nonspherical nuclei—will be discussed in this chapter. While there are many important and interesting aspects of these subjects, they will get us into deeper complexities than we have previously encountered. More elementary students may therefore find it expedient to bypass much, or even all, of this material, which can be done with little loss of continuity.

6-1 Odd-A Spherical Nuclei

A nucleus with odd A must have either even-Z odd-N or odd-Z even-N. For definiteness here we shall consider the former case, but all conclusions will apply equally well to the latter if we interchange the words neutron and proton.

The simplest states of an even-Z odd-N nucleus are the single-quasi-particle (SQP) states, which, from the definition of quasi particle given in Sec. 5-6, are states formed by adding a particle or a hole to the nearest even-even nucleus. *If there were no collisions* in nuclei, orbits would fill in order of increasing energy, as illustrated for the $\mathfrak{N} = 3$ shell in Fig. 5-5, and energies of states would be just the sum of the energies of the occupied orbits from Fig. 4-5. For nuclei in which the $\mathfrak{N} = 3$ neutron shell is filling, the energies E_j of the SQP states would then be as shown in Fig. 6-1. In order to understand it, let us consider some examples. If there were five neutrons in the shell, the lowest-energy state would be the one in which all were in $d_{5/2}$ orbits. If the odd neutron were in an $s_{1/2}$ rather than a $d_{5/2}$ orbit, the energy $E_{1/2}$ of this excited SQP state would be ϵ_s with the energy definitions of Fig. 5-2. Similarly if the odd neutron were in a $d_{3/2}$ orbit, we would have a SQP state with $E_{3/2} = \epsilon_d$. These energies are shown in Fig. 6-1, where the energy of the lowest SQP state is taken as zero.

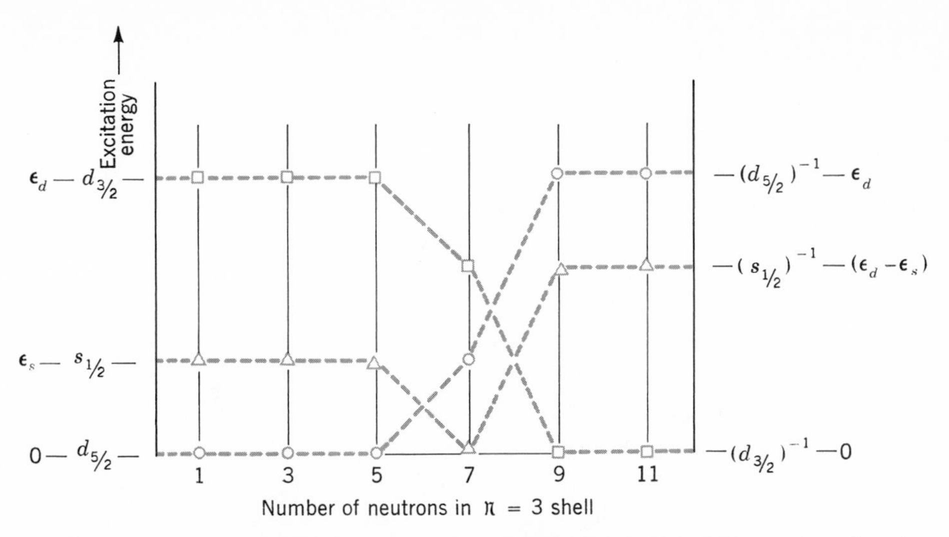

FIGURE 6-1 Energies of SQP states in the $\mathfrak{N}$ = 3 shell as a function of the number of neutrons in that shell under the assumption that there are no residual interactions.

For a nucleus with seven neutrons in the $\mathfrak{N} = 3$ shell, the lowest-energy SQP state under our assumption of no collisions would be the one with six neutrons in $d_{5/2}$ orbits and the odd neutron in an $s_{1/2}$ orbit; we would therefore have $E_{1/2} = 0$. If the odd neutron were in a $d_{3/2}$ orbit, this SQP state would have an energy higher than the ground state by $\epsilon_d - \epsilon_s$, whence $E_{3/2} = \epsilon_d - \epsilon_s$. A state with five neutrons in $d_{5/2}$ and two neutrons in $s_{1/2}$ orbits would have excitation energy ϵ_s since it would be higher than the ground state by the energy needed to excite a neutron from a $d_{5/2}$ to an $s_{1/2}$ orbit; this is also a SQP state according to the definition since it differs from the ground state of the nucleus with eight neutrons in the shell by having a single hole in the $d_{5/2}$ orbits. We therefore have a $d_{5/2}$ SQP state with $E_{5/2} = \epsilon_s$, as shown in Fig. 6-1.

Continuing with our assumption of no collisions, the ground state of a nucleus with 11 neutrons in the $\mathfrak{N} = 3$ shell would have 6 in $d_{5/2}$, 2 in $s_{1/2}$, and 3 in $d_{3/2}$ orbits; it differs from the ground state of a 10-neutron nucleus by the addition of a $d_{3/2}$ neutron particle and from the ground state of a 12-neutron nucleus by the "addition" of a $d_{3/2}$ neutron hole, so it is a SQP state. As it is the lowest-energy state of the 11-neutron nucleus, we have $E_{3/2} = 0$. Other SQP states in this nucleus are those in which the hole is in the $s_{1/2}$ and $d_{5/2}$ orbits. Their energies are $E_{1/2} = \epsilon_d - \epsilon_s$ and $E_{5/2} = \epsilon_d$ since these are the energies that must be added to the ground state to reach them; they may be reached by exciting a neutron from these orbits to the $d_{3/2}$ orbit, which is vacant in the ground state.

The behavior of the E_j under the assumption of no collisions is therefore as shown in Fig. 6-1. Where N is less than the value for which the odd neutron in the ground state is in that orbit, the SQP state is a *particle* state, differing from the ground state of the nucleus with one less neutron by the addition of a neutron particle. Where N is greater than the value for which the odd neutron in the ground state is in that orbit, the SQP state is a *hole* state, differing from the nucleus with one more neutron by the addition of a neutron hole. Ground states fulfill the definition of SQP states by being either particle or hole states. In the region where a SQP state is a particle state it moves down with increasing N, and the region where it is a hole state it moves up in energy with increasing N.

There are some odd-A nuclei in which there are no collisions, namely, the single-particle and single-hole nuclei. For these, then, Fig. 6-1 gives the correct result. The energies E_j of the nuclear states in these nuclei are therefore just the orbit energies of Fig. 4-5, so locating these states experimentally allows us to put Fig. 4-5 on a quantitative basis; methods for doing this will be described in Sec. 14-3. Energy levels in some single-particle and single-hole nuclei are shown in Fig. 6-2.

For all other odd-A nuclei, collisions do occur, so let us consider what changes they introduce. As we learned in Secs. 5-3 and 5-5, because of collisions the ground states of even-even nuclei are rather complicated. The ground state of the two- and four-neutron nuclei O^{18} and O^{20} already have the $s_{1/2}$ orbit filled part of the time (see Fig. 5-6), so to some extent the $s_{1/2}$ SQP state in O^{19} is a hole state rather than a particle state. And this is true in general; instead of being purely particle states or purely hole states, the SQP states are partly particle and partly hole states: that is the reason for the name quasi particle. Since all the orbits are filling smoothly, as shown in Fig. 5-6, the relative amounts of particle and hole nature in a SQP state change smoothly with neutron number. The fuller an orbit gets, the more holelike and the less particlelike its SQP state becomes. This is readily seen from the wave function for a SQP state of a nucleus with N neutrons

$$\psi_{N,j}(\mathrm{SQP}) = \sqrt{1 - V_j^2}\,\psi_{N-1}\psi(j) + V_j\psi_{N+1}\psi(j^{-1}) \tag{6-1}$$

Equation (6-1) says that the wave function of the SQP state is the sum of two terms, the ground state of the even-even nucleus $(N - 1)$ plus a *particle* in the state j and the ground state of the even-even nucleus $(N + 1)$ plus a *hole* in the state j. The fraction of the time it is a hole state is V_j^2, which was defined in Sec. 5-7 as the fullness of the single-particle state j, and the fraction of the time it is a particle state is $1 - V_j^2$, which is the degree to which the state j is empty.

As a consequence of collisions, the behavior of the SQP energies is modified from that given in Fig. 6-1 to what is shown in Fig. 6-3. The smooth transition from particle to hole states as N increases causes the energies E_j to shift smoothly rather than abruptly. Roughly speaking, we may say that the sharp angles of Fig. 6-1 are rounded off by the residual interactions.

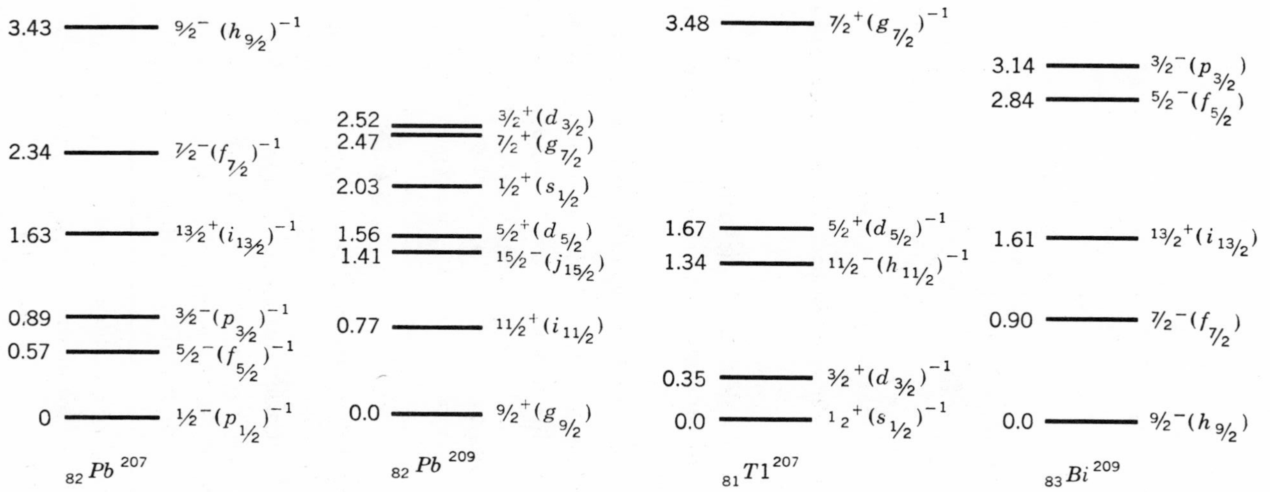

$_{82}Pb^{207}$		$_{82}Pb^{209}$		$_{81}Tl^{207}$		$_{83}Bi^{209}$	
3.43	$\frac{9}{2}^-(h_{9/2})^{-1}$	2.52	$\frac{3}{2}^+(d_{3/2})$	3.48	$\frac{7}{2}^+(g_{7/2})^{-1}$	3.14	$\frac{3}{2}^-(p_{3/2})$
2.34	$\frac{7}{2}^-(f_{7/2})^{-1}$	2.47	$\frac{7}{2}^+(g_{7/2})$	1.67	$\frac{5}{2}^+(d_{5/2})^{-1}$	2.84	$\frac{5}{2}^-(f_{5/2})$
1.63	$\frac{13}{2}^+(i_{13/2})^{-1}$	2.03	$\frac{1}{2}^+(s_{1/2})$	1.34	$\frac{11}{2}^-(h_{11/2})^{-1}$	1.61	$\frac{13}{2}^+(i_{13/2})$
0.89	$\frac{3}{2}^-(p_{3/2})^{-1}$	1.56	$\frac{5}{2}^+(d_{5/2})$	0.35	$\frac{3}{2}^+(d_{3/2})^{-1}$	0.90	$\frac{7}{2}^-(f_{7/2})$
0.57	$\frac{5}{2}^-(f_{5/2})^{-1}$	1.41	$\frac{15}{2}^-(j_{15/2})$	0.0	$\frac{1}{2}^+(s_{1/2})^{-1}$	0.0	$\frac{9}{2}^-(h_{9/2})$
0	$\frac{1}{2}^-(p_{1/2})^{-1}$	0.77	$\frac{11}{2}^+(i_{11/2})$				
		0.0	$\frac{9}{2}^+(g_{9/2})$				

FIGURE 6-2 Energies of SQP states in single-particle and single-hole nuclei adjacent to the closed-shell nucleus $_{82}Pb^{208}$. As demonstrated in the text, these are the orbit energies of Fig. 4-5 for the orbits shown in parentheses; for hole states, the energies are inverted from those in Fig. 4-5 as can be understood from Fig. 6-1.

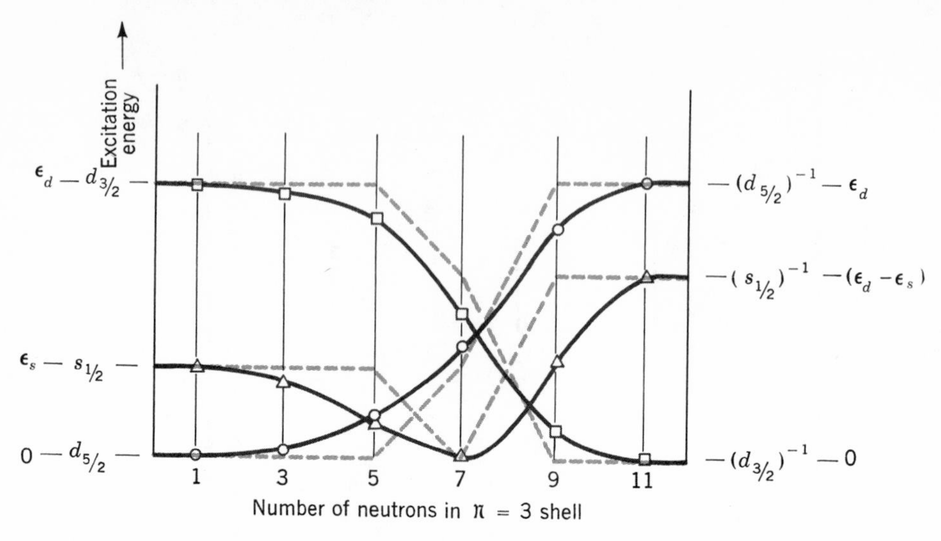

FIGURE 6-3 Actual energies of SQP states in the $\mathfrak{N}$ = 3 shell. Dashed lines are from Fig. 6-1.

Elaborate theoretical calculations make it possible to put Fig. 6-3 on a quantitative basis. One of the simplest of these is the pairing theory, introduced in Sec. 5-7, which gives

$$E_j = [(\epsilon_j - \lambda)^2 + \Delta^2]^{1/2} \tag{6-2}$$

To be consistent with the definition used above, E_j for the lowest-energy SQP state as obtained from this formula must be subtracted from (6-2). In this theoretical treatment, E_j can be expressed in terms of V_j^2 by solving (6-3) and (5-29) simultaneously; this gives

$$E_j = \frac{\Delta}{2V_j(1 - V_j^2)^{1/2}} \tag{6-3}$$

From Fig. 6-3 we see that each SQP state takes its turn at being the ground state as the shell fills. It is not necessarily the ground state in the same number of nuclei as it would be if the true situation were as in Fig. 6-1. For example, in Fig. 6-3 the ground state is $\frac{1}{2}^+$ in nuclei with both five and seven neutrons. In the $\mathfrak{N}$ = 5 shell, for which the energies of SQP states are shown in Fig. 6-4, the ground state is the $s_{1/2}$ SQP state when there are 13, 15, 17, or 19 neutrons in the $\mathfrak{N}$ = 5 neutron shell.

The low-energy states of an odd-A nucleus are a combination of terms formed by coupling each of the states of the nearest even-even nuclei with the various SQP states. Note that this use of the word "term" is somewhat different from our previous definition, but its relationship to "states" is the same, and they have other similar properties. The origin of these terms is illustrated in Fig. 6-5 for Sn^{117}. The states of Sn^{116} or Sn^{118} are shown at the left in that figure; from

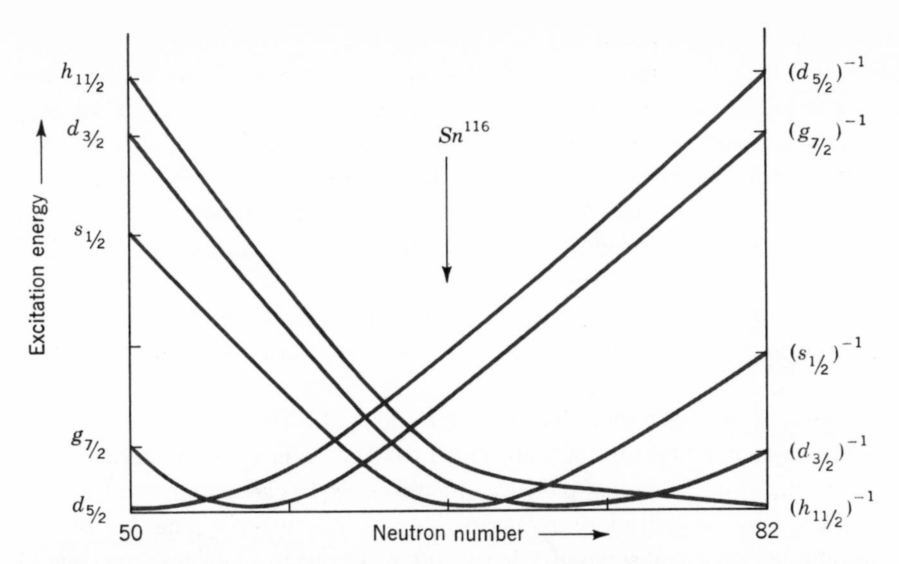

FIGURE 6-4 Energies of SQP states in the $\mathfrak{N} = 5$ shell vs. number of neutrons. This is based on the orbit energies in the shell-theory potential for a Sn^{116} nucleus. Orbit energies shift somewhat as a function of mass, so this plot would not be valid for nuclei far removed from Sn^{116} as explained in Sec. 5-12.

FIGURE 6-5 The formation of terms in wave functions for states of Sn^{117}.

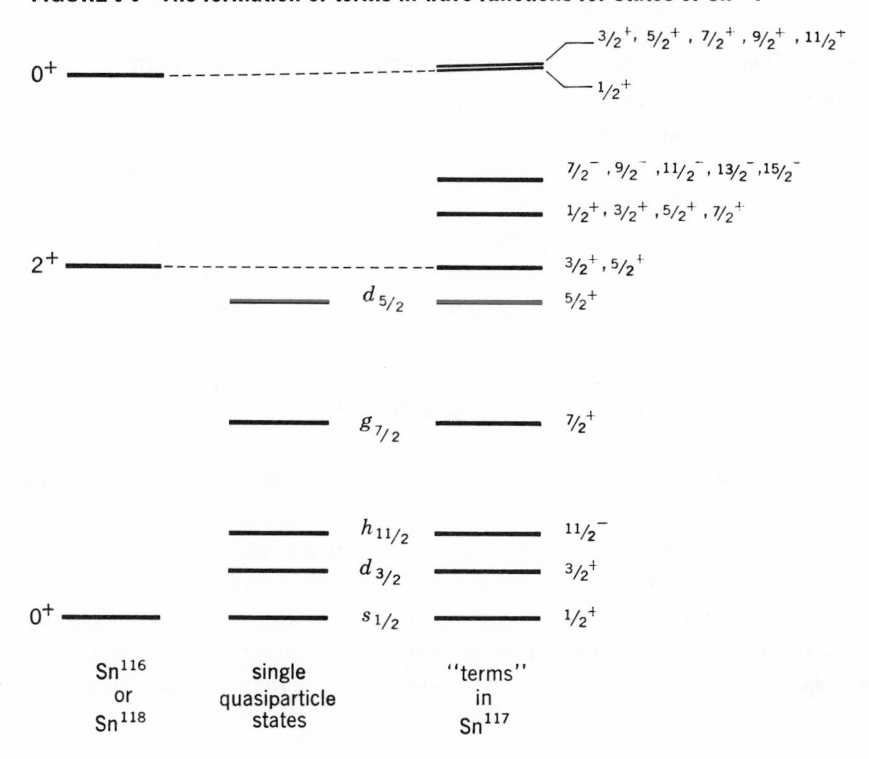

the discussion in Chap. 5, the ground state and vibrational states of these two nuclei should be very similar. The SQP states as obtained from Fig. 6-4 are shown in the center portion of Fig. 6-5, and the terms, shown in the right column, represent all combinations of the levels in the left and center columns. The energies of the terms are equal to the sum of the two energies in the combination, their angular momenta are given by all possible sums of the two angular momenta, and their parities are the product of the two parities. For example, the highest-energy term shown is from coupling the 2^+ vibrational state of the even-even nuclei with the $g_{7/2}$ SQP state; this gives terms with $I = 3/2, 5/2, 7/2, 9/2$, and $11/2$, all with positive parity since the $g_{7/2}$ ($l = 4$) and the 2^+ state both have positive parity.

The *states* of Sn^{117} are then linear combinations of the terms in Fig. 6-5 with the same I^π. The residual interactions which cause mixing of terms here are not nearly as strong as in the case of pairing, which was used as an example in Chap. 5, so terms mix only if rather close in energy. To put this into the language of Sec. 5-1, the time between collisions of a type that will take the nucleus from one of these terms to another is so long that only relatively small violations of energy conservation are consistent with (5-4). Of course all these terms contain a number of paired nucleons, and pairing collisions occur here as frequently as elsewhere.

Since they are not near in energy to terms of the same I^π, the lowest four terms in Sn^{117} are almost completely unmixed; these terms are essentially the complete wave functions of nuclear states at that energy, so those states are nearly pure SQP states. However, it is readily seen from Fig. 6-5 that the $d_{5/2}$ SQP state is close in energy to two other $5/2^+$ states, those formed from coupling the 2^+ vibration to the $s_{1/2}$ and $d_{3/2}$ SQP states. There are therefore three states in which the $d_{5/2}$ SQP state has an appreciable amplitude; in each of these states, an appreciable fraction of the time is spent as Sn^{116} (ground state) plus a $d_{5/2}$ particle and as Sn^{118} (ground state) with a $d_{5/2}$ hole in the ratio prescribed by (6-1). In those three nuclear states, an appreciable fraction of the time is also spent as the 2^+ vibrational state of Sn^{116} plus an $s_{1/2}$ or $d_{3/2}$ particle or as the 2^+ vibrational state of Sn^{118} with an $s_{1/2}$ or $d_{3/2}$ hole. The system goes from one to another of these possibilities as a result of collisions.

The low-lying energy levels of Sn^{117} and some of the other odd-A Sn isotopes are shown in Fig. 6-6, which also shows for each nuclear state the fractions of the time the nucleus spends in the corresponding SQP state. The method of determining these will be discussed in Sec. 14-3.

It should not be inferred from this discussion that all spherical odd-A nuclei have as simple an energy-level spectrum as the Sn isotopes. As shown in Fig. 5-11, the first excited state of the even-even Sn isotopes lies at a much higher energy than in neighboring nuclei because Sn has a closed shell of 50 protons. In neighboring nuclei like $_{48}Cd$, $_{46}Pd$, etc., the first excited state, and hence also the next

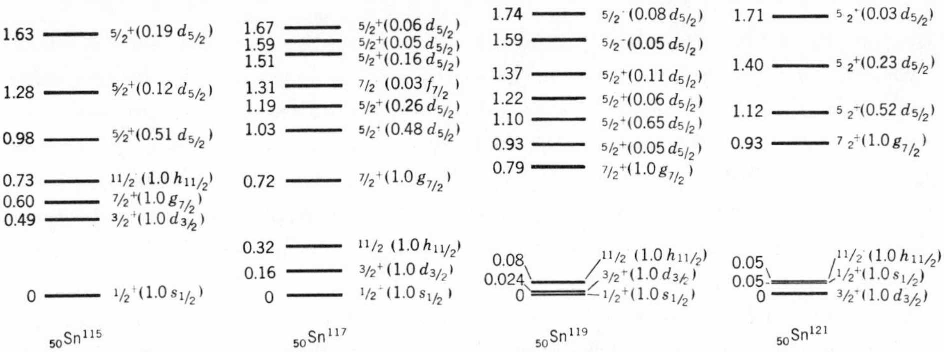

FIGURE 6-6 Energy levels in some odd-A isotopes of $_{50}$Sn. Numbers in parentheses are the portion of the SQP state contained in that nuclear state; for example, $(0.48 d_{5/2})$ means that 48 percent of the $d_{5/2}$ SQP state is in that level, whence the other 52 percent must be mixed into other levels. [*These numbers were obtained by deuteron-stripping experiments as explained in Sec.* **14-3** *and are from Phys. Rev.,* **156: 1315 (1967).**]

three excited states, which are at about twice the energy of the first according to Fig. 5-9, are at a much lower energy. It is easy to see from Fig. 6-5 how many more low-energy terms this would introduce, so almost all the SQP states are mixed into several nuclear states. As an example the equivalent of Fig. 6-6 for Pd^{107} and Pd^{109} is shown in Fig. 6-7.

The lowest-energy states of even-even nuclei are the ground state and the one- and two-phonon vibrational states, and we see from Figs. 5-9 and 5-12 that

FIGURE 6-7 Energy levels in some odd-A isotopes of $_{46}$Pd. [*Numbers in parentheses have the same meaning as in Fig.* **6-6** *and were obtained in the same way; from Phys. Rev.,* **161: 1257 (1967).**]

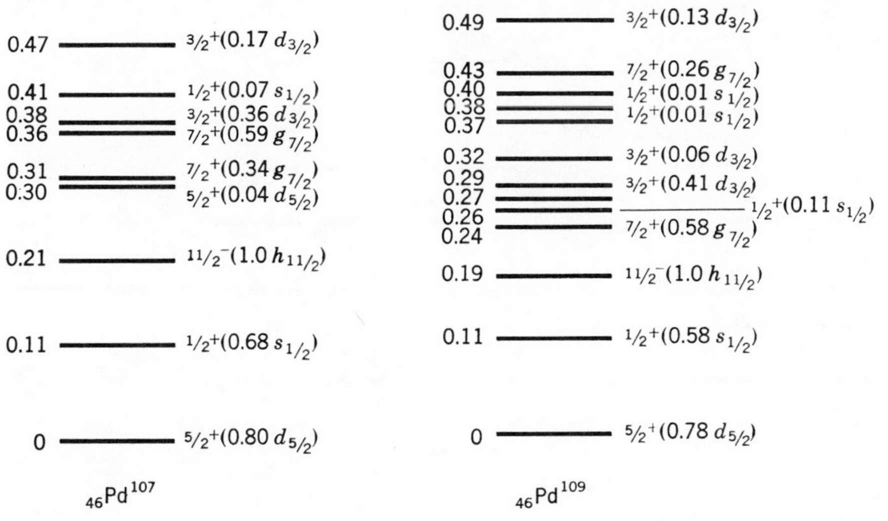

their energies are generally quite similar among isotopes of the same element. From Fig. 6-4 we see that the energies of the SQP states are also quite similar in neighboring nuclei. Since the states of odd-A nuclei are composed of combinations of these two, we expect them to be quite similar for neighboring isotopes, and from Figs. 6-6 and 6-7 we see that this is indeed the case. As noted above, the large difference between the Sn and Pd isotopes is explained by the difference in the energies of the vibrational states in their even-even isotopes due to the fact that Sn has a closed proton shell, but we see from Fig. 5-11 that in all other cases there is little difference. We therefore expect the states of odd-A isotopes of neighboring elements to be similar. For example, the low-energy states of ${}_{44}Ru^{105}$ are very similar to those of ${}_{46}Pd^{107}$.

The above description of states in odd-A nuclei is only an approximation. For example, the $s_{1/2}$ SQP state in Sn^{117} cannot even part of the time be exactly the same as the ground state of Sn^{116} plus a single $s_{1/2}$ neutron since there are sometimes two $s_{1/2}$ neutrons in the Sn^{116} ground state and the Pauli exclusion principle does not allow three $s_{1/2}$ neutrons. Many other contradictions of this type may be seen.

A more accurate approach is to proceed in a manner similar to the treatment for O^{18} in Sec. 5-8. For the three-neutron case of O^{19}, the beginning of the treatment is shown in Fig. 6-8. By placing the three nucleons in the three available

FIGURE 6-8 Configurations and terms in O^{19}.

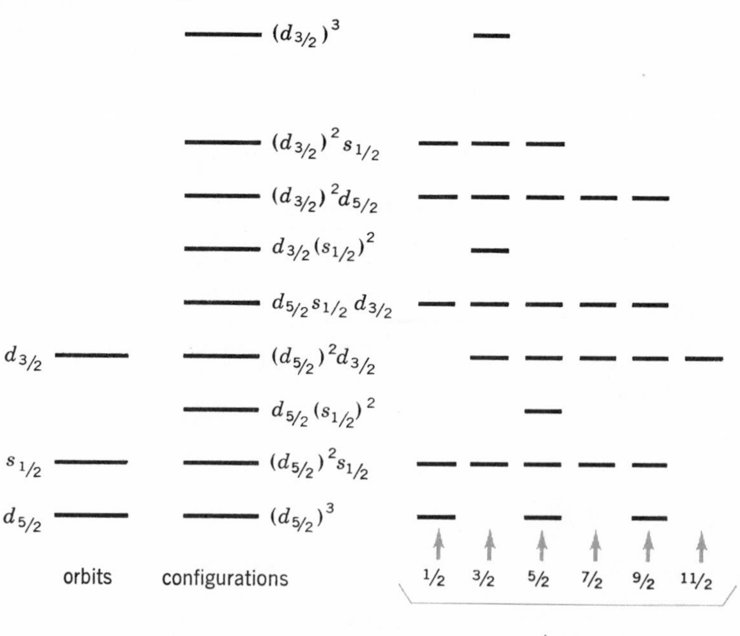

orbits, nine configurations are possible, each with an energy equal to the sum of the orbit energies. With the various possibilities for adding orbital angular momenta, the terms obtained from these configurations (we are now back to our original use of the word "term") are shown at the right in Fig. 6-8. All these terms have positive parity since the sum of three even l values is always even. The angular momenta obtainable from $(d_{5/2})^3$ can be shown by the methods of Sec. 5-13 to be limited to the values listed, and the single I from $(d_{3/2})^3$ is readily understood if we think in terms of holes.

The states of O^{19} are linear combinations of these terms with the same I^{π}. For example, there are five $\frac{1}{2}^+$ states, each being a linear combination of $(d_{3/2})^3$, $(d_{5/2})^2 s_{1/2}$, $d_{5/2} s_{1/2} d_{3/2}$, $(d_{3/2})^2 d_{5/2}$, and $(d_{3/2})^2 s_{1/2}$ terms. Those who have not omitted Sec. 5-4 will recognize that the energies of these five states and the coefficients of the five terms in each of them can be calculated by finding the eigenvalues and eigenfunctions of a 5×5 matrix with the same structure as the one in (5-23). The ϵ_j in that matrix are the energies of the terms in Fig. 6-8. The v_{ij} are calculated from (5-18) with V' chosen by one of several prescriptions; one such prescription is to use the nucleon-nucleon force developed in Chap. 3. Similar calculations must be done for the $\frac{3}{2}^+$ states (involving seven terms and hence a 7×7 matrix), the $\frac{5}{2}^+$ states, etc. The number of v_{ij} (referred to as *matrix elements*) to be computed is rather large, and finding eigenvalues and eigenfunctions of the several matrices is laborious, so calculations of this type can be very lengthy. Moreover it is clear that their length increases rapidly as the number of orbits included and the number of particles filling them increases. Computer codes have been developed for these calculations, and the scientific literature contains an abundance of their results.

At higher excitation energies, there are states in odd-A nuclei with all the complicated excitations outlined in the third paragraph of Sec. 5-11, and at sufficiently high energies, the statistical treatment discussed in that section becomes valid.

6-2 Spherical Odd-Odd Nuclei

In nuclei with odd numbers of both neutrons and protons, there must be at least one neutron quasi particle and one proton quasi particle. The low-energy states are then formed from the various couplings of these two. For example, in $_{51}Sb^{118}$, the neutron SQP states are the same as in Sn^{117}, as shown in the second column of Fig. 6-5. The SQP proton states are just the orbits in the $\mathfrak{N} = 5$ shell, as listed in Fig. 4-5 and shown in Fig. 6-9. The low-energy configurations of Sb^{118} are the 25 possible combinations of these. Each of these configurations gives terms of several total angular momenta, and the various terms of the same I^{π} mix due to

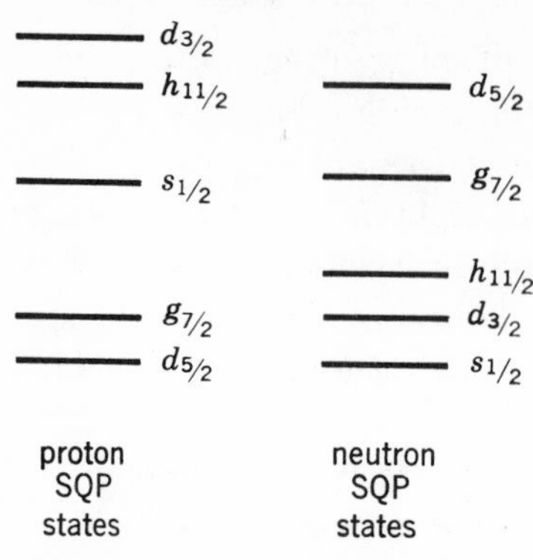

FIGURE 6-9 Proton and neutron SQP states in isotopes of $_{51}$Sb. The lowest-energy states in odd-odd Sb nuclei are combinations of these neutron and proton SQP states.

residual interactions. Clearly the situation is very complex, and there is no energy gap to simplify things as in even-even nuclei.

In some cases where the lowest-energy SQP states of both neutrons and protons are well separated from the others, the lowest-energy states of the odd-odd nuclei can be taken as combinations of the lowest SQP states only. In Y^{90}, for example, the lowest SQP states are $p_{1/2}$ for protons and $d_{5/2}$ for neutrons. The lowest-energy nuclear states of Y^{90} therefore have the configuration $d_{5/2}p_{1/2}$, which leads to a 2^- and a 3^- state. The forces between the $d_{5/2}$ neutron and the $p_{1/2}$ proton are different in these two states, so the energies of the two states are slightly different. There are simple, although not always reliable, rules for determining which has the lower energy and is therefore the ground state.[1] More details of the Y^{90} states are discussed in connection with Fig. 14-8.

6-3 Isobaric Spin and Isobaric Analog States

While the states of odd-odd nuclei are generally very complex, there is one very simple aspect that is important to consider here. It arises from the fact that, as we found in Sec. 3-11, nuclear forces are charge-independent. We have already seen an example of this simple aspect in the two-nucleon problem, for which the lowest-energy states of the three possible systems, n^2 (the dineutron), H^2 (the deuteron), and He^2 (the diproton) are shown in Fig. 6-10. Since the interactions are exactly the same (ignoring coulomb forces) in the three $T = 1$ states, their

[1] M. H. Brennan and A. M. Bernstein, *Phys. Rev.*, **120:** 927 (1960).

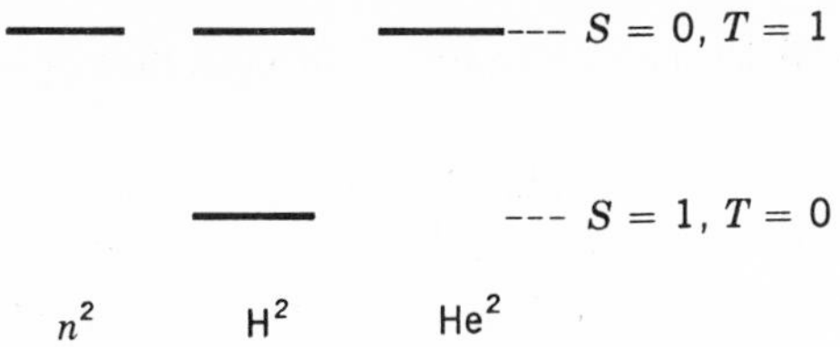

$S = 0, T = 1$

$S = 1, T = 0$

n^2 H^2 He^2

FIGURE 6-10 A simple example of isobaric analog states, the lowest-energy states of the two-nucleon system from Chap. 3.

energies and wave functions are the solution of the same Schrödinger equation and must therefore be identical. These three states are called *isobaric analog states;* isobar is the general term for "same number of nucleons," so the name means that these are analogous states with the same A.

To understand the general importance of this concept, we must go through a rather lengthy development. We begin by introducing a new method of representing terms in wave functions, illustrated in Fig. 6-11 for O^{18}. Each space in a diagram there represents an orbit for a nucleon, and a circle in that space indicates that the orbit is occupied. In the term shown, the top orbits are occupied by neutrons and not by protons. These orbits are not necessarily arranged in order of increasing energy, so this diagram can represent any term in a wave function for O^{18} in which all orbits occupied by protons are also occupied by neutrons.

FIGURE 6-11 Effects of T_- and T_+ operations. Diagrams are representations of wave functions, showing which orbits are occupied; each space represents an orbit, and a circle in the space indicates that it is occupied. Vertical dashed lines indicate that all lower energy orbits are occupied by both neutrons and protons. The T_+ operator converts the left diagram to the center diagram and the center diagram to the right diagram; the T_- operator does the inverse.

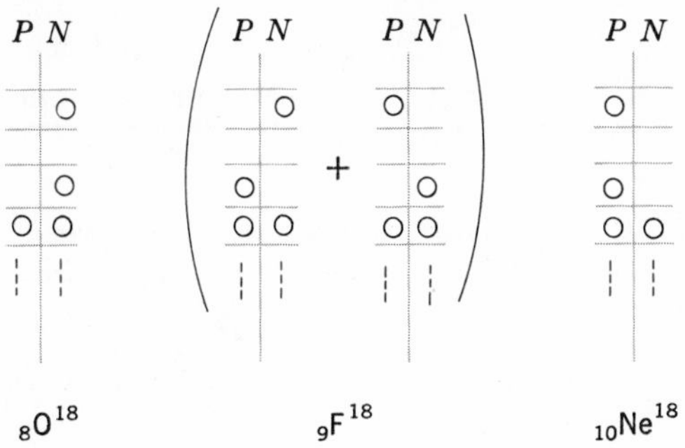

P N P N P N P N

${}_8O^{18}$ ${}_9F^{18}$ ${}_{10}Ne^{18}$

Each of the terms in our wave functions (5-16) can be represented by a diagram of this type.

Next we introduce two operators, T_+ and T_-. The T_+ operation converts a term written in the above way into a sum of terms in each of which one of the neutrons is converted into a proton in the same orbit so long as that orbit is not already occupied by a proton.[1] For example, it converts the O^{18} diagram to the F^{18} diagram in Fig. 6-11. When it operates on the F^{18} diagram, the operation on each term leads to the diagram for Ne^{18} in that figure, so the result is 2 times the Ne^{18} diagram. A T_+ operation on the Ne^{18} diagram gives zero since there are no neutrons that can be changed to protons in the same orbit. We shall ignore multiplicative constants.

These results can be summarized as

$$T_+(O^{18}) = (F^{18}) \qquad T_+(F^{18}) = (Ne^{18}) \qquad T_+(Ne^{18}) = 0$$

where (O^{18}) refers to the O^{18} diagram in Fig. 6-11, and similarly for the others.

The T_- operation does the opposite of the T_+, changing protons into neutrons, with the same rules, so its effect here can be summarized as

$$T_-(Ne^{18}) = (F^{18}) \qquad T_-(F^{18}) = O^{18} \qquad T_-(O^{18}) = 0$$

The reason this is important is that we shall presently prove the following theorem: If the nucleon-nucleon force is charge-independent, and if some wave function ψ_1 is a wave function for an actual nuclear state, $T_+\psi_1$ and $T_-\psi_1$ are also wave functions for actual nuclear states (unless they are zero), and the energies of these states are the same except for the coulomb energy. All states that can be generated from one another by successive T_+ and T_- operations are called isobaric analogs, and the complete group is called an *isobaric spin multiplet.*

In order to deal with this multiplet, we introduce two quantum numbers, T and T_z. The latter is defined as[2]

$$T_z = \frac{Z - N}{2} \tag{6-4}$$

This is just one-half of the difference between the numbers of protons and neutrons in the nucleus, which is readily obtained from the atomic number and

[1] We ignore the signs of terms and multiplicative constants here, although these are necessary in most quantitative applications. This method and several of its applications are discussed in C. D. Goodman, *Nucl. Phys.*, **55:** 449 (1964).

[2] For historical reasons, T_z is sometimes defined as the negative of (6-4). When this is done, the roles of T_+ and T_- are interchanged.

atomic weight; for O^{18}, F^{18}, and Ne^{18} it is -1, 0, and $+1$, respectively. The quantum number T is then equal to the largest T_z in the multiplet; we show later that T is the isobaric spin quantum number we have used previously. From the symmetry of our scheme between protons and neutrons, it is evident that any state has an isobaric analog with equal but opposite T_z in which all orbits occupied by protons in the original state are occupied by neutrons in the analog state and vice versa. We may therefore conclude that isobaric analogs of a state with a given T occur in nuclei with $T_z = -T, -T + 1, \ldots, +T$.

Some of the states of $A = 14$ nuclei are shown in Fig. 6-12*a*. The situation is the same as in the $A = 18$ nuclei described above. The $T_z = 0$ nucleus N^{14} has states with $T = 0, 1, 2, \ldots$, the $T_z = \pm 1$ nuclei C^{14} and O_{14} have states with $T = 1, 2, \ldots$, and the $T_z = \pm 2$ nuclei B^{14} and F^{14} have states with $T = 2$ or more.

The concept of isobaric analog states depends on the assumption of charge independence of the force between nucleons. In using it, we have ignored coulomb forces, which act only between protons and therefore tend to destroy the simple ideas we have been using. However, we have seen that coulomb forces are too weak to have important effects in the short-range nucleon-nucleon interactions which determine the nuclear wave functions, so our discussions of wave functions is still basically valid. On the other hand, the coulomb force has a long range, allowing a proton to interact with all other protons in the nucleus. The combined effect of all these small interactions is appreciable, and it causes energies of states containing more protons to be raised. This is shown in Fig. 6-12*b*.

FIGURE 6-12 Isobaric analog states in $A = 14$ nuclei: *(a)* without coulomb forces; *(b)* the effect of coulomb forces. States are classified according to the T quantum numbers.

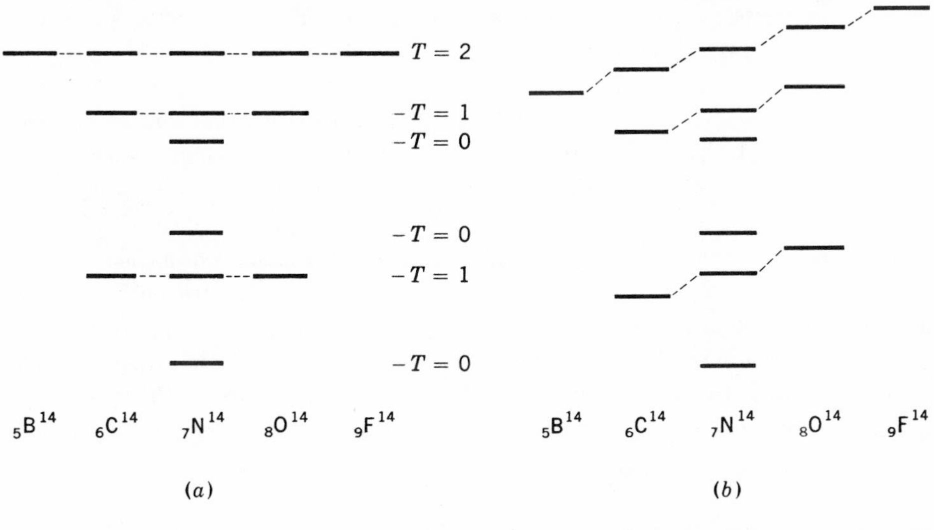

$T = 2$
$T = 1$
$T = 0$
$T = 0$
$T = 1$
$T = 0$

${}_5B^{14}$ ${}_6C^{14}$ ${}_7N^{14}$ ${}_8O^{14}$ ${}_9F^{14}$

(*a*)

${}_5B^{14}$ ${}_6C^{14}$ ${}_7N^{14}$ ${}_8O^{14}$ ${}_9F^{14}$

(*b*)

Another interesting aspect of Fig. 6-12 is that the 0^+ state with maximum pairing is the isobaric analog of the ground state of C^{14} and O^{14}, and we see that it is *not* the ground state in N^{14}. In the ground state, the two odd nucleons are in the same orbit ($1p_{1/2}$) but moving in the same direction, rather than in opposite directions as in the paired situation. Being in the same orbit, they spend a maximum amount of time close to each other, so this gives an energy lowering similar to that given by pairing. In this situation they can interact in the $T = 0$ two-nucleon states, and as we know from the case of the deuteron, this gives a stronger attractive force than does an interaction in the $T = 1$ two-nucleon states experienced by paired nucleons (the fact that paired nucleons must interact in $T = 1$ two-nucleon states is obvious from the fact that two neutrons or two protons can be paired). Consequently, the unpaired state has a lower energy and is the ground state of N^{14}. It should not, however, be concluded that this is always the case in nuclei with equal numbers of neutrons and protons ($N = Z$). The energy lowering of the paired situation is enhanced by the coherence effect explained in Sec. 3-4, so where there are several orbits available to contribute to this coherence (i.e., when p in (5-19) is large), the paired state becomes lower in energy. Consequently, in some of the heavier $N = Z$ nuclei like ${}_{25}Mn^{50}$ and ${}_{27}CO^{54}$, the $T = 1$, 0^+ state is the ground state.

For nuclei with odd A, T_z from (6-4) and hence also T, is half-integer. As explained above, the interactions are stronger when T is a minimum, and there is no energy advantage to be gained from pairing if T is chosen otherwise, as was the case in N^{14}, so the lowest-energy states do have the minimum value of T, namely $|T_z|$. For example, the low-energy states of ${}_3Li^7$ and ${}_4Be^7$, shown in Fig. 6-13, have $T = \frac{1}{2}$. On this basis, all states of these two nuclei should be isobaric analogs of one another and hence should have similar energies and wave functions. From the figure, we see that this correspondence is rather clear. The absolute energies of these states are, of course, shifted due to coulomb effects in accordance with the previous discussion.

The isobaric analog state in F^{18} shown in the center portion of Fig. 6-11 contains only two terms since there are only two neutrons that can be changed into protons in the same orbit, but in heavy nuclei things are much more complex. For example, in the state of ${}_{83}Bi^{208}$ which is the isobaric analog of the ground state of ${}_{82}Pb^{208}$, there are $126 - 82 = 44$ neutrons whose places in the Pb^{208} wave function must be taken by a proton, one in each term. In a nucleus as heavy as Pb, replacing a neutron by a proton requires a great deal of extra energy (about 18 MeV) because of the coulomb repulsion between this proton and all the other protons in the nucleus (see the calculation in Sec. 1-3). On the other hand, in accordance with the discussion of Sec. 4-6, the ground states of Pb^{208} and Bi^{208} are at nearly the same energy since it takes roughly equal energies to remove a

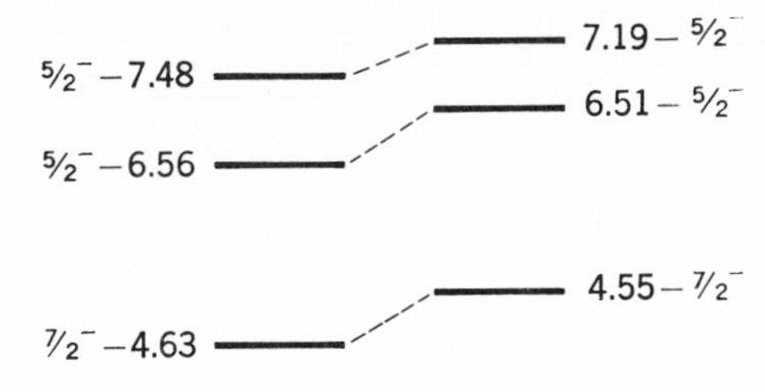

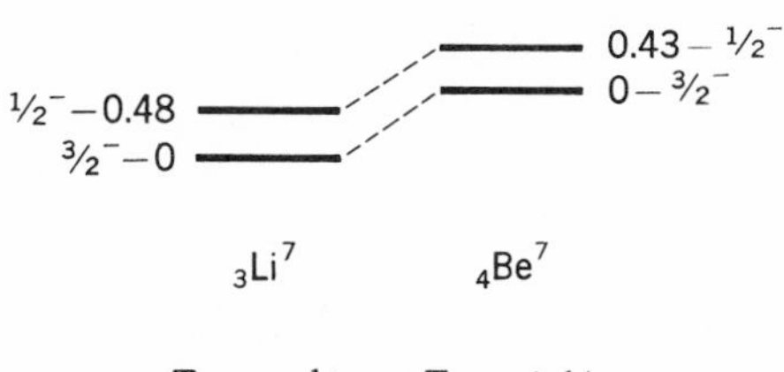

$_3Li^7$ $\qquad$ $_4Be^7$

$T_z = -\frac{1}{2}$ $\qquad$ $T_z = +\frac{1}{2}$

FIGURE 6-13 States of Li^7 and Be^7. Horizontal lines show the energies of states, and the numbers attached to them show their I^π and excitation energy above the ground state. Dashed lines connect isobaric analogs. Note that they have the same I^π and nearly equal excitation energies.

neutron from the former or a proton from the latter to reach the ground state of Pb^{207}. We may therefore conclude that the isobaric analog state under consideration here is a highly excited state of Bi^{208}, as shown in Fig. 6-14.

From (6-4), $T_z = -22$ for Pb^{208} and -21 for Bi^{208}. Since the lower-energy states in Bi^{208} do not have isobaric analogs in Pb^{208}, they must be $T = 21$ states. We therefore see that the low-energy states of any heavy nucleus must have $T = |T_z|$. The ground state of Pb^{208} must then be $T = 22$, whence its isobaric analog state in Bi^{208} is $T = 22$, $T_z = -21$.

In the remainder of this section, which may be skipped by less advanced students, we shall prove the basic theorem upon which all our discussion has been based, that if ψ_1 is a nuclear state, $T_+\psi_1$ and $T_-\psi_1$ are also nuclear states with the same energy. The proof goes easily once we can establish that T_+ (and T_-) commutes with the hamiltonian H; or

$$HT_+ = T_+H$$

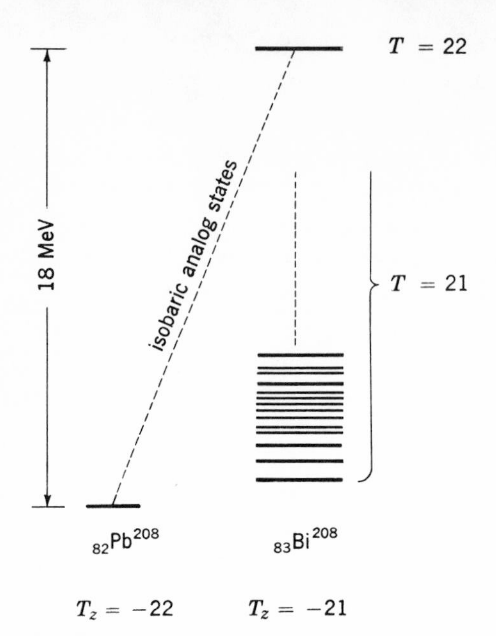

FIGURE 6-14 Ground state of $_{82}Pb^{208}$ and its isobaric analog in $_{83}Bi^{208}$. Also shown are other states of Bi^{208} and their T values.

Let us first assume this to be so and say that some state ψ_1 with energy E_1 satisfies the Schrödinger equation

$$H\psi_1 = E_1\psi_1$$

Operating on both sides with T_+ gives

$$T_+H\psi_1 = E_1T_+\psi_1$$

and using the commutation relation gives

$$H(T_+\psi_1) = E_1(T_+\psi_1)$$

This is a Schrödinger equation, so $T_+\psi_1$ is the wave function for a nuclear state with energy E_1. This proves our theorem if we can prove the commutation relation.

To do this, it is necessary to establish the mathematical equivalence of isobaric and regular spin. The logical procedure is to start at the beginning and define an operator $\boldsymbol{\tau}$ which behaves mathematically like $\boldsymbol{\sigma}$. We drop all other distinctions between neutrons and protons but we require that wave functions have a τ-spin part just as they have a regular spin part, and we interpret τ spin *up* and *down* as a proton and a neutron respectively. In analogy with spin, we define

$$\mathbf{T} = \sum_i \boldsymbol{\tau}_i$$

where the sum is over all nucleons. As demonstrated for S in Sec. 3-4, in a two-nucleon system, $T = 1$ and $T = 0$ states are symmetric and antisymmetric, respectively, in τ-spin space on interchange of the two nucleons. If we require that the total wave function be antisymmetric with respect to interchange, we automatically satisfy the Pauli exclusion principle since a system of two protons or two neutrons has $T = 1$ ($T_z = \pm 1$) and hence a symmetric τ-spin wave function, so the product of the space and ordinary spin wave functions must be antisymmetric. Moreover, this product is symmetric in $T = 0$ states, which justifies our use of $T = 0$ and $T = 1$ in Sec. 3-4. Our other uses of T in Chap. 3 depended only on its use there plus its mathematical equivalence to S. From the above definition of T,

$$T_+ = \sum_i \tau_{+i}$$

and it is easily shown that this operation acts as described in connection with Fig. 6-11. Our definition (6-4) of T_z follows directly from the above definitions, and the properties of the T quantum number relative to those of T_z are familiar from the spin analogy.

Now we can proceed with our proof of the commutation relation between H and T_+. The hamiltonian for a nucleus includes a sum of all the two-nucleon interactions between pairs of nucleons in the nucleus. We found in (3-20) that the two-nucleon potential between nucleons j and k depends on T only through terms of the form $\mathbf{T}\cdot\mathbf{T}$, where $\mathbf{T} = \boldsymbol{\tau}_j + \boldsymbol{\tau}_k$. These terms may be written as $\boldsymbol{\tau}_j \cdot \boldsymbol{\tau}_k$ plus constants since $\boldsymbol{\tau}\cdot\boldsymbol{\tau} = \tau(\tau + 1) = \frac{1}{2} \times \frac{3}{2} = \frac{3}{4}$ for each τ. Hence the entire hamiltonian depends on $\boldsymbol{\tau}$ as

$$H = A + B \sum_j \sum_{k \neq j} \boldsymbol{\tau}_j \cdot \boldsymbol{\tau}_k$$

where A and B are not functions of the $\boldsymbol{\tau}$. This double sum may be written

$$\sum_j \sum_{k \neq j} \boldsymbol{\tau}_j \cdot \boldsymbol{\tau}_k = \Big(\sum_j \boldsymbol{\tau}_j\Big) \cdot \Big(\sum_k \boldsymbol{\tau}_k\Big) - \sum_k \boldsymbol{\tau}_k \cdot \boldsymbol{\tau}_k = \mathbf{T}\cdot\mathbf{T} - \text{const}$$

where now T is for the entire nucleus, as we have been using it in this section. As is well known from the analogy with spin, T_+ commutes with $\mathbf{T}\cdot\mathbf{T}$ ($= T^2$), so we find that T_+ commutes with H. This is the result we were seeking. In this treatment we used the result (3-20) that the two-nucleon potential can be written as a single expression which depends on $\mathbf{T}$ only through $\mathbf{T}\cdot\mathbf{T}$; this would clearly be impossible if the nuclear force were not charge-independent.

6-4 Spheroidal Nuclei—The Shell-theory Potential

Before getting into our discussion of nonspherical nuclei, let us pause to clarify a few quantum-mechanical concepts. In quantum theory, only a few physical quantities can be known accurately; i.e., they are not subject to uncertainty principles. They must be conserved quantities, i.e., constants of the motion, in classical mechanics, and in quantum mechanics they are represented by quantum numbers. The energy and total angular momentum of a system are examples. As regards angular momentum, there is a further limitation in that only the total angular momentum and *one* of its components (conventionally designated as the z component) can be known accurately. If a quantity is not conserved in classical mechanics in a particular problem, the quantum number that represents it is not a good quantum number in the corresponding quantum-mechanical problem even though it may be good in other problems. For example, we know from atomic physics that m_l and m_s are good quantum numbers in the absence of a spin-orbit interaction, but when a spin-orbit interaction is present, they are not good quantum numbers and we must use j and m. This is because the directions of the orbital and spin angular momenta are no longer conserved in the classical problem. Classically, the motion is sometimes described as a precession of the l and s vectors about the j vector, which must have a fixed direction since its components are conserved quantities.

In this example only the directions of vectors are not conserved, but there are also cases where even the magnitude of some angular momenta are not conserved. A familiar example is a spinning top, where the total angular momentum is the sum of the spin and precessional angular momenta. Neither of the latter two is conserved, and angular momentum is continually exchanged between these two motions. However, their vector sum, the total angular momentum, and its components are conserved. In the quantum-mechanical analogy, only the total angular momentum and its z component are represented by good quantum numbers.

It was pointed out in Sec. 4-8 that many nuclei are ellipsoidal rather than spherical and that in the great majority of cases since two of the principal axes of the ellipsoid are equal, the shape is spheroidal. There is therefore rotational symmetry about the unequal principal axis, which we designate as the z direction. From the discussion of Sec. 4-2, the shell-theory potential V is approximately proportional to the nucleon density, so in these nuclei it is also spheroidal, whence V is a function of θ, where θ is measured from the z axis, as is customary. On the other hand, due to the rotational symmetry about the z axis, V is *not* a function of ϕ.

We have pointed out in Sec. 3-5 that in a potential where V is a function of θ, orbital angular momentum is not conserved. Hence l, and consequently j $(= |\mathbf{l} + \mathbf{s}|)$, are not good quantum numbers. However, since the potential is not a function of ϕ, there is no torque in the z direction, whence the z component of angular momentum is conserved, so m *is* a good quantum number. It is called Ω rather than m because the latter is reserved for the component of j along an

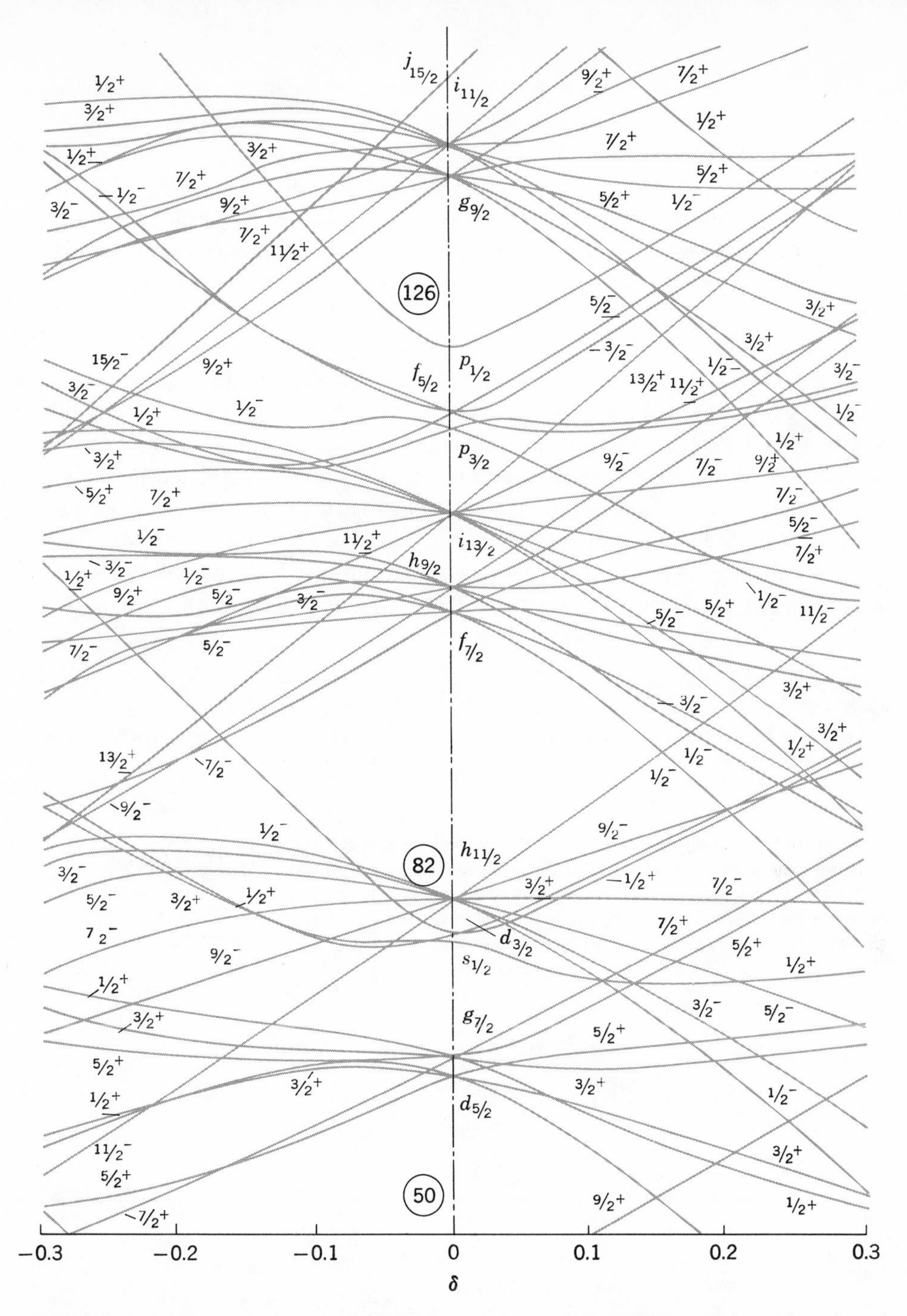

FIGURE 6-15 Orbit energies in the shell-theory potential of spheroidal nuclei as a function of their deformation δ. Positive δ is prolate and negative δ is oblate deformation. [*From S. G. Nilsson, Kgl. Danske Videnskab. Selskab, Mat. Fys. Medd.,* **29: 16 (1960).**]

axis fixed in space, whereas there are situations, as we shall see, where the symmetry axis of the nucleus is not fixed in space. In such cases, Ω is the component of j along this symmetry axis even though it is not stationary. The allowed values of Ω are the same as those previously used for m, namely, $\pm\frac{1}{2}, \pm\frac{3}{2}, \ldots, \pm j$.

The behavior of the orbit energies in the shell-theory potential as a function of its shape is shown in Fig. 6-15. The deformation δ is plotted horizontally (it will be more precisely defined later), with positive values indicating prolate, and negative values oblate, spheroidal deformations. When the nucleus is spherical ($\delta = 0$), the orbits are designated by the j quantum number, and since the energy does not then depend on the spatial orientation, all Ω states corresponding to that j are at the same energy. In fact the $\delta = 0$ part of Fig. 6-15 should be identical to Fig. 4-5.[1] As the nucleus becomes nonspherical, j loses its meaning, so states are designated only by Ω and the parity (parity is still conserved and is unchanged by the deformation). We see from Fig. 6-15 that states of different Ω separate in energy as the nucleus becomes deformed; let us explain this effect.

The shell-theory potential in a spheroidal nucleus is shown schematically in Fig. 6-16. Let us say the dashed and solid lines are where the depth of the potential

[1] There are slight differences between the ordering of orbit energies within shells. These are largely due to the fact that Fig. 6-15 was calculated before detailed energies of orbits were experimentally determined. In any case, the ordering of orbit energies within shells is subject to variations depending on what other orbits are occupied (see Sec. 5-12).

FIGURE 6-16 The shell-theory potential in spheroidal nuclei. Dashed and solid lines show schematically where the potential is 90 and 10 percent of its maximum depth, respectively. The diagram has rotational symmetry about the z axis. AA, BB: two orbits viewed edge on whose energies are discussed in the text.

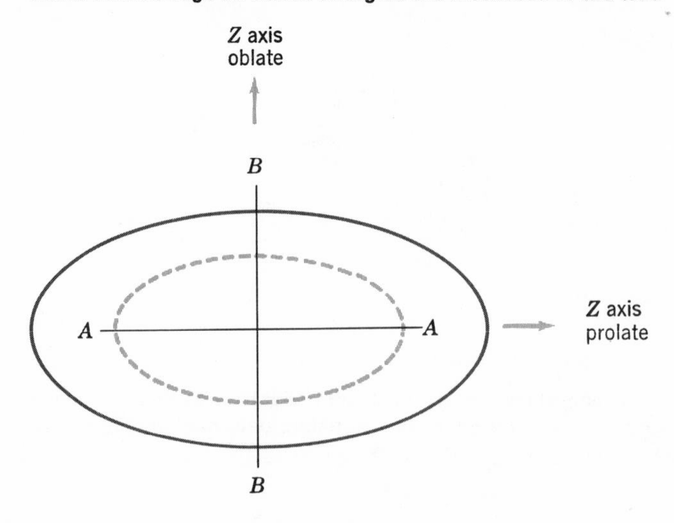

is 90 and 10 percent of its value at the center respectively. An orbit oriented as AA in Fig. 6-16 will experience a wider and deeper (on an average) potential than will an orbit oriented as BB, whence, as explained in Chap. 2, the energy of the former will be lower. If the nucleus is an oblate spheroid, the angular momentum vector of orbit AA is parallel to the z axis while that of BB is perpendicular to it, so AA is an orbit of large $|\Omega|$ while BB is an orbit of small $|\Omega|$. This explains why, for an oblate deformation in Fig. 6-15, the energy increases with decreasing $|\Omega|$. We note that the sign of Ω, which gives the sense of the angular momentum vector or, more mundanely, the direction in which the nucleon traverses its orbit, does not affect the energy.

If the nucleus is prolate, the z axis, being the axis of symmetry, is horizontal in Fig. 6-16. Now the higher-energy orbit BB has its angular-momentum vector parallel to the z axis, whence it corresponds to a large $|\Omega|$; hence for prolate deformations, small-$|\Omega|$ orbits lie lower in energy than large-$|\Omega|$ orbits, as we see in Fig. 6-15.

The curves in Fig. 6-15 were first calculated by S. G. Nilsson, whence it is known as a *Nilsson diagram*. No discussion of the general subject of spheroidal nuclei could be complete without mentioning the names of A. Bohr and B. R. Mottelson, who originated and contributed heavily to all aspects of the theory as developed in this and the following sections.

6-5 States of Even-Even Spheroidal Nuclei—The Ground-state Rotational Band

The ground state of an even-even spheroidal nucleus is in many ways similar to that of an even-even spherical nucleus. Each nucleon is paired with another nucleon, always moving in the same orbit but in opposite directions, that is, Ω for one equals $-\Omega$ for the other, and the collision pattern by which pairs go from one orbit to another is the most coherent and rhythmic one. The only difference is that the orbits here are those from Fig. 6-15 with the appropriate deformation rather than those from Fig. 4-5. The I^π of this state is clearly 0^+; the total angular momentum is zero since each pair consists of two nucleons with equal and opposite angular momenta, and the parity is plus because each pair contributes $2l$ to the sum of l values, whence that sum is even.

We have already learned in modern-physics courses that molecules are capable of rotation in quantum mechanics but atoms are not. The reason for this derives from the fact that a spherically symmetric system has no preferred direction in space, so a rotation, which by definition changes its orientation in space, is meaningless. It is closely related to the fact that one cannot see whether or not a perfect sphere is rotating; only by watching imperfections on the surface, i.e., deviations from spherical symmetry, can rotations be detected. For the same reason, an ellipsoid with two equal principal axes cannot rotate about the third axis (the symmetry axis), but there is no difficulty in its rotating about one of the equal axes just as there is no difficulty in seeing whether a football is rotating end over end.

A spheroidal nucleus is therefore capable of rotating about one of the equal principal axes.[1] Its rotational angular momentum **R** is

$$\mathbf{R} = \mathcal{I}\boldsymbol{\omega}$$

where $\mathcal{I}$ is its moment of inertia. Its energy of rotation is

$$E = \tfrac{1}{2}\mathcal{I}\omega^2 = \frac{R^2}{2\mathcal{I}} \tag{6-5}$$

Since the motion of the nucleons contributes nothing to the angular momentum (they are in a 0^+ state), **R** is the total angular momentum of the nucleus, whence, from the usual properties of angular momentum in quantum mechanics,

$$R^2 = I(I+1)\hbar^2 \tag{6-6}$$

where I is integer.

The part of the wave function for the nucleus which describes the rotational motion by giving the probability of various orientations in space is similar to that developed in Secs. 2-3 and 2-4 to describe the angular motion of particles in orbits. It is

$$\psi \propto P_{I,m_I}(\theta)$$

where P_{I,m_I} are the associated Legendre polynomials listed in Table 2-2. We found in Sec. 2-7 that this wave function has parity $(-1)^I$. If the nucleus is symmetric with respect to reflections about the origin, as is clearly the case here since the parity is positive (recall that the nucleons are in a 0^+ state), nothing is changed if the nucleus is rotated through 180°. However, a negative-parity rotational wave function implies that there is a meaning to rotation through 180°. Such wave functions are therefore unacceptable, whence I must be even.

By inserting (6-6) into (6-5) and applying this last rule, we obtain the final result for the energies of these rotational states as

$$E = \frac{\hbar^2}{2\mathcal{I}} I(I+1) \qquad I = 0, 2, 4, 6, \ldots \tag{6-7}$$

In all these states, the motion of the particles relative to the shell-theory potential is the same as in the ground state, so the group of states represented by (6-7)

[1] Since there is rotational symmetry about the unequal axis, any two mutually perpendicular lines in the plane perpendicular to the symmetry axis and passing through the center of the nucleus can be taken as principal axes. For ease of reference, we shall refer to them as the *equal principal axes*.

is called the *ground-state rotational band.* Since $I(I+1)$ for the values of I given in (6-7) are 0, 6, 20, 42, 72, . . . , the energies of excitation above the ground state are in the ratio 1, $\frac{20}{6} = 3.33$, 7.0, 12.0, . . . for the states of I^π equal to 2^+, 4^+, 6^+, 8^+, . . . , respectively. A low-energy-level spectrum of this type is characteristic of spheroidal even-even nuclei, and the accuracy with which (6-7) is fulfilled, at least for $I = 2$ and 4, is almost unique in the theory of nuclear structure. Some examples are shown in Fig. 6-17, and a compilation of results is given in Fig. 6-18.

FIGURE 6-17 Experimentally determined ground-state rotational bands in typical even-even spheroidal nuclei. The numbers at the left are excitation energies in MeV ***(top row)*** **or keV** ***(bottom row)*****, and I^π for the states is shown at the right. These states were determined in coulomb excitation experiments described in Sec. 14-7.**

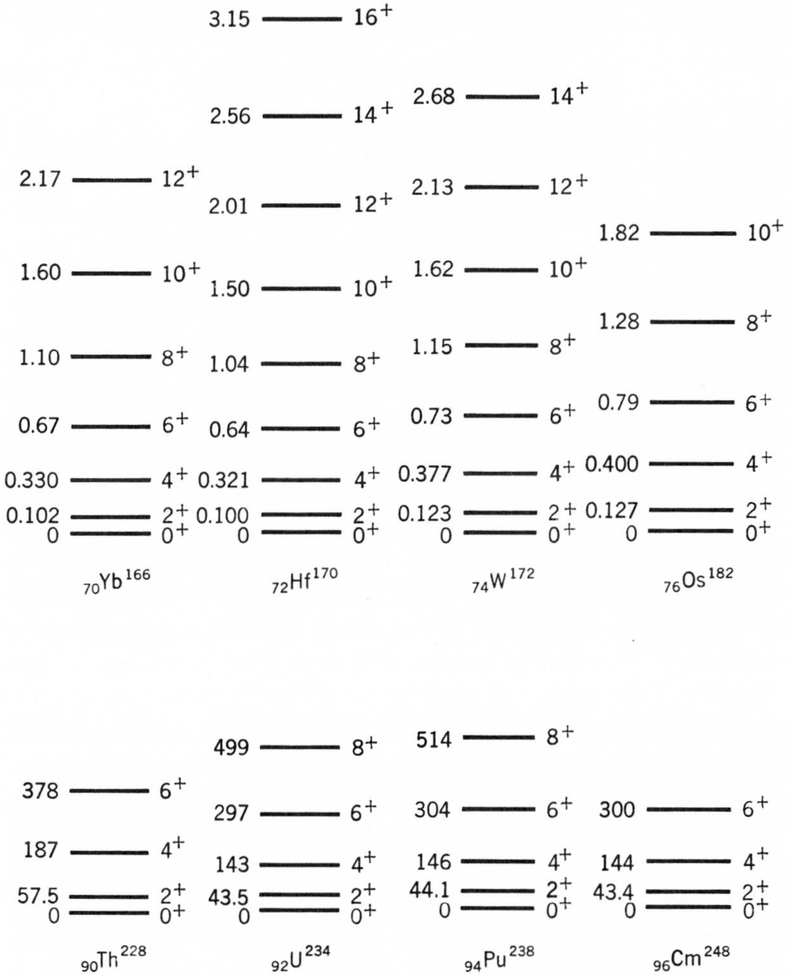

I^π	$_{70}Yb^{166}$	$_{72}Hf^{170}$	$_{74}W^{172}$	$_{76}Os^{182}$
16^+		3.15		
14^+		2.56	2.68	
12^+	2.17	2.01	2.13	
10^+	1.60	1.50	1.62	1.82
8^+	1.10	1.04	1.15	1.28
6^+	0.67	0.64	0.73	0.79
4^+	0.330	0.321	0.377	0.400
2^+	0.102	0.100	0.123	0.127
0^+	0	0	0	0

I^π	$_{90}Th^{228}$	$_{92}U^{234}$	$_{94}Pu^{238}$	$_{96}Cm^{248}$
8^+		499	514	
6^+	378	297	304	300
4^+	187	143	146	144
2^+	57.5	43.5	44.1	43.4
0^+	0	0	0	0

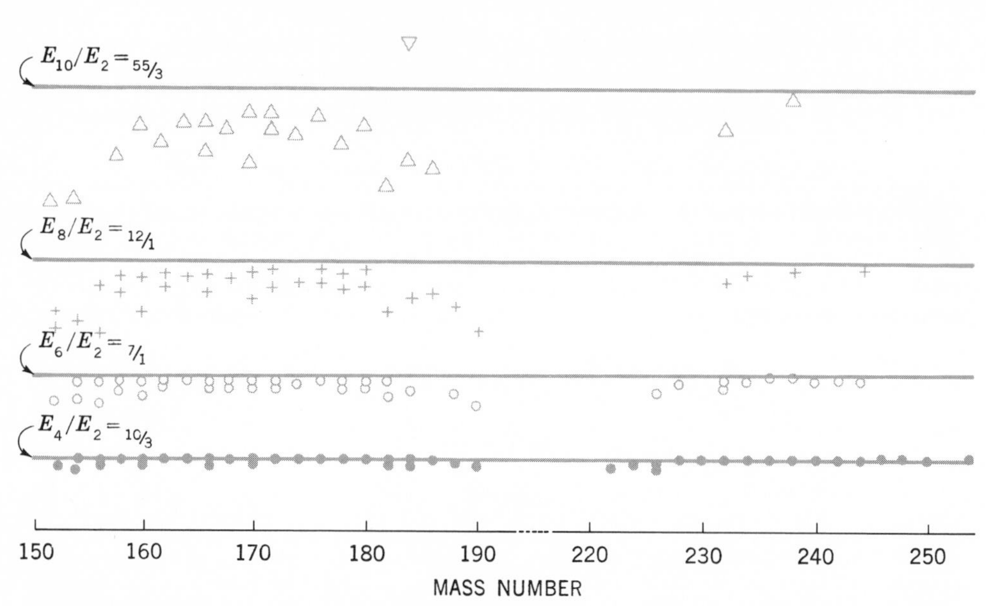

FIGURE 6-18 Ratio of energies of various members of ground-state rotational bands in spheroidal even-even nuclei to the energy of the lowest (2^+) member. Subscripts are I of the states, and the horizontal lines are the predictions of (6-7). [*From O. Nathan and S. G. Nilsson in K. Siegbahn (ed.), "Alpha, Beta, Gamma Ray Spectroscopy," North-Holland Publishing Company, Amsterdam,* **1965;** *by permission.*]

By comparing the experimentally determined energy levels from Fig. 6-17 with (6-7) we can determine the moment of inertia $\mathscr{I}$. In the simplest interpretation of the above discussion we might get the impression that $\mathscr{I}$ can be calculated by the methods of elementary physics considering the nucleus to be a rigid body. If this were the case, $\mathscr{I}$ would not be very dependent on the deformation, since the moment of inertia of a sphere is not much different from that of an ellipsoid with the relatively small deformations encountered in nuclei. However, this is too simple a picture; it turns out that $\mathscr{I}$ is considerably less than the rigid-body value and is quite sensitive to the amount of deformation.

In Fig. 6-18 we see that deviations from (6-7) become appreciable for large I. This is partly because, in that formula, $\mathscr{I}$ appears as a constant whereas when the nucleus rotates with increasing angular velocity, it is stretched by the centrifugal force causing the deformation, and hence $\mathscr{I}$, to increase. This causes the energies of the higher states in Fig. 6-18 to fall below the predictions of (6-7). This *centrifu-*

gal stretching as well as other small effects can be taken into account by modifying (6-7) to

$$E = \frac{\hbar^2}{2\mathscr{I}}[I(I+1) - \alpha I^2(I+1)^2] \tag{6-7a}$$

where α and $\mathscr{I}$ are constant for all I.

6-6 Collective Vibrations and Other States in Spheroidal Even-Even Nuclei

If a spherical rubber ball is squeezed along a diameter and then let go, there is only one type of $\lambda = 2$ vibration that can be set up, namely, that shown in Fig. 5-10*a*. However, a spheroidal ball squeezed at opposite ends of its principal axes and then let go has two different $\lambda = 2$ vibrations depending on which principal axis is squeezed. If it is the unequal axis, the vibrations in the nuclear case are known as β vibrations; they are shown schematically in Fig. 6-19*a*. If it is one of the equal axes, the vibrations are as shown in Fig. 6-19*b*. Vibrations very similar to these are known as γ vibrations; a more accurate description of γ vibrations will be given in Sec. 6-7.

In considering the angular momentum carried by these vibrations, we must again face the problem encountered in Sec. 6-4 for particle angular momenta, namely, that only its component parallel to the symmetry axis (which we desig-

FIGURE 6-19 The two types of $\lambda = 2$ vibrations in spheroidal nuclei: *(a)* β vibrations and *(b)* γ vibrations. Two perpendicular views are given for both; curves are shapes of the nucleus at the two extremes of the vibration.

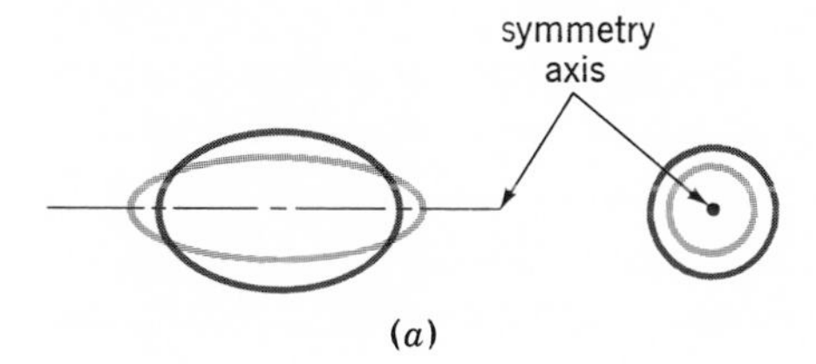

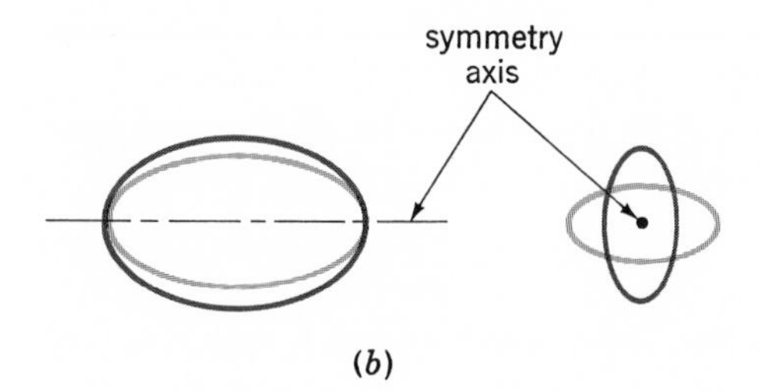

nate as the z axis) is represented by a good quantum number. We call it K. In β vibrations, we see from Fig. 6-19*a* that the motion relative to the symmetry axis has no θ dependence, so the deformations about the symmetry axis, in the parlance of Fig. 5-8, are $\lambda = 0$; that is, they are represented by the Legendre polynomial from Table 2-2 with no θ dependence, namely, P_{00}. We found in Sec. 5-9 that the angular momentum involved in a vibration is equal to λ, so the component of angular momentum from β vibrations along the symmetry axis is 0; they therefore have $K = 0$. For γ vibrations, on the other hand, we see from Fig. 6-19*b* that the motion relative to the symmetry axis is very similar to that shown in Fig. 5-10 for $\lambda = 2$, whence $K = 2$.

In addition to their vibrational motion, these nuclei may also rotate, so for each vibration there is a band of states. However, we cannot clearly distinguish between the angular momentum derived from vibrations and that derived from rotation since neither is by itself a conserved quantity. All we can know about a state is the total angular momentum I and its z component K. We do know that the rotational angular momentum must be perpendicular to the symmetry axis. It turns out, in fact, that what may be considered loosely as the component of the total angular momentum in that direction plays the role of the rotational angular momentum R in (6-5). That is, if we define R as the third side of a right triangle whose hypotenuse is the total angular momentum and whose other side is its z component, the pythagorean theorem gives

$$R^2 = I(I + 1) - K^2$$

whence the energy of the states, from (6-5), is

$$E = \frac{\hbar^2}{2\mathcal{I}} [I(I + 1) - K^2] \qquad \textbf{(6-8)}$$

While this "derivation" leaves very much to be desired, the result (6-8) can be derived rigorously and is correct. Note that in it the energies of states are determined by the only quantities having definite meaning here, namely, I and K. Since K is the z component of I, I must be equal to or greater than K. The lowest-energy state is then the one with $I = K$.

The states of the "rotational band" based on a vibration have energies given by (6-8). For γ vibrations, $K = 2$, so the values of I^π are $2^+, 3^+, 4^+, \ldots$. The parity must be even since rotations cannot change the parity, as was established in the discussion leading to (6-7). For β vibrations, $K = 0$, whence the values of I^π are $0^+, 2^+, 4^+, \ldots$. The odd values of I are not allowed here for the same reasons as those given in connection with the ground-state rotational band. Examples of β and γ vibrational bands are shown in Fig. 6-20.

We can see from the "derivation" of (6-8) that all states in the vibrational bands, including the lowest, combine both vibrational and rotational motion—indeed it is not really meaningful to separate the two. In states with $I \gg K$

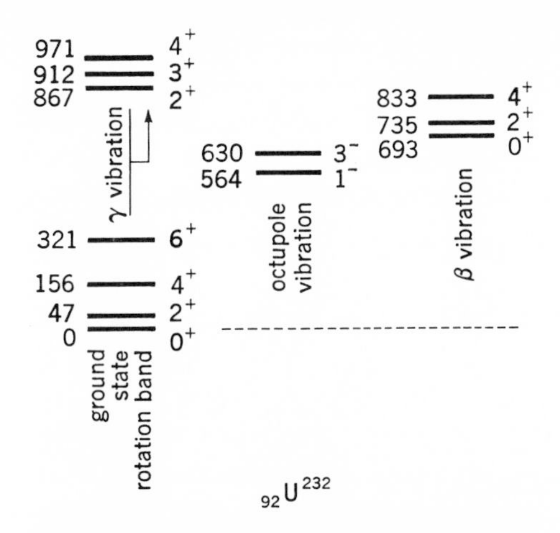

FIGURE 6-20 Experimentally known states of $_{92}U^{232}$. States belonging to different types of rotational bands are displaced horizontally for clarity. Numbers at the left are excitation energies in keV, and I^π are given at the right. Data were obtained mostly by beta-ray and gamma-ray decay studies.

it is clear that most of the angular momentum is rotational; in low-energy states, however, things are not simple. For example, in a β vibration it is evident from the similarity with the corresponding vibration in spherical nuclei that the vibration is predominantly of angular momentum 2. The fact that the lowest-energy β vibration has $I = 0$ must therefore be interpreted as meaning that there is rotation in this state which opposes the vibrational angular momentum. Such interpretations must of necessity be crude, since neither vibrational nor rotational angular momentum are conserved in the motion.

In addition to the $\lambda = 2$ vibrations we have been discussing there are $\lambda = 3$ (octupole) vibrations (see Fig. 5-10*b*) known in spheroidal nuclei. For reasons closely akin to those we have been discussing, the I^π of the states belonging to this rotational band are $1^-, 3^-, 5^-, \ldots$. Note that here only odd-parity states are acceptable since the vibration is an odd-parity one [recall that parity is $(-1)^\lambda$ for vibrations]. Their energies are given by (6-8) with $K = 1$. There is an example of an octupole vibrational band in Fig. 6-20.

In addition to vibrational excitations, even-even spheroidal nuclei can have all the types of excited states exhibited by spherical nuclei. As in the latter, states can arise from noncoherent collision patterns, from broken pairs, from higher-energy configurations, from exciting one or more nucleons to higher-energy shells, etc. In all cases, these states can be described only by the projection of their total angular momentum on the symmetry axis K and their parity. Each then gives rise to a rotational band, the energies of which are given by (6-8). The lowest-energy state in this band then has $I = K$. Where there are no vibrations involved, K is just the sum of the Ω values of occupied orbits.

6-7 Generalized Treatment of Nuclear Shapes

Up to this point we have been working under the tacit assumption that the shape of a nucleus in a given state is fixed, either spherical or ellipsoidal with a certain deformation. However, the nuclear shape merely represents the density distribution of the nucleons, so it changes as the nucleons go through their collision patterns. As a result, the shape of the nucleus fluctuates, and calculations can be made of the fraction of the time a nucleus spends in various shapes. Here we shall discuss some of the results of these studies.[1]

Under the approximation that the nuclear shape is an ellipsoid with a fixed volume, the shape can be specified in general by two variables, e.g., the length of any two principal axes, the length of the third being then determined by the volume. A more convenient choice of variables is β and γ, defined by Fig. 6-21.

[1] K. Kumar and M. Baranger, *Nucl. Phys.*, **A92:** 608 (1967); **A110:** 490, 529 (1968); **A122:** 241, 273 (1968).

FIGURE 6-21 Shape of an ellipsoid as a function of the variables β and γ. Curves give the lengths of the three principal axes of the ellipsoid as a function of γ. The difference between these lengths and the average radius R_0 is βR_0 times the ordinates of the curves.

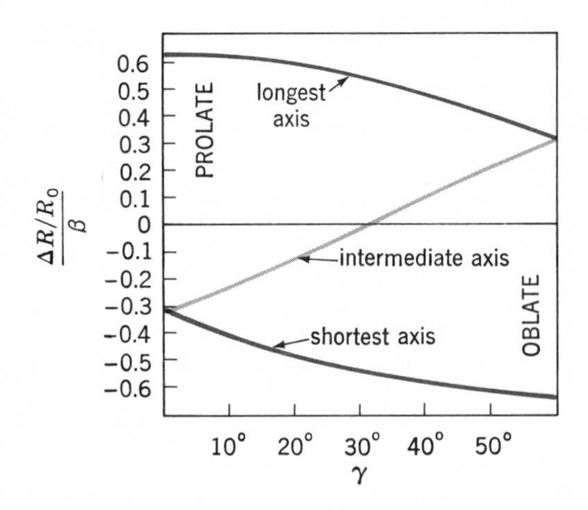

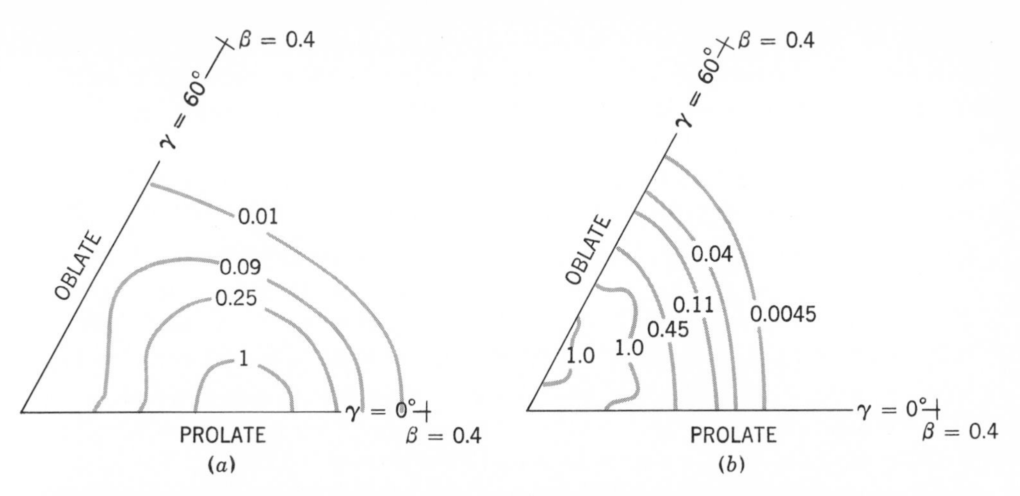

FIGURE 6-22 Probability of various shapes of *(a)* **W^{184} and** *(b)* **Pt^{196} as specified by β and γ. Lines are contours of relative probabilities as given by the attached numbers.** [*From K. Kumar and M. Baranger, Nucl. Phys.,* **A122: 273 (1968);** *by permission from North-Holland Publishing Co.*]

We see there that γ is an angular variable which can take values between 0 and 60°. It determines the ratio of the lengths of the principal axes; $\gamma = 0$ and 60° correspond to prolate and oblate spheroids respectively, $\gamma = 30°$ corresponds to an ellipsoid with the intermediate axis equal in length to the average of the longest and shortest, etc.

The variable β can take values between zero and infinity. It determines the magnitude of the deviation from spherical shape. For example if $\gamma = 0°$, the ratio of the longest axis to the average radius is $1 + 0.63\beta$ and the ratio of the two shorter axes to the average radius is $1 - 0.31\beta$. If $\beta = 0$, the shape is spherical, while large values of β correspond to ratios of axis lengths very different from unity. The β and γ vibrations discussed in the last section can now be described more accurately; they correspond to periodic variations of β and γ, respectively. The deformation δ used as the abscissa in Fig. 6-15 is approximately equal in magnitude to β.

The shape of an ellipsoidal nucleus of a given volume is determined by a point in $\beta\gamma$ space. Such spaces are shown in Fig. 6-22. The diagrams there show the relative probability for various nuclear shapes. The lines are contours of equal probability specified by the attached numbers. From these contours, one can visualize the probability as a function of β and γ. This probability may also be thought of as the fraction of its time the nucleus spends in the corresponding shape.

Figure 6-22*a* shows the results of calculations for the ground state of W^{184}, which is what we have previously been calling a spheroidal nucleus. We see from it that W^{184} spends most of its time as a prolate ellipsoid with $\beta \simeq 0.2$. From Fig. 6-21 we see that $\gamma = 0°$, $\beta = 0.2$ implies that the long axis is 12.5 percent longer than the average radius R_0 and the other two principal axes are 6.25 percent shorter than R_0; hence the long axis is about 20 percent longer than the shorter ones. We also see from Fig. 6-22*a* that W^{184} spends a small fraction of its time in various other shapes, including oblate spheroidal ($\gamma \simeq 60°$) and near spherical ($\beta \simeq 0$). Figure 6-22*b* shows the results for the ground state of Pt^{196}, which is what we have been previously calling a spherical nucleus. It does indeed spend a large fraction of its time with β near zero, but the fraction of its time spent with $\beta \lesssim 0.15$ is by no means negligible. All values of γ are about equally likely, so there is no strong preference for near prolate or near oblate shapes.

From Fig. 6-22, we may conclude that our previous assumption that ellipsoidal nuclei have a fixed spheroidal shape was reasonably accurate. The probability for shapes very different from the most probable one is really quite small. However, our previous assumption of a strictly spherical shape for nuclei like Pt^{196} was much less accurate. In the rather considerable fraction of their time such nuclei spend in an ellipsoidal shape, they may rotate. The first excited (2^+) state of "spherical" even-even nuclei is therefore not a simple vibration but includes part-time rotations. The same may also be said for the two-phonon vibrations; this accounts in a large measure for the deviations from the simple energy ratio of these states predicted by vibration theories (cf Sec. 5-10). It also causes the *average* shape of the nucleus in these states to deviate from spherical symmetry. Experimental determinations of these shapes by measuring electric quadrupole moments (see Sec. 7-5) have played an important role in elucidating these matters.

6-8 States of Spheroidal Odd-A Nuclei

Just as in the case of spherical nuclei, the simplest states in odd-A spheroidal nuclei are the single-quasi-particle (SQP) states, those which differ from the ground state of neighboring even-even nuclei by the addition of one particle or one hole. Since the ground states of even-even nuclei are 0^+, all the angular momentum comes from the added particle or hole. If it is a SQP state designated by Ω in Fig. 6-15, the nuclear state has $K = \Omega$. This is the lowest state of a rotational band which has members with $I = K, K + 1, K + 2, \ldots$ and energies given by (6-8). An exception occurs for $K = \frac{1}{2}$, but we shall not consider it here.

There are rotational bands of this type for each SQP state, i.e., for each Ω state shown in Fig. 6-15 in the energy region where filling occurs. The energies of these SQP states are governed by considerations similar to those in spherical nuclei discussed in connection with Fig. 6-3. As an Ω state begins to fill, the corresponding SQP state moves closer to the ground state and becomes less of a particle state and more of a hole state; when it is about half full, it is the ground state; as it becomes still fuller, it moves up in energy above the ground state and becomes less and less of a particle state and more and more of a hole state. The wave function is again given by (6-1). It should be noted that each Ω state can accommodate only two particles ($\pm\Omega$), but this does not hamper the gradual filling aspect, just as it did not hamper it for the $s_{1/2}$ state in spherical nuclei.

At higher energies there are β, γ, and octupole vibrations based on SQP states and states arising from particle excitations of various types; each of these leads to a rotational band. States of the same I^π and similar energies mix, leading to energy shifts and mixed wave functions. Nevertheless, in spite of all this complexity, bands due to individual SQP states can be distinguished by the way the various members are excited in certain types of nuclear reactions (see Sec. 14-3). The energy levels from a typical odd-A spheroidal nucleus are shown in Fig. 6-23. They are arranged into bands to reduce the complexity of the diagram, and many of these bands can be identified with SQP states from Fig. 6-15.

FIGURE 6-23 Experimentally determined energy levels of $_{72}Hf^{177}$. Different rotational bands are displaced horizontally for clarity. The K^π of each band is identical with I^π of its lowest member as shown at the right. Excitation energies in keV are shown at the left. Note that $K = \frac{1}{2}$ bands do not follow (6-8). *(From R. K. Sheline, "Nuclear Structure, Dubna Symposium 1968," International Atomic Energy Agency, Vienna,* **1968.**)

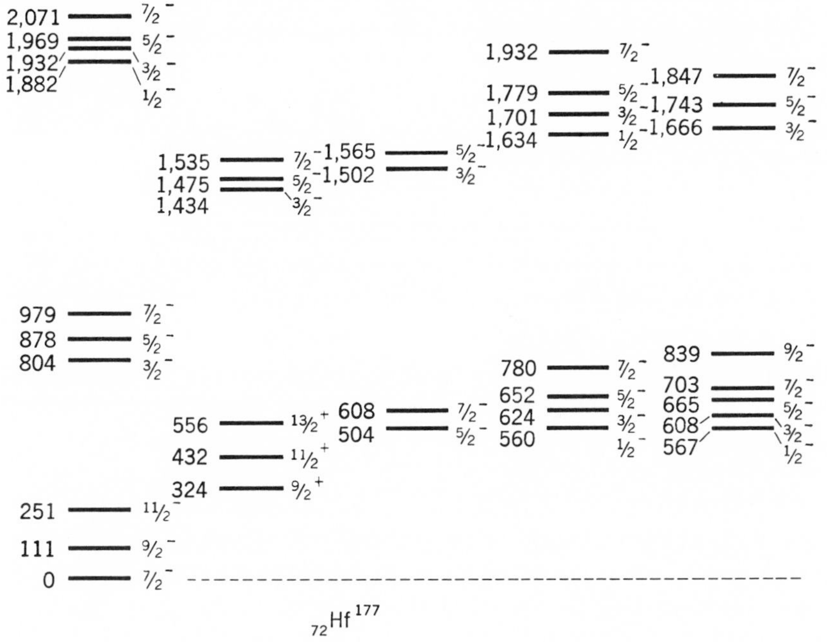

Problems

6-1 Construct the analogy of Fig. 6-1 for the $\mathfrak{N} = 6$ shell.

6-2 Under the assumption that Fig. 6-3 is correct, draw energy-level diagrams showing the energies and I^π for the three lowest-energy states of Mg^{25}, Mg^{27}, and Mg^{29}.

6-3 By use of (6-2) and the ϵ's given in Prob. 5-5, calculate the curves in Figs. 5-6 and 6-3. [*Hint:* Calculate the E_j, V_j^2, and N from (5-29) for various λ.]

6-4 From (6-3), show that a SQP state is the ground state when it is half full.

6-5 Assuming that Fig. 6-4 is accurate, draw a diagram showing the energies and I^π of the lowest energy states of Sn^{113} and Sn^{125}.

6-6 If the center part of Fig. 6-5 were changed by increasing the energy of the $h_{11/2}$ state by a factor of 3, how would it change the states of Sn^{117}? Consider the same question if the $d_{3/2}$ state in the center part were raised to that energy. Discuss the difference.

6-7 Explain the difference among the low-energy states of the different Sn isotopes in Fig. 6-6. Explain the differences between these and the low-energy states in the Pd isotopes shown in Fig. 6-7.

6-8 Construct the equivalent of Fig. 6-5 for the Pd isotopes and use it to explain some of the features of Fig. 6-7.

6-9 If the configuration of the ground state of F^{18} is $(d_{5/2})_\nu(d_{5/2})_\pi$, what are the possible values of I^π?

6-10 Construct a diagram similar to Fig. 6-8 for Ca^{37}.

6-11 Construct the configuration part of a diagram like that in Fig. 6-8 for the nucleus O^{23}.

6-12 If the method given at the end of Sec. 6-1 is used to study O^{18} and all the v_{mn} are taken as -1, write out all the matrices whose eigenvalues are the energies of states in O^{18}.

6-13 If forces between two neutrons were the same as those between two protons but not the same as those between a neutron and a proton (this is called *charge symmetry* as opposed to *charge independence*) what relationships would be expected between the states of Li^7 and Be^7? Of C^{14}, N^{14}, and O^{14}?

6-14 In the representation used in Fig. 6-12, find the wave functions for the isobaric analogs of the ground state of B^{14}.

6-15 What are the T values for the ground states of Cu^{63}, Ag^{109}, and U^{238}?

6-16 Compare the energy levels of Hf^{170} and Pu^{238} from Fig. 6-17 with (6-7) and with (6-7a) by choosing $\mathcal{I}$ and α to fit the lowest- and highest-energy levels shown.

6-17 Look up the energy levels of Tb^{173}, Re^{185}, and Ac^{223} and compare with the predictions of (6-8) for the lowest-energy levels. Explain the differences between these and the low-energy states in the Pd isotopes shown in Fig. 6-7.

6-18 A nucleus of $A = 180$ is ellipsoidal in shape with $\beta = 0.15$, $\gamma = 15°$. What are the lengths of the three principal axes?

6-19 An odd-A nucleus has an $\Omega = \frac{7}{2}^+$ SQP state as its ground state. What are I^π and K for the first two excited states? If the energy of the first excited state is 100 keV, what is the energy of the second excited state?

6-20 If the nucleus in Prob. 6-19 has an $\Omega = \frac{11}{2}^+$ SQP state at an excitation energy of 180 keV, how will the energy-level diagram be affected?

Further Reading

See General References, following the Appendix.

Bohr, A., and B. R. Mottelson, *Kgl. Danske Videnskab. Selskab, Mat. Fys. Medd.*, **27**(16) (1953).

deShalit, A., and I. Talmi: "Nuclear Shell Theory," Academic, New York, 1959.

Eisenbud, L., and E. P. Wigner, "Nuclear Structure," Princeton University Press, Princeton, N.J., 1961.

Feenberg, E.: "Shell Theory of the Nucleus," Interscience, New York, 1959.

Kisslinger, L. S., and R. A. Sorenson: *Kgl. Danske Videnskab. Selskab. Mat. Fys. Medd.*, **27**(16) (1953).

Lane, A. M., "Nuclear Theory," Benjamin, New York, 1964.

Mayer, M. G., and J. H. D. Jensen: "Elementary Theory of Nuclear Shell Structure," Wiley, New York, 1955.

Chapter 7

Miscellaneous Aspects of Nuclear Structure

In our development of the structure of complex nuclei in the last three chapters, we have bypassed a few important subjects, which we return to consider here. The first subject is binding energies of nuclei; it has many practical aspects and in addition leads us into the basic question of how complex nuclei arise from the nucleon-nucleon force and how these forces lead to the shell-theory potential which was the basis of our previous developments. Of the three observable properties of nuclei discussed in Sec. 1-6, we have so far treated only the angular momentum. The other two, the magnetic dipole moment and the electric quadrupole moment, will be discussed later in this chapter.

7-1 Masses and Binding Energies of Nuclei

The binding energy of a nucleus B is defined as the energy required to break it up into free neutrons and protons. One method of breaking a nucleus up into its constituents would be to separate one nucleon at a time. Since we found in Sec. 4-6 that nucleon separation energies are approximately constant at about 8 MeV, we may estimate that the binding energy of a complex nucleus is about 8 MeV times the number of nucleons, or

$$\frac{B}{A} \simeq 8 \text{ MeV} \tag{7-1}$$

Here we consider methods of determining this quantity more accurately and of understanding its detailed behavior.

In accordance with relativity theory, the mass of a system bound by an energy B is less than the mass of its constituents by B/c^2. If a nucleus of mass M contains Z protons and N neutrons, we may therefore express its binding energy as

$$B = (ZM_p + NM_n - M)c^2 \tag{7-2}$$

When the mass is expressed in atomic mass units (amu), the conversion factor c^2 works out to be 931.5 MeV/amu. If masses are measured with sufficient accuracy, (7-2) can be used to determine B. An equation similar to (7-2) can be written for any bound system, such as the sun with its planets, a magnet holding an iron nail, a molecule composed of two or more atoms, an atom with its electrons bound to the nucleus, etc. In all cases, the mass of the bound system is less than the sum of the masses of its constituents by B/c^2. However, in all these problems the percentage change in mass is so small as to make the effect of only academic interest. In one of the more favorable ones, an electron bound in a hydrogen atom, $B \simeq 13$ eV while $Mc^2 \simeq 10^9$ eV, so the change in mass is only about 1 part in 10^8. In nuclei, however, we see from (7-1) that the fractional change in mass is about 8/931.5 MeV, or close to 1 percent. This is readily measurable.

The technology of accurately measuring masses of nuclei is known as *mass spectroscopy*. We shall discuss some of its aspects in Chap. 9, and detailed treatments of the subject are given in many textbooks.[1] Over the years it has developed into a highly refined art, and it is now capable of measuring masses to an accuracy of about 1 part in 10^7, which means that B can be determined from (7-2) to an accuracy of about 1 part in 10^5.

Another method of measuring nuclear masses is by determining the energy release Q in nuclear reactions. In most cases, the reaction can be written

$$a + X \rightarrow b + Y + Q \tag{7-3}$$

where a and b are nucleons or light nuclei and X and Y are heavy nuclei. If the target nucleus X is initially at rest, the energy of the incident particle a is known, and the energy and angle of emission of b are measured, the energy of Y can be calculated from conservation of momentum, whence Q is determined. From conservation of energy,

$$M_Y = M_a + M_X - M_b - \frac{Q}{c^2} \tag{7-4}$$

where the M's are the masses of the particles in (7-3) given by the subscripts. If the masses of a, b, and X are known, the determination of Q then gives the mass of Y from (7-4). Nuclear decay processes can be written in the form (7-3) with a omitted, whence the same method is applicable.

As a result of many decades of mass spectroscopy and accurate measurements of Q values from nuclear reactions and decay processes, the masses of a great many nuclei have been determined. Since mass spectroscopic techniques directly measure atomic rather than nuclear masses, it is conventional to present results

[1] Consult Further Reading lists at the end of this chapter and Chap. 9.

in the form of atomic masses. They differ from nuclear masses principally by the rest mass of their Z electrons, although the binding energy of the electrons is not negligible. A list of atomic masses is given in Table A-3 of the Appendix. Since the mass of an atom is very close to its mass number A, listings are given only of deviations from A. For convenience in use, these deviations are converted into energy units in Table A-3.

Tables of atomic masses have many practical uses. Nucleon separation energies, which were discussed in Sec. 4-6, are just the difference in binding energies of the nucleus in question and the one with one less nucleon. This is readily calculated from the masses of these nuclei with (7-2). Mass tables are useful for calculating Q values for nuclear reactions with (7-4) or for calculating the minimum energy necessary to induce a nuclear reaction. But one of the more interesting applications of masses is in the determination of binding energies. The explanation of these binding energies is our next topic.

A plot of B/A for nuclei found in nature is shown in Fig. 7-1. We see that it is very crudely a constant as given by (7-1) but that it falls off at small and at large values of A. The latter may be explained as a coulomb effect; the coulomb repulsion acts between every pair of protons and hence increases as Z^2, which increases more rapidly than A for naturally occurring nuclei. The drop at small A in the curve of Fig. 7-1 is due to the fact that full binding is achieved by a nucleon only when it is completely surrounded by other nucleons. This, of course, is not the situation for nucleons on the surface of the nucleus; since light nuclei have a greater fraction of their nucleons on the surface, their binding energy per nucleon is reduced.

The curve in Fig. 7-1 is an important one from a practical standpoint. The fact that it is peaked at medium-mass nuclei means that the binding energy can

FIGURE 7-1 Binding energy per nucleon B/A for nuclei found in nature.

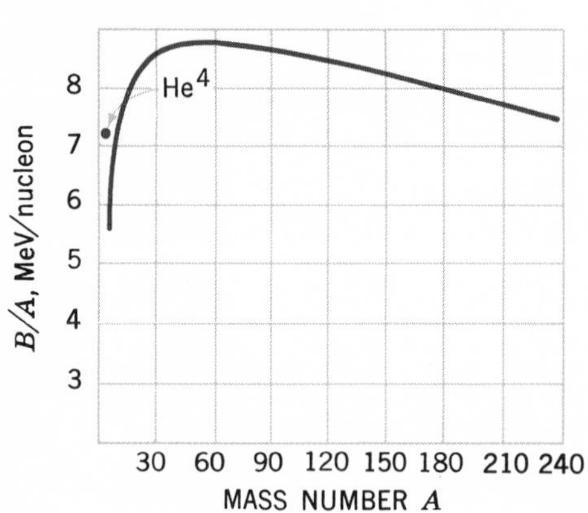

be increased, and hence energy in other forms can be released, by either fusing light nuclei together or by breaking heavy nuclei apart. As we shall see, these are the bases of the two methods for producing nuclear energy. It may be noted that the curve in Fig. 7-1 is peaked at Fe; we shall find in Chap. 15 that this is the reason for the large abundance of that element in nature.

7-2 The Semiempirical Mass Formula

In the last section we obtained a qualitative understanding of the principal features of Fig. 7-1. However, B/A is known to a very high accuracy, and it is known not only for nuclei found in nature but for a much larger number of unstable nuclei, so we might hope to do much better—to achieve a quantitative and more basic understanding of the binding energies of all of these nuclei. In this section we use a semiempirical approach which leads to a simple equation for binding energies of nuclei. As such, it has many practical applications, but in addition we shall find it to be useful in the next section, where we approach the problem from a more basic standpoint.

The semiempirical approach is usually discussed under the title *semiempirical mass formula* or *Weizsacher formula*. In it, the nucleus is treated as an assemblage of interacting particles similar in some ways to a drop of liquid (this is sometimes called a liquid-drop model) with coulomb forces, effects of the Pauli exclusion principle, and details arising from the considerations of Chaps. 5 and 6 added as corrections. We now consider the various contributions to the binding energy in such a model.

Volume Term

For a drop of liquid to evaporate, a certain amount of heat must be supplied; as we all know, it is the product of a constant Q_v, known as the heat of vaporization, and the mass of the material, which is just the number of molecules A times the mass of each molecule M_m. This heat is the energy required to overcome all the interactions between molecules, which is therefore just equal to the binding energy of the liquid drop B. Hence we have

$$B = Q_v M_m A \qquad \text{(7-5a)}$$

Since Q_v and M_m are constants, this says that the binding energy per molecule B/A is independent of the number of molecules in the system. The reason for this simple relationship is that the binding energy is the sum of all the interactions between molecules, and each molecule interacts only with its neighbors; since the number of neighbors each molecule has in a liquid is independent of the overall size of the system, B/A is independent of A. This is clearly a characteristic of any system in which the range of interactions between particles is small compared with the dimensions of the system. For nuclei, therefore, we expect a term in the expression for binding energy

$$B_v = c_v A \tag{7-5b}$$

where c_v is a constant which corresponds to $Q_v M_m$ for the liquid drop.

Surface Term

In our analogy with a liquid drop, the proportionality between B and A depends on the assumption that the size of the drop is so large that nearly every molecule is surrounded by its full complement of neighbors with which to interact. This is *not* true for molecules on the surface since they are not surrounded on all sides. If a surface molecule is surrounded, on an average, by a fraction f_1 of the number of molecules surrounding one inside the drop, and if the fraction of the molecules that are on the surface is f_2, expression (7-5a) should therefore be corrected to

$$B = Q_v M_m A(1 - f_2) + Q_v M_m A f_2 f_1$$

The first term is the binding due to the $A(1 - f_2)$ molecules that are not on the surface, and the second term represents the binding of the Af_2 molecules that are on the surface and are consequently bound only f_1 as strongly. This expression simplifies to

$$\begin{aligned} B &= Q_v M_m A - Q_v M_m A f_2(1 - f_1) \\ &= c_v A - c_v A f_2(1 - f_1) \end{aligned}$$

where the second expression applies to nuclei as in the analogy between (7-5a) and (7-5b). The first term is just B_v from (7-5b), whence we expect an additional term in our expression for the binding energy of nuclei

$$-B_a = c_v A f_2(1 - f_1) \tag{7-6}$$

due to surface effects. Let us now estimate f_1 and f_2.

If we picture the system as an assemblage of balls packed together, we see that f_1 is somewhat greater than $\frac{1}{2}$ but closer to $\frac{1}{2}$ than to unity; we take it to be

$$f_1 \simeq \tfrac{2}{3}$$

In estimating f_2, we may interpret (1-2) to mean that a nucleon is "on the surface" if it is within a shell of thickness about 1.07 F centered at the radius R. The volume V_S of this shell is then

$$V_S \simeq 4\pi R^2 \times 1.07 \text{ F} = 4\pi \times 1.07^3 A^{2/3}$$

while the total volume of the nucleus V_T is

$$V_T = \tfrac{4}{3}\pi R^3 = \frac{4\pi}{3}\, 1.07^3 A$$

The fraction of the nucleons that is on the surface, f_2, is roughly

$$f_2 \simeq \frac{V_S}{V_T} \simeq \frac{3}{A^{1/3}}$$

By inserting these estimates of f_1 and f_2 into (7-6) we obtain

$$-B_a = c_v A^{2/3} \tag{7-7}$$

Quite aside from our estimates of f_1 and f_2, it is clear that B_a must be proportional to the surface area of the nucleus, which, in view of (1-2), means

$$-B_a = c_a A^{2/3} \tag{7-8}$$

where c_a is a constant. By comparing this with (7-7), we estimate

$$c_a \simeq c_v \tag{7-9}$$

Coulomb Term

The amount of work that must be done against repulsive coulomb forces in order to "assemble" a nucleus represents a negative term in the binding energy $-B_c$. In the approximation that a nucleus may be considered to have a constant charge density ρ out to a radius R with no charge for $r > R$,

$$\rho = \frac{Ze}{\frac{4}{3}\pi R^3} \tag{7-10}$$

The work dW required to bring a thin spherical shell of this charge up to its radius r is obtained from the coulomb law with one charge equal to the total charge inside r and the other charge equal to the charge of the shell, whence

$$dW = \tfrac{4}{3}\pi r^3 \rho \times 4\pi r^2\, dr \frac{\rho}{4\pi\epsilon_0 r}$$

A first estimate of B_c, which we denote B_c', is obtained by integrating this from $r = 0$ to R, which gives, after insertion of (7-10),

$$B_c' = \frac{\frac{3}{5} Z^2 e^2}{4\pi\epsilon_0 R} \tag{7-11}$$

In this calculation, even if there were only one proton in the nucleus, we would obtain

$$(B_c')_{\text{proton}} = \frac{\frac{3}{5} e^2}{4\pi\epsilon_0 R} \tag{7-11a}$$

whereas it is clear that no work is done against coulomb forces in assembling such a nucleus. The term (7-11*a*) is part of the energy required to assemble a proton, and our basic equation (7-2) is not meant to include such terms; it is based on the assumption that protons already exist. B_c is therefore obtained by subtracting from (7-11) a term equal to (7-11*a*) for each proton. Since there are Z such terms and $Z^2 - Z = Z(Z - 1)$, this gives

$$-B_c = \frac{\frac{3}{5}Z(Z-1)e^2}{4\pi\epsilon_0 R} \tag{7-12}$$

Symmetry Energy

In the absence of coulomb forces, the binding energy of a nucleus of a given A is a minimum when it has equal numbers of neutrons and protons. The increase in energy required to have unequal numbers is known as the *symmetry energy*, and this gives a decrease in the binding energy $-B_s$. In order to calculate this, we must first see how the kinetic and potential energies of nucleons, T_i and V_i, respectively, contribute to the total energy of the system E_T. In terms of the two-nucleon interaction v_{ij} these are

$$E_T = \sum_i T_i + \frac{1}{2}\sum_i\sum_j v_{ij}$$

$$V_i = \sum_j v_{ij}$$

where the sums exclude the term $j = i$. The factor $\frac{1}{2}$ in the first equation is required to avoid counting the interaction between each pair of nucleons twice, but it does not appear in the second equation since the potential of particle i is just equal to the sum of the potentials arising from its interaction with each of the other nucleons. Inserting the second into the first gives

$$E_T = \sum_i T_i + \frac{1}{2}\sum_i V_i \tag{7-13}$$

Any change in E_T is a change in the binding energy, whence

$$-B_s = \Delta(\Sigma T_i) + \tfrac{1}{2}\Delta(\Sigma V_i) \tag{7-14}$$

where the Δ represents the difference between these quantities for a given nucleus and one of the same A but with $N = Z$.

The determination of the first term in (7-14) may be understood with the help of Fig. 7-2. The horizontal lines there are the energies of the allowed orbits, and a circle on the line indicates that the orbit is occupied. The left side of the diagrams gives the orbits occupied by protons, and the right side shows the occupancy by neutrons. In the figure, the energy spacing between adjacent allowed orbits is taken to be a constant which we call ϵ; we shall return to con-

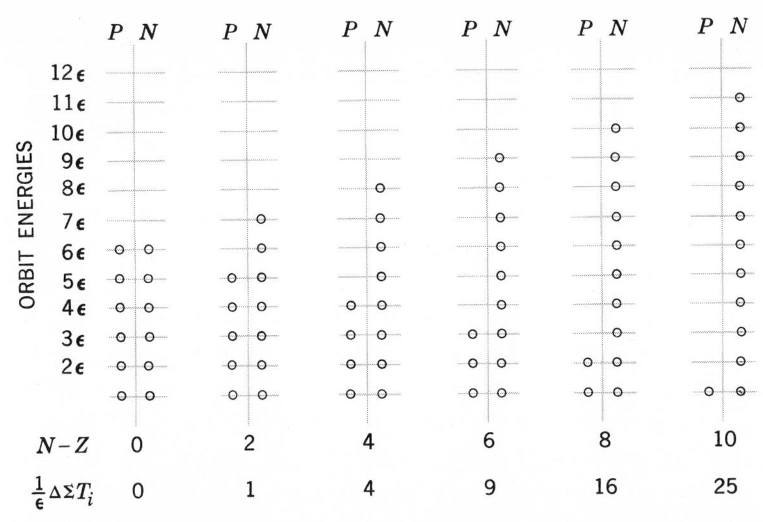

FIGURE 7-2 Dependence of ΣT_i in (7-14) on $N - Z$. Each diagram represents a nucleus with the same A $(= N + Z)$ but with different $N - Z$ as shown below. Horizontal lines designate orbits, and circles on them indicate that they are occupied. Occupation by protons and by neutrons is indicated in this way to the left and right respectively of the vertical lines. The energies assumed for the orbits, ϵ, 2ϵ, . . . are shown at the left. The row labeled $\Delta\Sigma T_i/\epsilon$ is the difference (in units of ϵ) between the total energy of occupied orbits for the nucleus in question and the one with $N - Z = 0$.

sider this assumption in the next paragraph. In the lower part of Fig. 7-2 are shown the values of $N - Z$ and $\Delta\Sigma T_i$, the increase in the total kinetic energy of occupied orbits above its value for the $N = Z$ nucleus. (Note that the potential wells, and hence the potential energies are not changed in Fig. 7-2.) For example, in the $N = Z$ nucleus, the sum of the energies of occupied orbits is

$$2\epsilon(1 + 2 + 3 + 4 + 5 + 6) = 42\epsilon$$

while in the $N - Z = 8$ nucleus it is

$$\epsilon(1 + 2 + 1 + 2 + 3 + 4 + 5 + 6 + 7 + 8 + 9 + 10) = 58\epsilon$$

which represents an increase of 16ϵ. It is readily seen from the values of $\Delta\Sigma T_i$ that

$$\Delta\Sigma T_i = \frac{\epsilon}{4}(N - Z)^2 \tag{7-15a}$$

This result may be obtained in general by noting that in any of the diagrams of Fig. 7-2, $(N - Z)/2$ nucleons are higher in energy than in the $N = Z$ nucleus by an amount $(N - Z)\epsilon/2$.

In obtaining (7-15a) we have made the approximation that each allowed orbit can accommodate only one neutron and one proton. This is not the situation for the orbits shown in Fig. 4-5. For example, a $d_{5/2}$ orbit can accommodate six nucleons of each type, so we might think that nuclei of a given A with $N - Z = 0$, ± 2, ± 4, and ± 6 can all be reached by changing protons to neutrons (or vice versa) in the same $d_{5/2}$ orbit, and therefore with no cost in energy. However, it was shown in Sec. 5-7 that orbits do not fill one at a time, but rather several orbits fill simultaneously; that in going from $N - Z = 0$ to 6, neutrons would be filling other orbits than $d_{5/2}$, and increasingly higher-energy orbits, on an average, as $N - Z$ increases. Similarly the proton holes being made in the process would go into several orbits simultaneously, and increasingly lower-energy orbits on an average as $N - Z$ increases. Every time a proton is changed to a neutron, the former come from an increasingly lower-average-energy orbit and the latter goes into an increasingly higher-average-energy orbit, which is just what was assumed in our derivation. Moreover these changes in average energy are smooth and regular. This justifies our assumption of equally spaced orbits. On the basis of these arguments, ϵ should be taken as the average spacing between orbits in the region of filling. It is shown in Table 7-1 that this is well approximated by

$$\epsilon \simeq \frac{25 \text{ MeV}}{A}$$

Using this formula in (7-14) gives

$$\Delta \Sigma T_i = 6.3 \text{ MeV} \frac{(N - Z)^2}{A} \tag{7-15b}$$

TABLE 7.1: CALCULATION OF ϵ IN FIG. 7.2 FOR VARIOUS A The second column is the energy difference between shells from (4-19). The average spacing between orbits is obtained by taking the number of orbits in the shell that is filling (or in the case of closed shells, the average of these numbers for the shell above and the shell below) to be spread over this energy. These numbers for neutrons and protons are averaged and the numbers in the second column are divided by them to obtain ϵ. The last column shows that $\epsilon \approx 25$ Mev/A.

A	ΔE per shell $= 33$ MeV/$A^{1/3}$, MeV	Orbits per shell: Neutron	Proton	Av	$\epsilon = \Delta E$ per orbit, MeV	ϵA, MeV
208	5.6	50	38	44	0.126	26.2
120	6.7	32	27	30	0.22	26.3
70	8.0	22	22	22	0.36	25.5
16	13.0	9	9	9	1.44	23.0

The dependence of the potential energy on $N - Z$ is given by (4-3). For each of the N neutrons V_i is increased by ΔV_s, while for each of the Z protons it is decreased by ΔV_s, whence

$$\Delta \sum_i V_i = (N - Z)\,\Delta V_s = 27\text{ MeV}\,\frac{(N - Z)^2}{A}$$

Inserting this along with (7-15*b*) into (7-14)

$$-B_s = 19.8\text{ MeV}\,\frac{(N - Z)^2}{A} \qquad \textbf{(7-16a)}$$

It is conventional to write

$$-B_s = c_s\,\frac{(N - Z)^2}{A} \qquad \textbf{(7-16b)}$$

whence from (7-16*a*) we expect

$$c_s \simeq 19.8\text{ MeV}$$

Pairing Term

Up to this point in our discussion of binding energy, we have paid no explicit attention to the details of nuclear structure as developed in Chaps. 4 to 6. The most important aspect of nuclear structure as regards ground states of nuclei (only ground states are considered in mass tables and in the empirical mass formula) is *pairing*, which lowers the energy of the ground state of even-even nuclei by 2Δ below the energy of the 2QP states (cf. Figs. 5-3 and 5-4). In heavy odd-odd nuclei, as we learned in Sec. 6-2, the ground state is a 2QP state; there must be two unpaired nucleons, namely, the odd neutron and odd proton. There is therefore a difference in binding energy of 2Δ between even-even and odd-odd nuclei if all other things are the same. According to Sec. 6-1, the ground states of odd-A nuclei are SQP states—they have one unpaired nucleon—whence the effect of pairing on their binding energy is halfway between that for even-even and odd-odd nuclei. The effect of pairing on the binding energy is therefore given by a term B_p defined as

$$B_p = \delta = \begin{cases} +\Delta & \text{for even-even nuclei} \\ 0 & \text{for odd-}A\text{ nuclei} \\ -\Delta & \text{for odd-odd nuclei} \end{cases} \qquad \textbf{(7-17)}$$

Equation (7-17) gives us a method for obtaining Δ. We can calculate it from neutron separation energies if we consider three nuclei with a given even number of protons and with mass numbers $A - 1$, A, and $A + 1$, which are even, odd, and even, respectively. Consider the expression on the right side of the following tentative equation:

$$\Delta = \frac{B(A+1) + B(A-1)}{2} - B(A) \tag{7-18}$$

All the other terms we have developed as contributors to B have equal and opposite values for the two terms in this expression, so the only contribution is from (7-17), which gives Δ. Thus Eq. (7-18) is established. In terms of neutron separation energies S_n we can write

$$B(A) = B(A+1) - S_n(A+1)$$
$$B(A-1) = B(A) - S_n(A) = B(A+1) - S_n(A+1) - S_n(A)$$

Inserting these into (7-18), we find

$$\Delta = \frac{S_n(A+1) - S_n(A)}{2} \tag{7-19}$$

For example, the neutron separation energies for Cd^{114} and Cd^{113} are 9.048 and 6.538 MeV, respectively, whence from (7-19) we obtain

$$\Delta = \frac{9.048 - 6.538}{2} = 1.255 \text{ MeV}$$

Formulas analogous to (7-19) could be obtained from odd-proton nuclei or from considering nuclei with the same number of neutrons but with different numbers of protons; roughly the same results would be obtained. (Actually there are small consistent differences, but we ignore them here.) These formulas can be used as the working definition of Δ.

The magnitude of Δ depends on the strength of the pairing interaction, which, in turn, depends on the frequency of collisions of the type described in Sec. 5-3. The frequency of these collisions is inversely proportional to the volume of the nucleus, which would imply $\Delta \propto 1/A$. However, there are effects which accentuate the pairing effect somewhat in heavy nuclei, e.g., the larger number of nucleons involved. Empirically it turns out that $\Delta \propto A^{-3/4}$, and it is fairly well represented by

$$\Delta = 33 \text{ MeV} \times A^{-3/4} \tag{7-20}$$

The Complete Formula

The total binding energy of a nucleus should be equal to the sum of B_v, B_a, B_c, B_s, and B_p. From (7-5*b*), (7-7), (7-12), (7-16*b*), and (7-17) this is

$$B = c_v A - c_a A^{2/3} - \frac{3}{5}\frac{Z(Z-1)e^2}{4\pi\epsilon_0 R} - c_s \frac{(N-Z)^2}{A} + \delta \tag{7-21}$$

By use of (7-2) this can easily be transformed into an expression for the mass of nuclei, and by the addition of Z electron masses it can be converted into an expression for the mass of atoms. Since the mass of a hydrogen atom M_{H} is equal to $M_p + M_e$, this may be written

$$M(Z,A) = NM_n + ZM_{\mathrm{H}} - \frac{1}{c^2}\left[c_v A - c_a A^{2/3} - \tfrac{3}{5}Z(Z-1)\frac{e^2}{4\pi\epsilon_0 R} - c_s\frac{(N-Z)^2}{A} + \delta\right] \tag{7-21a}$$

This equation is known as the *semiempirical mass formula.*

It should be noted that (7-21) is actually a function of only two independent variables, for example, N and Z. In order to show this explicitly, the value of R from (1-2) may be inserted, and all A's may be replaced by $N + Z$. It can also be written as a function of A and Z by replacing N by $A - Z$.

Since Table A-3 gives a large amount of accurate information on binding energies of nuclei with various Z and A, the constants c_v, c_a, and c_s can be evaluated from it. The values obtained for them are[1] typically

$$\begin{aligned} c_v &= 14 \text{ MeV} \\ c_a &= 13 \text{ MeV} \\ c_s &= 19 \text{ MeV} \end{aligned} \tag{7-22}$$

Note that the values of c_s and c_a agree quite well with the estimates given in the discussion.

Equation (7-21) gives binding energies of all nuclei for which M is known with average errors of a few MeV. Where higher accuracies are desired, more accurate formulas are available which treat various nuclear-structure effects in greater detail. The most important of these effects are deviations at closed shells. In these treatments, additional terms are introduced so that when the expanded formula is fit to the data, somewhat different values of c_v and c_a are obtained, as

$$\begin{aligned} c_v &\simeq 16.1 \text{ MeV} \\ c_a &\simeq 20.2 \text{ MeV} \end{aligned} \tag{7-23}$$

Formula (7-21) and its improved versions have many important applications. It can be used to estimate masses for nuclei where no measurements are available and hence to estimate Q values of nuclear reactions through (7-4). If the energy release in a decay process is positive, that decay can proceed, so the initial nucleus is unstable; this method can therefore be used to estimate the limits of nuclear stability. The magnitude of the surface term is used to calculate the *surface ten-*

[1] J. Orear, A. H. Rosenfeld, and R. A. Schluter, "Nuclear Physics: A Course Given by E. Fermi," The University of Chicago Press, Chicago, 1950.

sion, which is defined as the increase in energy per unit increase in surface area; this was used in our discussion of vibrations in Sec. 5-9. But the application of most basic interest is in the study of *nuclear matter*, which will be discussed in the next section.

7-3 Hartree-Fock Calculations and Nuclear Matter

In this section, we seek a more fundamental understanding of the binding energy B. Naïvely one might suppose that B is just the negative of the sum of the energies of all occupied orbits, but that is *not* correct; under that assumption the average binding energy per nucleon B/A would be much larger than the separation energy of the least bound nucleon S_n, whereas we know experimentally that the two are both about equal to 8 MeV. The fallacy is that the potential energy of a nucleon, which is essentially the depth of the shell-theory potential, arises from the sum of its interactions with all other nucleons, so if we assume that the sum of these potential energies (minus the kinetic energies) must be supplied to remove all nucleons from the nucleus, we are counting the interactions of each nucleon twice, once when it is removed and then again when all the other nucleons are removed. This point is demonstrated in the derivation of (7-13), and in fact E_T in (7-13) is essentially the binding energy B.

If one accepts the shell-theory potential as given in Secs. 4-1 and 4-2, a calculation of B from (7-13) is straightforward; however, as explained in Sec. 4-3, that potential is correct only for orbits near the top of the well. The velocity dependence of the nuclear force makes the forces on a nucleon depend on its kinetic energy, which is much higher for these than for deeply bound nucleons, so the value of V_0 we have been using is grossly incorrect for the latter. Obviously we need something better than the shell-theory potential to calculate binding energies of nuclei.

It should be clearly understood that these problems are not present in calculating the energy levels of a given nucleus. For this purpose, as we found in Chaps. 5 and 6, only the relative energies of the various orbits need be known, and, as is apparent from (2-5), these are essentially independent of the depth of the potential well. This means that the energy-level structure of nuclei is not very sensitive to the nuclear force! In support of this conclusion, it is interesting to point out that most of the information presented in Chaps. 5 and 6 was developed before the nucleon-nucleon force was known in any detail, and it was developed essentially without use of that force.

Far from being discouraged by the inability of shell theory to explain binding energies, we should look upon this as a great opportunity. In order to understand binding energies, we must truly understand the shell-theory potential in all detail, and this can only be done by learning to derive it from the nucleon-nucleon force. In proceeding with a theoretical problem of this type, it is important to have numbers to calculate as a check on the approximations used. In this case we have not only the parameters introduced in (4-1) but the binding energy, which is more sensitive to the details of the force and of the calculation. Calculations of this type are carried out by the Hartree-Fock method, which may be familiar to more advanced students as a technique developed for treating the quantum-mechanical problem of complex atoms.

Hartree-Fock calculations of the shell-theory potential have had good qualitative success and some degree of quantitative success. One result from them is a calculation of the effective mass for nucleons arising from the velocity dependence of the nucleon-nucleon force as discussed in Sec. 4-3. For nucleons deeply bound in the well, the result is $M^* \simeq 0.5M$. If this is assumed to be valid for all nucleons in the nucleus, application of (4-6) and (4-5) leads to

$$V_0 \simeq V_{00} - \frac{p^2}{2M}$$

For the least bound nucleons which are effective in determining the details of nuclear structure, we know from (4-2) and (4-11) that $V_0 = 57$ MeV and

$$\frac{p^2}{2M} = V_0 - 8 \text{ MeV}$$

whence

$$V_{00} \simeq 106 \text{ MeV}$$

Since Hartree-Fock calculations are very complex, another approach that has been used is to introduce the concept of infinite nuclear matter. Ordinary nuclei are limited in size by the coulomb term in (7-21), which clearly makes nuclei unstable for large A. However, if the coulomb forces could somehow be turned off, nuclei would be stable in any size. If we then let A go to infinity, we have a hypothetical *infinite nuclear matter* in which the volume term of (7-21) is completely predominant, whence

$$B = c_v A \tag{7-24}$$

Just as physicists have had great success in calculating the properties of normal matter in the solid and liquid state, they have succeeded in calculating the properties of nuclear matter from the nucleon-nucleon force of Chap. 3; the fact that the system is unbounded introduces many simplifications. In the calculation, a relation between B and the average distance between nucleons is obtained; the stable situation is where B is a maximum, and this choice gives both c_v (as B/A) and the actual average distance between nucleons. The former may be compared with the value from (7-23), and the latter is simply related to ρ_0 in

(1-1), for which the experimentally determined value is given in (1-2). The agreement is reasonably good, and discrepancies can be accounted for by the approximations used in the calculations. It is interesting to point out that in these calculations, three-body forces contribute about 2.5 MeV of the 16 MeV ($= c_v$) binding energy per nucleon; this is the origin of the 15 percent estimate given at the end of Sec. 3-13.

Another interesting problem arises if we introduce a surface into the nuclear matter problem; this is sometimes referred to as *semi-infinite nuclear matter.* Analogous calculations have been done with solids and liquids, allowing the estimation of such properties as the surface tension, the reflectivity of light, and the work function. Nuclear-matter calculations of this type yield values for c_a in (7-21) and for a in (1-1). Again comparisons with the empirically determined values from (7-23) and (1-2) give reasonable agreement considering the approximations used in the calculations.

As a result of the various calculations discussed in this section, it seems fair to state that there is a reasonable degree of understanding of how the binding energy, size, and shape of complex nuclei are determined by the nucleon-nucleon force and of how this force leads to the shell-theory potential from which the energy levels of nuclei were explained in Chaps. 5 and 6.

7-4 Magnetic Dipole Moments

It was pointed out in Sec. 1-6 that nuclei have a magnetic dipole moment μ, the vector sum of contributions from orbital motion, (1-7), and from spin, (1-8), with

$$\left.\begin{aligned} g_l &= 1 \\ g_s &= 5.585 \end{aligned}\right\} \quad \text{for protons} \qquad \left.\begin{aligned} g_l &= 0 \\ g_s &= -3.826 \end{aligned}\right\} \quad \text{for neutrons} \tag{7-25}$$

In a closed shell, all directions in space are equally represented by these vectors, so the total magnetic moment is zero. When two nucleons are paired, both their orbital and spin angular momenta are equal and opposite, so their total magnetic moment is again zero. Hence $\mu = 0$ for all even-even nuclei in their ground states,[1] and the total magnetic moment of a nucleus is derived from the unpaired nucleons.

If a state is a pure single-particle state, its magnetic moment is just that for a single nucleon in one of the orbits of Fig. 4-5; we call this μ_1. In a pure single-hole state, both the magnetic moment and the angular momentum are the negative of those for a single particle, but since the magnetic moment is defined by its value in the direction of the angular momentum, once again $\mu = \mu_1$. A single-quasi-particle (SQP) state spends part of its time as a single-particle state and

[1] Actually it can be proved rigorously that μ must be zero for any state with $I = 0$.

the rest of its time as a single-hole state, and in either case the magnetic moment is μ_1, so again for a SQP state we expect $\mu = \mu_1$. We found in Sec. 6-1 that the ground states of odd-A nuclei are principally SQP states, so their magnetic moments should be close to μ_1.

To calculate μ_1, we assume that the nucleus is in a state of $m = j$, so the total angular momentum is in the z direction; μ_1 is then given by

$$\mu_1 = \frac{e\hbar}{2M_p}(g_l l_z + g_s s_z) \tag{7-26}$$

where l_z and s_z are the z components of **l** and **s**. Unfortunately, as we know from Sec. 4-4, l and s are not good quantum numbers, so l_z and s_z are not constants of the motion; only j and m are good quantum numbers here. However, by use of (2-31) we can determine the expectation value $\langle s_z \rangle$ of s_z. It is[1]

$$\langle s_z \rangle = \begin{cases} \dfrac{m}{2l+1} & \text{for } j = l + \frac{1}{2} \\ -\dfrac{m}{2l+1} & \text{for } j = l - \frac{1}{2} \end{cases} \tag{7-27}$$

By taking $m = j$ and expressing l in terms of j we convert (7-27) to

$$\langle s_z \rangle = \begin{cases} \frac{1}{2} & \text{for } j = l + \frac{1}{2} \\ -\frac{1}{2}\dfrac{j}{j+1} & \text{for } j = l - \frac{1}{2} \end{cases} \tag{7-28}$$

In order to find μ_1 we can manipulate (7-26) as

$$\begin{aligned} \mu_1 &= \frac{e\hbar}{2M_p}(g_l l_z + g_s s_z) \\ &= \frac{e\hbar}{2M_p}[g_l(l_z + s_z) - g_l s_z + g_s s_z] \\ &= \frac{e\hbar}{2M_p}[g_l j + (g_s - g_l)s_z] \end{aligned}$$

The expectation value of μ_1, $\langle \mu_1 \rangle$, is then found by use of (7-28) as

$$\langle \mu_1 \rangle = \frac{e\hbar}{2M_p}[g_l j + (g_s - g_l)\langle s_z \rangle]$$

$$\langle \mu_1 \rangle = \begin{cases} \dfrac{e\hbar}{2M_p}[g_l(j - \frac{1}{2}) + \frac{1}{2}g_s] & \text{for } j = l + \frac{1}{2} \\ \dfrac{e\hbar}{2M_p}\left[g_l j - \frac{1}{2}(g_s - g_l)\dfrac{j}{j+1}\right] & \text{for } j = l - \frac{1}{2} \end{cases} \tag{7-29}$$

[1] The derivation of this result may be found, for example, in M. G. Mayer and J. H. D. Jensen, "Elementary Theory of Nuclear Shell Structure," p. 231, Wiley, New York, 1955.

Where the single nucleon is a proton, (7-29) with (7-25) gives (dropping the angle brackets)

$$\mu_{1p} = \begin{cases} \dfrac{e\hbar}{2M_p}(j + 2.29) & j = l + \frac{1}{2} \\ \dfrac{e\hbar}{2M_p}\left(j - 2.29\dfrac{j}{j+1}\right) & j = l - \frac{1}{2} \end{cases} \tag{7-30a}$$

If it is a neutron they give

$$\mu_{1n} = \begin{cases} -1.91\dfrac{e\hbar}{2M_p} & j = l + \frac{1}{2} \\ +1.91\dfrac{j}{j+1}\dfrac{e\hbar}{2M_p} & j = l - \frac{1}{2} \end{cases} \tag{7-30b}$$

Plots of (7-30) are shown by the lines in Fig. 7-3 for protons and in Figs. 7-4 for neutrons; they are known as *Schmidt lines* after their originator. Also shown

FIGURE 7-3 Magnetic moments of odd-Z, even-N nuclei in units of nuclear magnetons $e\hbar/M_p$. The lines are calculated from (7-30*a*), and the points are experimental determinations. *(From L. R. B. Elton, "Introductory Nuclear Theory," W. B. Saunders Company Philadelphia,* **1966;** *by permission.)*

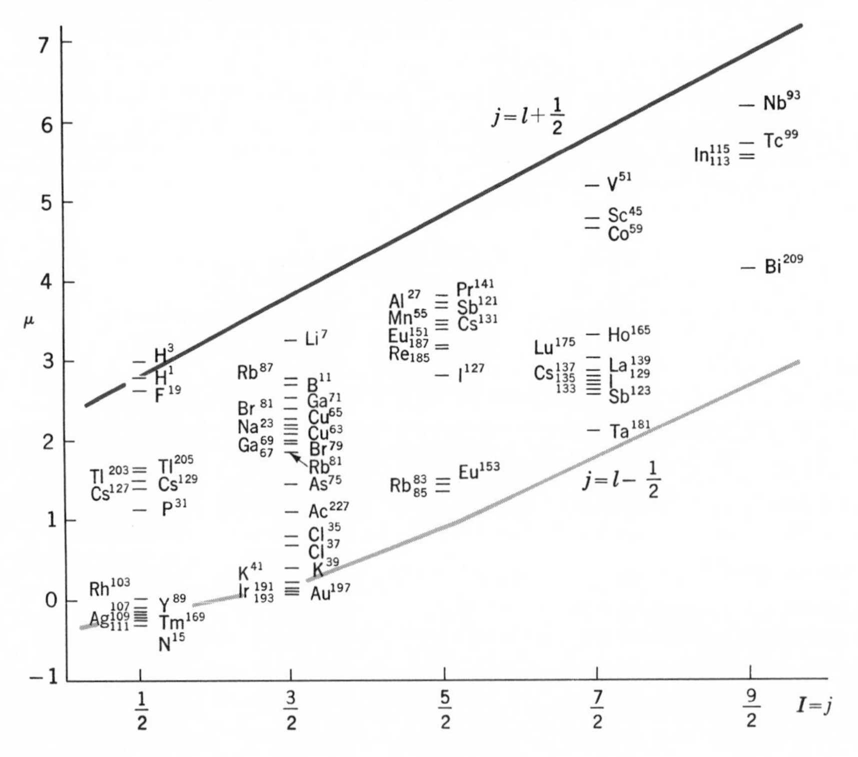

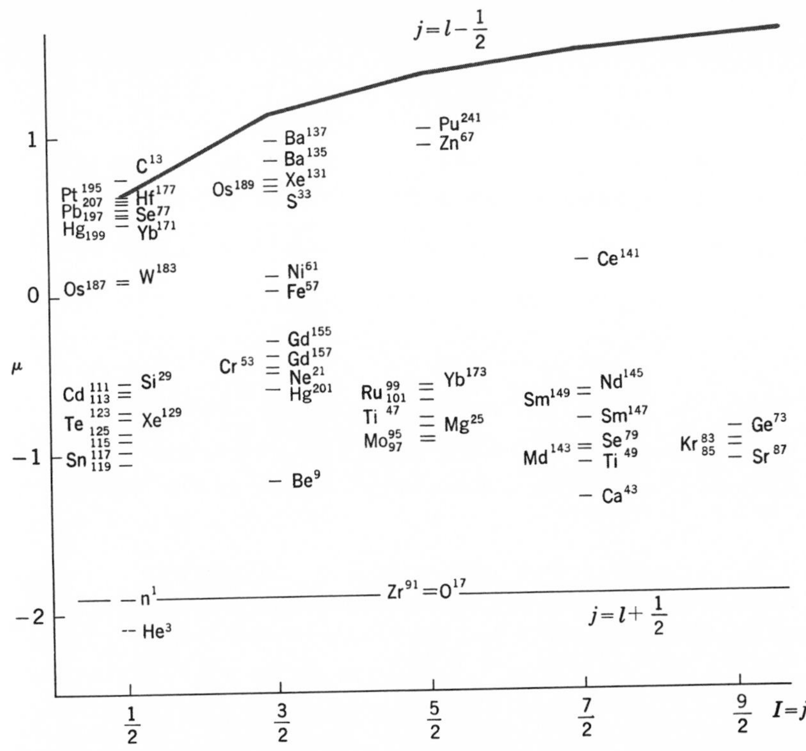

FIGURE 7-4 Magnetic moments of even-Z, odd-N nuclei in units of nuclear magnetons $e\hbar/M_p$. The lines are calculated from (7-30*b*), and the points are experimental determinations. *(From L. R. B. Elton, "Introductory Nuclear Theory," W. B. Saunders Company, Philadelphia,* **1966;** *by permission.)*

there are measured magnetic moments for the ground states of many odd-A nuclei. It is seen that the agreement is only qualitative, which indicates that the ground states of these nuclei are not pure SQP states. The measured values do lie closer to the predicted line than to the other line in all cases; for example, μ for Ca^{43}, whose ground state is principally an $f_{7/2}$ SQP state, lies closer to the $j = l + 1/2$ line ($7/2 = 3 + 1/2$), whereas μ for Sb^{123}, whose ground state is principally a $g_{7/2}$ SQP state, lies closer to the $j = l - 1/2$ line ($7/2 = 4 - 1/2$). Moreover, we do expect deviations from the predictions of (7-30) because we know from Sec. 6-1 that ground states of odd-A nuclei are not pure SQP states; Fig. 6-7 is an example of this type. However, the magnitude of the deviations in Fig. 7-3 and 7-4 may seem surprisingly large.

The explanation for this may be understood by advanced students through the following calculation. By use of (2-31) we can calculate the expectation value of the magnetic moment as

$$\langle \mu \rangle = \int \psi^* \left(\sum_k \mu_1{}^{(k)} \right) \psi \, d\tau \tag{7-31}$$

where[1] the sum is over all nucleons. If ψ is expanded as

$$\psi = c_0 \phi_0 + \sum_i c_i \phi_i$$

where ϕ_0 is the SQP state, substitution into (7-31) gives

$$\langle \mu \rangle = c_0{}^2 \int \phi_0^* (\Sigma \mu_1) \phi_0 \, d\tau + c_0 \Sigma c_i [\int \phi_0^* (\Sigma \mu_1) \phi_i \, d\tau + \int \phi_i^* (\Sigma \mu_1) \phi_0 \, d\tau] + \cdots \tag{7-32}$$

The first term in (7-32) is the contribution from the SQP state; for $c_0{}^2 = 1$, $c_i = 0$, (7-32) reduces to (7-30). For the case of Pd^{107} shown in Fig. 6-7, the nucleus spends 80 percent of its time as a SQP state, which implies $c_0{}^2 = 0.8$, whence, from (5-10), $\sum_i c_i{}^2 = 0.2$. Within these restrictions, the second term in (7-32) could be quite large. For example if it consists of 20 subterms with each $c_i = 0.1$ (note that this satisfies $\Sigma c_i{}^2 = 0.2$), the coefficient of the second term in (7-32) is much larger than that of the first (SQP) term. In practice, all c_i are not generally of the same sign, so the effect is not so extreme, but it is clear that large deviations from the SQP prediction are easily explained even though 80 percent of the time Pd^{107} is a SQP state.

One reason why the measured values lie between the Schmidt lines in Figs. 7-3 and 7-4 is that the dominant ϕ_i term is often one in which the odd nucleon is in the other state of the same l. For example, in Pd^{107}, which is largely a $d_{5/2}$ SQP state, the principal other term is the one from the coupling of the 2^+ state of the neighboring even-even nuclei with a $d_{3/2}$ SQP state. Since $d_{5/2}$ is on the $j = l + \frac{1}{2}$ and $d_{3/2}$ is on the $j = l - \frac{1}{2}$ Schmidt line, the actual value lies between the two lines. This effect has been verified by detailed calculations.[2]

For cases where the angular momentum is due to rotation of a nucleus which has no intrinsic angular momentum, as for the rotational states based on the ground states of ellipsoidal even-even nuclei, the magnetic moment can be derived from (1-6) with M becoming AM, the total mass of the nucleus, and e becoming Ze, the total charge. If all angular momentum is due to rotation, $L = \hbar I$ and $g = 1$. Using these in (1-6) gives

$$\mu = \frac{e\hbar}{2M} I \frac{Z}{A} \tag{7-33}$$

[1] $\int d\tau$ here is over the coordinates of all the nucleons in the nucleus and their spins.

[2] R. J. Blin-Stoyle and M. A. Perks, *Proc. Phys. Soc. (London)*, **A67:** 885 (1954).

In the few cases where μ has been measured for rotational states, the values obtained agree reasonably well with (7-33). In fact, (7-33) seems to apply also to magnetic moments of the first 2^+ state of spherical even-even nuclei. This is partially understandable in terms of the treatment of Sec. 6-7, where these states were found to be rotational to a considerable extent.

It may be recalled from Sec. 6-8 that the ground states of odd-A spheroidal nuclei owe some of their angular momentum to rotation of the nucleus as a whole. In determining their magnetic moments, (7-33) applies to these contributions, and it turns out that this causes their magnetic moments to fall between the Schmidt lines. Examples of this type are included in Figs. 7-3 and 7-4.

7-5 Electric Quadrupole Moments

The electric charge distribution in nuclei is of general interest because it is a measurable quantity and is rather simply interpreted as the density distribution of protons in the nucleus. It was shown in connection with (1-10) that this charge distribution can be described by a series of terms consisting of the total charge, the electric dipole moment, the electric quadrupole moment, etc.

From (1-10) the electric dipole moment D is

$$D = \int \rho z \, dV \tag{7-34}$$

Since the charge distribution in a nucleus arises from the probability distribution of the protons, ρ is proportional to $|\psi|^2$ for the protons. But since nuclear states have a definite parity, $|\psi|^2$ is an even function of z (that is, it has the same value for $+z$ as for $-z$) so multiplying it by z in (7-34) makes the integrand an odd function of z. The integral from 0 to $+\infty$ will therefore be equal and opposite to the integral from $-\infty$ to 0, so the entire integral vanishes. A nucleus cannot therefore have an electric dipole moment. Similar arguments require that all odd-numbered electric moments be zero, and it may similarly be shown that even-numbered magnetic moments are zero.

The lowest-order deviations from spherical symmetry in a nucleus are therefore detected by measurements of the electric quadrupole moment Q. From (1-10) it is defined as

$$\begin{aligned} Q &= \int \rho(3z^2 - r^2)\, dV \\ &= e \sum_{i=1}^{Z} (3z_i^2 - r_i^2) \end{aligned} \tag{7-35}$$

where the sum is over all the protons in the nucleus; or quantum-mechanically from (2-31)

$$\langle Q \rangle = e \sum_{i=1}^{Z} \int \psi^*(3z_i^2 - r_i^2)\psi \, d\tau \tag{7-35a}$$

In the quantum-mechanical treatment there is a complication in determining Q (we drop the angle brackets) because it must be related to some particular direction. The only externally detectable quantity having a definite direction in a nucleus is its total angular momentum I, so the measured value of the quadrupole moment depends on the orientation of the charge distribution relative to that of I. When $I = 0$, there is no direction that can be specified; all directions must be given equal weight, whence $Q = 0$. This quantum-mechanical "peculiarity" has manifestations for all values of I; it turns out, in fact, that also for $I = \frac{1}{2}$, Q must be 0.

The simplest situation where Q is interesting is for the ground state of a single-particle or single-hole nucleus. Since closed shells have spherical symmetry, the entire value of Q must be derived from the orbit of the odd nucleon, so the sum in (7-35) reduces to a single term. To get some estimate of Q due to a single nucleon, consider the latter to be moving in a circular orbit of radius r in the plane $z = 0$. Inserting $z = 0$ into (7-35) then yields

$$Q_{SP} = -er^2 \tag{7-36a}$$

The orbits of nucleons in the nucleus are not like classical orbits in being confined to a plane. This would correspond to the wave function's being restricted to a single value of ϕ, whereas we see from (2-25) that this is not the case. The accurate treatment of the problem by use of (7-35a) gives[1]

$$Q_{SP} = -\frac{2j-1}{2j+2}\,e\overline{r^2} \tag{7-36b}$$

where $\overline{r^2}$ is the average value of r^2 for the orbit; it is somewhat less than R^2 from (1-2). For large values of j, (7-36b) approaches (7-36a); for smaller j it is always less in magnitude as expected from the fact that orbits are not planar whence the z^2 term in (7-35) makes a contribution of opposite sign. For $j = \frac{1}{2}$, (7-36b) gives $Q = 0$, in agreement with the general rule mentioned above.

The expression (7-36b) should be valid for a single-proton nucleus, like $_{83}Bi^{209}$; for that nucleus, (7-36b) gives -0.25 e-barn (barn is the unit for 10^{-24} cm^2) whereas the measured value from Table A-2 is -0.35. For a single-neutron nucleus like $_8O^{17}$, Q should be zero since the electric charge of the neutron is zero. It turns out, however, that a neutron passing through a nucleus has a small effect on the motions of all the protons in the nucleus, tending to "drag them along," and as a result of their accumulated motions, a neutron in an orbit induces a quadrupole moment not much smaller than that calculated from (7-36b). For example,

[1] See footnote page 175.

for O^{17} (7-36*b*) gives −0.038, whereas the measured value is −0.026 *e*-barn. This effect also explains why a single proton gives a somewhat larger Q than is predicted by (7-36*b*), as we found above for Bi^{209}.

For single-hole nuclei, the nucleus is spherical except for a missing nucleon, whence the sign is reversed in (7-36*b*). An example of this is $_{19}K^{39}$, for which (7-36*b*) with sign reversed gives +0.046, which compares well with the measured value, +0.055 *e*-barn. From this effect we see that the sign of the quadrupole moment changes from plus to minus as the nucleon number goes through closed shells. It may be recalled from Sec. 4-8 that when either the neutron or proton number is near a closed shell, the nucleus is basically spherical, so in odd-A nuclei in this region, Q is determined principally by the odd nucleon. We therefore expect nuclei in which either the neutron or proton number differs by 1 from a closed shell to be governed by (7-36*b*), with the sign positive if there is a single hole in the closed shell and negative if there is a single particle.

In order to test this conclusion, measured quadrupole moments for odd-A nuclei from Table A-2 are plotted in Fig. 7-5 vs. the number of nucleons of the type which is odd; i.e., for odd-Z nuclei the abscissa is Z, and for odd-N nuclei the abscissa is N. We see there that the change in Q at closed shells from positive for holes to negative for particles is strikingly evident.

The largest and most easily interpreted electric quadrupole moments are those of spheroidal nuclei, where the relationship between Q and deformation was illustrated in Fig. 1-5*b*. If the electric charge is assumed to be distributed uniformly through the volume of the spheroid, a straightforward integration of the first of (7-35) leads to a classical quadrupole moment Q_0 as

$$Q_0 = \frac{3}{\sqrt{5\pi}} ZeR^2\beta \tag{7-37}$$

where β is the deformation parameter defined in Sec. 6-7. Due to the quantum-mechanical peculiarities mentioned above, the expectation value for a measurement of Q is

$$Q = \frac{3K^2 - I(I+1)}{(I+1)(2I+3)} Q_0 \tag{7-38}$$

For the ground state of an odd-A nucleus, $I = K$, whence (7-28) reduces to

$$Q = \frac{I}{I+1}\frac{2I-1}{2I+3} Q_0 \tag{7-39}$$

As an example, for $I = 5/2$, the coefficient in (7-39) is 0.36, whence, with (7-37), (7-39) becomes

$$Q \simeq 0.27Z\beta eR^2$$

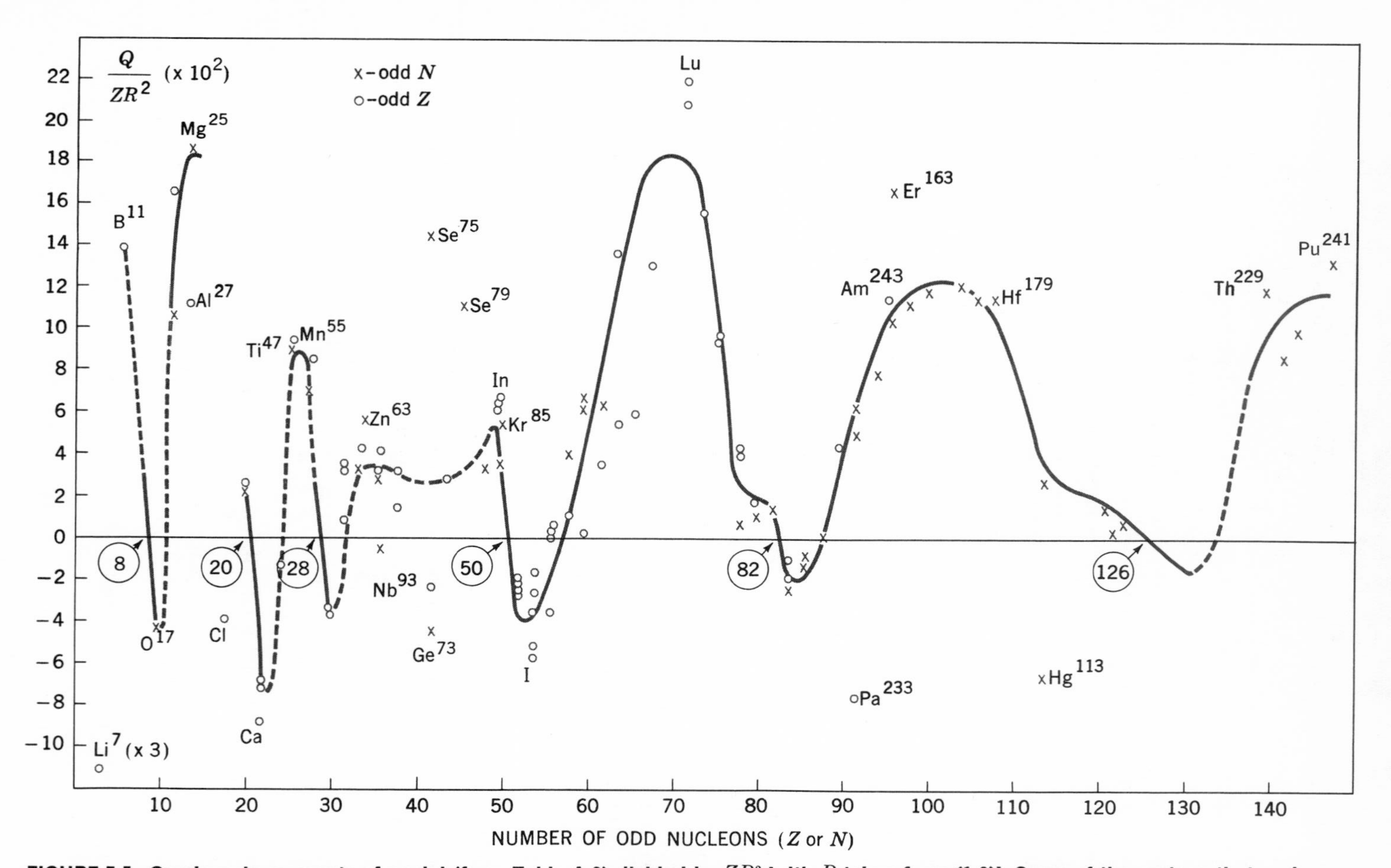

FIGURE 7-5 **Quadrupole moments of nuclei (from Table A-2) divided by ZR^2 [with R taken from (1-2)]. Some of the systematic trends are shown by the solid lines. Note the switch from positive to negative values at closed shells and the regions of large spheroidal deformation near odd nucleon numbers 70, 100, and 140. From (7-37) and (7-39) it is seen that the ordinate here, Q/ZR^2, is proportional to the deformation for large I.**

For a typical nucleus in the rare-earth region, $Z \simeq 65$, $\beta \simeq 0.3$, whence[1]

$$Q \simeq 5.3eR^2 \simeq 2.7 \text{ } e\text{-barns}$$

This is much larger than the single-particle value, (7-36*b*), which for this case is $\sim 0.5eR^2$. All the large quadrupole moments in Fig. 7-5 are from this type of nucleus. One can easily check that they occur in the regions specified in Sec. 4-8.

There are techniques available for studying the electric quadrupole moment of excited states, and much attention has been paid to the results for the first excited (2^+) states of spherical even-even nuclei; these are the states which we described in Sec. 5-9 as vibrations of the type shown in Fig. 5-10*a*. In that figure, the average shape is spherical, so $Q = 0$ is expected. However, measurements revealed that they have nonnegligible quadrupole moments, typically about 0.5 *e*-barn. This dilemma was resolved by calculations of the type discussed in Sec. 6-7; it was found that the midpoints of these vibrations differ appreciably from a spherical shape.

Problems

7-1 From the masses given in Table A-3, calculate B and B/A in MeV for Sn^{116}. Calculate the separation energies for a neutron and a proton.

7-2 Calculate the various terms in (7-21) for Sn^{116}. From them, determine B, B/A, and the mass. Compare the results with the values obtained in Prob. 7-1.

7-3 From (7-21), calculate the separation energy of a neutron from Sn^{116} and compare it with the result of Prob. 7-1.

7-4 Derive a formula for the surface tension of a nucleus and obtain a numerical estimate.

7-5 Compare the measured and calculated values of μ for Ge^{71}.

7-6 Estimate the magnetic moment for the three lowest-energy states of Er^{166}.

7-7 What is the electric quadrupole moment of Pb^{207}; of Sn^{117}?

7-8 What sign is expected (and why) for the electric quadrupole moment of K^{36}, $In,^{117}$ Sb^{125}, Co^{57}, and Cu^{61}?

[1] The nucleons at large radii have a heavily disproportionate influence on the quadrupole moment because, in (7-35), there is an r^2 in the integrand and dV contains a factor r^2. Hence the R in (7-37), which is the radius of the equivalent uniform distribution of nucleons, is larger than that in (1-2): it is conventional to use $R = 1.3A^{1/3}$ F for this purpose.

7-9 Estimate the quadrupole moment of Ca^{39} ($I = \frac{3}{2}$).

7-10 For Ta^{181}, $Q = 4.20$ e-barns. Estimate β for this nucleus ($I = K = \frac{7}{2}$).

7-11 With the help of Fig. 6-23 and Table A-2, estimate β for Hf^{177}.

Further Reading

See General References, following the Appendix.

Kopferman, H.: "Nuclear Moments," Academic, New York, 1958.
Nuclear Data, periodical published by Academic Press, New York.
Ramsey, N. F.: "Nuclear Moments," Wiley, New York, 1963.
Segre, E.: "Experimental Nuclear Physics," Wiley, New York, 1953.

Chapter 8

Nuclear Decay and Reaction Processes

In previous chapters, we have described a great many states in which a nucleus can exist. From what we have said so far, there is no reason why a nucleus should not remain in any state indefinitely. In fact, however, it does not, because there are processes in nature that induce transitions from one nuclear state to another. These fall into two categories: (1) those which occur spontaneously, referred to as *decay*, and (2) those which are initiated by bombardment with a particle from outside, called *reactions*. In this chapter we discuss some of the simple properties of these processes and develop a background for more detailed discussions of them in Chaps. 10 to 14.

When a process occurs spontaneously, changing a system from one state to another, conservation of energy requires that the final state be of lower energy than the initial state and that the difference in these energies be sent out of the system in some way. In all cases this is accomplished by energetic particles being emitted. These particles are rather easy to observe experimentally, and in fact it was their experimental discovery in the 1890s that initiated research in nuclear physics.

There are three basic types of decay: electromagnetic processes, beta decay, and nucleon emission. We discuss each of these in turn and consider some of their common properties. In the latter part of the chapter we shall introduce the subject of nuclear reactions.

8-1 Electromagnetic Decay Processes

When a nucleon is in a high-energy orbit while a low-energy orbit is unfilled, it can jump to the lower-energy orbit with the energy thereby released coming off as a quantum of electromagnetic radiation, which we call a *gamma ray*. This process is the same as the one in atoms where an electron jumps from a higher- to a lower-energy orbit with the emission of a quantum of light. The only distinction is that the energies of nucleon orbits, and therefore the differences in their energies, are much larger, of the order of MeV in nuclei vs. eV in atoms. Hence the energy of the emitted quantum is much larger, and from the familiar relationships between energy, frequency, and wavelength of quanta,

$$E = h\nu = \frac{hc}{\lambda}$$

the frequency is larger and the wavelength is proportionately shorter.

It is important to distinguish between the type of orbit change involved here and the orbit changing through collisions introduced in Chap. 5. The latter are part of the description of a state, whereas the former involve transitions from one such state to another. In the latter there is no loss of energy by the system; what energy changes do occur can be either positive or negative, but they are temporary and are governed by the limitations of the uncertainty principle (5-4). In the process under discussion here, on the other hand, the energy of the nucleus must decrease, and this energy loss is permanent. Since a nucleus must always be in a definite quantum state (as defined in Sec. 5-3), the only way for it to lose energy is to make a transition to a lower-energy state, e.g., from the first excited state to the ground state.

Gamma-ray emission results from an interaction of the nucleus with the electromagnetic radiation field, just as light emission results from an interaction of atomic electrons with that field. The existence of such fields, pervading all space and unobservable except when interactions occur, is an aspect of all quantum field theories (see Sec. 3-9). In the interaction, energy is transferred from the nucleus to the field, and this excitation of the field appears as a gamma ray. The nucleus can interact via the electromagnetic force not only with this field but also with atomic electrons. These electrons and the protons in the nucleus have electric charges, and both of these, as well as neutrons, have magnetic moments, so the usual coulomb and Biot-Savart force laws apply, allowing energy to be exchanged. One way for this energy to be exchanged is for a nucleon to drop from a higher-energy to a lower-energy orbit, with the energy lost being given to an atomic electron. In general, this is more than enough energy to overcome the binding of the electron to the atom, so the electron is emitted with a kinetic energy equal to the energy lost by the nucleus minus the binding energy of the electron. As in the case of gamma-ray emission, when the nucleus loses energy it must make a transition to a lower-energy state.

This process is called *internal conversion*. The name derives from the fact that when the phenomenon was first discovered, it was believed to occur when a gamma ray emitted from the nucleus of an atom knocked out an orbital electron from that atom by the photoelectric effect; the energy of the gamma ray was thus believed to be converted internally (within the atom) to energy of an electron. An interaction similar to internal conversion is familiar in atomic physics. Sometimes when an electron jumps from a higher-energy to a lower-energy orbit, instead of light being emitted, the energy is transferred to one of the outer electrons, causing it to be emitted. This is known as the *Auger effect.*

A third electromagnetic decay process is one in which, instead of a gamma ray being emitted, an electron-positron pair comes off. This is known as *internal pair formation;* it is important in only a very few cases, so we shall pay little attention to it here.

The electromagnetic decay processes, principally gamma-ray emission and internal conversion, do not change the number of neutrons or protons in the nucleus and therefore cause transitions between different states of the same nucleus. Through them, a nucleus in an excited state decays to a lower-energy excited state, which in turn decays to a still lower-energy excited state, etc., until eventually the ground state is reached. As we shall see, these transitions ordinarily take place within a very small fraction of a second, so as a consequence of electromagnetic decay processes nuclei are almost always found in their ground states.

8-2 Beta Decay

Consider the nucleus ${}_7N^{16}$, which contains seven protons and nine neutrons. In its lowest-energy state, according to Fig. 4-5, there must be a neutron in the $\mathfrak{N} = 3$ shell while there is a hole in the $\mathfrak{N} = 2$ shell for protons. It is apparent that the energy of the system could be very much lowered if the neutron could be changed into a proton, as it could then be in an $\mathfrak{N} = 2$ rather than in an $\mathfrak{N} = 3$ orbit. But nothing we have said up to this point would allow a neutron to change into a proton. Transformations must be induced by forces, and none of the forces of nature we have discussed so far—gravitational, electromagnetic, or nuclear—can change a neutron into a proton. However, it turns out that there is a force in nature that does cause such transitions. It is therefore the fourth basic force known to man; it induces transitions of the type

$$n \leftrightarrows p + e^- + \bar{\nu} \tag{8-1}$$

where e^- is a negative electron and $\bar{\nu}$ is a massless, chargeless particle called the *antineutrino.* This interaction also can induce any process derivable from (8-1) by exchanging a particle for its antiparticle (or vice versa) and putting it on the other side of the equation. One example of this type is

$$p \leftrightarrows n + e^+ + \nu \tag{8-2}$$

where e^+ (the positive electron) and ν (the neutrino) are the antiparticles of the e^- and $\bar{\nu}$, respectively. Either (8-1) or (8-2) can proceed to the right as a decay process provided it is energetically possible. (They can proceed to the left only in the extremely unlikely event that the three particles on the right arrive at the same place simultaneously with the proper energy and momentum to conserve those quantities.) Since the electron and neutrino (or their antiparticles) carry away energy, the energy of the nuclear system must be lowered in the process; energy conservation requires that it be lowered by at least an amount equivalent to the rest mass of the electron. In the case of N^{16}, the energy of the nucleus is lowered by (8-1), so we have the decay

$${}_7N^{16} \rightarrow {}_8O^{16} + e^- + \bar{\nu}$$

If we consider the nucleus ${}_9F^{16}$, which has nine protons and seven neutrons, the situation is reversed: here there is a proton in an $\mathfrak{N} = 3$ orbit while there is a vacancy for a neutron in an $\mathfrak{N} = 2$ orbit, so the energy of the nuclear system can be lowered by a decay of the type (8-2), which gives

$${}_9F^{16} \rightarrow {}_8O^{16} + e^+ + \nu$$

For historical reasons, the electrons emitted in these processes are known as *beta rays* and the process is called *beta decay*. When negative electrons are emitted, the process is designated β^-, and when positive electrons are emitted, the designation is β^+.

A third process that can be deduced from (8-1) by use of the stated prescription is

$$p + e^- \rightarrow n + \nu \tag{8-3}$$

Since orbital electrons are nearly always attached to nuclei and, as we know from atomic physics, spend some fraction of their time inside the nucleus, this process can proceed spontaneously if the energy of the system is thereby lowered. It is called *electron capture* (EC). From the standpoint of nuclear structure, (8-3) causes a proton to change into a neutron, which is the same as the effect of (8-2). These processes therefore are generally in competition with one another. However in (8-2) an electron must be created, so the energy of the nuclear system must be lowered by at least an amount equal to the rest mass of the electron, whereas in (8-3) an electron is destroyed, so the process can take place even if the energy of the nucleus is thereby *raised* by as much as the rest mass of the electron! This is understandable if we consider the whole atom as a single system; the energy of this complete system is still lowered by the decay.

The beta-decay processes do not change the number of nucleons in the nucleus, and they change the number of protons by 1, so they connect nuclei of the same A and of adjacent Z values. If we consider nuclei of the same A, the conditions for electron capture and β^- decay in terms of nuclear masses M' are

EC: $\quad M'_{Z+1} + M_e > M'_Z$

β^-: $\quad M'_Z > M'_{Z+1} + M_e$

These are simpler when written in terms of atomic masses M, where they become simply

EC: $$M_{Z+1} > M_Z$$

β^-: $$M_{Z+1} < M_Z$$

Thus, if we consider two atoms of the same A and adjacent Z, the one with the larger mass can always decay into the one with the smaller mass by either (8-1) or (8-3); only one of the two can be stable.

If the semiempirical mass formula (7-21a) is written as a function of A and Z, for a given A it takes the form $aZ^2 + bZ + c + \delta$, where a, b, and c are constants and δ is a different constant for odd-A, for even-even, and for odd-odd nuclei, as given by (7-17). These are the equations of parabolas, so we expect plots of M vs. Z for each of these three types of nuclei to be parabolic in shape with a minimum at some value of Z. Such plots are shown in Fig. 8-1a for $A = 135$ and in Fig. 8-1b for $A = 136$. Since the latter includes both odd-odd and even-even nuclei, there are two parabolas displaced in binding energy by 2Δ in accordance with (7-17).

Since beta decays change Z by one unit and lead to a decrease in the mass of the system, all the solid lines with arrows in Fig. 8-1 correspond to beta decays. Those which lead to an increase in Z correspond to a neutron changing into a proton by (8-1) and are labeled β^-, whereas those which lead to a decrease in Z represent the change of a proton into a neutron by (8-2) or by (8-3) and are labeled β^+ or EC; where the mass change is less than twice the electron rest mass (1.022 MeV), only (8-3) is possible, whence the latter label is used.

From Fig. 8-1a it is apparent that there can be only one stable nucleus for each odd A. All other nuclei of that A decay into it by one or more beta decays. From Fig. 8-1b, however, we see that there can be more than one stable nucleus for a given even A because beta decays here go successively from one parabola to the other. In the example shown there are three stable nuclei with $A = 136$, namely, Xe^{136}, Ba^{136}, and Ce^{136}. For lighter nuclei, the parabolas become steeper due to the A in the denominator of (7-16), so there is usually only one stable nucleus for each even A. An example of this type is shown in Fig. 8-2.

Figure 8-3 is a plot of Z vs. N for all nuclei indicating which nuclei are stable and which decay by various processes. The line through the stable nuclei in this figure is known as the *line of beta stability*. Its approximate equation could be obtained by expressing (7-21a) as a function of A and Z and then imposing the condition

$$\left[\frac{\partial M(Z,A)}{\partial Z}\right]_{A=\text{const}} = 0$$

From Fig. 8-3 we see that it follows the line $N = Z$ for light nuclei, where the symmetry energy term is dominant in (7-21), but then deviates from it as the

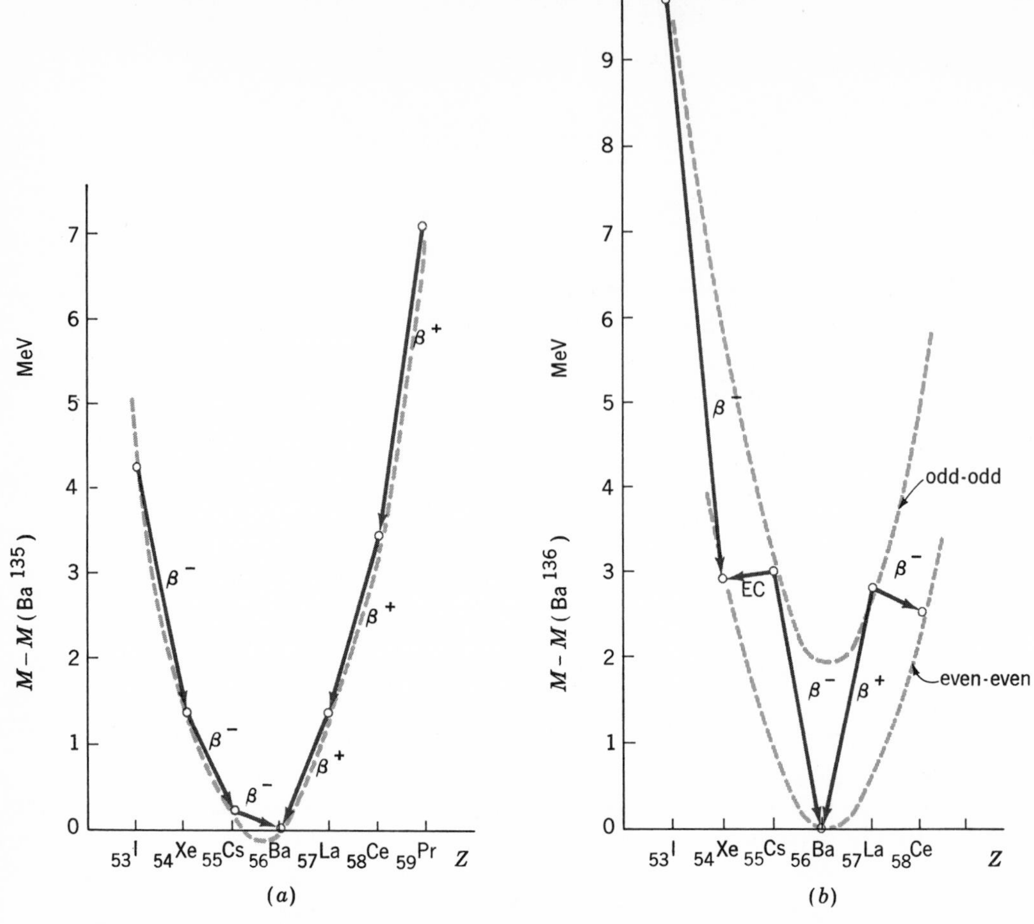

FIGURE 8-1 Masses of various nuclei of the same A plotted vs. their proton number Z: *(a)* $A = 135$; *(b)* $A = 136$. Solid lines with arrows indicate beta-decay transitions. Dashed parabolas are predictions of a formula like (7-21).

coulomb energy term becomes increasingly more important, causing stable nuclei to have more neutrons and fewer protons. Nuclei to the left and above the line of beta stability decay by β^+ or EC, and those to the right and below that line undergo β^- decay.

From Fig. 8-1*b* it is apparent that Xe^{136} and Ce^{136} would not be stable if there were a process available in nature by which two neutrons could change into two protons and vice versa. There apparently is no such direct process, but it should be possible for this to take place through (8-1) or (8-2) in two steps; this is called *double beta decay*. In view of the uncertainty principle, Xe^{136} can decay by (8-1) into Cs^{136} in spite of the fact that such a decay violates energy conservation provided that the violation lasts for only a very short time as given by (5-4).

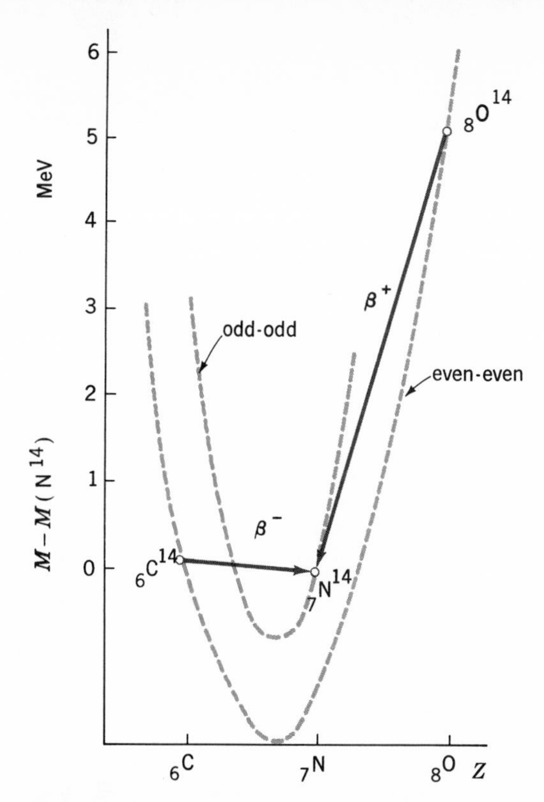

FIGURE 8-2 Masses of various nuclei of $A = 14$. Solid lines with arrows indicate beta-decay transitions. Dashed parabolas are predictions of a formula like (7-21).

This will be the case provided Cs^{136} decays into Ba^{136} by (8-1) within this time. The time required for a normal beta decay is generally many seconds, so the probability for a decay to take place in a time of the order of 10^{-21} s is very small. The decay rate in double beta decay is therefore so slow that it has not yet been observed experimentally.

There is another interesting process that can be derived from (8-1) by the prescription given after that equation, namely,

$$\bar{\nu} + p \rightarrow n + e^+ \tag{8-4}$$

This indicates that antineutrinos incident upon protons can induce a reaction in which a neutron and a positive electron are produced. This reaction was used in the first experiments in which neutrinos were detected experimentally.[1]

[1] F. Reines, C. L. Cowan, Jr., F. B. Harrison, H. W. Kruse, and A. D. McGuire, *Phys. Rev.*, **117**: 159 (1960).

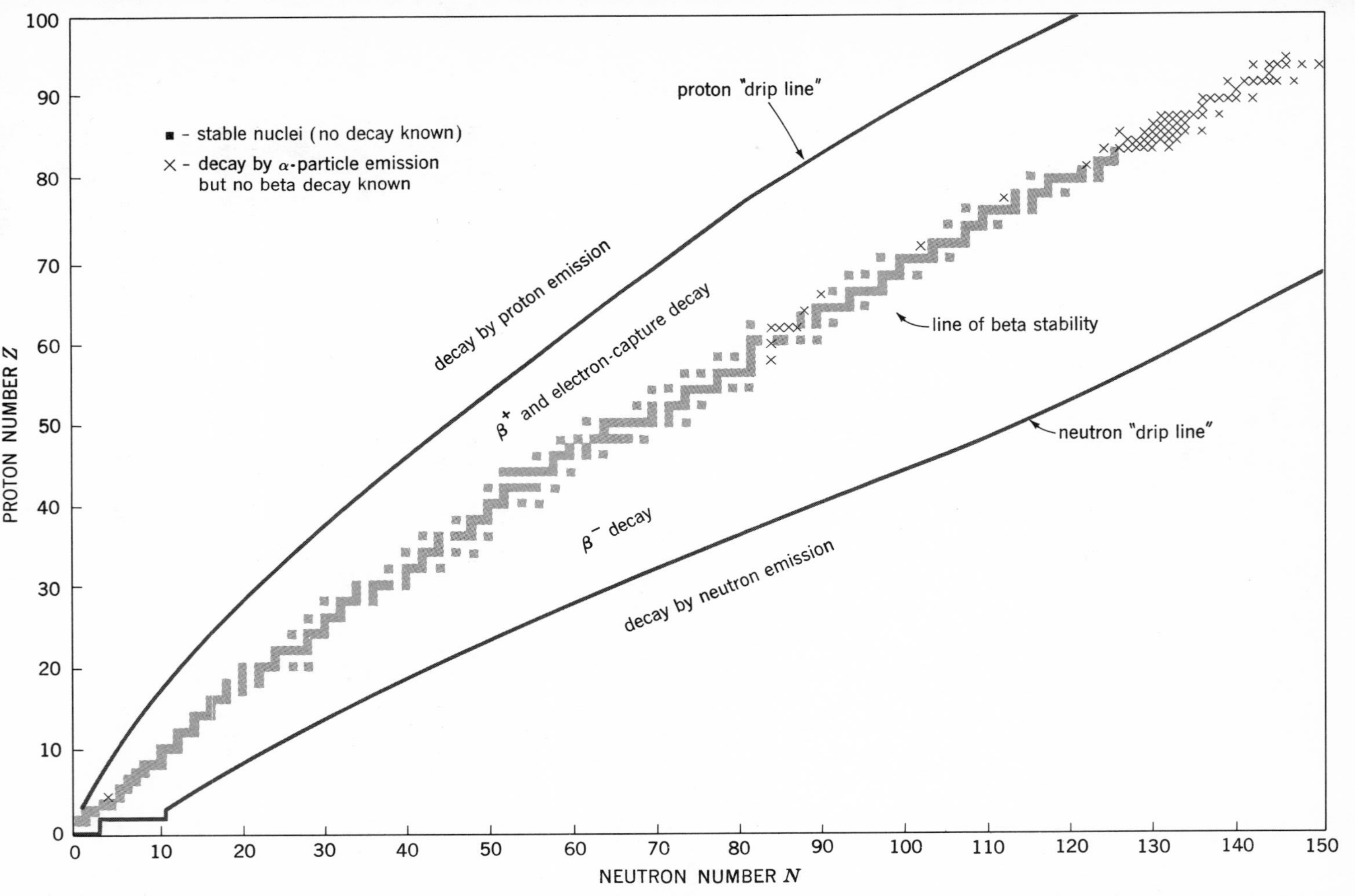

FIGURE 8-3 **Plot of Z vs. N showing which nuclei are stable** *(shaded)* **and which decay by various processes. The neutron and proton drip lines show roughly where nucleon emission becomes energetically possible and precedes beta decay. This figure is to a large extent a reduced version of the chart of nuclides discussed in Sec. 8-7, which may be consulted for more detailed information.** [*Drip lines from W. D. Myers and J. Swiatecki, Nucl. Phys.*, **81: 1 (1966)** *with odd-even differences averaged out.*]

8-3 Nucleon Emission

If a neutron or a proton has enough energy to escape from the nucleus, i.e., if it is in an orbit whose kinetic energy is larger than the depth of the shell-theory potential, it can come out of the nucleus. This is the process of *nucleon emission.* As shown in Sec. 4-6, the separation energy for even the least bound nucleons in stable nuclei is generally about 8 MeV, so emission of a single nucleon is usually energetically possible only when the nucleus is excited by about 8 MeV or more. The 8-MeV separation energy, however, is valid only in reasonably heavy nuclei near the line of beta stability. In light nuclei there are large variations; e.g., we found in Sec. 3-2 that the separation energy in the H^2 nucleus (the deuteron) is only 2.2 MeV. In $A = 5$ nuclei the first four nucleons are in $\mathfrak{N} = 1$ orbits while the fifth nucleon must be in an $\mathfrak{N} = 2$ orbit, so we expect its separation energy to be small; it turns out, in fact, to be negative (the energy of the lowest $\mathfrak{N} = 2$ orbit is above the top of the well), whence the beta-stable nucleus He^5 decays by neutron emission. There are hence no stable nuclei with $A = 5$, a fact which has important astrophysical consequences, as we shall see in Chap. 15.

From the discussion in Chap. 7, the separation energy S_n of a neutron from a nucleus in its ground state may be expressed as

$$\begin{aligned} S_n &= [M(Z, N-1) + M_n - M(Z,N)]c^2 \\ &= \left\{ -\left(\frac{\partial M(Z,N)}{\partial N}\right)_{Z=\text{const}} + M_n \right\} c^2 \\ &= \left(\frac{\partial [\cdot]}{\partial N}\right)_{Z=\text{const}} \end{aligned}$$

where $[\cdot]$ is the term enclosed by these brackets in (7-21*a*) written as a function of Z and N. When S_n becomes negative, neutron emission becomes energetically possible so these nuclei are unstable. From (7-21*a*) we see that the symmetry energy term increases rapidly with increasing N, causing $\partial[\cdot]/\partial N$ to become negative for nuclei far to the right of the line of beta stability in Fig. 8-3, whence these nuclei can decay by neutron emission. A rough estimate of the Z and N values where this first occurs is shown in Fig. 8-3; it is called the neutron *drip line.* It is clear from the above treatment and (7-21*a*) that this line should have odd-even differences, but they have been averaged out. For analogous reasons, nuclei far above the line of beta stability are unstable relative to proton emission; a rough estimate of the Z and N values where this first occurs is shown by the proton drip line in Fig. 8-3.

In addition to the emission of single nucleons, it is also possible for groups of nucleons to come off. When large amounts of excitation energy are available, e.g., after a nucleus has been struck by a high-energy particle, practically every type of light nucleus has been observed to be emitted. In most cases, emission of these nuclei requires even more energy than emission of a single nucleon; e.g., if a deuteron is to be emitted, the excitation energy must be at least the sum of the separation energies of a neutron and a proton less the binding energy of the deuteron. Since the former are 8 MeV each and the latter is 2.2 MeV, this is about $2 \times 8 - 2.2 \simeq 14$ MeV.

For emission of a He^4 nucleus, which, from Table A-3 has a binding energy of about 28 MeV, a similar calculation gives $4 \times 8 - 28 \simeq 4$ MeV as the energy required. This relatively low value, caused by the rather large binding energy of the closed-shell He^4 nucleus, allows the emission of He^4 nuclei, known for historical reasons as *alpha particles*, to occur much more frequently than the emission of deuterons or of other complex nuclei. It turns out, in fact, that in heavy nuclei, the energetics of alpha-particle emission is even more favorable than indicated by the above estimate. A more accurate estimate than (4-11) for the nucleon separation energy S_n may be obtained as

$$\begin{aligned} S_n &= B(A) - B(A-1) \\ &= \frac{dB}{dA} \\ &= A\,\frac{d(B/A)}{dA} + \frac{B}{A} \end{aligned} \tag{8-5}$$

The advantage of this form is that it can be evaluated directly from Fig. 7-1 by measuring the slope of the curve in that figure. For $A = 160$ and 240, $d(B/A)/dA$ are about $-1/120$ and $-1/100$, and B/A are 8.2 and 7.6, whence from (8-5), S_n and S_p are about 6.9 and 5.2 MeV, respectively. As the energy available for alpha-particle emission is approximately 28 MeV $- 2S_n - 2S_p$, this indicates that it is sometimes energetically possible for $A \simeq 160$ nuclei to emit alpha particles, and for $A \simeq 240$ nuclei, alpha-particle emission is energetically possible by typically 7 MeV for nuclei in their ground states. (Since alpha decay consists of the emission of two neutrons and two protons, it deviates appreciably from the line of beta stability in very heavy nuclei, as a consequence of which the last estimate should be reduced by about 2 MeV.) Alpha decay is consequently a common process among heavy nuclei. Nuclei which are stable against beta decay but which have been observed to decay by alpha-particle emission are indicated in Fig. 8-3 by crosses. We see there that the number of stable nuclei is limited at large A by this process.

There is one very important example of alpha decay in light nuclei, namely, ${}_4Be^8$, which decays into two alpha particles. It turns out that this is the only beta-stable nucleus of $A = 8$, so as a result of its alpha decay there are no stable nuclei with $A = 8$. This fact, as we shall see in Chap. 15, has very important consequences for the nature of our universe.

Another type of complex nucleus emission which is energetically possible even in the ground states of many nuclei is a breakup into two approximately equal parts, known as *fission*. As an example, let us consider the decay process

$$_{92}U^{238} \rightarrow {}_{46}Pd^{119} + {}_{46}Pd^{119} \tag{8-6a}$$

According to Fig. 7-1, B/A is about 7.6 MeV in U^{238} while it is about 8.4 MeV in Pd^{119}, so this process results in an energy *release* of $238 \times (8.4 - 7.6) = 190$ MeV. It is readily seen that similar (and in fact even greater) energy releases are obtained if the two decay products are not of equal mass, as, for example, in

$$_{92}U^{238} \rightarrow {}_{56}Ba^{144} + {}_{36}Kr^{94} \tag{8-6b}$$

Fission is energetically possible in any nucleus where the binding energy per nucleon is less than in a nucleus with half its mass. From Fig. 7-1 it is seen that this is the situation in every nucleus with A greater than about 100! The only reason nuclei with $A > 100$ exist in nature is that the *rate* of the fission decay process is very slow in them. This is but one example of the great importance of considering decay rates; they will be treated in considerable detail in Chap. 10.

8-4 Combinations of Decays

We shall find in Chaps. 10 to 12 that typical times required for nucleon emission, electromagnetic processes, and beta decay are about 10^{-18}, 10^{-12}, and 10^{2} seconds respectively. These times, which are the inverse of the decay rates, are indicative of the relative strengths of the basic forces causing them, the nuclear, electromagnetic, and beta-decay interactions respectively. They indicate that the nuclear force is the strongest of the three and that the beta-decay interaction is by far the weakest. For this reason, these two are commonly called the *strong interaction* and the *weak interaction*, respectively.

From the practical standpoint the difference in decay rates means that when a nucleus is highly excited, it will usually decay by nucleon emission until this is no longer energetically possible; it will then decay by gamma-ray emission (or internal conversion) until the ground state of the nucleus is reached; if this ground state is not stable, it will then beta-decay. If, as frequently happens, the beta decay goes to a state other than the ground state of the product nucleus, it is followed very shortly by further gamma-ray emission from transitions leading to that ground state. These gamma rays cannot come off until the beta decay occurs, so they are emitted at the rate of the beta decay.

In view of this time scale, beta decay, for all practical purposes, is limited to the ground states of nuclei (a few exceptions will be noted in Sec. 12-6). Similarly, gamma-ray emission does not occur very frequently when a nucleus is excited by more than about 10 MeV. When nuclei decay with an observable half-life, they are said to be *radioactive* because a sample of material containing these nuclei is active in emitting radiation. For example, U^{238} is radioactive since alpha particles are constantly being emitted from it.

8-5 Decay Laws

A fundamental property of transitions in any area of quantum physics is the transition rate λ, defined as the probability per unit time for the transition to occur. The estimation of transition rates will be the central theme of Chaps. 10 to 12; in this section we assume that transition rates are known and use them to derive some useful decay laws. Let us say we have a sample containing N nuclei that have a probability λ per unit time to decay. The number decaying per unit time, $-dN/dt$, is then

$$-\frac{dN}{dt} = \lambda N \tag{8-7}$$

This is a differential equation which is readily solved (by multiplying both sides by dt/N and integrating) to give

$$N = N_0 e^{-\lambda t} \tag{8-8}$$

where N_0 is the value of N at $t = 0$. By differentiating this we find the number of decays per unit time, and hence the number of particles emitted per unit time, to be

$$-\frac{dN}{dt} = \lambda N_0 e^{-\lambda t} \tag{8-9}$$

From (8-8) and (8-9) we see that both the number of radioactive nuclei and the rate of emission of decay particles decrease exponentially in time, approaching zero asymptotically.

From the definition of a logarithm, any number b can be written

$$b = e^{\ln b}$$

Raising both sides to the power x,

$$b^x = (e^{\ln b})^x = e^{x \ln b} \tag{8-10}$$

Taking x to be

$$x = -\frac{\lambda t}{\ln 2}$$

application of (8-10) gives

$$e^{-\lambda t} = e^{x \ln 2} = 2^x = 2^{(\lambda/\ln 2)\,t}$$

If we define

$$T_{1/2} = \frac{\ln 2}{\lambda} = \frac{0.693}{\lambda} \tag{8-11}$$

(8-8) and (8-9) become

$$N = N_0 2^{-t/T_{1/2}} \tag{8-8a}$$

$$\frac{dN}{dt} = \lambda N_0 2^{-t/T_{1/2}} \tag{8-9a}$$

When t increases by $T_{1/2}$, N and $-dN/dt$ decrease by half, whence $T_{1/2}$ is called the *half-life*. Each time a half-life passes, the number of radioactive nuclei, as well as the rate of decay-particle emission, is cut in half. The assumptions on which (8-8) and (8-9) are based, and therefore the decay laws, are valid for all nuclear decays regardless of half-life and for light emission by atoms and other atomic decay processes as well.

If a nucleus can decay by several different processes for which the probabilities per unit time are $\lambda_1, \lambda_2, \lambda_3, \ldots$, clearly the total probability λ per unit time for decay is

$$\lambda = \lambda_1 + \lambda_2 + \lambda_3 + \cdots$$

In view of (8-11), the half-life $T_{1/2}$ is then

$$\frac{1}{T_{1/2}} = \frac{1}{(T_{1/2})_1} + \frac{1}{(T_{1/2})_2} + \cdots$$

where $(T_{1/2})_1$ is what the half-life would be if only process 1 were available, etc. These latter are called *partial half-lives*. If one of them is much shorter than the others, it is dominant in determining $T_{1/2}$. The fraction of the decays going by process 1 is clearly λ_1/λ and similarly for the other processes, and this ratio is constant with time.

In some cases the residual nucleus after a decay is also radioactive with some different half-life. If the transition rates for the two decays are λ_1 and λ_2, the number of these nuclei at any time, N_1 and N_2, respectively, are clearly related by

$$-\frac{dN_1}{dt} = \lambda_1 N_1$$

$$\frac{dN_2}{dt} = \lambda_1 N_1 - \lambda_2 N_2$$

If $N_1 = N_{10}$ at $t = 0$, the solution of these coupled differential equations is

$$N_1 = N_{10} e^{-\lambda_1 t}$$

$$N_2 = N_{10} \frac{\lambda_1}{\lambda_2 - \lambda_1} (e^{-\lambda_1 t} - e^{-\lambda_2 t}) \tag{8-12}$$

From these solutions we can see that if the half-life of the *daughter* is much shorter than that of the *parent* ($\lambda_2 \gg \lambda_1$), after a time $t \gg 1/\lambda_2$,

$$N_2 = N_{10} \frac{\lambda_1}{\lambda_2} e^{-\lambda_1 t} \qquad t \gg \frac{1}{\lambda_2} \tag{8-13}$$

From (8-13) we see that the daughter eventually decays with the half-life of the parent. By comparing (8-13) with (8-12) we find that the ratio of the two types of nuclei stays constant at

$$\frac{N_1}{N_2} = \frac{\lambda_2}{\lambda_1} \tag{8-14}$$

If the half-life of the daughter is much longer than that of the parent ($\lambda_1 \gg \lambda_2$), the two decays are effectively decoupled; the first decay is finished before much happens in the second. An important example of (8-13) occurs when gamma-ray emission follows a beta decay. As we have stated above, the former ordinarily has a much shorter half-life than the latter, whence the gamma-ray emission occurs with the beta-decay half-life.

An important practical application of (8-14) is found in certain medical situations where it is advantageous to use short-half-life beta-decay activities. Since these ordinarily decay away very rapidly after they are produced, their use is impractical in most cases. However, if they are available as daughters of long-lived parent activities, the parent can be conveniently stored. Upon demand, the daughter can be chemically separated from the parent to give a pure short-lived radioactive source. In this application, the parent material is referred to as a *cow*, and the separation of the daughter is called *milking.*

In some situations, such as the decay of the naturally occurring uranium isotopes, there is a long series of decays from one nucleus to another. Treatments of this problem are found in standard textbooks.[1] If the first decay has a half-life much longer than the others, (8-13) and (8-14) are valid for any one of the subsequent decays.

Another closely related situation of interest is when radioactive nuclei are being produced at some constant rate P, as, for example, by nuclear reactions in an accelerator or a nuclear reactor. The rate of change of the number of radioactive nuclei, dN/dt, is equal to the difference between the production rate and the decay rate from (8-7), or

$$\frac{dN}{dt} = P - \lambda N$$

The usual initial condition is $N = 0$ at $t = 0$. The solution of this differential equation is

$$N = \frac{P}{\lambda}(1 - e^{-\lambda t}) \tag{8-15}$$

N asymptotically approaches a limiting value of P/λ; it reaches half of this value after 1 half-life, three-fourths of it after 2 half-lives, etc. In producing radioactive isotopes, it clearly does not pay to extend the production period over more than a

[1] See for example, R. D. Evans, "The Atomic Nucleus," McGraw-Hill, New York, 1955.

few half-lives. From (8-9) and (8-15) with $N = N_0$, the strength of a radioactive source immediately after the end of its production is

$$\frac{dN}{dt} = P(1 - e^{-\lambda t})$$

We see from this that the decay rate of source cannot exceed the rate at which the radioactive nuclei were produced. It approaches this rate if the production time is extended for several half-lives.

The strength of a radioactive source is traditionally given in curies (Ci); 1 Ci is 3.70×10^{10} disintegrations per second. Such units as the microcurie, millicurie, kilocurie, and megacurie, are also used with the obvious meanings.

8-6 Nuclear Reactions—Some Basic Properties

When a particle strikes a nucleus, the resulting interaction is spoken of as a *nuclear reaction.* Since in general the incoming particle brings in energy, energy is available for one or more particles to be emitted. It is standard notation to designate a nuclear reaction by enclosing the incoming and outgoing particles in parentheses separated by a comma. For example, an (n,p) reaction is one in which the incoming particle is a neutron and the outgoing particle is a proton. Neutrons incident on a nucleus can also induce (n,γ), (n,α), (n,d), etc., reactions. If the outgoing particle is identical with the incident particle, as in (n,n), or (p,p), the reaction is referred to as *scattering.* If the nucleus is left unchanged in the process it is called *elastic scattering,* whereas if the nucleus is left in an excited state it is called *inelastic scattering* and usually designated (n,n'), (p,p'), (α,α'), etc. When enough energy is available, more than one particle can be emitted in a nuclear reaction. For example, a $(p,2n)$ reaction is one in which two neutrons are emitted, a $(p,\alpha 2p3n)$ reaction is one in which an alpha particle, two protons, and three neutrons are emitted, etc. Reactions in which several particles are emitted, as in the last example, are called *spallation.*

In Sec. 7-1 we defined the energy release in a nuclear reaction Q and discussed how measurements of Q can be used to determine masses. In dealing with other aspects of nuclear reactions, the problem is turned around: one usually wants to calculate the Q value of the reaction from mass tables. This can be done by solving (7-4) for Q. For example, by use of Table A-3 we can calculate Q for the $Fe^{56}(d,p)Fe^{57}$ reaction as

$$\begin{aligned} Q &= [M(Fe^{56}) + M(H^2)] - [M(Fe^{57}) + M(H^1)] \\ &= [M(Fe^{56}) - 56 \text{ amu} + M(H^2) - 2 \text{ amu}] \\ &\qquad - [M(Fe^{57}) - 57 \text{ amu} + M(H^1) - 1 \text{ amu}] \\ &= (-60.605 + 13.136) \text{ MeV} - (-60.175 + 7.289) \text{ MeV} \\ &= 5.417 \text{ MeV} \end{aligned}$$

where the numbers followed by amu are the A values; they must cancel out to conserve the number of nucleons, so the second step can be omitted. It is therefore only necessary to replace each mass in the first equation by the entry opposite it in Table A-3. Note that we have used atomic masses rather than nuclear masses, but these differ only by the masses of the atomic electrons and there are equal numbers of electron masses on each side of the reaction equation. The result indicates that the total energy of all particles emitted in this reaction is larger by 5.417 MeV than the energy of the incident deuteron.

The Q value of some reactions is negative; an obvious example is the $Fe^{57}(p,d)Fe^{56}$ reaction, for which the Q-value calculation includes the same numbers as in the above example except that all signs are reversed, whence $Q = -5.417$ MeV. To some approximation this means that the reaction can be induced by protons of any energy above 5.417 MeV. However this is not exactly true since momentum must also be conserved in the reaction; this would not be possible if the incident proton had 5.417 MeV, as the emitted deuteron and Fe^{56} nucleus would then come off with zero energy and thus necessarily with zero momentum (see Prob. 8-7).

In nuclear reactions which lead to the emission of one outgoing particle plus a residual nucleus, the division of the available energy between these two is determined by momentum conservation. The problem is most easily handled in a coordinate system fixed to the center of mass of the interacting particles. This has the added advantage that the theory of nuclear reactions and formulas derived from it are developed for a system in which the center of mass is at rest. Treatments of the transformation between the center-of-mass and laboratory coordinate systems are given in many standard textbooks.[1]

When a particle is incident normally upon a thin sheet of material containing target nuclei with which it can react, the probability of a reaction is proportional to the number of target nuclei per unit area in the sheet. The proportionality constant has the units of area and is called the *cross section* σ. It may be visualized as the effective area a target nucleus presents to the incident particle for undergoing the reaction; the probability of the reaction is just equal to the probability that the incident particle strikes within this effective target area. Since the number of target nuclei per unit area is $n_t x$, where n_t is the number of target nuclei per unit volume in the material and x is the thickness of the sheet, the probability of a reaction is

$$\text{Probability} = n_t \sigma x \tag{8-16}$$

This result is valid only if the probability of a reaction is sufficiently small to ensure that the incident beam is not appreciably attenuated in passing through

[1] See page 198, for example, *ibid.*

the sheet. This attenuation effect can be taken into account and put to good use when dealing with the total cross section σ_T for all processes which remove particles from the incident beam. From (8-16), the beam intensity I is decreased by $-dI$ in going through a thickness of material dx according to

$$-dI = In_t\sigma_T\, dx$$

This is a differential equation whose solution is

$$I = I_0 \exp(-n_t\sigma_T x) \tag{8-17}$$

If the factors in the exponent are known, (8-17) can be used to compute the attenuation I/I_0; or σ_T can be determined experimentally by measuring this beam attenuation.

Since cross sections have the units of area, they may be expressed in square centimeters; a more convenient unit is the barn, defined as 10^{-24} cm^2. A rather extensive discussion of cross sections for various nuclear reactions will be presented in Chaps. 13 and 14.

8-7 Chart of Nuclides

A convenient way of displaying information about nuclei is in a chart form first used by Segre and lately kept current by Knolls Atomic Power Laboratory of the General Electric Co., under the sponsorship of the U.S. Atomic Energy Commission. Copies of this chart are available in most physics laboratories, and a personal copy may be obtained free of charge by writing to Educational Relations, General Electric Co., Schenectady, N.Y., 12305.

The chart is a plot of N vs. Z like the one in Fig. 8-3 except that each square denoting an individual value of N and Z is expanded in size, and in it is printed some of the most important information about the nucleus with that N, Z, including its mass, natural isotopic abundance, the I^π of its ground state, its half-life and the energies of its radiations if it is unstable, similar information on its isomers, etc. In addition to making this information readily available, the chart is useful in rapidly determining the radioactivities produced by nuclear reactions. For example, an (n,α) reaction decreases Z by two units and N by one unit, so it corresponds to moving down two squares and to the left one square. By use of the chart one can therefore very rapidly determine what radiations are produced by an (n,α) reaction on any nucleus. Analogously, the activities produced by an (n,γ) reaction on any target nucleus can be determined by shifting one square to the right, and similar use may be made of the chart for other types of reactions.

From an inspection of the chart, one quickly observes that beta decays generally have half-lives ranging from tenths of seconds to tens of years. This is a range that fits in well with times characteristic of everyday human experience. As a result, beta decays and the gamma decays that frequently follow them have many practical applications in human affairs. Some of them will be discussed in Sec. 15-1.

Problems

8-1 Construct diagrams like those in Fig. 8-1 for $A = 116$ and $A = 117$.

8-2 Use Table A-3 to determine how Be^7 decays.

8-3 List all processes that can be derived from (8-1) by use of the prescription given after that equation. Discuss their possible physical interest.

8-4 From the masses listed in Table A-3 calculate the energy of alpha particles emitted by Th^{232}.

8-5 Calculate the energy that would be released by fission into two equal parts of Zr^{90}, Sn^{116}, Er^{164}, and Pb^{200}.

8-6 Calculate the Q value for the $O^{16}(p,\alpha)$ reaction.

8-7 Calculate the minimum-energy protons that can induce the $Fe^{57}(p,d)Fe^{56}$ reaction. (*Hint:* This energy is a minimum when the deuteron comes off in the direction of the incident proton. Why?)

8-8 A source of Co^{60} originally had a strength of 1 Ci. What is its decay rate after 2 years?

8-9 A source of P^{32} originally contains 10 percent of P^{33}. How long must one wait to get a source from which 90 percent of the decays are from P^{33}?

8-10 A sample of Zr^{95} originally had a decay rate of 10^6 disintegrations per second. Plot the decay rate of its daughter, Nb^{95}, as a function of time.

8-11 A beam of 10^{13} protons per second produces Zn^{65} by the $Cu^{65}(p,n)$ reaction with a cross section of 0.5 barn. The target is pure Cu^{65}, 0.001 in. thick. How much Zn^{65} is produced in 1 h?

8-12 Suggest at least four nuclear reactions that can produce Co^{57}.

8-13 Neutrons of 15 MeV produce (n,p) $(n,2n)$, and (n,α) reactions. What activities are produced by 15-MeV neutrons on natural nickel?

8-14 If 1 g of cobalt is placed in a thermal neutron flux of 10^{12} neutrons/cm^2-s for 10 days, what is the strength in curies of the Co^{60} activity produced? The thermal neutron cross section for Co^{59} is 37 barns.

Further Reading

See General References, following the Appendix.

Feather, N.: "Nuclear Stability Rules," Cambridge University Press, Cambridge, 1952.

Mann, W. B., and S. B. Garfinkel: "Radioactivity and Its Measurement," Van Nostrand, Princeton, N.J., 1966.

Nuclear Data, periodical published by Academic Press, New York.

Segre, E.: "Experimental Nuclear Physics," Wiley, New York, 1953.

Chapter 9

Experimental Methods of Nuclear Physics

No treatment of nuclear reactions and decays can proceed far without reference to their experimental aspects. We therefore digress at this point to discuss the experimental methods of nuclear physics. This subject must be classed as one of the highlights of any nuclear-physics course; not only have these methods allowed us to investigate the nucleus, but because of their simplicity and great power they have found applications in many other fields of physics and in engineering, biology, medicine, etc. In fact it is probably in the area of experimental techniques that nuclear physics has made its greatest contribution to science.

9-1 Unusual Advantages in Nuclear Experimentation

There are few areas of physics where the "climate" is as favorable for experimentation as in nuclear investigations. This is due principally to three very convenient features: (1) nuclear energies are far higher than energies usually encountered in nature or in the laboratory, (2) nuclei generally carry an electric charge, and (3) half-lives for nuclear decays are often of convenient length.

One advantage from the high energy has already been mentioned in Chap 7: binding energies can be accurately determined by measurements of masses. Another advantage of dealing with energetic particles is that they pass easily through appreciable thicknesses of matter. In experiments where it is desirable for the particle energy to remain virtually unchanged in the traversal, as in determinations of cross sections at a certain energy or in accurate measurements of energies of emitted particles, the material can be made sufficiently thin without difficulty. Individual nuclear reactions and nuclear decays can therefore be studied in a very quantitative way; analogous experiments in other areas of physics are often extremely difficult.

Many other factors that traditionally lead to great experimental complication in other fields are negligible in nuclear physics because of the high energies involved. A photon of light typically carries an energy of a few electron volts, so light cannot disturb the nuclei under study. Because thermal energies are typically a small fraction of 1 eV, temperature effects are also negligible. The results obtained from experiments on nuclei in a dark room at subzero temperatures are no different than when they are in a glowing furnace. Stray electric and magnetic fields and gravity also have entirely negligible effects in most nuclear experiments. The behavior of a nucleus does not depend on whether it is in a free state or bound in a molecule or in a crystal; and it is not affected by atomic collisions. While the importance of the simplifications mentioned in this paragraph may not be fully appreciated by the student, they would be classed as extremely relevant by atomic, solid-state, or plasma physicists, who spend much of their working lives worrying about complicating effects from these sources.

The advantages deriving from the fact that most nuclei carry an electric charge embrace virtually all methods of accelerating or detecting nuclear particles. Their electric charge allows them to be accelerated by simply applying an electric field and allows their mass or energy to be determined by passing them through a magnetic field and measuring the resulting deflection. Thanks to this charge they exert electrical forces on the electrons in atoms as they pass; in the interaction, energy is exchanged, causing the nuclear particle to be slowed down and the electrons to be knocked out of the atoms. The slowing-down aspect gives an easy method of changing energies or of stopping particles, and the electrons that are knocked out provide a basis for many methods of detecting the nuclear particle and measuring its mass, charge, and energy. The fact that most nuclear particles have an electric charge means that a beam of these particles used in an experiment constitutes an electric current; since electric currents are easily and accurately measurable, the determination of beam intensity is a trivial problem.

The wide range of half-lives encountered in nuclear processes is a unique feature of this field of physics. In Chaps. 11 and 12 we shall explain how half-life measurements can be used to obtain information about nuclear structure. Most beta-decay and some alpha-decay half-lives, being measurable with clocks or even calendars, are an extra very easy measurement which can be used for identification purposes. In addition, the long lifetimes often allow chemical separations to be made; these have been invaluable in simplifying experimental situations and were, for example, responsible for the discovery of fission and for much of the early work on that process. Residual nuclei from nuclear reactions usually decay by beta emission, whence the easily measurable half-lives serve as indicators for the reactions and can be used to determine their cross sections. On the other hand, if the cross section for a reaction is known, a measurement of the amount of induced beta activity gives a determination of the intensity of the incident-particle beam. We shall see in Chap. 13 that because different nuclear reactions are induced by different energies of incident particles, a measurement of the relative numbers of different reactions, made by determining the amount of beta decay induced by each, can be used to find the energy distribution of a beam of particles. These techniques are especially useful for neutron beams, which are not easily measured or analyzed by other methods. By use of long-lived activities, sources of alpha particles, electrons, gamma rays, and fission fragments are readily made available to experimenters; by combining an alpha-particle or gamma-ray source with materials with reasonable (α,n) or (γ,n) cross sections, neutron sources are easily made. All these sources are sometimes the core of an experiment, but in nearly all experiments they are of inestimable value in testing and adjusting experimental apparatus.

In the remainder of this chapter we describe the physical principles behind the most important experimental techniques of nuclear physics. We avoid technical and engineering details and omit techniques which are basically nonnuclear (e.g., atomic beam measurements) but even within these limitations, our discussion must necessarily be brief and incomplete. Interested students are therefore encouraged to do supplementary reading in areas not covered here in sufficient depth to suit their particular interests.

9-2 Interaction of a Charged Particle with Matter

In the last section we pointed out the importance of the fact that an energetic electrically charged particle passing through matter loses energy by collisions with electrons, and we shall see more applications of it in Chap. 15. In this section, we treat this process in more quantitative detail. We consider a particle of kinetic energy E, velocity v in the x direction, and charge ze passing through matter composed of n_1 atoms per unit volume of atomic number Z. If the trajectory of the particle passes an electron at a minimum distance q, the distance between them in the position shown in Fig. 9-1 is $q/(\sin\theta)$, whence the force is $F = (ze^2 \sin^2\theta)/4\pi\epsilon_0 q^2$. The net impulse given to the electron as the particle passes is in a direction perpendicular to its path and is therefore equal to the force

FIGURE 9-1 Interaction between an energetic charged particle (solid circle) and an electron. The former is moving in the x direction, as indicated by the arrow.

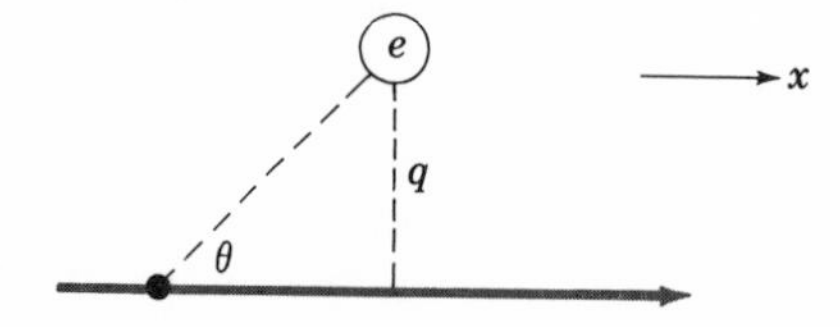

in this direction times the time of interaction, or $F \sin \theta \, dt$. This impulse is the momentum p_e transferred to the electron, whence we have

$$p_e = \int_{-\infty}^{\infty} F \sin \theta \, dt \tag{9-1}$$

But $dt = dx/v = q \, d\theta/(v \sin^2 \theta)$. Using this and the above expression for F in (9-1), the integral can be evaluated to give

$$p_e = \frac{ze^2}{2\pi\epsilon_0 qv}$$

The kinetic energy T_e given to the electron is then

$$T_e = \frac{p_e^2}{2M_e} = \frac{z^2e^4}{8\pi^2\epsilon_0^2q^2M_ev^2} \tag{9-2}$$

This energy must come from the energy of the charged particle, whence the energy of the latter is reduced. In traveling a path length dx ($dx \gg q$), the number of electrons it passes at a distance between q and $q + dq$ is $2\pi q \, dq \, dx$ times the number of electrons per unit volume, which is n_1Z. Its energy loss due to interactions with these electrons is therefore

$$-dE = \frac{z^2e^4n_1Z}{8\pi^2\epsilon_0^2M_ev^2} \frac{2\pi q \, dq}{q^2} \, dx$$

This must be integrated over q to give, after some rearrangement of terms,

$$-\frac{dE}{dx} = \frac{z^2e^4n_1Z}{4\pi\epsilon_0^2M_ev^2} \ln \frac{q_{\max}}{q_{\min}} \tag{9-3}$$

where $q_{\max}$ and $q_{\min}$ are the upper and lower limits of the integration. The minimum distance that is meaningful in this problem is the electron wavelength as viewed by the passing particle, which is

$$q_{\min} = \frac{h}{M_ev}$$

The maximum distance at which energy can be transferred is that at which the time of interaction is of the order of the period of the electron in its atomic orbit. The former is about q/v (see Fig. 9-1), and the latter, from the usual solution of the time-dependent Schrödinger equation, is h/I, where I is the energy of the electron orbit, better known as the *ionization potential;* equating these gives

$$q_{\max} = \frac{hv}{I}$$

With these values, the argument of the logarithmic term becomes

$$\frac{q_{\max}}{q_{\min}} = \frac{M_ev^2}{\bar{I}}$$

where $\bar{I}$ is the average ionization potential of all the electrons. If this is inserted into (9-3), the result differs from that obtained from an accurate and complete quantum-mechanical treatment[1] by a factor of 2 in the argument of the logarithm. The correct result, including relativistic corrections, is

$$-\frac{dE}{dx} = \frac{z^2 e^4 n_1 Z}{4\pi\epsilon_0{}^2 M_e v^2}\left[\ln\frac{2M_e v^2}{\bar{I}} - \ln\left(1 - \frac{v^2}{c^2}\right) - \frac{v^2}{c^2}\right] \tag{9-4}$$

This is the rate at which a charged particle loses energy as it passes through matter. The average ionization potential $\bar{I}$ is approximately 13 eV times Z, except for the very lightest atoms, where it is somewhat higher. For a 5-MeV proton $2M_e v^2 = 10{,}000$ eV, whence the bracketed term varies from about 2.3 for heavy atoms to 4.5 for air and 6.2 for hydrogen; for a 20-MeV proton it is larger than these by $\ln 4 \simeq 1.4$.

Plots of (9-4) for protons as a function of their energy are shown in Fig. 9-2.

[1] For a more complete treatment see R. D. Evans, "The Atomic Nucleus," McGraw-Hill, New York, 1955.

FIGURE 9-2 Rate of energy loss of protons of various energies in passing through C, Cu, and Pb. Interpolations can be made for materials of other atomic numbers Z. These curves may be used for other particles if E is divided by their mass M and if dE/dx is divided by z^2, where z is their charge. Note that the distance is in milligrams per square centimeter, which is actually ρx, the distance times the density of the material. *(From H. A. Enge, "Introduction to Nuclear Physics," Addison-Wesley Publishing Company, Inc., Reading, Mass.,* **1966;** *by permission.)*

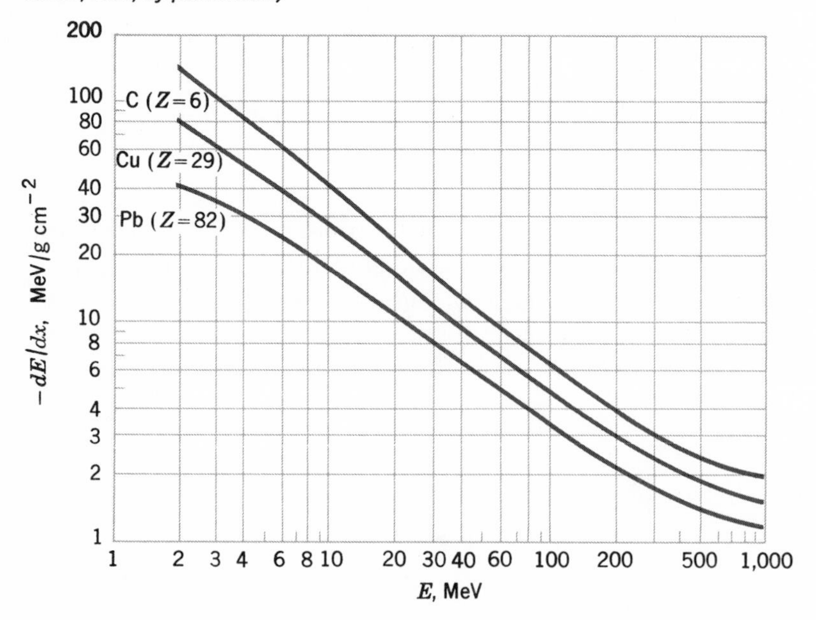

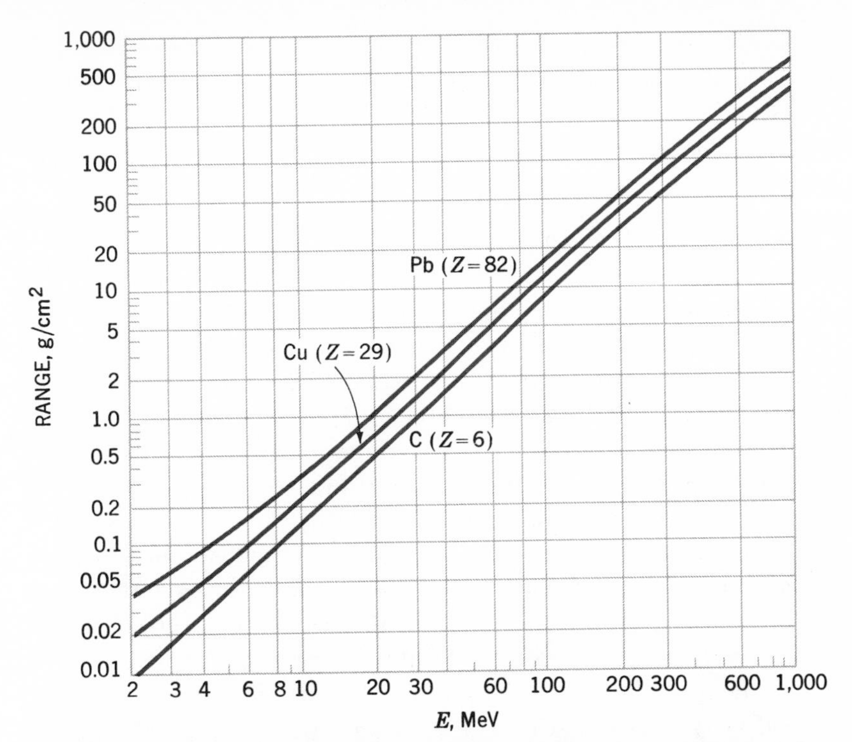

FIGURE 9-3 Range of a proton in C, Cu, and Pb. Interpolations can be made for materials of other atomic numbers. These curves may be used for other particles if their energy is divided by M and the range is divided by z^2/M. *(From H. A. Enge, "Introduction to Nuclear Physics," Addison-Wesley Publishing Company, Inc., Reading, Mass.,* **1966;** *by permission.)*

The rate of energy loss for other charged particles of the same velocity, i.e., with the same E/M, is larger by a factor z^2. If the energetic charged particle is an electron, a few other complications enter; treatments of that problem are available in standard textbooks.[1]

The factor n_1 in (9-4) can be expressed as

$$n_1 = \frac{N_A}{MW}\rho$$

where N_A is Avogadro's number, MW is the molecular weight, and ρ is the density. It is usually convenient to divide both sides of (9-4) by ρ; the right side then depends only on the material and not on its temperature or physical state, and x on the left side is converted to ρx, the mass per unit area, which is usually easier

[1] See, for example, *ibid.*

to measure experimentally (by a weight and area determination) than x itself, especially when x is very small. It may be noted that the ordinate in Fig. 9-2 is given in terms of ρx.

The *range* of a charged particle $R(E)$ is the distance it travels before losing all its energy, which may be expressed as

$$R(E) = \int_E^0 \frac{dE}{-dE/dx}$$

This integral can readily be evaluated numerically with (9-4), and curves of the results are widely available. Some examples are shown in Fig. 9-3.

It is important to note that the process of energy loss consists of a great many interactions. The average loss in a single interaction is a few times $\bar{I}$, which is typically hundreds of electron volts. Particles with several MeV of energy must therefore undergo thousands of such interactions before being stopped. Each of these interactions is, of course, subject to large statistical fluctuations in the energy transferred, and the rate at which interactions occur is also subject to statistical variations. However, the number of interactions is so large that these fluctuations average out, whence the range of all particles of a given charge, mass, and energy is quite similar. Measurements of range can therefore be used for energy determinations. There is, however, some straggling in these ranges due to statistical fluctuations, and it is large enough to be important in many experimental situations.[1]

9-3 Detectors for Energetic Charged Particles

As a result of the interactions with electrons discussed in the last section, the path of an energetic charged particle through a material is strewn with free electrons and positively charged ions (atoms with a missing electron). We pointed out that there are thousands of such *electron-ion pairs* formed directly by the charged particle, but many of the electrons have sufficient energy to produce further ionization by the same type of interaction, so eventually there is an electron-ion pair for about every 30 eV of energy loss by the original charged particle. If we accept this figure (which is typical for a gas; it varies considerably with the material), the path of a 3-MeV proton is lined with about 100,000 electron-ion pairs.

There are three principal methods of using this ionization to detect particles: (1) the ionization can be made visibly observable so that the tracks of particles can be seen or photographed; (2) light emitted when these electron-ion pairs recom-

[1] For an elaborate treatment of straggling see B. Rossi, "High Energy Particles," Prentice-Hall, New York, 1952.

bine can be detected and measured by a photosensitive device; and (3) by applying an electric field, the electrons and ions can be collected, thereby producing an electric signal.

Detectors Which Make Tracks Visually Observable

If air is supersaturated with water vapor, droplets of water form, but an interesting question is: Where? In clouds, it turns out that water droplets form on dust particles, but another favorable place for this condensation to take place is on ionized atoms. This is the basis for the *cloud chamber.* A vapor in a chamber is made supersaturated by cooling it either by suddenly expanding its volume, as in an *expansion chamber*, or directly with a coolant, as in a *diffusion chamber.* The supersaturated vapor condenses as droplets on the ions which lie along the path of the charged particle, and with proper illumination these droplets can be seen or photographed.

A closely related technique involves the use of a superheated liquid in a *bubble chamber.* A superheated liquid will boil, but an interesting question is: Where will the bubbles form? If dust is present, they will form on dust particles, but another favorable place is on ionized atoms. In a bubble chamber, a liquid is made ready to boil by suddenly releasing the pressure on it; bubbles form on the ions which are strewn along the paths of energetic charged particles, and they can be observed visually or photographed. Since a liquid is much denser than a gas, there is far greater opportunity for particles to have interactions in a bubble chamber than in a cloud chamber, whence the former is generally much more useful.

A third medium in which ionization can be made visually observable is a photographic emulsion. Ionization renders grains of silver halide developable (in fact this is the basis of ordinary photography, where the ionization is produced by light), whence upon development the tracks of energetic charged particles are marked out by developed grains which can be observed under a microscope.

A fourth device in which particle tracks can be made visually observable is a *spark chamber*, which consists of a number of parallel metal plates or wires. When a high voltage is applied between each pair of adjacent plates, sparks jump across; by far the most likely place for them to jump is where the gas that fills the space between them has been ionized by an energetic charged particle. The path of the latter is therefore marked by a series of sparks, which can be viewed or photographed, or the information about their positions may be coded into an electric signal suitable for computer analysis. Spark chambers have an advantage over the other methods mentioned: since the voltage across the plates or wires can be applied for a very short time ($\sim 10^{-7}$ s), only events originating at the same time, and therefore very probably from the same reaction, are observed.

The application of this voltage can also be accurately timed relative to events taking place in other detectors where this time relationship is of interest.

Scintillation Detectors

In a crystal of solid material, the interaction between an energetic charged particle and the electrons results in the latter being knocked out of their sites in the crystal lattice. When an electron falls back into these *vacancies*, light is emitted and some crystals are transparent to this light. The passing of an energetic charged particle is therefore signaled by a *scintillation* of light emitted from the crystal. In a scintillation detector (Fig. 9-4) this light is transformed into an electric pulse by

FIGURE 9-4 Schematic diagram of a scintillation detector. The glass envelope is sealed and evacuated. The resistor chain is to establish a voltage gradient between the photocathode and the first dynode, between each pair of consecutive dynodes, and between the last dynode and the anode; electrons are accelerated between each of these by the resulting electric fields. The external circuitry is explained in the text.

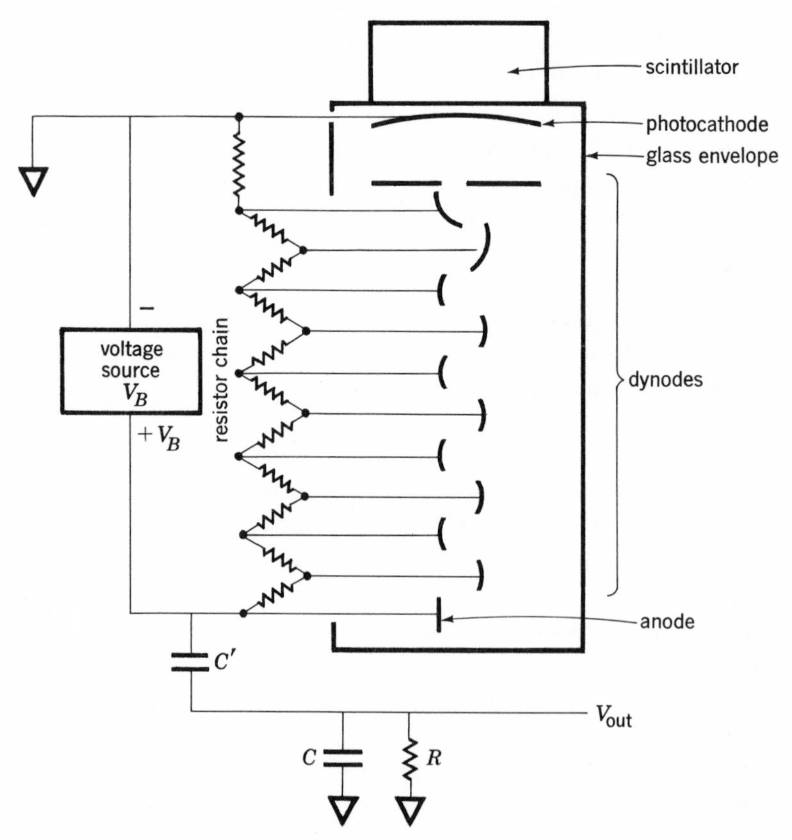

allowing it to strike a surface with a large cross section for a photoelectric effect, the process in which an incident photon knocks an electron out of an atom. This surface is the *photocathode* of a device known as a *photomultiplier*. In it, the electrons emitted by the photocathode are accelerated by an electric field and driven into a *dynode*, a plate with a surface from which electrons are easily knocked out. Each electron arriving at the dynode knocks out about three or four electrons, depending on the energy given to it by the electric field. The electrons emitted from this dynode are then accelerated toward a second dynode, from which each knocks out several electrons, and the process is repeated several times with the number of electrons being multiplied by three or four at each dynode; typical photomultipliers have 6 to 14 stages of this type. Electrons from the last dynode (total charge Q) are collected by a plate (called the *anode*), from which they flow onto capacitor C', thereby inducing an equal charge on capacitor C; this produces a voltage at the output, $V_{\text{out}} = Q/C$.

The process by which electrons drop into the vacancies in the scintillation crystal is a decay governed by (8-9), whence the number of emitted photons and hence eventually the charge Q is given by the integral of that equation as

$$\begin{aligned} Q(t) &\propto \int_0^t \lambda_v e^{-\lambda_v t}\, dt \\ &\propto 1 - e^{-\lambda_v t} \end{aligned}$$

where λ_v is the decay constant for vacancies in the crystal. Since Q is proportional to V_{out}, the output pulse is of this form, whence it has a rise time[1] of $2.2/\lambda_v$. The charge on capacitor C eventually flows off through resistor R with a time constant RC, so the output pulse is as shown in Fig. 9-5. The capacitor C' in Fig. 9-4 is inserted to remove the dc voltage from the output.

[1] The rise time of a pulse is conventionally defined as the time for it to rise from 10 to 90 percent of its final value. We shall adhere to this definition only roughly.

FIGURE 9-5 Pulse shape (output voltage vs. time) from a scintillation detector. The actual pulse, shown by the solid curve, is the product of the two dashed curves.

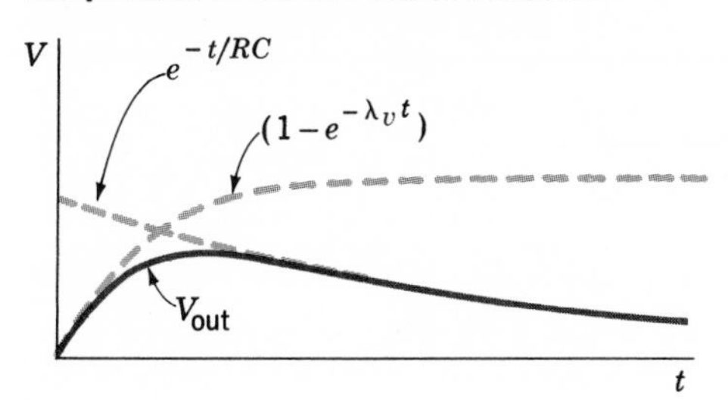

The size of the output pulse is proportional to the number of electrons emitted from the photocathode, which is proportional to the number of photons emitted in the scintillation, which is proportional to the number of electrons knocked out of their lattice sites, which is proportional to the energy lost in the crystal by the energetic charged particle. If the latter is stopped in the crystal, the size of the electric pulse produced is therefore proportional to its energy, whence we have here not only a method of detecting energetic charged particles but of measuring their energy.

There are two basic types of scintillators, organic and inorganic. The former include a variety of compounds (e.g., stilbene, *p*-terphenyl) which are available as crystals, in liquid solutions, or incorporated into plastics. They have the advantage of large λ_v, which gives fast rise times for output pulses (typically 3 to 30 ns†), but unfortunately the relationship between their light output and the energy deposited by the charged particle is nonlinear. This difficulty does not occur in most inorganic scintillators, of which NaI(Tl) is the most widely used example,[1] and these also have the advantage of containing high-atomic-number elements, which are most efficient for gamma-ray detection. Their principal disadvantage is their small λ_v, which causes slow output-pulse rise times [250 ns for NaI(Tl), longer for all others]. In most cases, NaI(Tl) is used for gamma rays, and plastic scintillators are used for neutrons.

Charge-collection Detectors

Detectors in which an electric field is applied to collect the charge are of two general types, *gas* and *semiconductor*. In the former, the energetic charged particle produces electron-ion pairs in a gas filling the region between two metal electrodes, across which a voltage is applied. The electric field produced by the voltage exerts a force on the electrons and ions, accelerating them toward the positive and negative electrodes, respectively. Their motion may be described as a *drift* because it is interrupted by frequent collisions with gas molecules, in which much of their velocity is lost; the average drift velocity v_d varies with the electric field strength, the type of gas, and its pressure but is typically about 10^7 cm/s.

The formation of the output pulse in a parallel-plate geometry (this type of detector is called an *ionization chamber*) may be understood with the help of Fig. 9-6, in which the detector itself is outlined by the dashed line. The voltage is applied by a voltage source V_B through a large resistor R_B; the electric field

† A nanosecond (ns) equals 10^{-9} s.

[1] NaI(Tl) is sodium iodide with a small amount of thallium, which absorbs the light and then emits light of a longer wavelength which is better matched to the photomultiplier response.

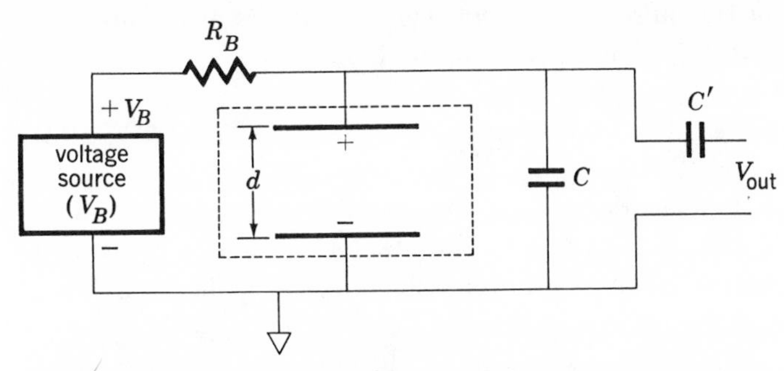

FIGURE 9-6 Schematic diagram of an ionization chamber or a semiconductor detector. The detector is the part surrounded by the dashed line, and the remainder is external circuitry explained in the text. In an ionization chamber, the region between the positive and negative electrodes is filled with a gas; in a semiconductor detector it is the depleted region in a back-biased p-n junction.

strength is V_B/d. In the steady state, the capacitor carries a charge $Q = CV_B$. When electrons drift toward the positive terminal of the detector, charge is induced on that plate. Because of the large value of R_B, very little charge per unit time, i.e., current, can come from V_B, so instantaneously this charge can only come from C. As its charge is reduced, the voltage across it, and hence the output voltage, changes by an amount dV. The quantitative aspects of the problem may be approached through energy conservation: the rate at which energy is given to the drifting electrons is the product of the force on them times their velocity, or

$$\frac{dW}{dt} = N_e e \frac{V_B}{d} v_d$$

where N_e is the number of electrons and e is their charge; this must come from the energy stored in the capacitor, whence

$$\frac{dW}{dt} = \frac{d}{dt}(\tfrac{1}{2}CV^2) = CV\frac{dV}{dt} \simeq CV_B\frac{dV}{dt}$$

where V is the voltage across the capacitor, which is never much different from V_B; equating these gives

$$\frac{dV}{dt} = \frac{N_e e v_d}{dC} \tag{9-5}$$

If the electron-ion pairs are created at a distance x from the positive plate, (9-5) continues for a time x/v_d, whence the total change in voltage at the output due to the motion of electrons ΔV_e is

$$\Delta V_e = \frac{N_e e}{C}\frac{x}{d}$$

The total time required for the pulse due to electrons to rise to its maximum t_R is just the time required for the electrons to be collected,

$$t_R = \frac{x}{v_d} \simeq \frac{d}{2v_d} \qquad \textbf{(9-6)}$$

where the last expression represents a rough average. For a typical chamber, $d \sim 4$ cm, $v_d \simeq 10^7$ cm/s, whence $t_R \simeq 2 \times 10^{-7}$ s $= 200$ ns. A similar treatment for the positive ions which drift at a much slower velocity v_d' over a distance $d - x$ gives a total voltage change due to ion motion, ΔV_i as

$$\Delta V_i = \frac{N_e e}{C} \frac{d - x}{d}$$

whence the total change in the voltage, $\Delta V = \Delta V_e + \Delta V_i$, is

$$\Delta V = \frac{N_e e}{C} = V_{\text{out}} \qquad \textbf{(9-7)}$$

The dc component of the voltage is prevented from reaching the output by C', so $\Delta V = V_{\text{out}}$, the output voltage.

Eventually, the original charge CV_B is restored to the capacitor by flow of charge from V_B through R_B with a time constant R_BC. If R_B is chosen so as to make this much longer than t_R for positive ions, the rise of the pulse is not much affected and the pulse is as shown in Fig. 9-7.[1]

[1] Since v_d' is so slow, R_B is often chosen to make R_BC shorter than t_R for the positive ions, whence the pulse decays much more rapidly than in Fig. 9-7 and the maximum never gets higher than the pulse due to the drift of the electrons. This leads to other complications, but we shall not pursue them here since gas ionization chambers are no longer widely used. They are discussed extensively in this section principally to facilitate explanation of other detectors.

FIGURE 9-7 Pulse shape from an ionization chamber. The actual pulse, shown by the solid curve, is the product of the dashed curve labeled "from positive ions + electrons" and the dashed decay curve. In a semiconductor detector, the rise of the second part of the curve, due to hole collection, is much more rapid.

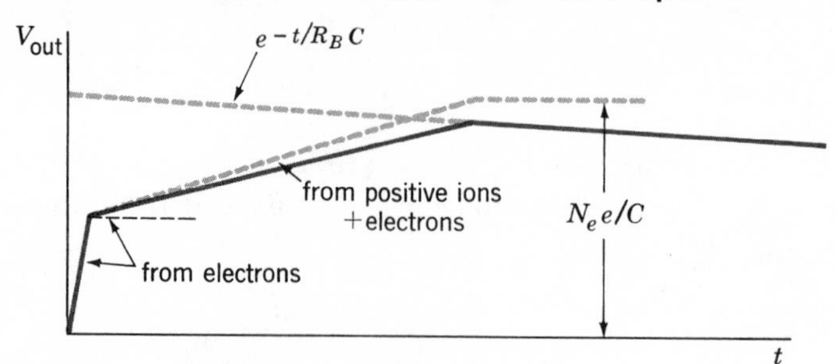

It should be noted from (9-7) that the largest pulse size is obtained with a minimum C. Hence no actual capacitor is used, and C consists of the capacitance between the detector plates and the stray capacitance in the wiring. In typical cases, this can be made as low as about 20×10^{-12} farad (F). From our previous estimate that for a 3-MeV energy loss by the energetic charged particle $N_e \simeq 10^5$ and from the well-known constant $e = 1.6 \times 10^{-19}$ C, (9-7) gives $V_{\text{out}} \simeq 10^{-3}$ volts. Since this is a rather small pulse, extensive amplification is necessary before it can be used for most purposes.

While the parallel-plate geometry of the ionization chamber has some advantages, larger pulse heights can be obtained by using a concentric-cylinder geometry in which the inner cylinder has a very small diameter; in practice it is a thin wire. The electric field near the wire in this geometry is very high so electrons can pick up enough energy between collisions in this region to produce further ionization through the interaction discussed in Sec. 9-2. We thereby obtain a multiplication of the number of electron-ion pairs by a factor M, whence the pulse size in (9-7) is multiplied by this factor; M can be as large as 1,000 or more. A detector of this type is called a *proportional counter*. In either a proportional counter or an ionization chamber the pulse size is proportional to N_e, which is proportional to the energy given up by the energetic charged particle, so they can be used for energy measurements.

A *Geiger counter* utilizes the same geometry as a proportional counter, but by increasing the applied voltage the electric field is made so high near the wire that even inner electrons of gas atoms are excited in collisions. As electrons drop back into these vacated orbits, the radiation emitted (ultraviolet light) is sufficiently energetic to produce ionization in other gas atoms, including those far from the original site: the discharge therefore spreads over the entire volume of the chamber. It is quenched only when the number of positive ions becomes so high that the electric field is greatly reduced. Since there is so much ionization and hence such a large number of electrons collected, the output pulse is quite large. Its size is limited only by the quenching action and is therefore independent of the number of electron-ion pairs produced by the energetic charged particle that initiated the pulse; a Geiger counter is therefore not useful for energy measurements. On the other hand, because the large size of the pulses it produces minimizes the need for expensive amplification circuitry, the price of an operating Geiger counter system is far below that of most other detectors.

The Geiger counter was the first detector developed in which the arrival of a particle leads to the output of an electric pulse, and since these pulses are the same for all particles, they are useful primarily for counting the number of particles reaching the detector. This device, developed by Geiger, therefore came to be called a Geiger counter. When other detectors were developed, even though they were used primarily for energy measurements, they were still often called *counters*. Thus we often hear of *scintillation counters*, referring to scintillation detectors, and *solid-state counters*, referring to semiconductor detectors.

The most widely used of the charge-collection type of detector is the semiconductor detector, commonly called *solid-state detector*. It consists of a *p-n* junction between *p*-type and *n*-type silicon or germanium. A reversed-bias voltage, i.e., a voltage in the direction in which the diode is *not* conducting, is applied, leaving a region near the junction depleted of charge carriers. Methods have been developed for extending the depleted region up to the surface of the detector and even through the entire crystal. When an energetic charged particle travels through this depleted region, its interaction with the electrons in the crystal leaves its path strewn with electron-hole pairs. It turns out, in fact, that there is an electron-hole pair for every 3.5 eV (in Si; in Ge, it is 2.8 eV) of energy lost by the charged particle. The electrons and holes are collected and give a pulse in the same manner as in a parallel-plate ionization chamber, whence Fig. 9-6 is applicable.[1] There is, however, a considerable quantitative difference in the pulse shapes from those in Fig. 9-7. The drift velocity in semiconductors is[1]

$$v_d \simeq \mu \frac{V_B}{d} \tag{9-8}$$

where μ is the *mobility*, which is about 1,350 and 480 cm^2/V-s for electrons and holes, respectively, in Si at room temperature. Since their mobilities are not very different, both electrons and holes are always collected, so (9-7) is valid and the pulse height can therefore be used for energy determinations. From (9-7) we can calculate that pulse heights produced by particles of several MeV are a few millivolts. V_B/d is typically 5,000 V/cm so from (9-8) for the slower hole collection $v_d \simeq 2.4 \times 10^6$ cm/s. For a typical thickness of about 0.05 cm (9-6) then gives $t_R \simeq 20$ ns. The initial rise, due to electron collection, is faster because of the higher mobility of electrons.

Semiconductor detectors can be much smaller physically than gas-filled detectors because the range of the particles to be detected in several orders of magnitude shorter in a solid than in a gas. They are also much smaller than a scintillation detector, which requires a photomultiplier and a resistor chain (see Fig. 9-4), both rather bulky. Some typically mounted silicon semiconductor detectors are shown in Fig. 9-8. Particles to be detected enter through the front in the top photograph and from the left side in the lower photograph. The area of silicon wafer which serves as the detector typically ranges from 25 to 1,000 mm^2 and is never more than a few millimeters thick. Germanium detectors, designated Ge(Li) (pronounced like "jelly") because their depleted region is greatly extended by drifting lithium ions through them, are usually made much larger (40 cm^3 and

[1] The electric field in a semiconductor detector is not uniform, but this is a relatively minor complication which we ignore here.

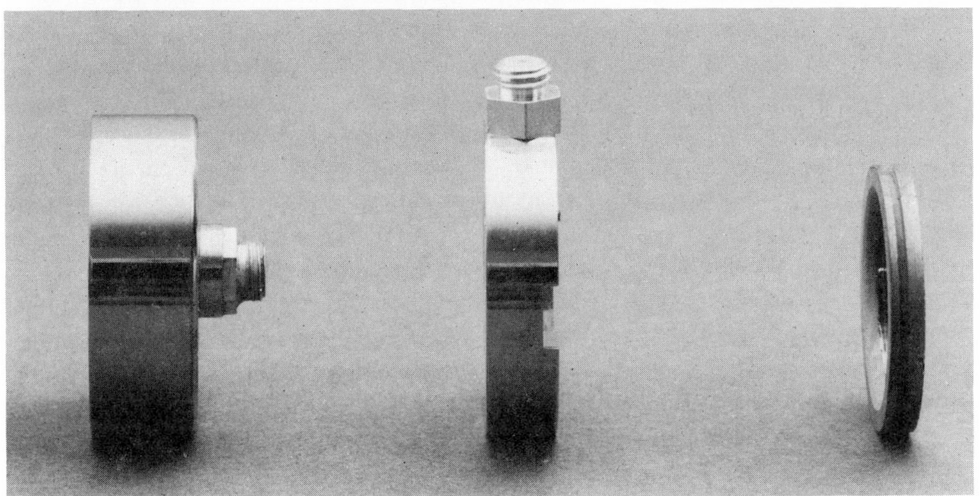

FIGURE 9-8 Two views of some typically mounted silicon semiconductor detectors. The diameters of these mounts are typically about 1 in. *(ORTEC, Oak Ridge, Tenn.)*

more) because they are used for gamma-ray detection (see Sec. 9-6). Since they must be kept at liquid-nitrogen temperature at all times to prevent deterioration, they are always in a bulky system which includes a liquid-nitrogen reservoir.

9-4 Energy Measurements and Identification of Energetic Charged Particles

In several of the detectors discussed in the last section, the size of the electric pulse was shown to be proportional to the energy given up by the charged particle being detected. If the size of the detector is sufficient to stop the particle, a measurement of the pulse height can be used to determine the particle's energy. From Fig. 9-3 we see that for protons of several MeV this requires detector thicknesses of the order of 1 mm of solid material or 1 m of gas; solid detectors such as scintillators or semiconductor detectors must therefore generally be used. The same is true for electrons of all but the lowest energies. Gas counters can be used for energy measurements only with short-range particles.

In a typical system, output pulses like those in Fig. 9-5 or 9-7 are amplified and shaped, principally by shortening their length, using standard electronics techniques. Their heights, i.e., the voltages at their peaks, are then digitized by an analog-to-digital converter (ADC). This device generates a voltage rising linearly with time until it reaches the height of the pulse. The number of cycles an auxiliary electronic oscillator goes through during this voltage rise is counted, and this total count is therefore a digital number whose size is proportional to the pulse height. Once the pulse height is digitized, it can be handled by various computer techniques; most commonly, a location in a magnetic core memory is reserved for each digital pulse height, and after each pulse is digitized in the ADC, the number in the corresponding memory location is increased by 1. The cumulative numbers in each memory location therefore give the number of pulses of the corresponding height, which is just the number of original charged particles with the corresponding energy. These numbers can be retrieved by any of several computer output devices, which print them, plot them, display them on an oscilloscope, transfer them to magnetic or punched paper tape for further computer processing, etc. These numbers give the energy distribution of detected particles; an example is shown in Fig. 9-9. A complex instrument for performing all the functions described in this paragraph is known as a *multichannel pulse-height analyzer;* at least one is available in nearly every nuclear-physics laboratory.

One of the most important properties of any energy-measuring device is its resolution ΔE, which may be defined as the full width at half maximum of the peak produced by a number of particles of identical energy. Its importance can be seen in Fig. 9-9; if the resolution is poor, individual peaks will not be resolved and information will be lost. When energies are measured through pulse heights, the quantity actually measured is N_e, and, as pointed out in Sec. 9-2, it is subject to statistical fluctuations in the energy-loss process.

These are the same type of fluctuations one encounters in drawing cards. If one card is drawn from a deck, the probability that it is an ace is 1/13. Therefore, if a card is drawn from a deck 1,300 times (reshuffling each time), the most probable number of aces drawn is 100, and from statistics theory the standard deviation is the square root of the most probable number, or 10. If 1,300 drawings are made a large number of times and the frequency for obtaining each number of aces is plotted vs. that number, one obtains a gaussian distribution centered at

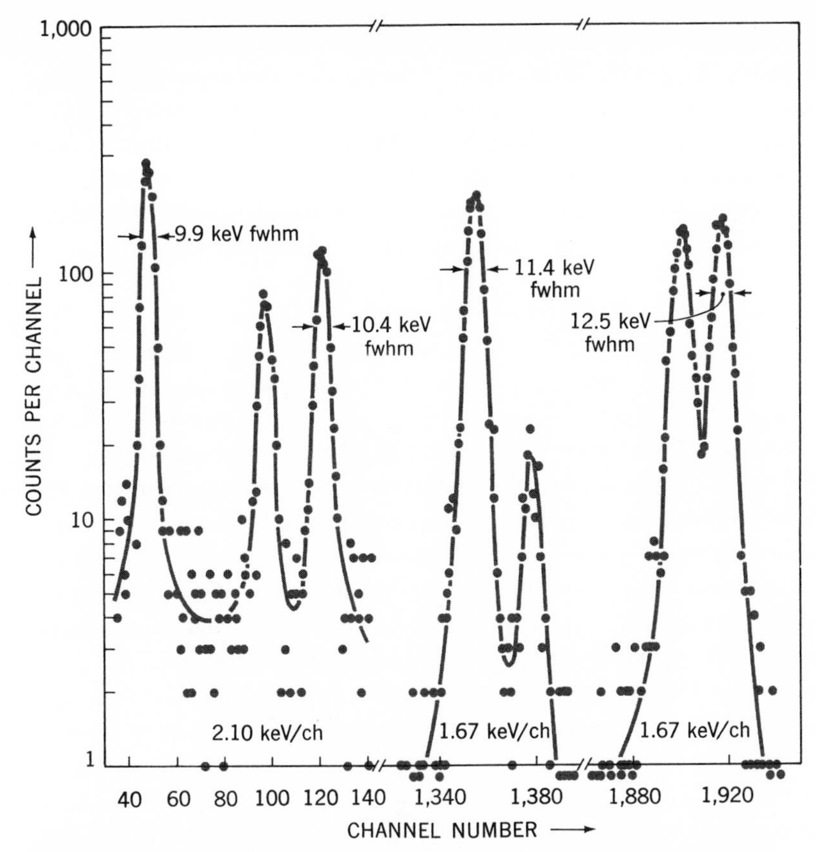

FIGURE 9-9 Energy spectrum of protons from Al^{27} (d,p) reactions as measured with a silicon semiconductor detector. The proton energies here vary from 8.9 to 12.3 MeV. [*From G. Andersson-Lindstroem, Nucl. Instr. Methods,* **56: 309 (1967);** *by permission from North-Holland Publishing Company.*]

100 and with a full width at half maximum of 2.3 standard deviations. In this case, this is 23, so the curve falls to half maximum at 88.5 and 111.5. This treatment applies so long as the probability for any individual event, like drawing an ace, is much less than unity.

In our problem the probability of any one electron's being knocked free by a passing charged particle is much less than unity, and so the treatment applies. If the most probable number of electrons affected is N_e, the width ΔN_e of the distribution of N_e values obtained for a large number of cases is $2.3\sqrt{N_e}$. Since our measurement of the energy is proportional to N_e, we find for the fractional energy resolution

$$\frac{\Delta E}{E} = \frac{\Delta N_e}{N_e} = \frac{2.3}{\sqrt{N_e}}$$

In the most favorable scintillation system there is one photoelectron produced at the photocathode for each 110 eV of energy lost, and in Sec. 11-3 we found that there is one electron set free for about each 30 and 3.5 eV of energy loss in gas counters and silicon semiconductor detectors, respectively. On this basis we may expect for a 5-MeV particle[1]

$$\frac{\Delta E}{E} = \begin{cases} \dfrac{2.3}{(5 \times 10^6/110)^{1/2}} = 1.1\% & \text{scintillator} \\ \dfrac{2.3}{(5 \times 10^6/30)^{1/2}} = 0.58\% & \text{gas counter} \\ \dfrac{2.3}{(5 \times 10^6/3.5)^{1/2}} = 0.19\% & \text{semiconductor} \end{cases}$$

We see from this calculation that semiconductor detectors have a very distinct advantage over the other types, and for this reason gas counters are rarely used for energy measurements. Scintillators, on the other hand, being more than an order of magnitude cheaper and much more rugged than semiconductor detectors and available in much larger sizes, find many applications where good energy resolution is not crucial. An example of the difference in the energy resolutions of a scintillator and a semiconductor detector can be seen in Fig. 9-15.

Where very high energy resolution is desired, the best technique is to measure the deflection of particles in a magnetic field. The radius of curvature ρ of a particle with charge ze, mass M, momentum p, and kinetic energy E in a field of strength B is

$$\rho = \frac{p}{zeB} = \frac{\sqrt{2ME}}{zeB} \tag{9-9}$$

where the second expression ignores relativistic corrections. There are many designs for magnetic field shapes (see Sec. 9-5) which focus particles to positions that are functions of ρ, so a measurement of ρ involves only a measurement of position. Any of the detectors discussed in Sec. 9-3 can be used for this purpose, but efficiency is improved if particles with a broad range of ρ values are detected in the same experiment while retaining the information on their positions. One widely used technique is to mount a photographic emulsion on the focal plane and after exposure for a suitable time and development, to count the tracks at various positions on it. An example of data obtained in this way may be seen in Fig. 14-4. Several other detection schemes are available.

It is generally important to know the identity of the particles being detected. Where magnetic analysis is employed, this is often simple: for electrons, ρ is so

[1] For reasons that are not understood, energy resolutions from scintillators are at least a factor of 2 poorer than given by this calculation. Our statistical treatment was not very thorough: actually, resolutions can be somewhat better than given here, by the so-called *Fano factor*.

much smaller than for other particles that there is no chance for confusion; for protons, an absorber (a sheet of material to reduce the energy of charged particles or stop them) placed in front of the detector can remove all other particles with the same ρ; etc. When two or more particles with the same ρ are present, rough measurements of their energy with a detector can give the necessary information.

Particle identification is a common problem in nuclear-reaction studies when energies are determined by pulse heights. For example, in the study of protons from (d,p) reactions leading to the data shown in Fig. 9-9, deuterons from (d,d') reactions, tritons from (d,t) reactions, and alpha particles from (d,α) reactions are simultaneously emitted. Particle identification can be achieved here by employing a thin detector (thickness t_1) in front of the thick detector. The particle loses an energy $(-dE/dx)t_1$ in the thin detector and the remainder in the thick detector. The pulse height from the thin detector is proportional to $-dE/dx$, and by adding the outputs from both detectors one gets a pulse proportional in height to E. Several methods have been developed for combining these to give particle identification. One such method can be understood by multiplying the numerator and denominator of (9-4) by $\frac{1}{2}M$ and ignoring the relatively small variations in the logarithmic term; this gives

$$-\frac{dE}{dx} \propto \frac{Mz^2}{E}$$

Electronic multiplication of the pulses proportional in height to $-dE/dx$ and E therefore gives a product pulse proportional in height to Mz^2, which is 1, 2, 3, and 16 for protons, deuterons, tritons, and alpha particles, respectively. Most multichannel pulse-height analyzers include provisions for sending the digitized information on energy described earlier in this section into different portions of the memory as indicated by an accompanying signal. This accompanying signal can be the product pulse which is proportional to Mz^2, and separate energy spectra of all particles can therefore be measured simultaneously.

In cloud chambers and bubble chambers particle identification and energy measurements can be accomplished by applying a magnetic field. The tracks of the particles are then curved, and ρ can be measured to give ME/z^2. The number of water droplets (in a cloud chamber) or of bubbles (in a bubble chamber) along the path is proportional to $-dE/dx$, whence from (9-4) one can determine Mz^2/E. Multiplying these two pieces of information gives M, which ordinarily is sufficient to identify the particles.

9-5 Magnetic Instruments

In many experimental situations it is advantageous to deflect or focus beams of particles or to measure their mass, velocity, or electric charge. In general these require application of a force, and forces on charged particles are most conveniently applied through electric and magnetic fields, as indicated by the Lorentz force law

$$\mathbf{F} = e[\boldsymbol{\mathcal{E}} + (\mathbf{v} \times \mathbf{B})]$$

where **F** is the force, $\boldsymbol{\mathcal{E}}$ is the electric field, and **B** is the magnetic field. Typical field strengths that can be *easily* obtained are $\mathcal{E} \simeq 3 \times 10^6$ V/m and $B \simeq 2$ Wb/m^2; the ratio of magnetic to electric forces in these fields is about 200 v/c. This is unity for 0.01-MeV protons, whence for particles in the MeV energy range, a given force is more easily applied with magnetic than with electric fields. There are applications which take advantage of the lower cost and/or more accurately calculable field strength vs. position of electric fields, and there are special situations in which combinations of electric and magnetic fields are useful, but in most cases, magnetic fields are used. Although a large variety of magnetic instruments have been developed for various applications, here we shall discuss only two of the most common types, the sector-magnet spectrograph and the quadrupole focusing lens.

A sector-magnet spectrograph is a device for measuring the radius of curvature of a charged particle in a magnetic field, whence, from (9-9), it can be used for measuring mass, velocity, or charge if the other two of these is known. It can also be used for changing the direction of a beam of particles with the same mass, velocity, and charge. The simplest type consists of a magnetic field in the shape of a wedge, or a slice of pie, as shown in Fig. 9-10. (Parts of the wedge not used by the particles may be cut off, as is done there.) Particles starting from the source and passing through the center of the slit follow the path labeled c; those leaving the source at an angle α to this ray follow paths labeled a (b), in which they travel a longer (shorter) distance in the magnetic field and hence are deflected more (less), whence they again intersect the path c. If the expression for the distance q at which this intersection occurs is expanded in powers of α, the condition that the term linear in α vanishes turns out to be[1]

$$\tan \phi = \frac{\tan \gamma + \rho/p + \tan \delta + \rho/q}{1 - (\tan \gamma + \rho/p)(\tan \delta + \rho/q)} \qquad \textbf{(9-10)}$$

where all symbols are defined in Fig. 9-10. If this condition is fulfilled, all particles emitted with small α intersect path c at the same point, whence, in the language of geometric optics, these particles are said to be *focused* to form an *image*. The terms in the expansion proportional to α^2, α^3, etc., cause this focusing to be imperfect, these imperfections being known as *second-order aberrations*, *third-order aberrations*, etc., respectively. Since α is much less than 1 radian, the α^2 term is

[1] For derivations of Eqs. (9-10) and (9-12) see M. Camac, *Rev. Sci. Instr.*, **22**:197 (1951).

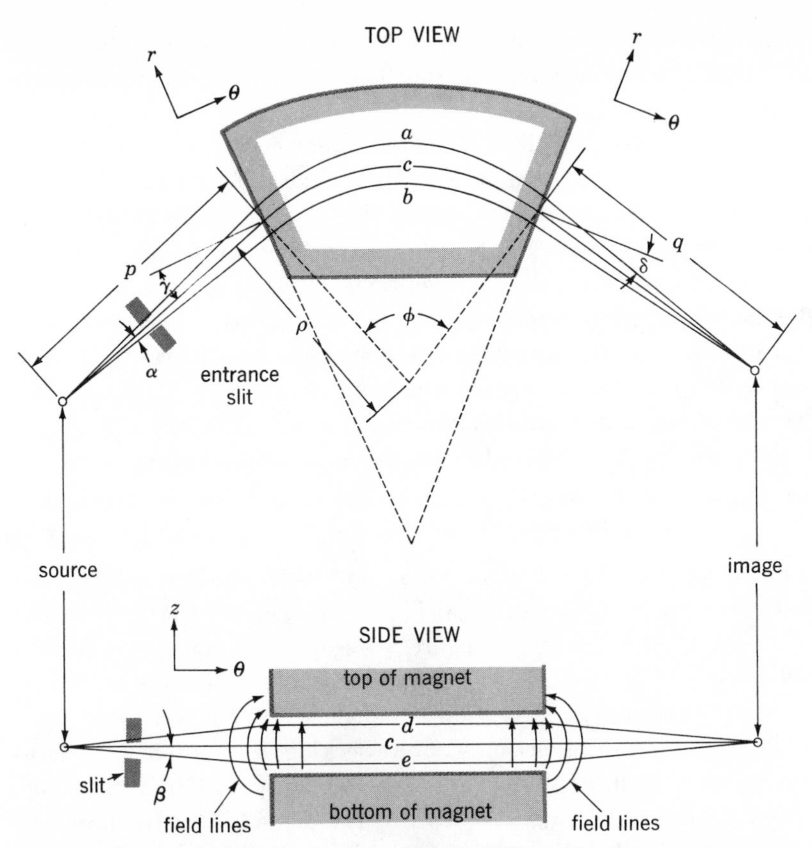

FIGURE 9-10 A uniform-field wedge-magnet spectrometer. The path c passes through the center of the entrance slit, and the other paths deviate from it by a small angle as shown. In the top view, the shaded region is the magnetic field. The angles γ and δ are between path c and the normal to the magnet edge.

ordinarily the most serious. It is clearly independent of the sign of α, whence paths a and b intersect at a point displaced slightly from c at the image point. This displacement, it turns out, is approximately $\rho\alpha^2$ in the direction of increased deflection. It can be eliminated by rounding the magnet edges, removing material on both sides of the place where the path c crosses the pole boundaries; this causes particles following paths a and b to be deflected slightly less without changing the deflection of path c. Third-order aberrations have opposite signs when the sign of α is opposite, which means that for large α, paths a and b intersect in front of (or behind) the focus, which we have effectively defined as that intersection for small α. This can be corrected by choosing the magnet edges to be third-order curves. Where third-order aberrations are troublesome, the detector can be positioned between the intersection of rays for small and larger α; in doing so, one partially compensates the third-order aberration by introducing a first-order aberration.

In accordance with (9-9), particles with a different energy have a different radius of curvature ρ, whence they are deflected through different angles ϕ, and from (9-10) they are focused at different points. The locus of these points is the focal plane, whose use was discussed in Sec. 9-4. An instrument using an appreciable length of this focal plane is called a *broad-range spectrograph.*

In most situations it is desirable for the spectrograph to accept particles leaving the source over as wide a range of directions as possible. This can be done by increasing the accepted range of α as much as possible without introducing excessive aberrations and also by achieving focusing in the vertical (z) direction, as shown in the lower part of Fig. 9-10. The force in this direction, from the Lorentz law, is

$$F_z = e(v_\theta B_r - v_r B_\theta) \tag{9-11}$$

In the magnetic field of Fig. 9-10, B_r is always zero in the region traversed by the particles, but B_θ is nonzero in the fringing field near the magnet edges. From the field lines shown in the side view of that figure, we see that F_z is negative (downward) above the median plane and positive (upward) below the median plane, provided v_r is positive or, equivalently, that the angle γ is in the direction shown in the figure. This is the focusing action we were seeking, and similar considerations easily lead to the conclusion that there is also a focusing action in the fringe field as the particles leave the magnet, provided the angle δ is in the direction shown. As a result of this focusing action, particles following paths like those labeled d and e in Fig. 9-10 are deflected toward the median plane. If the position at which they cross it is expanded as a power series in the angle β by which their original direction deviates from that of c, the condition that the term linear in β vanishes turns out to be[1]

$$\phi = \frac{1}{\tan\gamma - \rho/p} + \frac{1}{\tan\delta - \rho/q} \tag{9-12}$$

This gives the distance at which a focus in the z direction is achieved for small β; if the magnet can be designed such that (9-10) and (9-12) are simultaneously satisfied, focusing of particles leaving the source over a sizable range of directions is achieved.

Unfortunately, simultaneously satisfying (9-10) and (9-12) often leads to large angles γ and δ which are not easily and accurately obtainable with real magnets. An alternative approach is to choose these angles to be 0° (whence

[1] See footnote page 224.

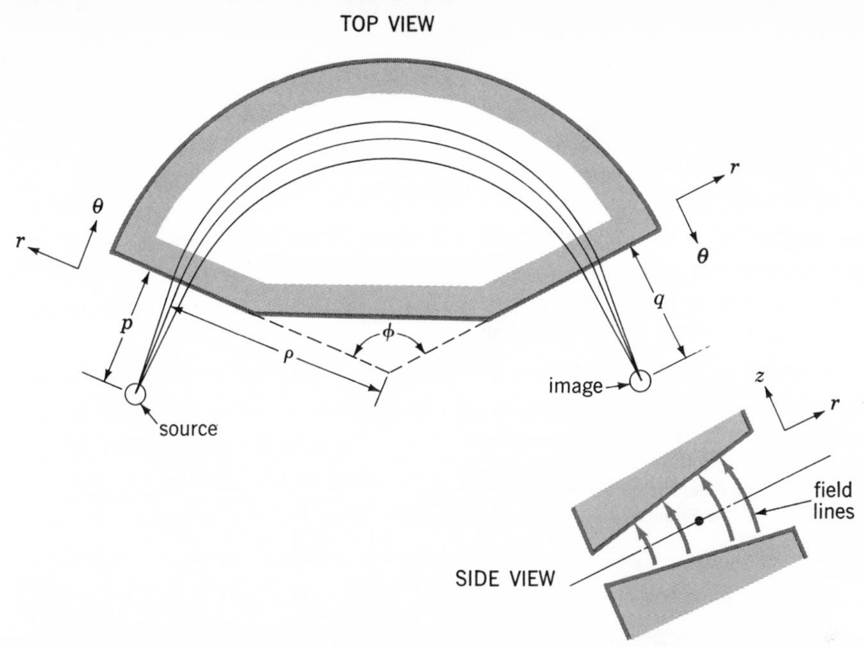

FIGURE 9-11 A nonuniform-field wedge-magnet spectrometer. The side view shows the shape of the poles looking in from the edge. In the region where the particles travel, the field falls off linearly with increasing r, and the resulting B_r gives a vertical focusing.

ϕ = wedge angle) and obtain the focusing in the z direction by introducing a nonuniform magnetic field as shown in Fig. 9-11. Since in this case B_r is nonzero, there is a contribution to F_z from the first term of (9-11). It is readily seen from the rules learned in elementary physics that if B decreases with increasing r as in Fig. 9-11, F_z is always directed toward the median plane. If the magnetic field in the region traversed by the particles ($r \simeq \rho$) is represented by

$$B = B_0[1 - n(r - \rho)]$$

the focusing conditions in the horizontal (r,θ) plane and in the vertical (z) direction are, respectively,[1]

$$q = \begin{cases} -\dfrac{\rho}{(1-n)^{1/2}} \tan\left[(1-n)^{1/2}\phi + \tan^{-1}(1-n)^{1/2}\dfrac{p}{\rho}\right] & \text{horizontal} \\ -\dfrac{\rho}{n^{1/2}} \tan\left[n^{1/2}\phi + \tan^{-1} n^{1/2}\dfrac{p}{\rho}\right] & \text{vertical} \end{cases}$$

It is immediately clear that these equations become identical and can therefore be simultaneously satisfied if $n = \frac{1}{2}$.

[1] These formulas are derived in D. L. Judd, *Rev. Sci. Instr.*, **21**:213 (1950).

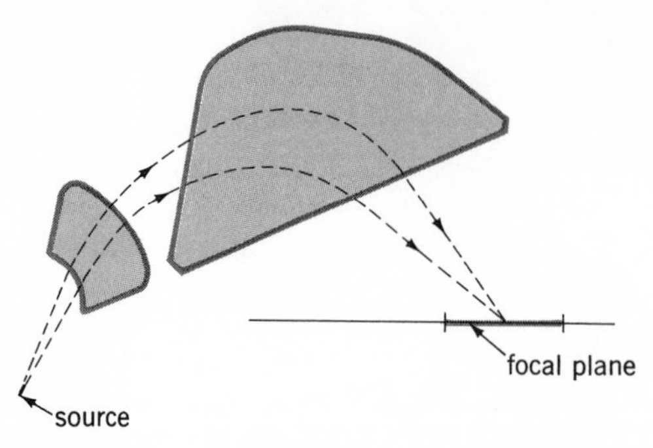

FIGURE 9-12 The magnetic field shape in a split-pole magnetic spectrograph designed by H. A. Enge. The dashed lines show two typical trajectories. This spectrograph is described in J. E. Spencer and H. A. Enge, *Nucl. Instruments Methods,* **49: 181 (1967).**

This method of obtaining vertical focusing would lead to great complications in a broad-range spectrograph since the broad range requires use of an extensive area of the magnetic field; the first method, the one shown in Fig. 9-10, is therefore used. In order to achieve sufficient vertical focusing without excessively large angles γ and δ, particles can be successively passed through two magnetic wedges, giving four fringe fields from which to obtain contributions to F_z. The use of four edges also gives more parameters, those describing the shapes of these edges, which can be adjusted to reduce aberrations. A spectrograph design employing these principles is shown in Fig. 9-12.

The second type of magnetic instrument to be discussed is the quadrupole lens, which focuses particles without deflecting them. Half of such a lens is shown in cross section in Fig. 9-13. The poles are shaped such that

$$B_x = ky$$
$$B_y = kx$$

where k is a constant depending on the dimensions and the number of ampere-turns in the coils. The particles go in the z direction ($v_x \simeq 0 \simeq v_y$; $v_z \simeq v$), whence the Lorentz force law gives

$$\begin{aligned} F_x &= v_yB_z - v_zB_y = -vkx \\ F_y &= v_zB_x - v_xB_z = +vky \end{aligned} \tag{9-13}$$

They therefore experience a focusing force in the x direction and an equal defocusing force in the y direction. A quadrupole lens consists of two such configurations,

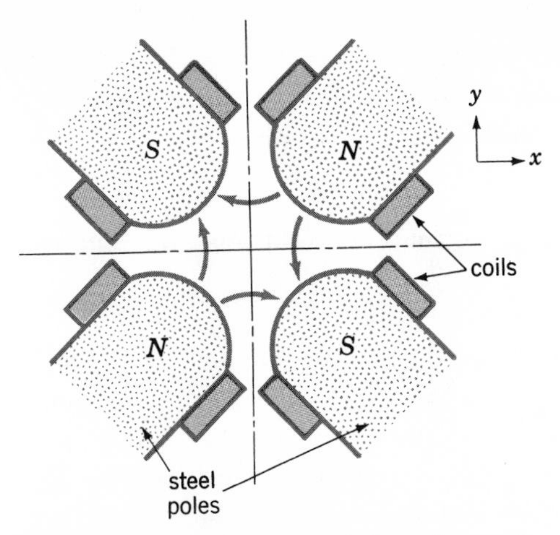

FIGURE 9-13 Cross-sectional view of a quadrupole lens. The curved lines with arrows are magnetic field lines.

Q_1 and Q_2, with fields in opposite directions; therefore if the one described in (9-13) is Q_1, in Q_2

$$\begin{aligned} F_x &= +vkx \\ F_y &= -vky \end{aligned} \tag{9-13a}$$

which gives focusing in the y direction and defocusing in the x direction. The motion is then as shown in Fig. 9-14 for the xz and yz planes. In both cases, there is a force acting in each direction, but the force *toward* the central axis occurs

FIGURE 9-14 Focusing action of a quadrupole pair. A parallel beam of particles is focused in the x direction *(top diagram)* in Q_1 and defocused in Q_2, while in the y direction *(bottom diagram)* it is defocused in Q_1 and focused in Q_2. In both directions the net result is a focusing.

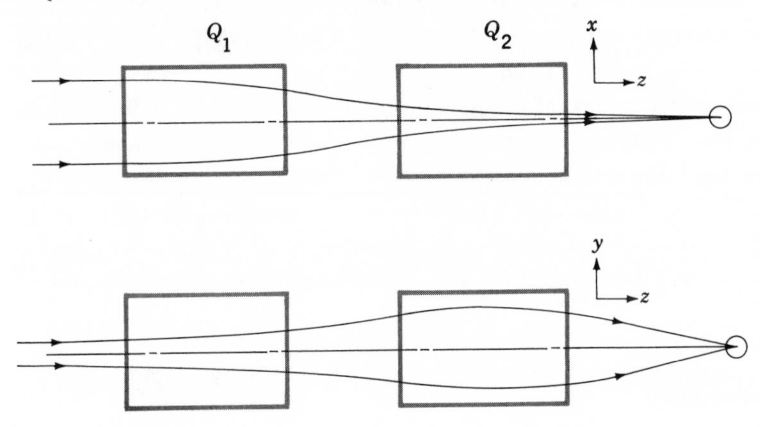

when the displacement is larger, whence, from (9-13), that force is the stronger of the two. The pair of lenses therefore gives focusing in both directions. It can be used in much the same way as lenses are used in optical systems.[1]

9-6 Detection, Energy Measurement, and Stopping of Neutrons and Gamma Rays

All the discussion of energy loss, detection, and energy measurement in the last few sections is applicable only to electrically charged particles. Since neutrons and gamma rays are uncharged, these subjects must be discussed separately for them. Unfortunately, the techniques used for these are less standardized than for charged particles, and there are many tricks and variations. We shall therefore confine our discussion to a few of the most widely used techniques.

Neutrons can induce nuclear reactions and scattering events, many of which result in the emission of charged particles; virtually all methods of neutron detection depend on detecting these charged particles. For detection of low-energy neutrons, the (n,α) reactions on B^{10} and Li^6 and the $U^{235}(n,f)$ reaction produce energetic particles which can readily be detected; it will be shown in Sec. 13-8 that these reactions have large cross sections for low-energy neutrons. Gas counters using boron trifluoride, BF_3, are most frequently used for this purpose; the BF_3 serves simultaneously as the counter gas and the source of boron. Scintillators of LiI and gas counters containing uranium sheets are also used for detection of low-energy neutrons.

Higher-energy neutrons ($\gtrsim 0.2$ MeV) are most frequently detected by taking advantage of their large cross sections for scattering from protons; this is the one of the nucleon-nucleon scattering experiments mentioned in Chap. 3. Since the two are of nearly equal mass, on an average the proton receives half of the original neutron energy. An organic scintillator provides both a target of protons and an efficient detector for them. Unfortunately, the pulse height does not give a measurement of the neutron energy since the fraction of this energy given to the proton depends on the scattering angle.

The most widely used method for determining neutron energies is to measure their time of flight over a known distance; this gives their velocity and hence their energy. Their starting time can be determined from the time of arrival of a beam pulse from an accelerator, by detecting a charged particle emitted simultaneously, or, for low-energy neutrons, by the opening of a mechanical shutter. Techniques for accurately measuring the time interval between these events and the arrival

[1] For formulas and curves for calculating the properties of quadrupole lenses see H. A. Enge, *Rev. Sci. Instr.*, **30:** 248 (1959).

of the neutron at its detector will be discussed in the next section. Most other methods for measuring energies of "fast" neutrons depend on determining both the energy and the angle of the protons from *n-p* scattering events. Because very slow neutrons have wavelengths comparable to spacings between the planes in a crystal, their energies can be measured by diffraction techniques similar to those used for x-rays; these determine their wavelength, which is related to their velocity by (2-1).

When neutrons are produced by nuclear reactions, their energy can be calculated from the energy of the incident particle and the reaction energetics, provided the residual nucleus is known to be left in its ground state. In a few reactions in light nuclei, such as $d + H^3 \rightarrow \mathrm{He}^4 + n$, there are no excited states in the residual nucleus for several MeV above the ground state, so this condition can be guaranteed over an appreciable energy range.

Gamma rays interact with the orbital electrons in ordinary matter through three processes well known from modern-physics courses, the photoelectric effect, Compton scattering, and electron-positron pair production. If these interactions occur in scintillators or in semiconductor detectors, the electrons are directly detected and the output pulses give a measure of their energy. Because in both the photoelectric effect and pair production the electron energy has a one-to-one relationship with the gamma-ray energy, the latter is obtained from a measurement of the former. A typical pulse-height spectrum from a scintillator and a semiconductor detector exposed to monoenergetic gamma rays is shown in Fig. 9-15; the shape of the spectrum is explained in the legend. Higher resolution and more accurate measurements of gamma-ray energies can be achieved by diffraction from a crystal lattice, as is done with x-rays; because of the very short wavelength of gamma rays, the deflection angle is quite small, but this difficulty is partially compensated for by bending the crystal into a curved shape to achieve a focusing.

The probabilities for interactions of gamma rays or neutrons with an atom or its nucleus, respectively, are expressed as total cross sections σ_T. For gamma rays σ_T is the sum of the cross sections for the three basic interactions mentioned above, and for neutrons it is the sum of the reaction and scattering cross sections to be discussed in Chap. 13. From the definition of cross sections, the probability for an event per unit distance traveled, μ, is

$$\mu = n_1 \sigma_T \tag{9-14}$$

where n_1 is the number of atoms per unit volume. The number of interactions per unit distance in a beam of intensity I is then μI, whence

$$-\frac{dI}{dx} = \mu I$$

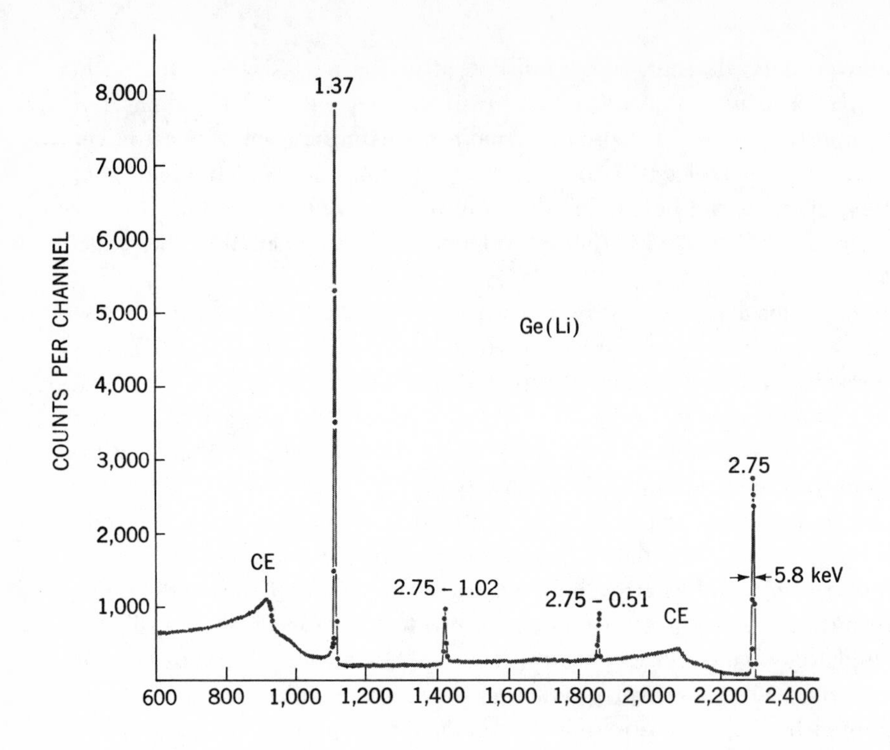

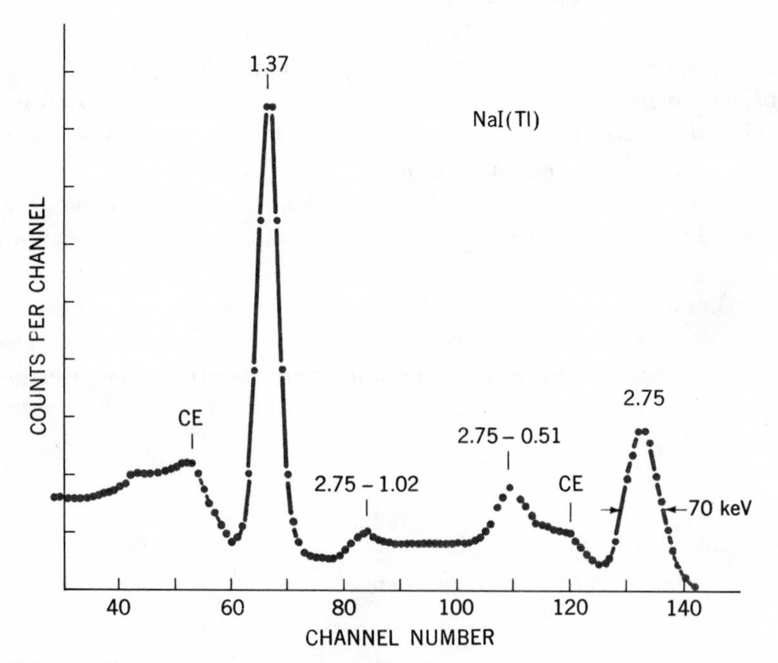

FIGURE 9-15 For legend see opposite page.

This is a differential equation whose solution is

$$I = I_0 e^{-\mu x} \tag{9-15}$$

From (9-15) we see that neutron and gamma-ray intensities are attenuated exponentially in passing through matter, in contrast with the situation for charged particles, which are stopped completely if the thickness of the material exceeds their range as given in Fig. 9-3. This difference arises from the fact that charged particles lose their energy gradually in a large number of very weak interactions, whereas neutrons and gamma rays are removed from a beam in a single catastrophic interaction.

Values of μ for gamma rays are shown in Fig. 9-16. From them we can estimate that for energies of a few MeV in solid materials, $1/\mu$ varies from a large fraction of an inch to several inches, depending mostly on the density of the material. For lower-energy gamma rays, the absorption rapidly becomes stronger. For neutrons, the cross sections we shall obtain in Chap. 13, in conjunction with (9-14), indicate that $1/\mu$ is a few inches for neutrons above 1 MeV in solid materials.

9-7 Timing Techniques

There are many situations in nuclear experiments where it is desirable to know the time relationship between two detected events. Perhaps the most common case is in studying processes in which two particles are emitted in the same event, as, for example, in a $(d,p\alpha)$ reaction. If one wants to study energy and/or angular correlations between the proton and the alpha particle, one must ascertain that they originated from the same reaction. This is done by assuring that they were emitted in *coincidence*, i.e., at the same time. Several electronic devices give an output pulse only when two (or more) input pulses are simultaneously present and hence can serve as the basis for a coincidence analyzer. The output pulse

---

FIGURE 9-15 Pulse-height spectrum from a Na^{24} source (2.75- and 1.37-MeV gamma rays) as measured with a Ge(Li) semiconductor detector and a NaI(Tl) scintillation detector. CE **denotes** *Compton edge,* **the highest-energy electrons that can be produced in a Compton scattering event; the continuous distributions extending below these are due to electrons from Compton scatterings. The peak labeled 2.75 is due to cases where all the energy of a 2.75-MeV gamma ray is absorbed in the crystal. Its principal interaction is electron-positron pair production, and in some cases one or both 0.51-MeV gamma rays from the annihilation of the positron escape from the crystal. These cause the peaks at 2.75 − 0.51 MeV and at 2.75 − 1.02 MeV. Note that the energy resolution from the semiconductor detector (5.8 kev) is much better than from the scintillator (70 kev), as explained in Sec. 9-4.** [*Ge(Li) data from SAIP, Malakoff, France; NaI(Tl) data from R. L. Heath, "Scintillation Spectrometry," USAEC.*]

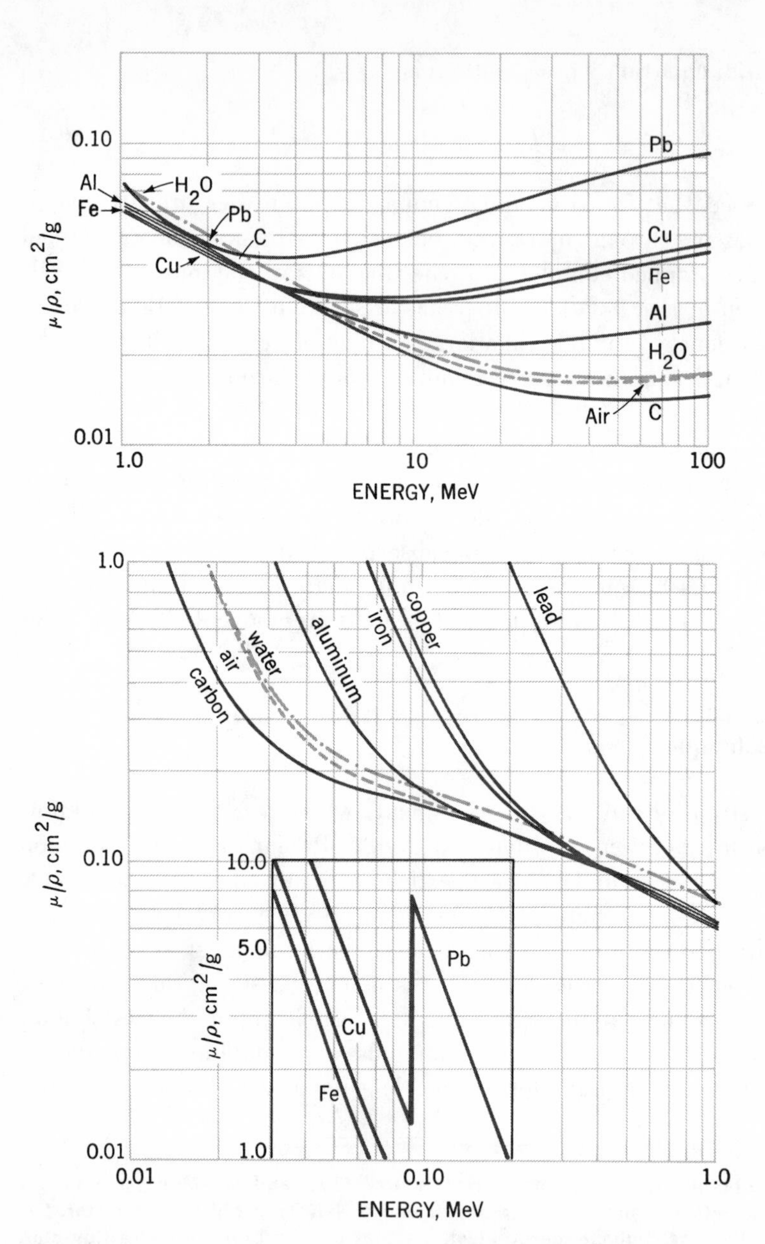

FIGURE 9-16 Attenuation coefficients for gamma rays at various energies and in various materials used in (9-15). *(From "Radiological Health Handbook," U.S. Department of Commerce.)*

from such a device can be used to "gate" other instruments used in the experiment, such as a multichannel pulse-height analyzer; i.e., these instruments will operate only when a pulse is present from the coincidence analyzer indicating that a coincidence has occurred.

The principal problem in an experiment of this type is accidental coincidences. To understand this, let us say that N_0 reactions occur per second, that the proton and alpha-particle detectors subtend solid angles Ω_1 and Ω_2, respectively, and that the angular distributions are isotropic. The number of coincidences detected per second N_c is then

$$N_c = N_0\Omega_1\Omega_2$$

If all particles detected come from these $(d,p\alpha)$ reactions, the total count rates in the two detectors N_1 and N_2 are

$$N_1 = N_0\Omega_1 \qquad N_2 = N_0\Omega_2$$

If particles are detected in the two detectors within the resolving time of the coincidence analyzer τ, they will be recorded as a coincidence. The chance that this will happen accidentally before or after each particle is received in the first detector is 2τ times the probability per second for a particle to reach the second detector, or $2\tau N_2$. Since this situation occurs N_1 times per second, the total number of accidental coincidences per second N_a is

$$N_a = 2\tau N_1 N_2$$

Combining these, we find for the ratio of true to accidental coincidences,

$$\frac{N_c}{N_a} = \frac{N_0\Omega_1\Omega_2}{2N_0{}^2\Omega_1\Omega_2\tau} = \frac{1}{2N_0\tau}$$

This ratio must clearly be kept reasonably large; it can be increased by reducing N_0, but this reduces N_c and hence the rate of data accumulation. It is therefore important to make τ as small as possible. For the best coincidence analyzers, τ is several nanoseconds, but a method of obtaining better time resolutions is described below. The contribution of accidental coincidences can be measured and approximately corrected for by inserting a time *delay* into the output of one of the detectors, as by requiring the output pulse to go through a long cable before reaching the coincidence analyzer; in this situation, only accidental coincidences can cause the latter to produce an output pulse.

In measuring short half-lives, one procedure is to use *delayed coincidence*. For example, if a gamma-ray emission follows a beta decay, the half-life of the former can be determined by measuring the probability for the pulse from the gamma-ray detector to follow that from the electron detector by various time intervals. This can be done by inserting various delays, e.g., cable lengths, between the gamma-ray detector and the coincidence analyzer and counting the number of coincidences as a function of the delay time. A more efficient procedure, however, is to use a time-to-pulse-height converter (TPHC), an instrument which produces an output pulse whose height is proportional to the time between the arrival of two pulses. This can be done (1) by causing a voltage to start rising linearly with time when the first pulse arrives and to stop rising when the second pulse arrives or (2) by converting the two input pulses into a standard shape and electronically integrating the area of their overlap. In our problem of measuring the half-life for a gamma-ray emission, we can start the TPHC with the pulse from the electron detector and stop it with the pulse from the gamma-ray detector, thereby obtaining output pulses whose pulse-height distribution is proportional to the distribution of time intervals between the emission of the electrons and the gamma rays. This pulse-height distribution can conveniently be measured with a multichannel pulse-height analyzer to give the entire decay curve (number of events vs. time interval) in one data accumulation. It is ironical that the TPHC converts time into pulse height and then the pulse height analyzer converts pulse height back into time to digitize it, but this is of no concern to the experimenter.

Another important application of a TPHC is in the measurement of flight times for velocity determinations, as, for example, in measuring neutron energies. Methods of deriving the two pulses were discussed in the last section. Since the length of the flight path is easily measured (with a meter stick) the distribution of pulse heights from the TPHC can be converted into an energy distribution of the neutrons by simple mathematical computations. The neutron energy resolution is clearly limited by the time resolution in the experiment. The time-of-flight method is sometimes used with charged particles for particle identification (see Sec. 9-4); since their energy is known from the pulse height in the detector, a measurement of their velocity gives a straightforward determination of their mass.

Coincidence experiments can also be done with a TPHC. An extra length of cable is inserted in the line carrying the pulse from one of the detectors so that the time interval corresponding to a coincidence is not zero, thus allowing one to measure true coincidences and accidental coincidences simultaneously and also giving better time resolutions—down to a few tenths of 1 ns in some cases.

Different charged particles produce ionization of different densities as they lose energy in passing through materials, and this leads to differences in the pulse shapes (voltage vs. time) produced in some detectors. These pulse-shape differences can be analyzed as a difference in the time interval between when the pulse reaches different fractions of its final height, to determine the particle type; this is often done with a TPHC. An important application of this technique is in distinguishing between pulses from neutrons and gamma rays in plastic scintillators. Neutrons are detected by the protons they produce, whereas gamma rays interact by producing electrons; electrons and protons of the same energy have very different velocities, whence, from (9-4), they produce very different ionization densities.

We have seen that good time resolution is important in most timing applications; let us consider the principal factors that affect it. The time at which an event is registered is determined by when the voltage pulse produced by a detector crosses some *discrimination level.* Electronic instruments based on fast tunnel diodes can detect the time of this crossing very accurately. The time resolution is therefore limited by how reproducibly various pulses of the same ultimate size reach this discrimination level. In charge-collection detectors this is determined by the ratio of electronic *noise* to the rate of rise of the pulse, dV/dt. This is illustrated in Fig. 9-17, which shows a random noise superimposed on two identical pulses, causing them to cross the discrimination level at times differing by δt. Since tunnel diodes require a few tenths of a volt for operation, whereas we have seen in Sec. 9-3 that pulses from semiconductor detectors are typically only a

FIGURE 9-17 The effect of noise on the time after initiation of the pulse when the discrimination level is crossed. *Upper diagram,* **the noise is positive, causing an early crossing;** *lower diagram,* **the noise is negative, causing a late crossing. The time difference between the two is δt. The small diagram at the top shows that the ratio of the noise to δt is dV/dt.**

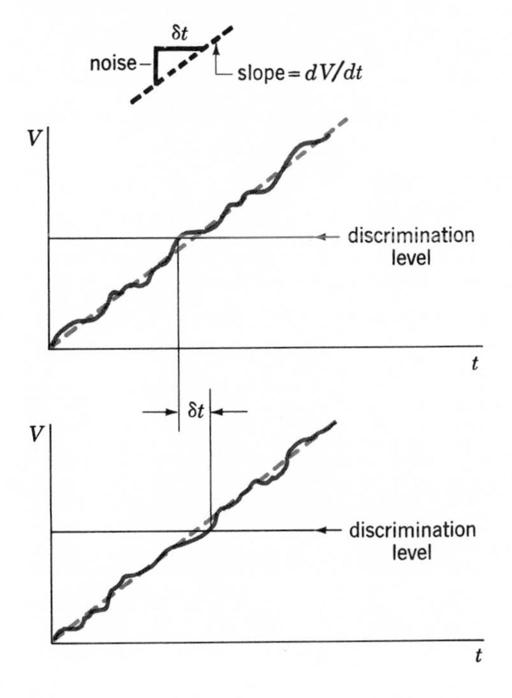

few millivolts in height, amplification is required. The need for a large-bandwidth (so as not to decrease dV/dt) low-noise amplifier is evident here, and the state of the art of producing such amplifiers limits the time resolution. For a given amplification, which implies for a given noise, dV/dt is determined by the ratio of pulse height to rise time; therefore from (9-6) to (9-8) low detector capacitance (including a short connecting cable to the amplifier), thin detectors, and high applied voltages improve the time resolution from gas-filled or semiconductor detectors.

In scintillators, the large noise-free amplification provided by photomultipliers eliminates the above problem; here the time resolution is limited by such things as (1) the differences in transit time through the photomultiplier of electrons originating from different parts of the photocathode and (2) statistical variations due to the small number of electrons produced there (if the pulse contains a total of five electrons and the discrimination level is crossed on the arrival of the third electron, there is a considerable statistical fluctuation in this arrival time). In favorable situations, scintillators can give time resolutions down to about 0.2 ns, and semiconductor detectors can do somewhat better.

No discussion of timing would be complete without mentioning a very widely used technique which does not involve the direct measurement of a time, namely, the *doppler-shift attenuation method* for measuring half-lives for emission of gamma rays in the range 10^{-11} to 10^{-14} s. It is based on the fact that when a nucleus moving with velocity v emits a gamma ray in its direction of motion, the wavelength λ of the gamma ray is shifted, due to the well known doppler effect, by an amount $\Delta\lambda$ given by

$$\frac{\Delta\lambda}{\lambda} = \frac{v}{c}$$

From the well-known relationships between wavelength and frequency ν, and between frequency and energy E_γ,

$$\lambda\nu = c \qquad E_\gamma = h\nu$$

this is easily converted to

$$\frac{\Delta E_\gamma}{E_\gamma} = \frac{v}{c}$$

If the gamma ray is emitted following a nuclear reaction such as $(p,p'\gamma)$, v can be determined from a knowledge of the energy and direction of the incident proton beam and a measurement of the energy and angle of the emitted proton (p') detected in coincidence with the gamma ray (we assume that the coincidence-resolving time here is much longer than the gamma-ray half-life). Hence the doppler shift ΔE_γ between different emission angles is readily detected by careful measurements of E_γ.

Up to this point we have been assuming that the nucleus which emits the gamma ray is traveling in a vacuum. But now suppose that after traveling a distance D, which requires a time D/v, it strikes some material which rapidly stops it. If there is a reasonably large probability for the nucleus to be stopped before the gamma ray is emitted, the doppler shift is *attenuated* and the extent of attenuation can be determined from a measurement of ΔE_γ. By attaching the stopping material to a *plunger* which allows D, and hence the time before stopping, to be varied, and then measuring ΔE_γ as a function of D, one can determine the probability for the gamma ray to be emitted at various times after the reaction. This then gives the half-life for gamma-ray emission. When half-lives are so short that the stopping time for the nucleus once it enters the stopping material is important, this must be taken into account by using the methods of Sec. 9-2. For very short half-lives, the stopping time in the material becomes predominant, and there is no need for a plunger; the stopping material is kept in direct contact with, or is an integral part of, the target in which the reaction takes place.

9-8 Accelerators

Nature as we experience it here on earth provides us with little opportunity for studying the nucleus. Nuclei are never naturally excited as atoms are when they emit light, and nuclear reactions never occur in bulk matter as chemical reactions do in such familiar forms as fire, fermentation, cooking, etc. There is some radioactive decay in nature, but this involves only a few nuclei in their lowest-energy states. Our principal information on nuclear structure has therefore come from nuclear reactions, from decay processes initiated by these reactions, and from the nucleon-nucleon scattering experiments discussed in Chap. 3. We shall find in Chap. 13 that all nuclear reactions except those induced by neutrons (which are not available in nature) require energies of several MeV to overcome the coulomb repulsion between nuclei, and there are similar energy requirements on most scattering experiments. With a few minor exceptions involving natural radioactivity and cosmic rays, particles with MeV energies became available only with the development of accelerators. Even the later development of nuclear reactors has done little to diminish the importance of accelerators as the key to research on nuclear structure.

Acceleration is accomplished by subjecting electrically charged particles to the forces of an electric field. The charged particles are generally derived from *ion sources*, in which an electrical discharge passing through a gas knocks electrons off (or in some cases, adds electrons to) atoms; electrons, on the other hand, are obtained by simply heating a metal, as is done in radio tubes. The method of applying the electric field divides accelerators into two categories, those in which a high dc voltage is used to give one or two accelerations, and those in which a radio-frequency voltage is used to give a large number of smaller accelerations.

In the first category, the dc voltage can be obtained by transformer-rectifier

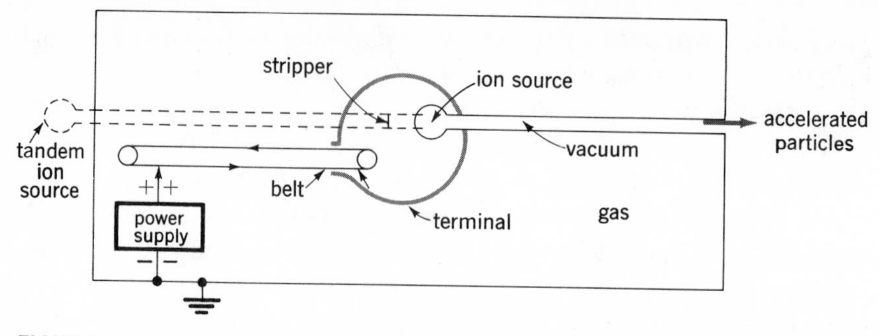

FIGURE 9-18 Schematic of a Van de Graaff accelerator. In an ordinary accelerator the ion source is in the terminal, as shown by the solid lines, but in a tandem accelerator it is as shown by the dashed line and a stripper is introduced in the terminal to change the charge from negative to positive. The tube down which the accelerated particles travel is at high vacuum, but all else is filled with high-pressure gas to hinder voltage breakdown.

methods or by the Van de Graaff method. In the latter, illustrated in Fig. 9-18, electric charge from a power supply is sprayed onto a belt made of insulating material, which conveys it to the inside of a metallic structure called the *terminal.* It is picked up there by a metallic contact and transferred to the outside of the terminal in accordance with the familiar principles of electrostatistics. The terminal thereby accumulates a large charge Q, which establishes a voltage difference $V = Q/C$ (where C is the capacitance) between it and ground. Electrically charged particles from the ion source are accelerated by this voltage. If they carry a charge ze, they come out with kinetic energy zeV.

The energy obtained with the same terminal voltage can be increased in a *tandem* Van de Graaff accelerator shown by the dashed lines in Fig. 9-18. Here negative ions (e.g., hydrogen atoms with two electrons) produced in an ion source at ground potential are accelerated to the terminal, reaching it with energy Ve (charges of negative ions are rarely larger than e). In the terminal they pass through a thin foil of material (*stripper*) which strips off two (or more) electrons, converting them into positive ions of charge ze. After passing through the terminal and out the other side, being positive ions they are again accelerated through the voltage V, whence their final energy is $(z + 1)eV$. The energy of a Van de Graaff accelerator is limited by the voltage that can be held without breakdown occurring. By use of large clearances and a high-pressure insulating gas, terminal voltages as high as 12×10^6 V have been used, giving proton energies up to 24 MeV.

Much larger energies can be obtained from the second category of accelerators, those using repeated accelerations with a high-frequency ac voltage. This category includes cyclotrons, linear accelerators, betatrons, and synchrotrons, each of several types, but we confine our discussion to cyclotrons. It is easily shown from (9-9) that a particle with a charge e and mass M in a magnetic field of strength B applied perpendicular to its velocity v will move in a circular path with angular frequency ω given by

$$\omega = \frac{eB}{M} \tag{9-16}$$

The radius of the circle r is

$$r = \frac{v}{\omega}$$

whence the energy E is

$$E = \tfrac{1}{2}Mv^2 = \frac{e^2B^2r^2}{2M} \tag{9-17}$$

In a cyclotron the accelerating field is applied across two hollow D shaped electrodes called *dees*, as shown in Fig. 9-19. If the frequency of the electric voltage is equal to ω from (9-16), the phase of the particles relative to this voltage stays constant and the particles are therefore accelerated each time they cross the gap between the dees. As their energies increase, they spiral outward, but according to (9-16), ω is independent of energy so the phase of the ions relative to the applied electric voltage does not change in first order. However, there are two-second order effects that alter this conclusion: (1) the mass increase due to relativistic effects, and (2) as we have seen in connection with Fig. 9-11, the magnetic field must decrease with increasing radius in order to achieve vertical focusing, i.e., to keep the force in the vertical direction in Fig. 9-19 directed toward the median plane, for otherwise the particles will strike the dees. According to (9-16), both these effects decrease ω with increasing radius, so the accelerated particles drift out of phase with the applied voltage. This limits the number of turns and hence the energy obtainable from an ordinary cyclotron.

In *spiral-ridge cyclotrons* the average magnetic field increases with radius so as to compensate the effect of the relativistic mass increase in (9-16) and thereby keep ω constant. This, of course, introduces vertical defocusing, but it is compensated for by introducing azimuthal variations into the magnetic field as shown in Fig. 9-20*a*. Particle paths then have a strong curvature in regions where the field is strong and a weak curvature where the field is weak, whence

FIGURE 9-19 The magnet and dees in a cyclotron.

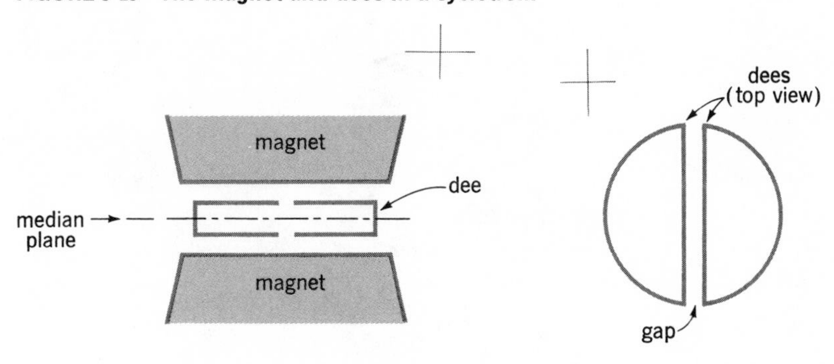

orbits are like the one shown in the figure. Effectively this is like passing the particles through three sector magnets, which gives a focusing in the vertical direction as discussed in connection with (9-12). This focusing can be enhanced by spiraling the azimuthal variations as shown in Fig. 9-20*b*. The path of a particle experiences this spiral as shown schematically in Fig. 9-20*c*. In view of the discussion leading to (9-12), this gives an alternately focusing and defocusing force, but due to the principle explained in Fig. 9-14, alternate focusing and defocusing gives a net focusing force.

Cyclotrons with magnetic field shapes like that in Fig. 9-20*b* allow particles to be accelerated to energies of several hundred MeV. When they reach their maximum energy, they can be deflected out of the magnetic field by various methods, one of which is by passing them through a region where there is a radial electric field.

Since cyclotrons and Van de Graaff accelerators are the most widely used machines in nuclear structure and reaction studies, it is interesting to compare their relative advantages. Cyclotrons can achieve much higher energies and can obtain any given energy much more cheaply. On the other hand, they produce a pulsed beam coming in short bursts with frequency ω; the beam is off between bursts, which is more than 90 percent of the time. This is a serious disadvantage in coincidence experiments since the ratio of true to accidental coincidences is determined by the instantaneous beam intensity. In a Van de Graaff accelerator, where the beam is steady in time, this same instantaneous beam intensity can be obtained without the beam's being off at all, whence the rate of data accumulation can be more than an order of magnitude faster. Another important advantage

FIGURE 9-20 Elements of a spiral-ridge cyclotron. *(a)* **Orbit of a particle. It is curved more than average in high-magnetic-field (B) regions and less than average in low-field regions, which leads to a focusing force in the vertical direction from the $v_r B_\theta$ term in (9-11), as explained in connection with Fig. 9-10.** *(b)* **Spiral of the field at large r. The additional focusing effect of this spiral is explained by** *(c)***; the particles cross the magnetic field edges at an angle (relative to the case of no spiral), which gives an alternate focusing and defocusing. This leads to a net focusing action as explained in connection with Fig. 9-14.**

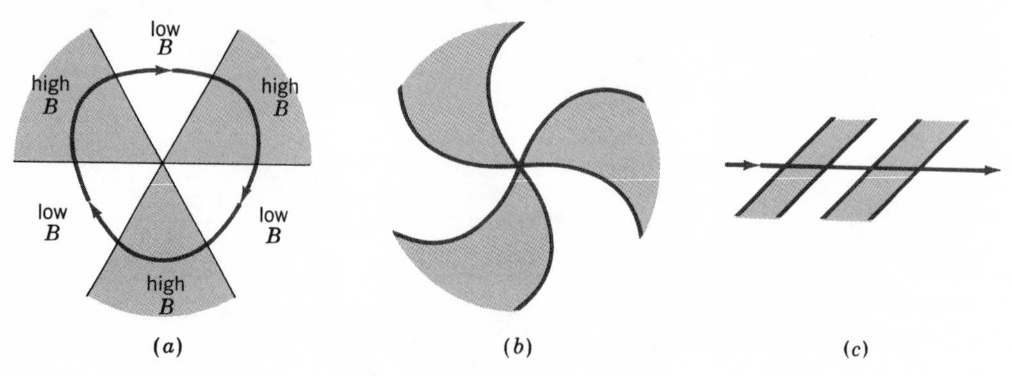

of the Van de Graaff is that it provides a beam with a very small energy spread, whereas in a cyclotron small variations in the positions of the orbit centers of different particles cause the emitted beam to have a rather large energy spread. This results in serious difficulties in experiments which require very high energy resolution, although methods of overcoming them have been developed.

9-9 Radiochemistry

Consider the problem of measuring the cross section for the reaction Cu^{65}-$(p,\alpha p)Co^{61}$ as a function of proton energy. One method would be to bombard a thin foil of Cu^{65} with protons of various energies and detect the emitted alphas and protons in coincidence at various combinations of angles. This would be an extremely difficult experiment. However, these cross sections can be measured much more easily by measuring the amount of Co^{61} produced. In fact, if a stack of Cu foils is used, the protons lose energy as they go through the stack at a rate accurately calculable from Fig. 9-3, whence by determining the amount of Co^{61} produced in each foil the cross section as a function of energy can be determined in a single bombardment. A determination of the amount N_0 of Co^{61} in a foil at the end of a bombardment is conveniently made by measuring the rate at which decay particles are emitted $(-dN/dt)$ as a function of time and fitting the results to (8-9) with the half-life of Co^{61}, which is 1.6 h. A difficulty arises, however, when there are other activities present. In bombarding copper with protons of high enough energy to induce $(p,\alpha p)$ reactions one induces many other reactions not only in Cu^{65} but in Cu^{63}, the other stable isotope of that element, and many of these also lead to radioactive nuclei. Some of their reactions and the half-lives of their products are

$$
\begin{aligned}
&Cu^{63}(p,2n)Zn^{62} && T_{1/2} = 9.2\text{ h}\\
&Cu^{63}(p,n)Zn^{63} && T_{1/2} = 38\text{ min}\\
&Cu^{65}(p,n)Zn^{65} && T_{1/2} = 245\text{ days}\\
&Cu^{65}(p,pn)Cu^{64} && T_{1/2} = 12.8\text{ h}\\
&Cu^{63}(p,\alpha n)Ni^{59} && T_{1/2} = 80{,}000\text{ years}
\end{aligned}
$$

In detecting the decay products from Co^{61}, one simultaneously detects products from these decays. In some cases, the separation can be made on the basis of gamma-ray energy or by analyzing the decay curve (count rate vs. time) into a sum of curves with the various half-lives. However, it is frequently impossible to do this with sufficient accuracy, especially when the activity of interest is present in only small amounts or when its half-life is not much different from some of the others. In such cases, the difficulty can be eliminated by performing a chemical separation of cobalt from the other elements present. This is an example of *radiochemistry*.

The procedure is to dissolve the Cu foils after the bombardment and add a known amount of nonradioactive Co to act as a *carrier*. The Co^{61} will mix with it thoroughly and follow it through all chemical processes. Nonradioactive carriers of Zn and Ni should also be added and chemical separations made of the cobalt from all these elements. If impurities are suspected, they can be similarly removed, and there are procedures for removing a large number of undesired elements with a single *scavenger carrier*. In the chemical procedures, some of the Co may be lost, but this can easily and accurately be corrected for by determining the amount of Co recovered at the end of the process and comparing it with the amount of carrier originally added (the Co^{61}, of course, has negligible mass). This allows extremely clean separations to be made without the painstaking care needed in ordinary quantitative chemical analysis.

Radiochemistry has been used extensively to determine the mass distribution of fission fragments. After bombardment, the uranium (or other fissionable material) is dissolved, and carriers of several of the product elements are added. After chemical processing to eliminate all but a single element from each sample, the number of nuclei of various radioactive isotopes is determined by counting decay particles. In bombardments with high-energy particles ($\gtrsim 50$ MeV), a large number of reactions take place, leading to isotopes of many different elements, so the problem is similar to that in fission and can best be approached with radiochemical techniques.

Radiochemistry is important in the preparation of radioactive isotopes; they must, of course, be made by nuclear reactions, but impurity elements with very large cross sections can sometimes contribute undesirable amounts of activity and must therefore be chemically removed. If necessary, separations can be made carrier-free, to eliminate mass from the sample; this must be done if the energy loss of the emitted particles in the material is important, in medical applications where the carrier material is chemically poisonous, etc.

Radiochemical techniques are almost always important when the amount of activity being studied is very small since impurities can then have important effects. This is the case in most applications of radioactivity to dating.

Problems

9-1 What thickness of aluminum is required to stop protons of 20 MeV? Protons of 5 MeV? Alpha particles of 5 MeV?

9-2 What thickness of aluminum will reduce the energy of 20-MeV protons to 19 MeV (use Fig. 9-2); to 10 MeV (use Fig. 9-3)?

9-3 If a scintillation detector yields one photoelectron for each 1,000 eV of energy deposited and the photomultiplier has 10 dynodes each of which emits four electrons for each electron striking it, how many electrons reach the anode of the photomultiplier when a 10-MeV proton strikes the scintillation crystal? If the capacitance from the anode to ground is 30 pF (1 pF = 10^{-12} F), what is the voltage of the output pulse?

9-4 A 10-MeV alpha particle loses all its energy in a proportional counter; one electron-ion pair is produced for each 30 eV of energy loss. The proportional counter has a multiplication $M = 500$, and the total capacitance between the wire and ground is 30 pF. What is the voltage of the output pulse?

9-5 What is the radius of curvature of a 20-MeV proton in a magnetic field of 1 Wb/m^2 (10,000 gauss)?

9-6 A wedge-magnet spectrograph is designed for normal entry and exit ($\gamma = \delta = 0$) and for equal object and image distances ($p = q$). If particles are deflected by 45° and $p = 60$ in., what magnetic field is needed to analyze 20-MeV protons?

9-7 Show that when a source is in a uniform magnetic field, the particles are focused after being turned through 180°. Show that the second-order aberration in this case is $\rho\alpha^2$.

9-8 If the angular distribution in neutron-proton scattering is isotropic in the center-of-mass system, describe the pulse-height distribution obtained when 10-MeV neutrons are incident on a plastic scintillator.

9-9 What thickness of lead reduces the flux of 1-MeV gamma rays by a factor of 10?

9-10 If σ_T for 20-MeV neutrons on Pb is 4 barns, what thickness of Pb is required to attenuate a neutron beam by a factor of 100?

9-11 A coincidence analyzer with a resolving time of 5 ns is used to study the decays from a source in which two gamma rays are emitted simultaneously. If one desires a true-to-accidental ratio of 5, how strong a source, i.e., how many decays per second, should be used? If each detector has a 1-in. diameter and is located 2 in. from the source, how many true coincidences per second are detected, and what is the count rate in each detector? If the efficiency of the detectors for detecting gamma rays which pass through them is only 10 percent, how are the results changed?

9-12 The time resolution in a neutron time-of-flight experiment is 1 ns, and the flight path is 1 m. Calculate the energy resolution for neutron energies between 0.1 and 20 MeV. Do the same for a 10-m flight path.

9-13 A cyclotron of 30-in. radius has a magnetic field of 1 Wb/m^2. To what energy can it accelerate protons? What frequency must be applied to the dees? If the voltage between the dees when the protons cross the dee gap is 50 kV, how long does the acceleration last?

9-14 Discuss the complications in trying to measure the $Cu^{63}(p,2n)$ cross section without radiochemical separations. Assume that the cross sections for (p,pn) and $(p,2n)$ are about equal.

Further Reading

See Elementary Textbooks, in General References, following the Appendix.

Ajzenberg-Selove, F.: "Nuclear Spectroscopy," Academic, New York, 1960.

Barkas, W. H.: "Nuclear Research Emulsions," Academic, New York, 1963.

Birks, J. B.: "The Theory and Practice of Scintillation Counting," Macmillan, New York, 1964.

Bleuler, E., and G. J. Goldsmith: "Experimental Nucleonics," Rinehart, New York, 1952.

Chase, R. L.: "Nuclear Pulse Spectrometry," McGraw-Hill, New York, 1961.

Faires, R. A., and B. H. Parks: "Radioisotope Laboratory Techniques," Newnes, London, 1964.

Livingood, J. J.: "Principles of Cyclic Particle Accelerators," Van Nostrand, Princeton, N.J., 1951.

Livingston, M. S., and J. P. Blewett: "Particle Accelerators," McGraw-Hill, New York, 1962.

Marion, J. B., and J. L. Fowler: "Fast Neutron Physics," Interscience, New York, 1960.

Price, J. W.: "Nuclear Radiation Detection," 2d ed., McGraw-Hill, New York, 1964.

Rossi, B. B.: "High Energy Particles," Prentice-Hall, New York, 1952.

—— and H. S. Staub: "Ionization Chambers and Counters," McGraw-Hill, New York, 1949.

Segre, E.: "Experimental Nuclear Physics," Wiley, New York, 1953.

Shutt, R. P.: "Bubble and Spark Chambers," Academic, New York, 1967.

Siegbahn, K.: "Alpha, Beta, and Gamma Ray Spectroscopy," North-Holland, Amsterdam, 1965.

Yuan, L. C. L., and C. S. Wu: "Methods of Experimental Physics: Nuclear Physics," Academic, New York, 1961.

Chapter 10
Nucleon Emission

In Chap. 8, we introduced the three basic types of nuclear decay in broad outline and showed how they limit the number of stable nuclei. The explanations there were largely on the basis of energy conservation. While conservation laws are often of great value in obtaining results, they hide many of the interesting details. For example, a frictionless roller coaster can be treated by energy conservation to find the velocity difference at two widely separated points, but in such a treatment one loses the interesting story of all the thrilling accelerations in between.

In this and the following two chapters we approach decay processes from a more basic viewpoint and treat them in some detail. In doing so, the principal new observable result we obtain is the decay rate λ, introduced in Sec. 8-5. However, the real payoff from a treatment of this type is not in predicting results but in achieving an understanding of what goes on in these processes.

In this chapter, we discuss the first of the three types of decay, nucleon emission. However, before launching into this treatment, we must divert briefly in Sec. 10-1 to develop two additional quantum-mechanical concepts.

10-1 Reflection and Transmission of Waves at Interfaces

In Chap. 2 we considered the quantum-mechanical problem of a particle trapped in a potential well. The results, as we have seen, have wide applications to the structure of the nucleus since, after all, the nucleus is a group of trapped nucleons. However, when we deal with decays and reactions, we no longer have trapped particles. This difference is easily stated from the wave point of view: a particle trapped in a potential is represented by a *standing wave*, whereas a particle involved in a decay or a reaction corresponds to a *traveling wave*.

In this section we treat two of these traveling-wave problems. In the first, we consider what happens to a particle when the potential in which it is moving changes. In the classical case this corresponds to a ball rolling up or down a hill, and we know the answer: it just changes velocity in accordance with

$$\tfrac{1}{2}Mv^2 = E - V$$

which can be solved to give

$$v = \left[\frac{2}{M}(E - V)\right]^{1/2} \tag{10-1}$$

However, when we consider the wave nature of matter and realize that the wavelength depends on the potential through (2-1), other effects arise. In this sense the problem is like that of a wave traveling down a string and encountering a place where the string is attached to a rope; in general, the wave will be partly reflected back up the string and partly transmitted down the rope. If the rope is very much heavier than the string, the wave will be almost completely reflected. This can be traced to the fact that the wavelength for a given frequency depends on the mass density, and there is a partial reflection when the wavelength changes. In analogy with this, we might expect that a particle which is represented by a wave will be partly reflected when it encounters a change in potential since this gives a change in wavelength.

Let us treat this problem in detail. A traveling wave in one dimension might be represented by

$$\psi \sim \sin(kx \mp \omega t)$$

where $k = 2\pi/\lambda$. The upper (minus) sign corresponds to a wave traveling to the right since, as t increases, the same argument of the sine occurs for increasing values of x; similarly, we deduce that the lower (plus) sign corresponds to a wave traveling to the left.

It is often convenient in mechanical and electromagnetic as well as quantum mechanics problems to introduce a complex notation and represent this by

$$\psi \sim \exp[i(kx \mp \omega t)]$$

or equivalently

$$\psi \sim e^{-i\omega t}e^{\pm ikx} \tag{10-2}$$

where the upper and lower signs retain the meaning indicated above. Students with experience in quantum mechanics will recognize that the time dependence must be chosen as in (10-2) in order to satisfy the time-dependent Schrödinger equation.

Now let us consider the situation illustrated in Fig. 10-1. A stream of particles of energy E $(E > 0)$ is traveling to the right when it encounters a change in potential from $-V_0$ to 0 at some value of x which for convenience we choose as $x = 0$. As we suspect that there may be some reflection, we take the wave function from (10-2) as

$$\psi = \begin{cases} (Ae^{ikx} + Be^{-ikx})e^{-i\omega t} & x < 0 \\ Ce^{ik'x}e^{-i\omega t} & x > 0 \end{cases}$$

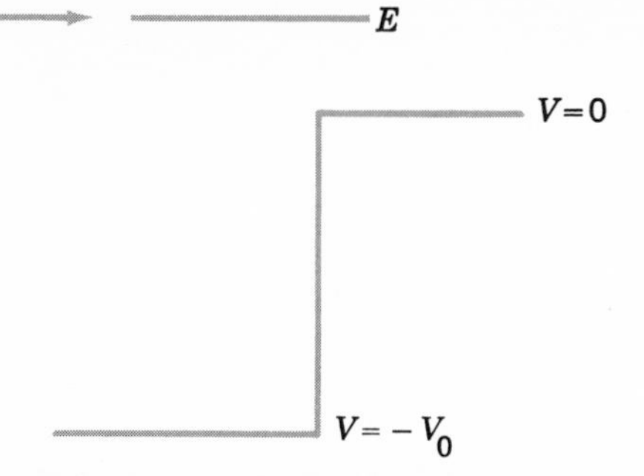

FIGURE 10-1 A change in the potential in which a particle is moving with energy E.

where, in accordance with (2-3),

$$
\begin{aligned}
k &= \frac{1}{\hbar\sqrt{2M(E - V)}} = \frac{1}{\hbar\sqrt{2m(E + V_0)}} \\
k' &= \frac{1}{\hbar\sqrt{2ME}}
\end{aligned}
\tag{10-3}
$$

From the continuity conditions on ψ and $d\psi/dx$ at $x = 0$, which were justified in Sec. 2-3, we find

$$
\begin{aligned}
A + B &= C \\
ikA - ikB &= ik'C
\end{aligned}
$$

These are readily solved to give

$$
\begin{aligned}
\frac{B}{A} &= \frac{k - k'}{k + k'} \\
\frac{C}{A} &= \frac{2k}{k + k'}
\end{aligned}
\tag{10-4}
$$

The probabilities for particles in some range of x values to be in the incident, reflected, and transmitted waves are proportional to A^2, B^2, and C^2, respectively. These probabilities are proportional to the *currents*, i.e., the number of particles per second passing each x value, times the length of time they spend in the range of x values, which is inversely proportional to their velocity. Since, from (10-1) and (10-3), velocities are proportional to k values, the incident $\mathcal{I}$, reflected $\mathcal{R}$, and transmitted $\mathcal{T}$ *currents* are proportional to kA^2, kB^2, and $k'C^2$, respectively. Thus from (10-4) we obtain finally

$$
\begin{aligned}
\frac{\mathcal{R}}{\mathcal{I}} &= \left(\frac{k - k'}{k + k'}\right)^2 \\
\frac{\mathcal{T}}{\mathcal{I}} &= \frac{k'}{k}\left(\frac{2k}{k + k'}\right)^2
\end{aligned}
\tag{10-5}
$$

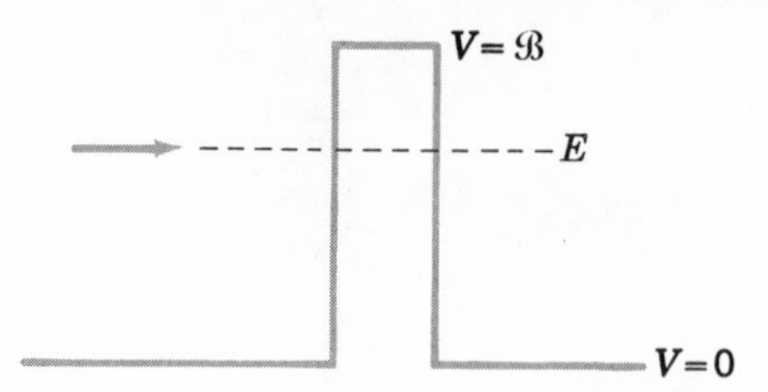

FIGURE 10-2 A one-dimensional barrier.

It is easily checked that the sum of these is unity, as expected from the fact that the sum of the reflected and transmitted currents must be equal to the incident current.

In our applications of (10-5) we shall be especially interested in the case where $k' \ll k$, for which the second of (10-5) becomes

$$\frac{\mathcal{T}}{\mathcal{I}} = \frac{4k'}{k} \tag{10-6}$$

As E goes to zero, k' from (10-3) and hence $\mathcal{T}$ from (10-6) go to zero, and the particle is certain to be reflected.

The second problem we consider in this section is one of barrier penetration, as illustrated in Fig. 10-2. A stream of particles is moving to the right with total energy E when it encounters a *barrier*, i.e., a region where $E < V$. In the classical case this corresponds to a ball rolling up a hill with a total energy less than its potential energy at the top of the hill. Clearly, it will stop and roll back down, never getting over the top of the hill: it will not penetrate the barrier.

For the quantum case, in analogy with the treatment of reflection above, we can write the wave function as

$$\psi = \begin{cases} Ae^{ikx} + Be^{-ikx} & x < 0 \\ De^{\kappa x} + Fe^{-\kappa x} & 0 < x < b \\ Ce^{ikx} & x > 0 \end{cases} \tag{10-7}$$

where

$$k = \frac{1}{\hbar\sqrt{2ME}} \qquad \kappa = \frac{1}{\hbar\sqrt{2M(\mathcal{B} - E)}}$$

Note that, in contrast to the situation in Sec. 2-2, we cannot disregard one of the two exponential solutions here because the region in which they apply does not extend to infinity. Matching values of ψ and $d\psi/d\psi$ at $x = 0$ and at $x = b$ leads to

$$\begin{aligned} A + B &= D + F \\ ikA - ikB &= \kappa D - \kappa F \\ De^{\kappa b} + Fe^{-\kappa b} &= Ce^{ikb} \\ \kappa De^{\kappa b} - \kappa Fe^{-\kappa b} &= ikCe^{ikb} \end{aligned}$$

Solving these simultaneously gives, after lengthy calculation, a barrier penetration $\mathcal{P}$ as

$$\mathcal{P} = \frac{\mathfrak{J}}{\mathfrak{g}} = \left|\frac{C}{A}\right|^2 = \left[1 + \frac{V_0^2(e^{\kappa b} - e^{-\kappa b})^2}{16E(V_0 - E)}\right]^{-1} \tag{10-8}$$

which is by no means zero. Thus we see that, because of their wave nature, particles can penetrate through barriers which are classically impenetrable. In fact, for small barrier thickness b, the right side of (10-8) approaches unity so we have 100 percent transmission. For large values of κb, $e^{\kappa b} \gg e^{-\kappa b}$, and the second term in the bracket is much larger than unity, whence (10-8) becomes

$$\mathcal{P} = \frac{16E(V_0 - E)}{V_0^2} e^{-2\kappa b} \tag{10-9a}$$

The coefficient of the exponential in (10-9*a*) is sensitive to details of the problem such as the potentials for $x < 0$ and $x > b$ and includes some of the reflection effects discussed above. The assumptions made here for the sake of simplicity are not realistic for nuclear problems in this regard, so we shall treat reflections by use of (10-6). The exponential, however, is a very important term here, as it can vary by many orders of magnitude; it gives the dominant effect of barrier-penetration considerations. We shall therefore use

$$\mathcal{P} \simeq e^{-2\kappa b} \tag{10-9b}$$

If the barrier does not have a rectangular shape so that κ is a function of x, it can be shown that (10-9*b*) is modified to

$$\mathcal{P} \simeq \exp\left(-2 \int_0^b \kappa \, dx\right) \tag{10-10}$$

The phenomenon of barrier penetration, while it may seem strange at first, can be easily understood in terms of the uncertainty principle. The energy of the particle may vary by an amount ΔE sufficient to make $E > \mathcal{B}$ provided this does not last much longer than a time Δt given by (5-4). If this time is long enough for the particle to traverse a distance b, it can penetrate the barrier. For example, let us say ΔE corresponds to an increase of energy as high above the barrier as it was originally below the barrier, or

$$\Delta E = 2(\mathcal{B} - E)$$

The velocity, from (10-1), is then

$$v = \left[\frac{2}{M}(\mathcal{B} - E)\right]^{1/2}$$

and the time Δt required to traverse the barrier is

$$\Delta t = \frac{b}{v} = b\left[\frac{2}{M}(\mathcal{B} - E)\right]^{-1/2}$$

Combining these and using (10-7) gives

$$\Delta E\, \Delta t = b[2M(\mathcal{B} - E)]^{1/2} = \hbar \kappa b$$

According to the uncertainty principle, then, the barrier can be penetrated provided

$$\kappa b \lesssim 1$$

This is crudely the result we get from (10-9*b*) although that equation is still valid for $\kappa b \gg 1$.

From this derivation it is easy to see why a particle with small mass can more easily penetrate barriers than a heavy particle. For a given ΔE, the former attains a higher velocity and hence takes a shorter time to cross the barrier region. Electrons therefore penetrate barriers with ease, whereas alpha particles have a considerably harder time than protons in penetrating barriers, and the difficulty is increased manyfold for fission fragments. That is why all the nuclei of $A \gtrsim 100$, which were shown in Sec. 8-3 to be unstable with respect to fission, are present on earth.

The phenomenon of barrier penetration is often called *tunneling through the barrier*, but stated in that way it runs contrary to a basic concept of physics. In view of our treatment with the uncertainty principle, it seems more palatable to call it *jumping over the barrier* during a fluctuation in its energy.

10-2 Decay Rates in Nucleon Emission—$l = 0$ Neutron Emission

We are now ready to embark on our treatment of nucleon emission. We shall progress in easy steps, starting with the simplest situation and then building up in complexity. To begin with, let us consider a possible state of the nucleus Fe^{57} whose wave function ψ_1 corresponds to the ground state of Fe^{56} plus a neutron in the $3s_{1/2}$ orbit, or

$$\psi_1(\mathrm{Fe}^{57}) = \psi[\mathrm{Fe}^{56}(\mathrm{GS})]\psi(3s_{1/2}) \qquad \textbf{(10-11)}$$

Since the ground state of Fe^{56} (an even-even nucleus) is 0^+, this state would have I^π of $\frac{1}{2}^+$. Since Fe^{57} has 31 neutrons, we see from Fig. 4-5 that in its low-energy states the $\mathfrak{N} = 4$ shell is just beginning to fill. The $3s_{1/2}$ orbit is in the $\mathfrak{N} = 5$ shell, which is at a considerably higher energy, so the state represented by ψ_1 is a highly excited state of Fe^{57}. In fact the $3s_{1/2}$ orbit is not even bound in this nucleus; let us say it is above the top of the well by 1 MeV, whence, from (4-11), the state ψ_1 has an excitation energy of about 9 MeV. Since the neutron is unbound, it is free to be emitted in a nucleon-emission process, written

$$\mathrm{Fe}^{57}(\tfrac{1}{2}^+) \rightarrow \mathrm{Fe}^{56}(\mathrm{GS}) + n(3s_{1/2}) \qquad \textbf{(10-12)}$$

It will reach the nuclear surface in a time of the order R/v_i, where R is the nuclear radius and v_i is its velocity inside the nucleus, but on reaching the surface it encounters a change of potential, so there is a good chance that it will be reflected back into the nucleus. In accordance with (10-6), it will probably be reflected a number of times equal to $k/4k'$; since k is proportional to velocity, this is $v_i/4v_e$, where v_e is its velocity outside the nucleus, which is the velocity with which it is eventually emitted. Hence, the average time τ_0 required for neutron emission in this case is

$$\tau_0 \simeq \frac{R}{v_i}\frac{v_i}{4v_e} = \frac{R}{4v_e} \tag{10-13a}$$

For a 1-MeV neutron use of the more accurate transmission formula (10-5) would change this to $\tau_0 \simeq R/3v_e$. Also, the distance traveled between reflections is closer to $2R$ than to R, and to get completely out of the nucleus (4-1) indicates that the neutron must reach a radius considerably larger than R. For these and other subtler reasons, a more accurate expression for τ_0 is[1]

$$\tau_0 \simeq \frac{R}{v_e} \tag{10-13b}$$

The relationship between the energy and velocity of a particle of mass M is

$$v = 1.4 \times 10^9 \text{ cm/s} \left[\frac{E(\text{MeV})}{M/M_p}\right]^{1/2}$$

whence (10-13*b*) becomes, with (4-2),

$$\tau_0 = 3.5 \left[\frac{E(\text{MeV})}{M/M_p}\right]^{-1/2} \left(\frac{A}{57}\right)^{1/3} \times 10^{-22} \text{ s} \tag{10-14}$$

The emission rate λ_0, or probability per unit time for nucleon emission, is $1/\tau_0$, whence, from (10-13*b*) and (10-14),

$$\lambda_0 \simeq \frac{v_e}{R} \simeq 3 \left[\frac{E(\text{MeV})}{M/M_p}\right]^{1/2} \left(\frac{A}{57}\right)^{-1/3} \times 10^{21} \text{ s}^{-1} \tag{10-15}$$

which, for the decay (10-12) is 3×10^{21} s^{-1}.

The situation we have been considering here, where the wave function is as given by (10-11), is most unrealistic for a state of 9 MeV excitation energy. There are many other configurations of the same I^π (in this case $\frac{1}{2}^+$) in this energy region, and there is every probability, in view of the discussions of Chaps. 5 and 6,

[1] This subject is usually discussed in terms of the *Wigner limit*, which is equivalent to $R/3v_e$. It was originally believed that τ_0 was of this order, but empirical evidence seems to indicate that (10-13*b*) is close to the correct value.

that these configurations will mix, i.e., that in any nuclear state, the nucleus will spend some time in each of these configurations, going from one to the other through collisions. Some of these other configurations are

$Fe^{56}(1^-)2p_{3/2}$	$Fe^{56}(2^+)2d_{5/2}$
$Fe^{56}(2^-)2p_{3/2}$	$Fe^{56}(3^+)2d_{5/2}$
$Fe^{56}(0^-)2p_{1/2}$	$Fe^{56}(3^+)1g_{7/2}$
$Fe^{56}(1^-)2p_{1/2}$	$Fe^{56}(4^+)1g_{7/2}$
$Fe^{56}(2^-)1f_{5/2}$	$Fe^{56}(1^+)3s_{1/2}$
$Fe^{56}(3^-)1f_{5/2}$	$Fe^{56}(0^+)3s_{1/2}$
$Fe^{56}(5^+)1g_{9/2}$	
$Fe^{56}(4^+)1g_{9/2}$	

As an example, the first entry in the list, $Fe^{56}(1^-)2p_{3/2}$, is a configuration consisting of a 1^- state of Fe^{56} plus a $2p_{3/2}$ neutron coupled to give a $\frac{1}{2}^+$ term. Since the $2p_{3/2}$ orbit is filling in the ground state of Fe^{57} (see Fig. 4-5), there is little excitation energy associated with the added neutron's being in this orbit, so for the excitation energy to be about 9 MeV, $Fe^{56}(1^-)$ must be a state of about 9 MeV excitation in Fe^{56}. But since from Sec. 5-11 we know that Fe^{56} has a large number of states of every I^π in this energy region, the first entry in the list may correspond to dozens of different configurations. Similar statements can be made about many of the other entries in the list, so there are perhaps a thousand terms in the wave function of any $\frac{1}{2}^+$ state in Fe^{57} at 9 MeV excitation, only one of which is (10-11). Typically, then, in any one of these states the configuration is (10-11) only about one one-thousandth of the time. But only when it is in this configuration can it decay by (10-12), so the decay rate is slower than (10-15) by a factor of about 1,000. This means $\lambda \simeq 3 \times 10^{18}\ s^{-1}$, which, from (8-11) corresponds to a half-life of 2.3×10^{-19} s.

To treat this problem more generally, let us say the amplitude of the term (10-11) in the wave function for any $\frac{1}{2}^+$ nuclear state i is θ_i. The fraction of the time the nucleus is in the configuration (10-11) is the square of its amplitude, θ_i^2, and during this time the decay rate is λ_0; hence the overall rate λ_i of decay of the state i is

$$\lambda_i = \lambda_0 \theta_i^2 \tag{10-16}$$

The quantity θ_i^2 is called the *reduced width* of the state i.†

Since amplitudes of terms in wave functions vary widely, θ_i^2 can vary by a large factor from one state to another; some states may have a larger and others

† The quantity that we call θ_i^2 is more often called θ_i^2/θ_0^2, the ratio of the reduced width to the single-particle reduced width. Historically λ_0 was taken to correspond to the Wigner limit, $\sim R/3v_e$, and θ_0^2 was later determined to be about $\frac{1}{3}$. By choosing $\lambda_0 = R/v_e$ in (10-13), we make $\theta_0^2 \simeq 1$.

a smaller piece of (10-11). But all the pieces must be in some nuclear state, so we have the *sum rule,*

$$\sum_i \theta_i^2 = 1 \tag{10-17}$$

where the sum is over all nuclear states of Fe^{57}. Of course only $\frac{1}{2}^+$ states can have this term; and since the configuration (10-11) will mix only with other configurations in the same energy region, the major contributions to the sum in (10-17) are from nuclear states in the energy region near 9 MeV of excitation.

For more advanced students, (10-17) can be derived as follows. The wave functions for all $\frac{1}{2}^+$ states of Fe^{57} can be written

$$\phi_i = \sum_k c_{ik}\psi_k \tag{10-18}$$

where ψ_1 is from (10-11) and the other ψ_k are all the other $\frac{1}{2}^+$ terms. All ϕ_i as well as all ψ_i must be orthonormal, whence

$$\begin{aligned} \int \phi_i^* \phi_j \, d\tau &= \delta_{ij} \\ \int \psi_i^* \psi_j \, d\tau &= \delta_{ij} \end{aligned} \tag{10-19}$$

Since c_{ik} is unitary, (10-18) implies

$$\psi_i = \sum_k c_{ki}^* \phi_k$$

Inserting this into the second of (10-19) and using the first of (10-19) gives

$$\sum_k c_{ki} c_{kj}^* = \delta_{ij}$$

For $i = j = 1$, this is

$$\sum_k |c_{k1}|^2 = 1 = \sum_i |c_{i1}|^2 \tag{10-20}$$

where the last is just a change of index designation. The c_{i1} are the amplitudes of ψ_1 in the wave functions ϕ_i of the nuclear states which is just our definition of θ_i, so (10-20) is the same as (10-17).

10-3 Decay Rates in Nucleon Emission—Penetration of Angular-momentum Barriers

In Sec. 10-2 we restricted our discussion to emission of $s_{1/2}$ neutrons. Neutrons in other unbound orbits can also be emitted, but in such cases there is the added complication of an angular momentum barrier. The existence of this barrier has already been seen in Fig. 2-4. For example, the potential well for $l = 3$ neutrons from that figure is reproduced in Fig. 10-3. A neutron with energy E, as shown there, must penetrate a barrier to be emitted.

This barrier can be understood physically if we consider a ball rolling around a circular "orbit" part way up the sides of a bowl-shaped depression. Its kinetic

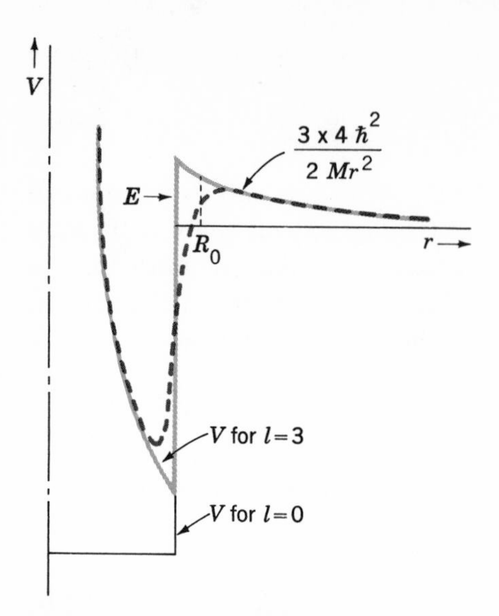

FIGURE 10-3 The angular-momentum barrier for an $l = 3$ neutron. In the square-well approximation, the total effective potential (thick solid line) is the sum of the nuclear potential (thin solid line) and the term representing the centrifugal force in (2-18). Due to the diffuse edge of the potential from (4-1), the actual potential is as shown by the dashed line. The barrier it introduces can be approximated by the shaded area. A particle with energy E and $l = 3$ must penetrate this barrier to be emitted.

energy is sufficient, if it is directed in a radial direction, to allow the ball to climb the side of the bowl and come out. However, its velocity cannot be spontaneously redirected to the radial direction since this would reduce its angular momentum and thereby violate conservation of that quantity. On the other hand, if it could somehow get out of the bowl and get far away, it could still have its original angular momentum in spite of its loss of kinetic energy in climbing the side of the bowl, because it would then have a large value of r, and angular momentum, for a given velocity, is proportional to r. It is therefore only a barrier rather than a deficiency in energy that prevents the ball from coming out.

To understand the barrier quantitatively, we note that if a nucleon at radius r is to have an angular momentum $\mathbf{r} \times \mathbf{p} = \sqrt{l(l+1)}\,\hbar$, its momentum p must be at least

$$p = \frac{\sqrt{l(l+1)}\,\hbar}{r}$$

Its kinetic energy must therefore be at least

$$E - V = \frac{p^2}{2M} = \frac{l(l+1)\hbar^2}{2Mr^2} \tag{10-21}$$

This is the equation of the curve shown in Fig. 10-3. Classically, the nucleon cannot get to a radius where its energy E is less than the value given by (10-21).

Returning to the problem of neutron emission, we recognize from our discussions of Sec. 10-1 that the barrier can be penetrated with a probability $\mathcal{P}$, calculable from (10-10). Thus, the decay rate (10-16) is modified to

$$\lambda_i = \lambda_0 \theta_i^2 \mathcal{P} \tag{10-22}$$

This is the complete expression for the decay rate in nucleon emission.

The calculation of $\mathcal{P}$ can be carried out exactly in the square-well approximation. Since the nuclear potential is zero in the region of the barrier (see Fig. 10-3), the barrier height is the right side of (10-21), whence, from (10-7)

$$\kappa = \frac{l(l+1)}{r^2} - \frac{2ME}{\hbar^2}$$

and the integral in (10-10) is between $r = R_0$, the radius of the square well, and the value of r where $\kappa = 0$. This integral is readily evaluated to give, finally,

$$\mathcal{P} = \exp\left[-2\sqrt{l(l+1)}\left(\log\frac{\alpha + \sqrt{\alpha^2 - R_0^2}}{R_0} - \sqrt{\frac{1 - R_0^2}{\alpha^2}}\right)\right] \tag{10-23}$$

where

$$\alpha^2 = \frac{l(l+1)\hbar^2}{2ME} = \frac{R_0^2 \mathcal{B}_l}{E} \simeq \frac{20l(l+1)}{E(\text{MeV})M/M_p} \qquad \text{F}^2 \tag{10-24}$$

In (10-24) $\mathcal{B}_l$ is the peak barrier height, which is its value at $r = R_0$, and the last expression is inserted for convenience in numerical calculations.

When we consider that the actual potential well (4-1) has rounded edges, the shape of the barrier is modified to that indicated by the dashed curve in Fig. 10-3. We see there that this modification eliminates the highest part of the barrier and therefore substantially reduces the integral in (10-10). Accurate calculations of $\mathcal{P}$ for this modified barrier can be done with computers, but for hand calculations it is more practical to retain (10-23) and mock up the effect by increasing the radius of the square well. This converts the barrier to the cross-hatched one in Fig. 10-3. The results of accurate calculations are then reasonably well approximated if we choose this radius to be

$$R_0 \simeq R + 2a \tag{10-25}$$

with R and a taken from (4-2).

To illustrate the effect of angular-momentum barriers in neutron emission we return to the example of Sec. 10-2. Let us say that one particular nuclear state i can decay by (10-12) with an emitted neutron energy of 1.5 MeV. It is also possible for this state to decay by

$$\mathrm{Fe}^{57}(\tfrac{1}{2}^+) \rightarrow \mathrm{Fe}^{56}(2^+) + n(d_{5/2}) \tag{10-26}$$

where $\mathrm{Fe}^{56}(2^+)$ is the first excited state of Fe^{56}, which has an excitation energy of 0.9 MeV; since the final nucleus has 0.9 MeV more excitation here than in the transition to the ground state, the neutron must have 0.9 MeV less energy, or 0.6 MeV. From (10-25) and (4-2), $R_0 = 6.1$ F, and from (10-24) with $l = 2$ (the neutron is in a $d_{5/2}$ orbit) and $E = 0.6$ MeV, $\alpha^2 = 200$ F^2. Inserting these into (10-23) gives $\mathcal{P} = \exp(-2.9) = \frac{1}{18}$; ignoring differences from the θ_i^2 in (10-22), the decay (10-12) occurs at a rate 18 times faster than (10-26).

There is now no reason to limit our consideration to $\frac{1}{2}^+$ states; states of all I^π in Fe^{57} can decay to states of all I^π in Fe^{56} by emission of neutrons with various angular momenta provided that angular momentum and parity are conserved. The rate of each of these decays is determined by (10-22). Very often, a decay from one state to another can take place in different ways. For example, a highly excited $\frac{9}{2}^+$ state (i) of Fe^{57} might well have terms in its wave function like

$$\begin{array}{ll} \theta_{i1}\mathrm{Fe}^{56}(4^+)3s_{1/2} & \theta_{i5}\mathrm{Fe}^{56}(4^+)2d_{5/2} \\ \theta_{i3}\mathrm{Fe}^{56}(4^+)2d_{3/2} & \theta_{i9}\mathrm{Fe}^{56}(4^+)1g_{9/2} \end{array}$$

where $\mathrm{Fe}^{56}(4^+)$ is the second excited state of Fe^{56}, which has an excitation energy of 2.1 MeV. The decay

$$\mathrm{Fe}^{57}(\tfrac{9}{2}^+) \rightarrow \mathrm{Fe}^{56}(4^+) + n$$

can then take place by emission of an $s_{1/2}$ neutron with reduced width θ_{i1}^2, a $d_{3/2}$ neutron with reduced width θ_{i3}^2, a $d_{5/2}$ neutron with reduced width θ_{i5}^2, and a $g_{9/2}$ neutron with reduced width θ_{i9}^2. For each of these there is a separate $\mathcal{P}$ and a separate λ_{ij} calculated from (10-22) with the appropriate θ_{ij}^2, and the total probability per unit time for decay λ_i is just the sum of the probabilities for decaying in each of these four ways, or

$$\lambda_i = \sum_j \lambda_{ij}$$

10-4 Decay Rates in Nucleon Emission—Penetration of Coulomb Barriers

In the emission of protons all considerations discussed in connection with neutrons are still valid, but there is an added complication, namely, a coulomb barrier. The existence of this barrier can be seen in Fig. 4-2, and it is shown in the square-well approximation in Fig. 10-4. It arises from adding the coulomb to the nuclear potential. It may seem strange that the coulomb force, which is repulsive, should prevent a proton from escaping from the nucleus. The reason

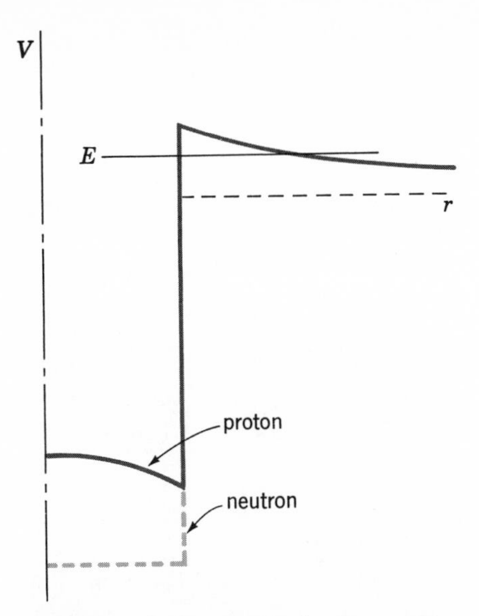

FIGURE 10-4 The coulomb plus nuclear potential in the square-well approximation, which gives a coulomb barrier. A proton with energy E must penetrate this barrier to be emitted.

for this is that the effects of the coulomb force have been discounted in advance in determining the energy. A proton with a given total energy E has much less *kinetic* energy than a neutron with that total energy, as we see from Fig. 10-4, because the bottom of the well is higher for protons. As a nucleon leaves the nucleus, it is slowed down by the nuclear forces of all the other nucleons pulling it inward (i.e., by "climbing" the side of the shell-theory potential which represents these forces) and a proton is slowed down by these forces just as much as a neutron. Since a proton has lower kinetic energy, it is more easily stopped.

The decay rate for proton emission is still given by (10-22) except that $\mathcal{P}$ now refers to penetration through the sum of the coulomb and angular-momentum barriers. For the emission of $s_{1/2}$ protons, only the coulomb barrier is present, and (10-10) gives

$$\mathcal{P} \simeq \exp\left[-2\sqrt{\frac{2M}{\hbar^2}}\int\left(\frac{Ze^2}{4\pi\epsilon_0 r} - E\right)^{1/2} dr\right] \tag{10-27}$$

where the range of integration is between $r = R_0$ and the radius where the integrand becomes zero. This integral can be evaluated to give, finally,

$$\mathcal{P} = \exp\left[-\beta f(x)\right] \tag{10-27a}$$

where

$$\beta = \frac{4zZe^2}{\hbar v} = 0.63zZ\left[\frac{M/M_p}{E(\text{MeV})}\right]^{1/2}$$
$$f(x) = [\cos^{-1} x^{1/2} - \sqrt{x(1-x)}]$$
$$x = \frac{E}{\mathcal{B}_c}$$
$$\mathcal{B}_c = \frac{zZe^2}{4\pi\epsilon_0 R_0}$$

We have used ze for the charge of the emitted particle to make these formulas valid for all charged particles; for protons, $z = 1$. $\mathcal{B}_c$ is the peak of the potential barrier which occurs at $r = R_0$, and $v = (2E/M)^{1/2}$, the velocity of the particle after emission.

For a medium-mass nucleus like Sn^{120} a calculation similar to (1-3*a*) with R_0 taken from (10-25) gives the maximum of the potential as 9.6 MeV. $\mathcal{P}$ can then be calculated from (10-27*a*) to be $e^{-6} \simeq 1/400$ for a 4-MeV proton, and $e^{-29} \approx 10^{-13}$ for a 1-Mev proton.

10-5 Reduced Widths for Emission of Alpha Particles and Fission

Let us consider the alpha decay

$$Po^{212}(GS) \rightarrow Pb^{208}(GS) + He^4(GS) \tag{10-28}$$

where GS refers to the ground state. Since all three of these nuclei are even-even and thus have $I = 0$, the orbital angular momentum l_α with which the alpha particle comes off must be zero to conserve total angular momentum. The reduced width $\theta_G{}^2$ for this decay is an expression of how much the wave function for Po^{212} in its ground state is like the product of the wave functions for Pb^{208} in its ground state, He^4 in its ground state, and the alpha particle moving relative to the Pb^{208} nucleus with $l_\alpha = 0$. This sentence will have to satisfy the less advanced students, who may skip to the paragraph leading up to (10-34), where they will see an example of its mathematical counterpart in (10-35); or if time is short, they may omit the remainder of this section.

Students with more experience in quantum mechanics will recognize that, in general, the wave function for Po^{212}(GS) can be written[1]

$$\psi[Po^{212}(GS)] = \Sigma\theta_{pq}\psi[Pb^{208}(p)]\psi[He^4(q)]\psi_\alpha \tag{10-29}$$

[1] In general the sum in (10-29) is also over l_α and its z component m_α, and θ should have subscripts corresponding to these, but this will not be necessary in the present example.

This is a linear combination of products of wave functions for all the states (p) of Pb^{208}, all the states (q) of He^4, and all the states of motion ψ_α of an alpha particle relative to a Pb^{208} nucleus given by the familiar expression for outgoing spherical waves

$$\psi_\alpha \propto (e^{-i\mathbf{k}_\alpha \cdot \mathbf{r}_\alpha}) \propto j_{l\alpha}(k_\alpha r_\alpha) Y_{l_\alpha m_\alpha}(\theta,\phi)$$

where j_l are the spherical Bessel functions and Y_{lm} are the spherical harmonics. By analogy with our definition of θ_i in Sec. 10-2, θ_G for the decay (10-28) is

$$\theta_G = \theta_{00} \tag{10-30}$$

where we have taken $p = 0$ and $q = 0$ as the ground states of Pb^{208} and He^4, respectively.

It is profitable at this point to consider the so called *overlap integral*

$$\int \psi_F^* \psi_I \, d\tau \tag{10-31}$$

where the subscripts I and F designate the initial and final states in the decay. ψ_I can be expressed as (10-29), and ψ_F is

$$\psi_F = \psi[Pb^{208}(GS)]\psi[He^4(GS)]\psi_\alpha(l_\alpha = 0) \tag{10-32}$$

When these are inserted into the overlap integral, all terms except the one with coefficient θ_{00} vanish because of the mutual orthogonality of all states of Pb^{208}, of all states of He^4, and of all states of ψ_α. The result is then

$$\int \psi_F^* \psi_I \, d\tau = \theta_{00}$$

By comparing this with (10-30) we see that

$$\theta_G = \int \psi_F^* \psi_I \, d\tau \tag{10-33}$$

It is readily seen that this derivation can be generalized to all cases of alpha decay, and it covers our use of θ_i in Secs. 10-2 and 10-3 (see Prob. 10-8). We may therefore take (10-33) as an alternative definition of θ_i. The square of the right side of (10-33) also serves as the mathematical counterpart of our phrase "how much ψ_F is like ψ_I."

From Chap. 5 we know that Pb^{208}(GS) is a closed-shell nucleus, and Po^{212}(GS) has two paired protons in the $\mathfrak{N} = 6$ shell and two paired neutrons in the $\mathfrak{N} = 7$ shell of Fig. 4-5. The lowest-energy orbits of these shells are $h_{9/2}$ and $g_{9/2}$ respectively, so a simple approximation to the wave function of Po^{212} is

$$\psi[Po^{212}(GS)] = \psi[Pb^{208}(GS)](h_{9/2})_0{}^2(g_{9/2})_0{}^2 \tag{10-34}$$

Unfortunately, since this is not of the form (10-29), we cannot obtain θ_G directly from it. We can make some progress by substituting (10-34) along with (10-32) into (10-33), which gives

$$\theta_G = \int \psi^*[He^4(GS)]\psi_\alpha^*(l_\alpha = 0)[(h_{9/2})_0{}^2(g_{9/2})_0{}^2] \, d\tau \tag{10-35}$$

This is an integral expressing how much two protons in $(h_{9/2})_0{}^2$ and two neutrons in $(g_{9/2})_0{}^2$ orbits about the *center of the nucleus* are like a He^4 nucleus—which means two neutrons and two protons in $1s_{1/2}$ orbits *relative to their own center of mass*—moving with $l_\alpha = 0$ relative to the rest of the system. There are straightforward quantum-mechanical procedures available for the numerical evaluation of integrals like (10-35).

As we know from Chap. 5, (10-34) is an oversimplification of the wave function for Po^{212}. Actually, all the proton orbits in the $\mathfrak{N} = 6$ shell and all the neutron orbits in the $\mathfrak{N} = 7$ shell are occupied to some extent in that nucleus, so the wave function is a linear combination of many terms like those in (10-34). θ_G is then a linear combination of integrals like the one in (10-35), each one of which can be evaluated in the same way. It is interesting to point out that the reduced width $\theta_G{}^2$ is much larger when there is a lengthy linear combination than when there is only a single term, as in (10-34). For example, if the wave function has 25 terms each with amplitude 0.2 (this satisfies the requirement that the sum of the squares of the coefficients be unity) and the integral in (10-35) which we designate θ_{G0} is the same for each term, we obtain

$$\begin{aligned} \theta_G &= 25 \times 0.2\theta_{G0} = 5\theta_{G0} \\ \theta_G{}^2 &= 25\theta_{G0}{}^2 \end{aligned} \qquad \textbf{(10-36)}$$

The reduced width $\theta_G{}^2$ is 25 times larger than for the wave function (10-34). It should be noted that this argument is valid only if the signs of all terms in the wave function are the same, so it is valid only for transitions between ground states of even-even nuclei. This is what is called a *coherence effect*. For transitions to other states, both plus and minus signs occur in the expression for θ^2, and a great deal of cancellation occurs.

In the more general case where some of the orbits are partially occupied in both the initial and final states, for a transition in which a pair of nucleons in the orbit j is emitted as part of the alpha particle, the reduced width is proportional to the number of orbit pairs that are occupied in the initial state, $(2j + 1)V_j{}^2$, and to the number of these orbit pairs that are unoccupied in the final state, $(2j + 1)(1 - V_j{}^2)$. If only the one neutron and one proton orbit, j_n and j_p, respectively, are involved, we therefore expect

$$\theta_G \propto [(2j_n + 1)V_{j_n}(1 - V_{j_n}{}^2)^{1/2}][(2j_p + 1)V_{j_p}(1 - V_{j_p}{}^2)^{1/2}]$$

When several j_n and j_p are involved, θ is the sum of terms of this type, whence θ^2 is

$$\theta^2 \propto \left\{\left[\sum_{j_n} (2j_n + 1)V_{j_n}(1 - V_{j_n}{}^2)^{1/2}\right]\left[\sum_{j_p} (2j_p + 1)V_{j_p}(1 - V_{j_p}{}^2)^{1/2}\right]\right\}^2 \qquad \textbf{(10-37)}$$

When this product of sums is expanded, there is a large number of terms all of the same sign, so θ^2 becomes rather large, as in (10-36). This again is the coherence effect in transitions between ground states of even-even nuclei.

The example (10-28) is simpler than the usual situation in that the wave function for the initial nucleus can be expressed as

$$\psi[\text{Pb}^{208}(\text{GS})]\phi$$

where ϕ is the wave function for the four additional nucleons. As an example of the more general situation, consider the alpha decay

$$\text{Po}^{213}(\text{GS}) \rightarrow \text{Pb}^{209}(\text{GS}) + \text{He}^4$$

The wave function for Po^{213} can be written

$$\psi(\text{Po}^{213}) = e_0\psi(\text{Pb}^{209}\text{-GS})\phi_0 + e_1\psi(\text{Pb}^{209}\text{-1})\phi_1 + e_2\psi(\text{Pb}^{209}\text{-2})\phi_2 + \cdots$$

where Pb^{209}-1 means the first excited state of Pb^{209}, etc., and the ϕ's are the wave functions for the four remaining nucleons. Only during the fraction of the time e_0^2 when the wave function is $\psi(\text{Pb}^{209}\text{-GS})\phi_0$ can this nucleus decay into Pb^{209}(GS) plus an alpha particle, so θ_G is e_0 times the expression analogous to (10-35). On the other hand, with this wave function Po^{213}(GS) can also decay into Pb^{209}-1 with θ_1 equal to e_1 times the expression analogous to (10-35), and similarly it can decay to the higher excited states of Pb^{209}.

For alpha-particle decays of nuclei in their ground states, typical values of θ_G^2 are about 0.1. For excited states, wave functions are more complex; therefore coefficients of individual terms are smaller, and θ_i^2 is much reduced. For very highly excited nuclei, θ_i^2 for alpha particle emission are about the same as those for nucleons where the same amount of energy is available.

In the case of fission, similar considerations apply; e.g., in the process (8-6*a*), θ^2 is a measure of how much U^{238} is like two Pd^{119} nuclei in particular states with some definite relative motion. This would be an extremely small number, but on the other hand there are perhaps a million different states in which each Pd^{119} nucleus can be left and there are about a hundred different fission breakups that can occur, as explained in connection with (8-6*b*). When the λ_i from (10-22) for all these, even with their extremely small θ_i^2 values, are added, the total decay rate exclusive of the factor $\mathcal{P}$ is about as large for fission as for neutron emission with a few MeV.

Actually the interactions that take place in fission are so complicated that it is impractical to treat them by the methods we have been discussing. More profitable approaches based on various detailed models of the fission process have been developed.

10-6 Barrier Penetration and Decay Rates in Alpha-particle Emission

We have seen that coulomb barrier penetration plays an important and effective role in the emission of protons from nuclei, but it has much more spectacular

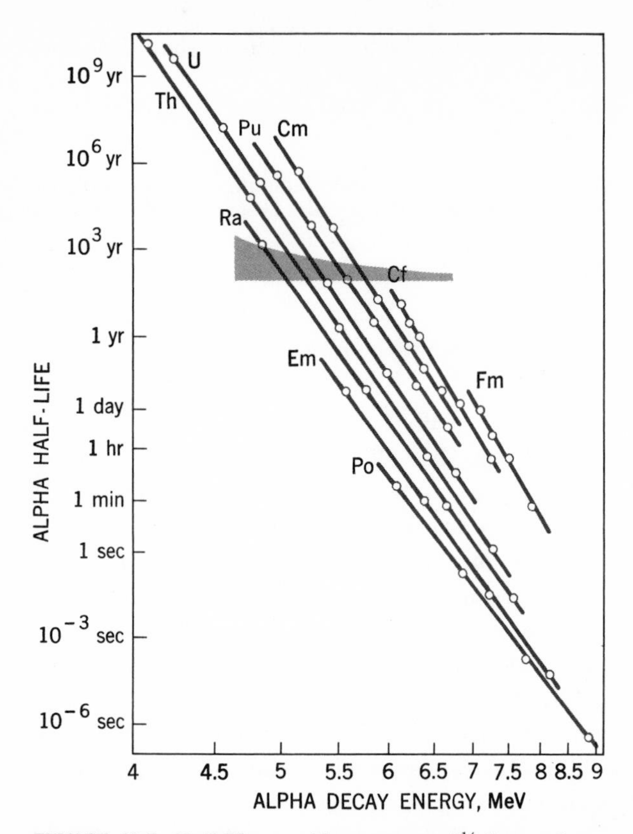

FIGURE 10-5 Half-life vs. (decay energy)$^{1/2}$ for various alpha decays. The ordinate scale is logarithmic. Cases for a given atomic number are connected by lines and labeled by the corresponding chemical symbol. [*From W. E. Stephens and T. Hurlimann, in E. U. Condon and H. Odishaw (eds.), "Handbook of Physics" McGraw-Hill Book Company. Used by permission.*]

effects in the emission of alpha particles from nuclei in their ground states. In many such cases there is no other way for the nucleus to decay, so it decays by alpha-particle emission regardless of how long it takes. As an example, consider the decay

$$_{92}U^{238} \rightarrow {}_{90}Th^{234} + He^4 + 4.2 \text{ MeV}$$

The maximum of the potential, $\mathcal{B}_c = 2Ze^2/4\pi\epsilon_0 R_0$ with R_0 taken from (10-25), is 28.7 MeV. Application of the formulas following (10-27*a*) then gives $x = 0.146$, $f(x) = 0.825$, $\beta = 112$; inserting these into (10-27*a*), we find $\mathcal{P} = e^{-92}$. A more accurate calculation without use of the square-well approximation gives $\mathcal{P} \simeq e^{-87} \simeq 2 \times 10^{-38}$. An alpha particle succeeds in coming out only twice in 10^{38} attempts! Inserting this value of $\mathcal{P}$ into (10-22) with $\theta^2 \simeq 0.1$ (from Sec. 10-5) gives

$\lambda = 5 \times 10^{-18}$ s^{-1}. From (8-11), the half-life is then about 1.4×10^{17} s $\simeq 5 \times 10^9$ years. This is of the order of the age of the elements composing our solar system; if it were much shorter than this, there would be no uranium left on earth.

The large value of the exponent makes the results of (10-27*a*) very sensitive to variations in Z and E. Changing the exponent by only 10 percent changes $\mathcal{P}$, and therefore the lifetime, by a factor of $e^{8.7} \simeq 7{,}000$. This fantastic sensitivity is exhibited in Fig. 10-5, where half-lives are plotted vs. E and Z for various nuclei which decay by alpha-particle emission. We see there that lifetimes vary from 10^{10} years to 10^{-6} s—by a factor of more than 10^{23}—for a variation of Z from 90 to 84 and of E from 4.1 to 8.9 MeV.

10-7 Barrier Penetration and Decay Rates in Fission

In comparing fission with alpha decay, we see immediately that the maximum of the potential is very much higher in fission because zZ is much larger; for U^{238}, for example, $zZ = 2 \times 90 = 180$ in alpha decay, and $46 \times 46 = 2{,}116$ in the fission decay (8-6*a*). As a consequence of this and of the inherently smaller penetrabilities of heavy particles mentioned in Sec. 10-1, fission can occur with appreciable probability only when x in (10-27*a*) is close to unity. When x is near unity, i.e., when E is near the maximum of the potential, it is important that the shape of this top be treated with some care, and, as demonstrated in Fig. 10-6, this is certainly not done in the square-well approximation. A better approximation would be to assume a rectangular top for the potential, as shown in the figure, and use (10-9*b*). For U^{238} we get the correct answer if we take this to be about 8 F wide with E about 6 MeV below its top. Then for the decay (8-6*a*) we find from (10-9*b*)

$$\mathcal{P} \simeq e^{-102} \simeq 10^{-44}$$

The fission decay rate for U^{238} is therefore about a million times slower than the rate of alpha-particle emission, so only one decay in a million is by fission. In some transuranic nuclei, fission occurs more rapidly than alpha decay, and the artificial preparation of new elements with higher and higher atomic numbers is limited principally by this process.

The fission decays we have been discussing are often called *spontaneous fission* to differentiate them from fission induced by nuclear reactions. There is no real difference between the two except for their half-lives; we see from our example of U^{238} that if only 6 MeV of excitation energy is added to that nucleus, as, for example, if it is struck by a 6-MeV gamma ray, E is above the maximum of the potential, whence $\mathcal{P}$ is of the order of unity and the half-life is of the order of 10^{-18} s. In the notation introduced in Sec. 8-7, this is a (γ,f) reaction.

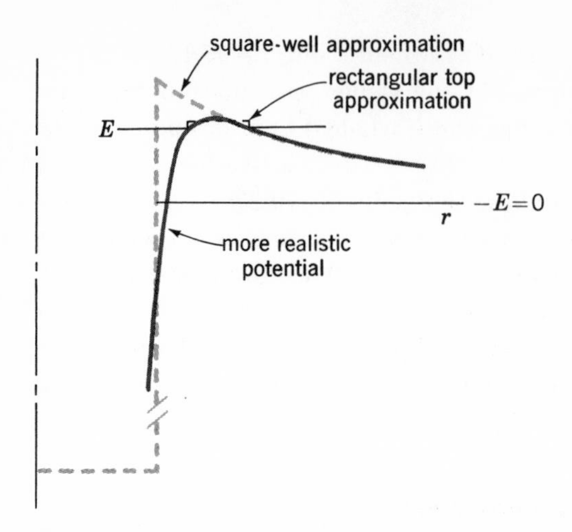

FIGURE 10-6 Potential between two fission fragments as a function of their separation r in the square-well approximation and in a more realistic approximation. When the energy available is E, very near the top of the barrier in the more realistic approximation, the barrier is better represented by the rectangular top, as shown, than by a square-well approximation.

An interesting and very important application of the remarkable sensitivity of the fission barrier penetrability to energy occurs in neutron-induced fission. Adding a neutron to U^{235} converts it into the even-even nucleus U^{236}, whereas adding a neutron to U^{238} converts it to the odd-A nucleus U^{239}. Due to the pairing term (7-17), the excitation energy is higher in the former case than in the latter, and this difference is enough to cause U^{236} formed in this way to decay by fission, whereas the U^{239} decays by gamma-ray emission. As a consequence, only U^{235}, a 0.7 percent abundant isotope, undergoes fission under bombardment with low-energy neutrons. If the maximum of the potential for fission of U^{239}, which is about 170 MeV high, were lower by even 1 MeV, the 99.3 percent abundant isotope U^{238} would also be fissionable with slow neutrons. If this were the case, construction of atomic bombs would have been very easy, and there is every likelihood that the Germans would have been the first to obtain them, early in World War II.

Fission fragments are much more complicated than simple particles like nucleons in that they can take on highly nonspherical shapes, undergo oscillations, and have various other types of excitations. As a result, the shape of the barrier is more complex than the one shown in Fig. 10-6. It turns out that it has two maxima as a function of r, with a dip in between. In some cases, there is a considerable time interval, often of the order of 10^{-9} s or longer, between the time when the two barriers are penetrated, so the half-life for fission even from some rather highly excited nuclei is of that order. We have seen in Secs. 10-2 and 10-3 that highly excited nuclei ordinarily emit neutrons with a half-life very much shorter than this, so we would expect neutron emission to occur first in such cases. However, once the first barrier is surmounted, since a considerable amount of the excitation energy is tied up in the potential energy of a highly deformed shape, neutron emission may no longer be possible. In some nuclei, fission has therefore been observed to occur with half-lives in the range 10^{-3} to 10^{-9} s. They are called *fission isomers*. The word isomer refers to two or more nuclei with observable half-lives that have the same number of neutrons and protons; we shall encounter another type of isomer in Sec. 12-6.

In all fission processes, once the barrier has been overcome, the two fragments are accelerated outward by the coulomb force so that their total final kinetic energies are close to the height of the barrier. In the fission of uranium this is about 170 MeV.

Problems

10-1 If the potential shown in Fig. 10-1 has $V_0 = 50$ MeV, calculate and plot the probabilities for transmission and reflection vs. E.

10-2 If a barrier is a right triangle with 10 MeV maximum height and a width at the base of 10 F, calculate and plot $\mathcal{P}$ for E between 0 and 9 MeV.

10-3 Consider the transition $Fe^{57}(\frac{5}{2}^+) \rightarrow Fe^{56}(GS) + n(2d_{5/2})$ and make a list of other configurations we would expect to find in the initial state.

10-4 If the neutron emitted in Prob. 10-3 comes off with an energy of $\frac{1}{2}$ MeV, and if $\theta_i^2 = 1/1{,}000$, calculate the emission rate.

10-5 List the ways in which a highly excited $\frac{9}{2}^-$ state of Pb^{209} can decay to a 5^- state of Pb^{208} by neutron emission.

10-6 Calculate the barrier penetration probability for three of the cases from Prob. 10-5 if the neutron is emitted with 3 MeV energy.

10-7 Calculate the barrier penetration for emission of protons of energies between 1 and 10 MeV from Pb^{208}.

10-8 Derive (10-33) for $l = 0$ neutron emission.

10-9 Why is alpha decay from an even-even nucleus in its ground state to excited states of the final nucleus not an important effect? With the chart of nuclides (see Sec. 8-7) look for cases where it is known.

10-10 Using the square-well approximation, calculate three well-separated points on one of the curves in Fig. 10-5 and compare the results with that figure.

10-11 In the rectangular-barrier-top approximation of Fig. 10-6, calculate $\mathcal{P}$ vs. E between 1 and 10 MeV below the barrier top.

Further Reading

See General References, following the Appendix.

Ajzenberg-Selove, F.: "Nuclear Spectroscopy," Academic, New York, 1960.

Hyde, E. K., I. Perlman, and G. T. Seaborg: "The Nuclear Properties of the Heavy Elements," Prentice-Hall, Englewood Cliffs, N.J., 1964.

Nuclear Data, periodical published by Academic Press, New York.

Segre, E.: "Experimental Nuclear Physics," Wiley, New York, 1953.

Siegbahn, K.: "Alpha, Beta, and Gamma Ray Spectroscopy," North-Holland, Amsterdam, 1965.

Wilets, L.: "Theories of Nuclear Fission," Clarendon Press, Oxford, 1964.

Chapter 11
Beta Decay

In this chapter we continue our more detailed analysis of the three types of nuclear decay with a discussion of beta decay. We have seen that this process arises from a basic interaction in nature, and we know from Sec. 3-9 that the basic interactions are treated by quantum field theory. In this case, it is the *beta-neutrino* field, which consists of an infinite density of negative energy, unobservable electrons and neutrinos pervading all of space. A β^- decay is then an exchange of energy between a nucleus and that field in which the nucleus contributes energy by changing a neutron into a proton, and this energy excites the field by changing a neutrino into an electron; the missing neutrino behaves physically as an antineutrino. Conversely, in a β^+ decay the nucleus converts a proton into a neutron, and the energy given to the field converts an electron into a neutrino, with the missing electron behaving as a positron. It is often more convenient in discussions to ignore the difference between particles and antiparticles and think of an electron and a neutrino as being created in the process.

In this chapter we consider the energy and angular momenta of the emitted particles, the effect of beta decay on the nucleus, and other related matters. Our treatment is based on a fundamental theorem of quantum mechanics which states that the transition rate λ_{if} between an initial (i) and a final (f) state is given by

$$\lambda_{if} = \frac{2\pi}{\hbar^2} |M|^2 \frac{dn_f}{dE} \tag{11-1}$$

where dn_f/dE is the number of final states per unit energy and M is the so-called *matrix element* for the transition, which can be formally written

$$M = \int \psi_f^* H \psi_i \, d\tau \tag{11-2}$$

where H is the interaction causing the transition. The matrix element will be further explained in Sec. 11-4, and pertinent examples will be discussed in detail there. Theorem (11-1) is conventionally derived in quantum-mechanics courses by use of time-dependent perturbation theory, but it is universally applicable and is widely used in various areas of physics.

11-1 Energy Spectrum of Electrons Emitted in Beta Decay

The total energy E_0 given up by the nucleus in the beta decay must be shared between the electron and the neutrino, and their energies are therefore related by

$$E_0 = E_e + E_\nu \tag{11-3}$$

Since the energy equivalent of the mass of the electron must be supplied in any case, this may be subtracted from both sides of (11-3) to give a relationship between the kinetic energies T as

$$T_0 = T_e + T_\nu = E_0 - 0.511 \text{ MeV} \tag{11-3a}$$

Note that because the neutrino has no mass, $T_\nu = E_\nu$. In general, only the electron can be observed experimentally; a typical energy spectrum for these electrons is shown in Fig. 11-1. From (11-3*a*), the energy spectrum for neutrinos must be the same curve reflected about $T_0/2$. Our first task will be to understand this spectrum.

In accordance with (11-1), the rate of beta decay is proportional to the number of final states per unit energy in the transition, dn_f/dE_0. When the final states are bound states of a system, there is no problem in counting them, but when they are particles emitted with some momentum vector **p**, the counting of states is not so obvious. The prescription from statistical physics for this situation is as follows. Consider a momentum space with axes p_x, p_y, p_z. Any momentum vector of the emitted particle can be represented by a point in this space. The

FIGURE 11-1 Energy spectrum of electrons emitted in the beta decay of Bi^{210}. [*From G. J. Neary, Proc. Phys. Soc. (London),* **A175: 71 (1940).**]

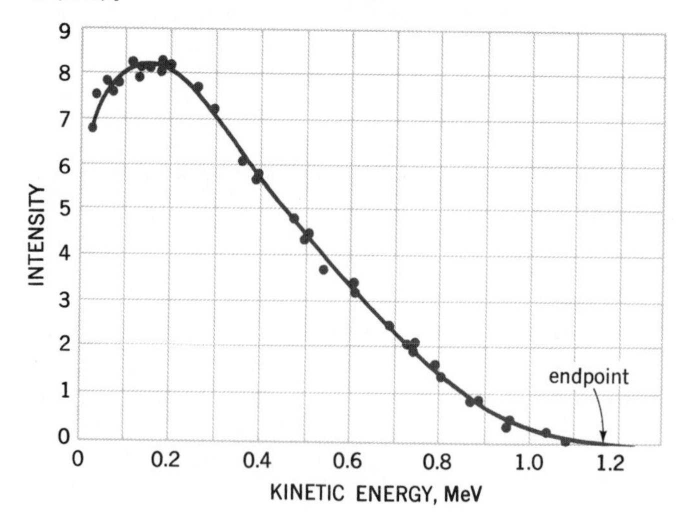

number of states is then proportional to the volume of this space occupied by the **p** vectors under consideration. In our case, we are interested in all momentum vectors of length between p and $p + dp$, regardless of their direction. They occupy a volume of this momentum space between two spherical shells of radii p and $p + dp$. This volume is $4\pi p^2\,dp$, so the number of states, dn, corresponding to momentum vectors between p and $p + dp$, is

$$dn \propto p^2\,dp$$

The total number of final states in a beta decay dn_f is just the product of the number of final states of the electron and the number of final states of the neutrino, or

$$dn_f \propto (p_e^2\,dp_e)(p_\nu^2\,dp_\nu) \tag{11-4}$$

where p_e and p_ν are the momenta of the electron and neutrino respectively.

Since the neutrino is not observable, we want to convert our variables from p_e and p_ν to p_e and E_0. As the neutrino has zero mass, the relativistic relationship between energy and momentum

$$E^2 = (M_0c^2)^2 + (pc)^2 \tag{11-5}$$

gives

$$p_\nu = \frac{E_\nu}{c} = \frac{E_0 - E_e}{c} \tag{11-6}$$

where the last expression is obtained from (11-3). In the transformation of variables, the prescription for transforming the differentials is

$$dp_e\,dp_\nu = J\,dp_e\,dE_0 \tag{11-7}$$

where J is the jacobian of the transformation,

$$J = \begin{vmatrix} \left(\dfrac{\partial p_e}{\partial p_e}\right)_{E_0} & \left(\dfrac{\partial p_e}{\partial E_0}\right)_{p_e} \\ \left(\dfrac{\partial p_\nu}{\partial p_e}\right)_{E_0} & \left(\dfrac{\partial p_\nu}{\partial E_0}\right)_{p_e} \end{vmatrix} = \begin{vmatrix} 1 & 0 \\ -\dfrac{1}{c}\dfrac{\partial E_e}{\partial p_e} & \dfrac{1}{c} \end{vmatrix} = \frac{1}{c} \tag{11-8}$$

(The subscripts outside the parentheses are kept constant in the differentiations.)

Inserting (11-6), (11-7), and (11-8) into (11-4), we obtain

$$dn_f \propto p_e^2(E_0 - E_e)^2\,dp_e\,dE_0$$

whence, from (11-1), the probability per unit time $d\lambda(p_e)$ for transitions in which the electron momentum is between p_e and $p_e + dp_e$ is

$$d\lambda \propto \frac{dn_f}{dE_0} \propto p_e^2(E_0 - E_e)^2\,dp_e \tag{11-9}$$

Another factor that must be considered here is the effect of the coulomb forces on the electrons. One such effect is to speed up positive electrons and slow down negative electrons after they leave the nucleus, thereby distorting (11-9), which is given as a function of p_e and E_e as observed after the coulomb forces have done their work. These effects are corrected for by multiplying (11-9) by the so-called *Fermi factor* $F(Z,E_e)$ for which tables and plots are widely available.[1] It is clearly less than unity for positive electrons since they are accelerated by the coulomb forces causing their E_e to be larger than the energy with which they left the nucleus, and greater than unity for negative electrons which are decelerated by the coulomb forces. Its deviation from unity increases with the strength of the coulomb force and hence with increasing atomic number Z, but for E_e more than a few tenths MeV, this deviation is not much more than a factor of 10 even for the highest-Z nuclei.

Inserting the Fermi factor into (11-9), we obtain

$$d\lambda \propto p_e^2(E_0 - E_e)^2F(Z,E_e)\,dp_e \qquad \textbf{(11-10)}$$

$d\lambda$ here is the rate at which beta decays occur in which the momentum of the emitted electrons is between p_e and $p_e + dp_e$. It is proportional to the intensity of electrons emitted within this momentum range, $I(p_e)\,dp_e$. Using this for $d\lambda$ in (11-10) gives

$$I(p_e) \propto p_e^2(E_0 - E_e)^2F(Z,E_e) \qquad \textbf{(11-11)}$$

This is the momentum spectrum for electrons emitted in a beta decay. It can be converted to an expression for the energy spectrum by multiplying it by dE_e/dp_e as obtained by differentiating (11-5).

The effect of the Fermi factor $F(Z,E_e)$ on momentum spectra is exhibited in Fig. 11-2, which shows spectra from a β^+ and a β^- decay with the same Z and nearly the same E_0. We see there that β^+ spectra are shifted toward higher energies and β^- spectra are shifted toward lower energies, as expected from the fact that the former are speeded up while the latter are slowed down by the coulomb force.

Actually one rarely sees data presented as in Fig. 11-1 and 11-2. It is much more conventional to divide $I(p_e)\,dp_e$ as obtained experimentally by $p_e^2F(Z,E_e)\,dp_e$ and plot the square root of the quotient vs. E_e. According to (11-11), this should give a straight line which extrapolates to zero at $E_e = E_0$. This is called a *Fermi-Kurie plot;* some examples are shown in Fig. 11-3. The advantages of this treatment are that it gives an accurate determination of E_0 and allows one to determine whether the spectrum is due to a single transition, as in Fig. 11-3*a*, or to transi-

[1] See, for example, H. A. Enge, "Introduction to Nuclear Physics," p. 316, Addison-Wesley, Reading, Mass., 1966.

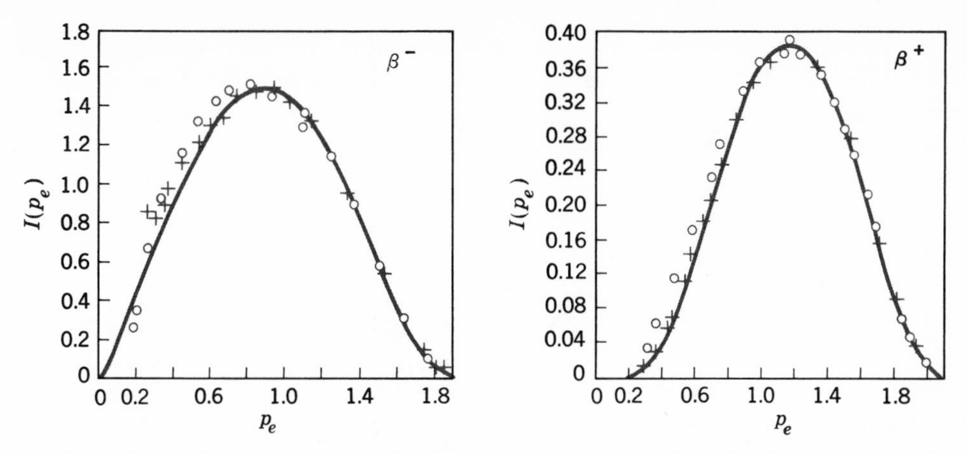

FIGURE 11-2 Momentum spectrum of electrons and positrons emitted in the beta decay of Cu^{64}. Data at left are for electrons from the β^- decay to Zn^{64}, and data at right are for positrons from the β^+ decay to Ni^{64}. Since the transition energies E_0 are nearly equal in the two decays, the difference is due principally to the Fermi factor $F(Z,E_e)$ in the theoretical formula (11-11). [*From J. R. Reitz, Phys. Rev.*, **77: 50 (1950).**]

tions to more than one final nuclear state, as in Fig. 11-3*b*. In addition it provides a sensitive test of the validity of (11-11) [we shall see in the next section that (11-11) is not always expected to be valid].

The total decay rate λ is proportional to the integral of (11-10), $f(Z,E_0)$, which can be written explicitly as

$$\lambda \propto f(Z,E_0) = \int_0^{p_0} p_e{}^2(E_0 - E_e)^2 F(Z,E_e)\, dp_e \qquad \textbf{(11-12)}$$

where p_0 is related to E_0 through (11-5). This is a completely calculable function, and plots of it are shown in Fig. 11-4.

11-2 Angular-momentum Considerations

Let us now consider the angular momentum carried off by the electron. Classically, this is $\mathbf{r} \times \mathbf{p}_e$, which cannot be larger than about $R_0 p_e$. For an electron kinetic energy of 1 MeV, (11-5) gives $p_e = 1.4$ MeV/c, and for a medium-size nucleus, $R_0 \simeq 6$ F. $R_0 p_e$ then comes out to be 2.8×10^{-23} MeV-s. But the smallest angular momentum other than zero allowed in quantum physics is $\sqrt{l(l+1)}\,\hbar$ with $l = 1$, which is 8.5×10^{-22} MeV-s, or 30 times larger. This means that a 1-MeV electron emitted even with $l = 1$ is hindered by a large angular-momentum barrier. Since the electron and the neutrino have, on an average, approximately equal momenta, emission of a neutrino with nonzero angular momentum is also hindered by a large barrier. The rate of emission is therefore by far the largest when the total orbital angular momentum of the

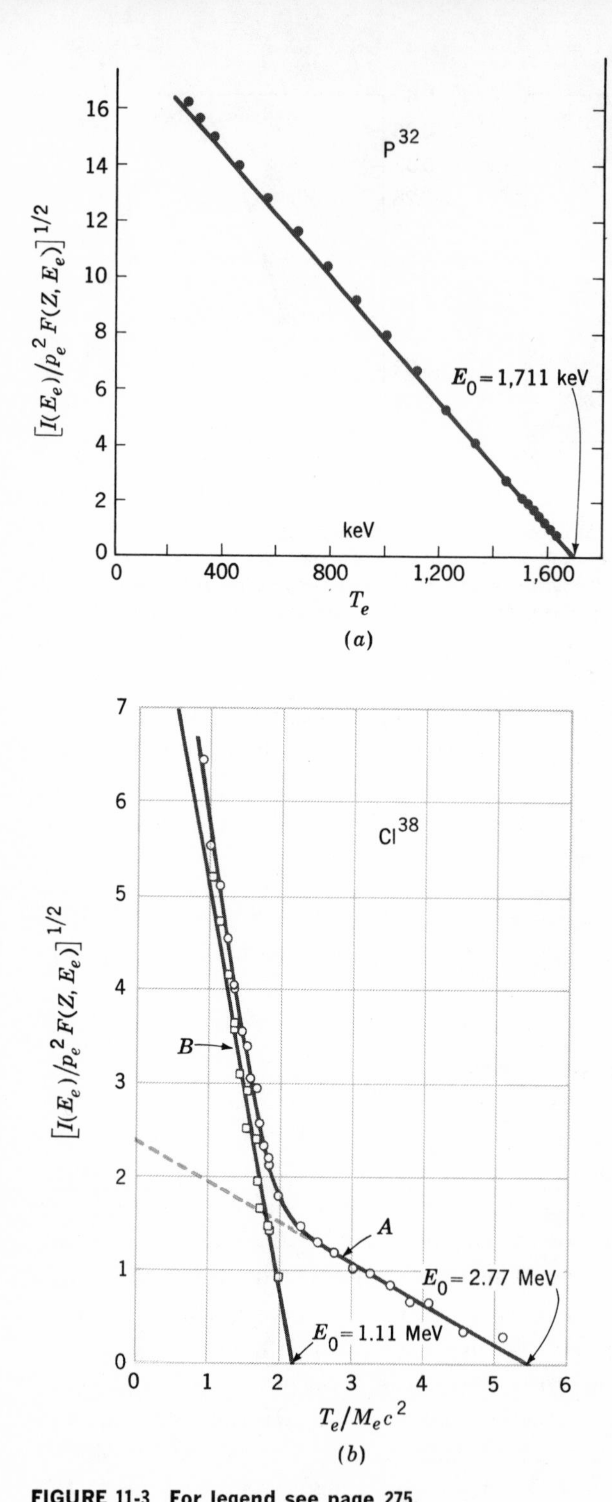

FIGURE 11-3 For legend see page 275.

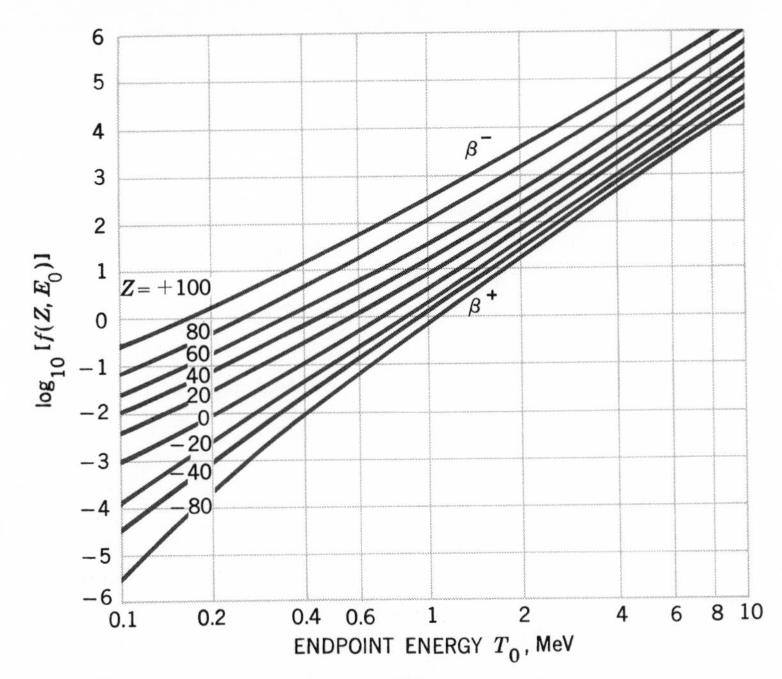

FIGURE 11-4 The function $f(Z,E_0)$ defined by (11-12). *(From R. D. Evans, "The Atomic Nucleus." Copyright 1955. McGraw-Hill Book Company. Used by permission.)*

electron and neutrino, L, is zero. Beta decays with $L = 0$ are referred to as *allowed*, while those with $L > 0$ are called *forbidden;* in particular, those with $L = 1$ are called *first forbidden*, those with $L = 2$ (which have an even larger angular momentum barrier to penetrate) are called *second forbidden*, etc.

The barrier penetrabilities can be estimated from (10-23) with α obtained from the second expression of (10-24). In all cases of interest, $\alpha \gg R_0$, whence (10-23) reduces to

$$\mathcal{P} = \exp\left[-2\sqrt{l(l+1)}\left(\log\frac{2\alpha}{R_0} - 1\right)\right] \tag{11-13}$$

For the case considered above (1-MeV electron, $l = 1$) this gives $\mathcal{P} = 1/80$; for $l = 2$, it gives $\mathcal{P} = 1/8{,}000$. For an 0.5-MeV electron, we similarly obtain

---

FIGURE 11-3 Examples of Fermi-Kurie plots. *(a)* **Decay of P^{32}, which is a single transition.** [*Data from F. T. Porter et al., Phys. Rev.,* **107: 135 (1957).**] *(b)* **Decay of Cl^{38}, which includes beta-decay transitions to two different states of Ar^{38} with $T_0 = 2.77$ and 1.11 MeV. When the straight line A (fitted to the high-energy data) is subtracted from the low-energy data, straight line B results, representing electrons from the 1.11-MeV transition.** [*From L. M. Langer, Phys. Rev.,* **77: 50 (1950).**]

$\mathcal{P} = 1/230$ for $l = 1$ and 1/44,000 for $l = 2$. In more detailed treatments of forbidden transitions, the barrier penetrability is not treated explicitly, as here, but its effects are incorporated into the wave functions.

Either the electron or the neutrino (or both for $L > 1$) may carry the orbital angular momentum with equal intrinsic probabilities. The factors $\mathcal{P}_e$ and $\mathcal{P}_\nu$ should be added to (11-11) and (11-12), whence the electron energy spectrum and $f(Z,E_0)$ are different for forbidden and for allowed decays. When measurements are made with good accuracy, the deviation from a straight line Fermi-Kurie plot can therefore be detected when transitions are forbidden. But the most important effect of $\mathcal{P}_e$ and $\mathcal{P}_\nu$ is to reduce the probability for all energies and hence to reduce the transition rate. If the average penetrability is $\mathcal{P}'$, (11-12) becomes

$$\lambda \propto \mathcal{P}' f(Z,E_0) \qquad \text{(11-12a)}$$

Aside from their orbital angular momenta, both the electron and neutrino have a spin, $s = \frac{1}{2}$ in both cases. Their total spin may therefore be $S = 0$ or $S = 1$, corresponding roughly to antiparallel or parallel spins; the more accurate meaning of these is outlined in Sec. 3-4. In the former case ($S = 0$) we have what is known as a *Fermi transition*, while the latter case ($S = 1$) is called a *Gamow-Teller transition*. As we shall see, both these types occur, and they have rather different effects on the nucleus.

11-3 Selection Rules

We have said that beta decay results from an interaction of the nucleus with the beta-neutrino field. The action of the nucleus on this field is to create the electron and neutrino, and we have been discussing some of their properties. In this section and the next, we consider the action of the field on the nucleus.

From conservation of angular momentum and parity, we can immediately deduce how I and π must change in the nucleus. In a Fermi transition the only angular momentum carried off by the electron and neutrino is their orbital angular momentum L, so the angular momenta of the initial and final nuclei, I_i and I_f, respectively, must satisfy the vector addition

$$\mathbf{I}_f = \mathbf{I}_i + \mathbf{L} \qquad \text{Fermi} \qquad \text{(11-14a)}$$

In a Gamow-Teller transition, the electron and neutrino carry off an additional spin angular momentum of one unit, whence

$$\mathbf{I}_f = \mathbf{I}_i + \mathbf{L} + \mathbf{1} \qquad \text{Gamow-Teller} \qquad \text{(11-14b)}$$

Conservation of parity requires that the product of the parities of the terms in the final wave function—the wave functions of the electron, neutrino, and final

TABLE 11-1: LOWEST AND NEXT LOWEST L VALUES FOR VARIOUS ΔI Those not possible if either I_i or I_f is zero are in parentheses

ΔI	0	1	2	3	4	5
Fermi	0, (1)	1, (2)	2, (3)	3, (4)	4, (5)	5, (6)
Gamow-Teller	(0), 1	0, 1	1, 2	2, 3	3, 4	4, 5

nucleus—be the same as the parity of the initial wave function. Since the parity of the electron-neutrino wave function is $(-1)^L$, this requires

$$\pi_i = \pi_f(-1)^L \tag{11-14c}$$

It is very often important to determine what type of beta transition occurs between two nuclear states. Since transition rates increase rapidly with decreasing L, this will always be the lowest L consistent with (11-14*a*) to (11-14*c*). Since parity switches with each consecutive L, this is either the lowest or next lowest L consistent with (11-14*a*) and (11-14*b*), whichever satisfies (11-14*c*). These are determined only by $\Delta I = |I_f - I_i|$ although there are exceptions where L values are not allowed in cases where either I_f or I_i are zero. The lowest and next lowest L values for each ΔI are shown in Table 11-1 for both Fermi and Gamow-Teller transitions; the exceptions mentioned above are in parentheses. The type of transition for a known ΔI is that characterized by the L value listed below it in Table 11-1 with the correct parity. A more conventional presentation of this same information is given in Table 11-2, where all cases corresponding to a given L are listed on a single line. The contents of these tables are known as *selection rules.*

We see from these tables that a beta decay between two $I = 0$ states of the

TABLE 11-2: SELECTION RULES FOR VARIOUS TYPES OF TRANSITION Those not possible if either I_i or I_f is zero are in parentheses. The yes or no under $\Delta\pi$ indicates whether the parity changes between the initial and final states

Transition type	L	*Fermi*		*Gamow-Teller*	
		ΔI	$\Delta\pi$	ΔI	$\Delta\pi$
Allowed	0	0	No	(0), 1	No
First forbidden	1	(0), 1	Yes	0, 1, 2	Yes
Second forbidden	2	(1), 2	No	2, 3	No
Third forbidden	3	(2), 3	Yes	3, 4	Yes
Fourth forbidden	4	(3), 4	No	4, 5	No

same parity can occur only through a Fermi transition. This will be important for our discussion in Sec. 11-5. It is also evident from the tables that there are many situations in which decays go in lowest order only through Gamow-Teller transitions; this includes all cases where $\Delta I = L + 1$. Such transitions are referred to as *unique*. The shape of the electron spectrum from them can be calculated accurately, whereas in cases where there is a mixture of Fermi and Gamow-Teller transitions the shape of the electron spectrum in forbidden transitions depends on the details of this mixture and therefore cannot be accurately calculated.

11-4 Matrix Elements in Beta Decay

The meaning of (11-2) can be understood as follows: H is an operator which operates on the wave function of the initial state ψ_i, converting it to $H\psi_i$. The matrix element M is then a measure of how much $H\psi_i$ is like the wave function of the final state ψ_f. As explained in Sec. 10-5, the mathematical answer to the question how much one wave function is like another is the overlap integral (10-31). This then leads to (11-2). In order to evaluate it, we must know the basic interaction H between the neutron which is changed into a proton (or vice versa) and the beta-neutrino field. This interaction was originally not known, but it was assumed that it was not velocity-dependent and this has been experimentally verified. Interactions between particles and fields occur at the position of the particle, so there is no function of distance involved, whence the only property of a nucleon the interaction can depend on is its spin σ_n. To make H a scalar, this interaction can only be of the form $\boldsymbol{\sigma}_n \cdot \boldsymbol{\sigma}_\nu$, where $\boldsymbol{\sigma}_\nu$ is the spin of the neutrino that is being changed into an electron (or vice versa). If the interaction includes this term, H is the Gamow-Teller operator H_{GT}, and if it does not include the spin term it is the Fermi operator H_F. In the latter case, no spin is given to the field so the electron has the same spin as the neutrino had, and the antineutrino which represents the missing neutrino has the opposite spin (a missing spin *up* is a spin *down*) whence their total spin is $S = 0$. In the Gamow-Teller interaction, the σ_ν operator changes the spin of the neutrino as it is converted into an electron, so the electron and the antineutrino come out with $S = 1$.

In the Fermi interaction, the effect of H_F on the nucleon is only to change a neutron into a proton (or vice versa) without otherwise affecting it. The effect of this operator on the nuclear wave function which appears in (11-2) is therefore to convert each term in it into a series of terms in each of which one of the neutrons is changed into a proton (or vice versa) in the same orbit. But this is just the operation T_+ (or T_-) described in connection with Fig. 6-11. As we learned in Sec. 6-3, these operators generate isobaric analog states, so the principal effect of the Fermi interaction is to induce transitions between isobaric analogs.

Two simple examples with which we have some familiarity are the decay of O^{14} to the first excited state of N^{14} (see Fig. 6-12*b*) and the decay of Be^7 to Li^7 (see Fig. 6-13). In these cases $|M_F|^2$ is close to unity. Actually it is given by[1]

$$M_F{}^2 = [T(T+1) - (T_z)_i(T_z)_f] \qquad \textbf{(11-15)}$$

where the subscripts i and f refer to initial and final nuclei, respectively. For the O^{14} decay this gives $|M_F|^2 = 2$ and for the Be^7 decay, $|M_F|^2 = 1$. Other examples of decays between isobaric analogs are $n \rightarrow H^1$, $H^3 \rightarrow He^3$, $O^{15} \rightarrow N^{15}$, $F^{17} \rightarrow O^{17}$, $Ca^{39} \rightarrow K^{39}$, and $Sc^{41} \rightarrow Ca^{41}$.

In heavy nuclei, however, we know from Sec. 6-3 that isobaric analogs are highly excited states. For example, let us consider the decay of Sb^{117} to Sn^{117}, for which the energy-level diagrams are shown in Fig. 11-5. The coulomb energy required to change a neutron into a proton in a Sn^{117} nucleus is readily calculated as 13.5 MeV, whereas the ground state of Sb^{117} is only 1.8 MeV higher in energy

[1] The expressions (11-15) and (11-18) are derived in M. G. Mayer and J. H. D. Jensen, "Elementary Theory of Nuclear Shell Structure," pp. 172 and 235, respectively, Wiley, New York, 1955.

FIGURE 11-5 Energy levels of interest in the beta decay of Sb^{117} (GS) to Sn^{117}. The low-energy levels of Sb^{117} are $T = 15/2$, with the $T = 17/2$ states beginning at an excitation energy of 11.7 MeV.

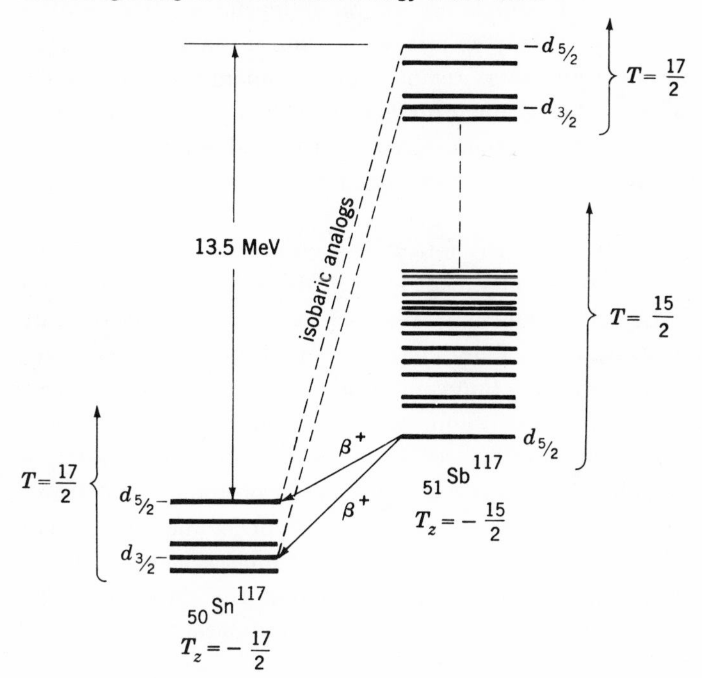

than the ground state of Sn^{117}, so the isobaric analogs ($T = {}^{17}\!/_2$ states) start at an excitation energy of 11.7 MeV. As we noted in Sec. 8-4, beta decay from highly excited states is of no practical interest because they decay very predominantly by gamma-ray or nucleon emission, so we are interested here only in the beta decay of the ground state of Sb^{117}. But Sb^{117}(GS) has $T = {}^{15}\!/_2$, so our conclusion would seem to be that it cannot beta-decay through the Fermi interaction.

While this conclusion is basically correct, it is not altogether so. An isobaric analog state is a true nuclear state only if all forces between nucleons are charge-independent. If they are not, this state is mixed into a number of nuclear states in the same energy region. The width of the energy region over which it mixes is proportional to the extent to which the forces are not charge-independent. This may be understood in analogy with our treatment of the shell-theory potential. If the shell-theory potential correctly represented all the forces in nuclei, the *terms* would be the states of nuclei. Since it is not so accurate, the terms are mixed into a number of states in the same energy region. The width of this energy region ΔE depends on the extent to which the shell-theory potential does not accurately portray the forces. This is the strength of the residual interactions, which determines the frequency with which collisions occur and hence the time between collisions Δt, and therefore it determines ΔE through (5-4).

We know, of course, that there are coulomb forces between protons, so the forces between nucleons are not completely charge-independent. This causes the isobaric analog states to be slightly spread out in energy; each one is mixed into several nuclear states over an energy region typically about 0.01 MeV wide. The low-energy states of Sb^{117}, being much further away than this, contain only a very small fraction of the isobaric analog states, perhaps about 1 part in 10^5. These states are therefore not pure $T = {}^{15}\!/_2$ states, but they spend about 1 part in 10^5 of their time as $T = {}^{17}\!/_2$ states.

During this time Sb^{117}(GS) can decay into a ${}^5\!/_2+$ state of Sn^{117} provided the wave functions of the two are the same except for a proton's being replaced by a neutron. For simplicity, let us assume that Sb^{117}(GS) is a $d_{5/2}$-proton SQP state and that the low-energy states of Sn^{117} are neutron SQP states. From (6-1) this implies that Sb^{117}(GS) has a wave function consisting of Sn^{116}(GS) plus a $d_{5/2}$ proton for a fraction of the time $(1 - V_{5/2}^2)_\pi$ and the state of Sn^{117} labeled $d_{5/2}$ in Fig. 11-5 has a wave function consisting of Sn^{116}(GS) plus a $d_{5/2}$-neutron for a fraction of the time $(1 - V_{5/2}^2)_\nu$. (The subscripts π and ν refer to protons and neutrons respectively.) These two wave functions are the same except for the replacement of a proton by a neutron, so the beta decay can take place between them. Combining these probabilities with the above probability that the states have the same T value, we find for the beta decay due to the Fermi interaction

$$Sb^{117}(GS) \rightarrow Sn^{117}(d_{5/2}): \quad |M_F|^2 \approx 10^{-5}(1 - V_{5/2}^2)_\pi(1 - V_{5/2}^2)_\nu \qquad \textbf{(11-16)}$$

Now let us turn to Gamow-Teller transitions. The operator H_{GT} performs the same operation as the T_+ (or T_-) operators, as explained in connection with Fig. 6-11, except that as a proton is converted to a neutron, its spin is changed. This change may or may not change the j of an orbit from $l + \frac{1}{2}$ to $l - \frac{1}{2}$ or vice versa, as, for example, by changing a $d_{5/2}$ orbit to a $d_{3/2}$ orbit.[1] If the nuclear force were not only charge-independent but also spin-independent, the states produced by this operation would be true nuclear states at the same energy as the isobaric analog states. However, we know that the nuclear force is quite strongly spin-dependent, so these states are mixed into nuclear states over a wide energy range centered at that energy. Since the ground state of Sb^{117} is rather far from that energy, it contains a relatively small share, typically about 1 percent, so the factor 10^{-5} in (11-16) is here about 10^{-2}. Moreover, Sb^{117}(GS) can now decay not only to the $d_{5/2}$ SQP state but also to the $d_{3/2}$ SQP state of Sn^{117}. For these transitions, in analogy with (11-16),

$$\begin{aligned} \mathrm{Sb}^{117}(\mathrm{GS}) \rightarrow \mathrm{Sn}^{117}(d_{5/2})&: \quad |M_{GT}|^2 \approx 0.01(1 - V_{5/2}{}^2)_\nu(1 - V_{5/2}{}^2)_\pi \\ \mathrm{Sb}^{117}(\mathrm{GS}) \rightarrow \mathrm{Sn}^{117}(d_{3/2})&: \quad |M_{GT}|^2 \approx 0.01(1 - V_{3/2}{}^2)_\nu(1 - V_{5/2}{}^2)_\pi \end{aligned} \qquad \textbf{(11-17)}$$

If, as we know to be true from Fig. 6-6, the lowest-energy $\frac{5}{2}^+$ state of Sn^{117} contains only a fraction of the SQP state, $|M|^2$ for that transition is reduced by this fraction and other $\frac{5}{2}^+$ states may be excited with $|M|^2$ proportional to the amount of the SQP state they contain.

In decays between odd-odd and even-even nuclei or in decays between odd-A nuclei involving paired nucleons, new complications enter, and the treatment of the V_j^2 in (11-16) and (11-17) must be modified. But the factors in front, $\sim 10^{-2}$ for Gamow-Teller transitions and $\sim 10^{-5}$ for Fermi transitions, are still applicable. There can, of course, be Gamow-Teller transitions between isobaric analog states in light nuclei if they are not $I = 0$ (cf. Table 11-1). In these cases, it can be shown that[2]

$$|M_{GT}|^2 = \begin{cases} \dfrac{I}{I+1} & I = l - \frac{1}{2} \\[2ex] \dfrac{I+1}{I} & I = l + \frac{1}{2} \end{cases} \qquad \textbf{(11-18)}$$

These are applicable in all the cases mentioned after (11-15) except for the O^{14} decay. For example, for the decay of H^3 into He^3, $I = \frac{1}{2}$ and $l = 0$ (they are in $1s_{1/2}$ states), so $|M_{GT}|^2 = 3$.

[1] More advanced students may understand H_{GT} as $\Sigma\tau_+(\sigma_+ + \sigma_- + \sigma_z)$, where the sum is over all the nucleons in the nucleus. The τ_+ changes a neutron to a proton, either the σ_+ or σ_- flips its spin while the other gives zero, and σ_z does not flip its spin. By contrast, H_F is simply $\Sigma\tau_+$ which is just T_+.

[2] See footnote page 279.

11-5 Decay Rate in Beta Decay

From (11-1) and (11-12a), we can write for the decay rate λ_β

$$\lambda_\beta = G' f(Z,E_0)|M|^2 \mathcal{P}' \tag{11-19}$$

The factor G' is the proportionality constant, and $|M|^2$ is

$$|M|^2 = (1 - x)|M_F|^2 + x|M_{GT}|^2 \tag{11-20}$$

where $x/(1 - x)$ is the ratio of the strengths of Gamow-Teller and Fermi interactions. If we use the relationship (8-11) between λ and the half-life, (11-19) can be written in its more familiar form

$$fT_{1/2} = \frac{1}{G|M|^2 \mathcal{P}'} \tag{11-21}$$

where $G = G'/(\ln 2)$. There is no way to predict G and x any more than we can predict the strength of any of the other three basic forces in nature such as the gravitational constant, the electron charge, or g^2 in (3-28). We can only determine them from experiment. This can be done here by measuring E_0 [to determine $f(Z,E_0)$] and $T_{1/2}$ for decays for which $|M_F|^2$ and $|M_{GT}|^2$ are known, namely, the ones mentioned after (11-15), for which $|M|^2$ can be calculated with that equation and (11-18). When this is done, the result is

$$G = \frac{1}{2{,}800} \qquad x = 0.56$$

If these values are used in (11-20) and (11-21), with $|M|^2$ taken from (11-16) and (11-17) and $\mathcal{P}'$ taken from the estimates of Sec. 11-2, values of $T_{1/2}$ come out as follows. For allowed transitions in light nuclei with $E_0 \simeq 1$ MeV, that is, the kinetic energy $T_0 \simeq 0.5$ MeV, $T_{1/2}$ is about 1 day. In heavy nuclei, due to the Z dependence of $f(Z,E_0)$, it is about an order of magnitude longer for β^+ and an order of magnitude shorter for β^- decay. For $T_0 \approx 10$ MeV, $T_{1/2}$ is about a million times shorter than these estimates (0.1 s in light nuclei), and if $T_0 \simeq 0.1$ MeV, $T_{1/2}$ is about 1,000 times longer (a few years in light nuclei). As T_0 gets still smaller, $T_{1/2}$ continues to get longer without limit. For first forbidden beta decays, half-lives are typically 100 times longer.

These results are more easily expressed in terms of the values of $fT_{1/2}$ in (11-21) or still more conveniently as log $fT_{1/2}$. As mentioned previously, this quantity can be determined experimentally by measuring E_0 and the half-life. Determinations have been made for a large number of decays where ΔI and $\Delta\pi$ are known, so classifications can be made from Table 11-1. The results are shown in Table 11-3, where we have included the classification *first forbidden unique*, since the results for those transitions are consistently larger than for other first forbidden decays. We have also included the classification *superallowed*, referring to cases where the matrix element is of the order unity, as in transitions between isobaric analog states. For them, $fT_{1/2} \simeq 1/G \simeq 2{,}800$.

TABLE 11-3: EMPIRICAL DETERMINATIONS OF $\log fT_{1/2}$

Type of transition	$\log fT_{1/2}$
Superallowed	~ 3.5
Allowed	5.5 ± 1.3
First forbidden	7.3 ± 1.3
First forbidden unique	8.5 ± 1.0
Second forbidden	~ 12
Third forbidden	~ 16
Fourth forbidden	~ 21

The variations indicated by the $\pm$ in Table 11-3 are due to variations in $|M|^2$ in (11-21) from one decay to another. We see that the overall variation is by a factor of 400 or so, although there are extreme cases well outside the limits listed in the table. The differences in the average values of $fT_{1/2}$ in Table 11-3 are due to the factor $\mathcal{P}'$ in (11-21) and are consistent with the discussion in Sec. 11-2. The value 5.5 for allowed transitions is easily understood from log 2,800 $\simeq$ 3.5 and the factor 0.01 in (11-17).

11-6 Operation of Selection Rules

Beta decay is not always (or even usually) a transition from one ground state to another. For example, consider the beta decay to an even-even vibrational nucleus for which energy-level diagrams are shown in Fig. 5-11. If the initial nucleus is 3^+, it can decay to the 2^+ first excited state by an allowed transition, whereas decay to the ground state would be second forbidden by the selection rules of Tables 11-1 and 11-2; decay to the 2^+ will therefore predominate.

If the initial state is 4^+ or 5^+, it can have an allowed transition to the 4^+ state ($\Delta I = 0, 1$; $\Delta\pi$: no) if there is sufficient energy, whereas transitions to the 2^+ states are second forbidden; the allowed transition will surely predominate. On the other hand, if the initial state is 4^-, transitions to either the 2^+ or the 4^+ are first forbidden ($\Delta I = 0, 1$, or 2; $\Delta\pi$: yes), so both may occur; in such situations, higher-energy transitions have larger values of $f(Z,E_0)$ and therefore usually occur more frequently. In most cases there is a low-lying state to which an allowed or a first forbidden transition can be made, so second or higher forbidden transitions are not very common.

Where I^π for a state is unknown, information can be obtained from beta-decay studies by measuring $fT_{1/2}$. This determines the degree of forbiddenness, which determines $\Delta I^{\Delta\pi}$ through the selection rules of Tables 11-1 and 11-2. If I^π for either the initial or final state is known, $\Delta\pi$ gives the parity and ΔI gives a few possible values of I for the other. For example, let us say an odd-odd nucleus decays to the ground state of an even-even nucleus (which must be 0^+) and $\log fT_{1/2}$ is measured to be 8.0. From Table 11-3, this indicates that the transition is first forbidden, whence from the selection rules we know $\Delta I = 0$, 1, or 2, and the parity is changed. The initial state must therefore be 0^-, 1^-, or 2^-.

In some cases, I^π can be determined uniquely. For example, if the above decay went to both the 0^+ and 2^+ states of the even-even nucleus with $\log fT_{1/2} \approx 5.0$, indicating that both transitions were allowed, the initial state could only be 1^+. While these methods are usually reliable, there is some risk in using $fT_{1/2}$ values to determine the degree of forbiddenness of a transition. From Table 11-3 we see that when $\log fT_{1/2}$ is in the range 6.0 to 6.8, the transition could easily be either allowed or first forbidden, and allowed transitions with $\log fT_{1/2}$ well above 7.0 are not unknown.

11-7 Decay Rates in Electron Capture

In Sec. 8-3 we pointed out that the beta-decay interaction can also induce the process (8-3) in which orbital electrons are captured. As the electron energy is fixed, the neutrino must carry all the energy released, so all neutrinos come out with the same energy, E_0 minus the binding energy of the electron. According to the derivation in Sec. 11-1, their emission probability is proportional to p_ν^2, which, ignoring the electron binding energy, is E_0^2/c^2. The decay rate is also proportional to the probability for the atomic electron to be at the position of the proton which changes into a neutron. This in inversely proportional to the volume occupied by the electron, and the latter is proportional to the cube of the Bohr radius, which is inversely proportional to Z. The net effect, therefore, is that the decay rate is proportional to Z^3. The operation of the field on the nucleus is described by the same H discussed in Sec. 11-4, so $|M|^2$ is the same as for electron emission. For forbidden transitions, the neutrinos come out with $l > 0$ and therefore must penetrate an angular-momentum barrier for which the penetrability is $\mathcal{P}$. Collecting these terms, we obtain

$$\lambda_{\text{EC}} = G_{\text{EC}} Z^3 E_0^2 |M|^2 \mathcal{P} \tag{11-22}$$

where G_{EC} is the proportionality constant which can be calculated in terms of G in a more thorough treatment.

In most cases, electron capture competes with emission of positive electrons, and since $|M|^2$ is the same for the two competing processes (it depends only on the nuclear states involved), the ratio $\lambda_{\text{EC}}/\lambda_\beta$ is independent of $|M|^2$ and hence of nuclear structure, and can be readily calculated from (11-19) and (11-22). The results for allowed transitions are shown in Fig. 11-6. From (11-22), we see that λ_{EC} increases rapidly with increasing Z, whereas a positive electron must pene-

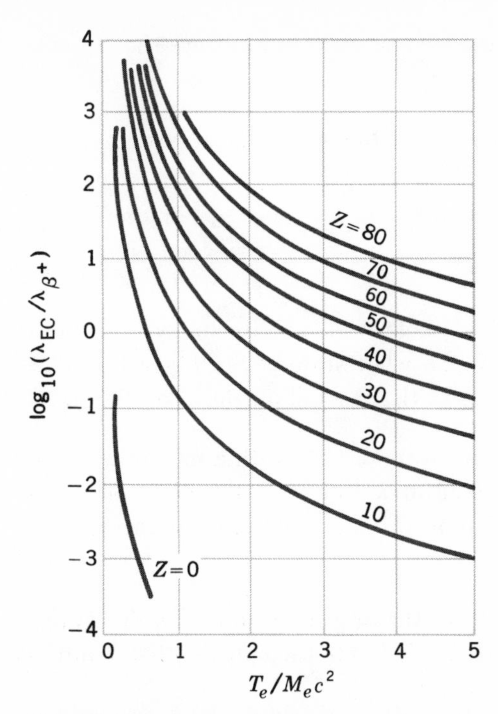

FIGURE 11-6 Ratio of electron capture to positron emission rates for allowed transitions as a function of atomic number Z and the maximum kinetic energy available in the β^+ decay (T_0). [*From E. Feenberg and G. Trigg, Rev. Mod. Phys.*, **22: 399 (1956).**]

trate a coulomb barrier so λ_β decreases with increasing Z. Electron capture therefore predominates in heavy nuclei, while emission of positrons generally predominates in light nuclei. Since $f(Z,E_0)$ increases more rapidly with energy than E_0^2, positron emission becomes relatively more probable with increasing E_0. At low positron energies, the fact that there is 1 MeV (that is, $2M_ec^2$) more energy available for electron capture than for positron emission and that the latter is strongly impeded by the coulomb barrier causes the curves in Fig. 11-6 to go up very fast.

Problems

11-1 From Table A-3, determine E_0 and T_0 for the beta decays of Mn^{56} and Co^{56}.

11-2 Calculate the electron spectrum from the beta decay of a neutron into a proton ($F = 1.0$). Make a plot of $I(p_e)$ vs. p_e and of $I(E_e)$ vs. E_e.

11-3 If a $\frac{5}{2}^+$ state decays by a first forbidden transition, what are the possible I^π of the final state?

11-4 What is the degree of forbiddenness in the following transitions?

(*a*) $\frac{1}{2}^+ \rightarrow \frac{3}{2}^+$ (*b*) $\frac{1}{2}^+ \rightarrow \frac{1}{2}^-$
(*c*) $\frac{5}{2}^+ \rightarrow \frac{7}{2}^-$ (*d*) $\frac{3}{2}^- \rightarrow \frac{9}{2}^-$
(*e*) $2^+ \rightarrow 1^+$ (*f*) $2^+ \rightarrow 5^-$

11-5 The ground state of In^{117} is a $g_{9/2}$ single-hole state, and it can decay to Sn^{117} with a total available energy of 1.5 MeV. To what state of Sn^{117} (see Fig. 6-6) does the transition lead? Check the result with the chart of nuclides (see Sec. 8-7).

11-6 Zn^{63} decays principally to the ground state of Cu^{63} with a maximum positron energy of 2.35 MeV. The half-life is 38 min. Calculate the $fT_{1/2}$ value and determine whether the transition is allowed or forbidden. What is the probability for electron capture?

11-7 A 1^+ state of an isotope of Cu decays to the 0^+ ground state of a Zn isotope with a maximum electron energy of 2.6 MeV. Estimate the half-life. If the initial state was 1^-, estimate the half-life.

11-8 In 1-MeV positron decays, what fraction of the decays are by electron capture if the decay is in Ca, Sn, and Hg?

11-9 A 1^+ state of a Tl isotope decays to a 0^+ state of a Hg isotope with a maximum positron energy of 1 MeV. Estimate the half-life. Do not ignore electron capture.

11-10 Suppose a state in a Bi isotope decays by beta decay only to the 2^+ state of an even-even Pb nucleus in which the lowest three states are 0^+, 2^+ and 4^+ in that order, the 4^+ state being at 1 MeV. The decay energy is 4 MeV, and the half-life is 5 s. What is I^π for the initial state?

Further Reading

See General References, following the Appendix.

Allen, J. S.: "The Neutrino," Princeton University Press, Princeton, N.J., 1958.

Ajzenberg-Selove, F.: "Nuclear Spectroscopy," Academic, New York, 1960.

Hyde, E. K., I. Perlman, and G. T. Seaborg: "The Nuclear Properties of the Heavy Elements," Prentice-Hall, Englewood Cliffs, N.J., 1964.

Kabir, P. K.: "The Development of Weak Interaction Theory," Gordon and Breach, New York, 1963.

Mayer, M. G., and J. H. D. Jensen: "Elementary Theory of Nuclear Shell Structure," Wiley, New York, 1955.

Nuclear Data, periodical published by Academic Press, New York.

Segre, E.: "Experimental Nuclear Physics," Wiley, New York, 1953.

Siegbahn, K.: "Alpha, Beta, and Gamma Ray Spectroscopy," North-Holland, Amsterdam, 1965.

Wu, C. S., and S. A. Moszkowski: "Beta Decay," Interscience, New York, 1966.

Chapter 12
Gamma-ray Emission

In contrast to nucleon emission and beta decay, the third type of nuclear decay, gamma-ray emission and other electromagnetic processes, is not peculiar to nuclear physics. It has nearly identical counterparts in atomic physics, and some of its aspects are closely related to electrical engineering.

As mentioned in Sec. 8-1, a gamma ray is an electromagnetic wave, basically no different from light or radio waves. An electromagnetic wave is an oscillating electric and magnetic field; the changing electric field induces a magnetic field, and the changing magnetic field induces an electric field, and so on. Such a wave can be generated by an oscillating electric charge which sets up an oscillating electric field, or by an oscillating electric current which sets up an oscillating magnetic field. In the first case we have what we call *electric multipole* radiation, designated E-l, and in the latter case it is *magnetic multipole* radiation, which we abbreviate M-l. We begin with a discussion of the former.

12-1 Electric Multipole Radiation from Quantum Systems

From classical electromagnetic theory we know that electric charges oscillating with a frequency ν emit electromagnetic radiation of that frequency. As we discussed in Sec. 1-6, the most general charge distribution can conveniently be expressed as the multipole expansion (1-10); the oscillations of a charge distribution can then be expressed as oscillations of each term in the expansion, and the resulting radiation may be thought of as being due to the various terms. This is especially convenient when one term in the expansion is predominant. For example, antennas used by radio and television stations are generally designed for electric dipole radiation. Electric dipole radiation is designated E1; electric quadrupole radiation, such as would result from symmetrical oscillations of the charge distribution shown in Fig. 1-5, is designated E2, etc.

The power radiated by an antenna in E-l radiation is found in the classical theory of electromagnetic radiation to be

$$P(\text{E-}l) = \frac{2(l+1)c}{l[(2l+1)!!]^2}\left(\frac{\omega}{c}\right)^{2l+2} Q_l^2 \qquad \textbf{(12-1)}$$

where $(2l + 1)!! = (2l + 1) \times (2l - 1) \times (2l - 3) \times \cdots 2$ or 1, $\omega = 2\pi\nu = 2\pi c/\lambda_r$ with λ_r being the wavelength of the emitted radiation, and Q_l is closely related to the maximum value (during the course of an oscillation) of the corresponding integral in (1-9). Since the charge in the nucleus is carried by the protons, it is often convenient to express these integrals in terms of the coordinates of the protons, x, y, and z. It is readily seen from (1-9) that this leads to

$$\begin{aligned} Q_1 &\sim Ze\bar{z} \\ Q_2 &\sim Ze(3\overline{z^2} - \overline{r^2}) = Ze(2\overline{z^2} - \overline{x^2} - \overline{y^2}) \\ &\cdots\cdots\cdots\cdots \end{aligned} \tag{12-2}$$

where bars indicate an average over all the protons in the nucleus of their coordinates at the extreme point of an oscillation. From (12-2) we see that Q_l is some fraction of ZeR^l. The magnitude of the various terms in (12-1) is therefore approximately proportional to $(\omega R/c)^{2l}$ or equivalently to $(R/\lambda_r)^{2l}$.

In quantum theory, the radiation is quantized into photons of energy $E = h\nu = \hbar\omega$, whence the rate of emission of photons is, from (12-1),

$$\lambda(\text{E-}l) = \frac{P(\text{E-}l)}{\hbar\omega} = \frac{2(l+1)}{\hbar l[(2l+1)!!]^2}\left(\frac{\omega}{c}\right)^{2l+1} Q_l^2 \tag{12-3}$$

This will allow us to calculate the half-life for gamma-ray emission once we know how to estimate Q_l.

Before facing that problem, let us consider the question of the angular momentum carried off in gamma-ray emission. In classical theory, the angular momentum in the electromagnetic radiation field L is related to the radiated energy E by

$$L = \frac{\sqrt{l(l+1)}}{\omega} E$$

Since the energy of a photon is $E = \hbar\omega$, its angular momentum L is then $\hbar\sqrt{l(l+1)}$, which is the angular momentum characterized by quantum number l.

Another way of understanding this result is to begin by pointing out that since energy is emitted by oscillating charges in discrete amounts, the energy in the oscillation must be decreased by equal discrete amounts; in the language of Sec. 5-9 the number of phonons in the oscillation is reduced by 1. While we can think of classical systems as emitting radiation continuously, a quantum system emits a photon only when its state of oscillation *changes*. In the oscillation shown in Fig. 5-10c, it is obvious from (12-2) that Q_1 has a large value, while in that shown in Fig. 5-10a, $Q_1 = 0$ but Q_2 is large. In transitions in which the number of phonons in these oscillations changes, E1 radiation is therefore emitted in the first case and E2 radiation in the second. In general, E-l radiation is emitted when a phonon of oscillation of $\lambda = l$ is given up by the nucleus. Since the phonon carries angular momentum λ and parity $(-1)^\lambda$, from conservation of these quantities a photon of E-l radiation must carry off angular momentum l and parity $(-1)^l$.

The momentum of a photon of energy E is E/c, which is somewhat less than that of an electron of the same energy, so a photon of a few MeV or less on being emitted from a nucleus must penetrate a large angular-momentum barrier, as discussed in Sec. 11-2. The penetrability of this barrier is not treated by a formula like (10-23) but is taken into account in (12-3) by the previously noted proportionality to $(R/\lambda_r)^{2l}$. The wavelength of a 1-MeV photon is about 1.1×10^{-12} m, whence if R is taken as the radius of a medium mass nucleus, $R/\lambda_r \simeq 1/200$. Since the other l-dependent factors in (12-3) vary slowly with l, λ(E-l) decreases by about four orders of magnitude for each successive value of l. It is interesting to point out here that in the emission of light by atoms, $R \simeq 0.5$ Å, $\lambda_r \simeq 5000$ Å, whence the emission rate decreases by about eight orders of magnitude for each successive l; as a result, emission of light with $l > 1$ is extremely rare.

12-2 Transitions between Nuclear States

Basically, electromagnetic radiation arises from a change in the state of oscillation of electric charges, but we have stated in Sec. 8-2 that gamma radiation is emitted when a nucleon jumps from one orbit to another. The consistency of these statements may be appreciated if we consider the time dependence of wave functions, a subject which was not discussed in Chap. 2 but was briefly mentioned in Sec. 10-1. This time dependence, which some may recognize as the solution of the time-dependent Schrödinger equation, is an oscillation with frequency $\omega = E/\hbar$. When a proton changes orbits, its energy and hence its frequency of oscillation changes, so we do have a change in the state of oscillation of an electric charge, which is just what is needed to emit electromagnetic radiation.

The quantum theory of radiation is based on (11-1), so the expression for Q_l is of the form (11-2) as

$$Q_l = \int \psi_f^* H(\text{E-}l)\psi_i \, d\tau \tag{12-4}$$

where H(E-l) represents the multipole moments (12-2). For elementary purposes, it may be thought of as an operator operating on ψ_i converting it to $Q_l^{SP}(H'\psi_i)$. Q_l^{SP} is a constant to be given by (12-6) below, and the operator H'(E-l)converts ψ_i into a sum of terms in each of which the orbit of one of the protons has its l value changed vector-addition-wise by l units. For example, if ψ_i and ψ_f differ only in that a proton is in an $f_{7/2}$ orbit in ψ_i and in a $p_{3/2}$ orbit in ψ_f, one of the terms in $H'(\text{E2})\psi_i$ will be identical with ψ_f, whence (12-4) gives $Q_2 = Q_2^{SP}$. The

SP superscript indicates that this is the value when the transition consists entirely of a *single particle* changing its orbit.

For students with experience in handling spherical harmonics $Y_{lm}(\theta,\phi)$—except for multiplicative constants, these are $P_{lm}(\cos\theta)e^{im\phi}$ from (2-25)—we shall treat this problem in more detail; others may go on to (12-6) without loss of continuity. It is easy to check that the terms in the multipole expansion (1-10) can be expressed as

$$\sum_k er_k{}^l Y_{lm}(\theta_k,\phi_k)$$

where the sum is over all the protons in the nucleus. These are the H(E-l) in (12-4) so it becomes

$$Q_l = \sum_k \int \psi_f^* er_k{}^l Y_{lm}\psi_i \, d\tau \tag{12-4a}$$

From (2-25), the θ,ϕ dependence of ψ_f and ψ_i are products of $Y_{l_{kf}m_{kf}}$ and $Y_{l_{ki}m_{ki}}$, respectively, whence (12-4*a*) includes integrals like

$$\int Y^*_{l_{kf}m_{kf}} Y_{lm} Y_{l_{ki}m_{ki}} \, d\theta_k \, d\phi_k$$

which are well known to be zero unless

$$\begin{aligned} \mathbf{l}_{kf} &= \mathbf{l}_{ki} + \mathbf{l} \\ m_{kf} &= m_{ki} + m \end{aligned}$$

That is, the only terms remaining are those in which l_{kf} and l_{ki} differ vector-addition-wise by l units; this explains the behavior described in the last paragraph. We are ignoring m quantum numbers although they must, of course, be taken into account in a complete treatment.

Now we consider the case cited where ψ_i and ψ_f differ only in that proton q is in an $f_{7/2}$ orbit in ψ_i and in a $p_{3/2}$ orbit in ψ_f. All terms in the sum (12-4*a*) other than the one for $k = q$ are a product of integrals, one of which is

$$\int \psi_q^*(p_{3/2})\psi_q(f_{7/2})r_q{}^2 \, dr_q \, d\theta_q \, d\phi_q$$

This is zero because the wave functions of any two different orbits are orthogonal, so all terms for $k \neq q$ are zero. In the term $k = q$, the integrals over the coordinates of particles other than q are unity because these particles are in the same orbit in ψ_i and ψ_f; the integral over the coordinates of q differs from the above integral in having an additional factor $er_q{}^2$, whence it is not zero. It can be written

$$|Q_2| = f_2 \int u_1(r_q) er_q{}^2 u_3(r_q) \, dr_q$$

where f_2 represents the integral over $d\theta_q \, d\phi_q$ and the effect of appropriate sums and averages over m_q quantum numbers, and $u_l(r_q)$ are r_q times the r-dependent part of the wave functions from (2-25) and Fig. 2-4. We use $|Q_2|$ rather than Q_2

because we are not paying attention to sign changes that may be induced by H(E-l) operators.

In general, then, when a transition consists of a single proton changing its orbit to one differing by l units of orbital angular momentum,

$$|Q_l| = Q_l^{SP} = |f_l \int u_f^* e r^l u_i \, dr| \tag{12-5}$$

This is called the single-particle value, Q_l^{SP}. In general, it is not a function only of l; for example, it is different for a $p_{1/2} \rightarrow f_{5/2}$ than for a $d_{5/2} \rightarrow g_{9/2}$ transition even though both are $l = 2$. However, the differences are not large, and V. F. Weisskopf has shown that a reasonable approximation for all cases is

$$Q_l^{SP} = \frac{3}{l+3} eR^l \tag{12-6}$$

As might be expected from (12-2) with $Z = 1$ (since only a single proton is involved), this is a fraction, somewhat less than unity, of eR^l. By inserting (12-6) into (12-3) we obtain the single-particle transition rates, also known as *Weisskopf rates*. They are shown in Fig. 12-1.

From what we have said, one might get the impression that E-l gamma-ray transitions occur only when protons change orbits since only protons have an electric charge. However, when neutrons change orbits, there are many subtle effects on the protons which cause a change in the oscillation of charge almost as large as that accompanying a proton orbit change. The lines in Fig. 12-1 are therefore approximately valid for neutron as well as proton orbit changes; for neutrons, $\lambda(E\text{-}l)$ is typically smaller by a factor of 2.

If ψ_i and ψ_f have several terms, those terms which differ in the orbit of one nucleon can contribute to the transitions. For example, if

$$\psi_i = a_1(g_{7/2})_{1/2}{}^3 + a_2(d_{3/2})_0{}^2 s_{1/2}$$
$$\psi_f = b_1(d_{5/2})_{5/2}{}^3 + b_2(d_{3/2})_0{}^2 d_{5/2}$$

a gamma-ray transition can lead from the second term of the first to the second term of the second by a nucleon's jumping from the $s_{1/2}$ to the $d_{5/2}$ orbit accompanied by E2 radiation. Q_2 for this transition is then close to $a_2 b_2 Q_2^{SP}$.

It is interesting to note from Fig. 4-5 that there are no cases where two orbits in the same shell differ in l and j by one unit. However we know from Chaps. 5 and 6 that all the low-energy states of nuclei are composed of different arrangements of nucleons in the same shell. There cannot therefore be E1 transitions between these states. On the other hand, there are a great many cases where two orbits in the same shell differ in l by two units, so E2 transitions are quite common among the low-energy states of nuclei. In states with enough excitation energy to have nucleons excited into the next shell, which according

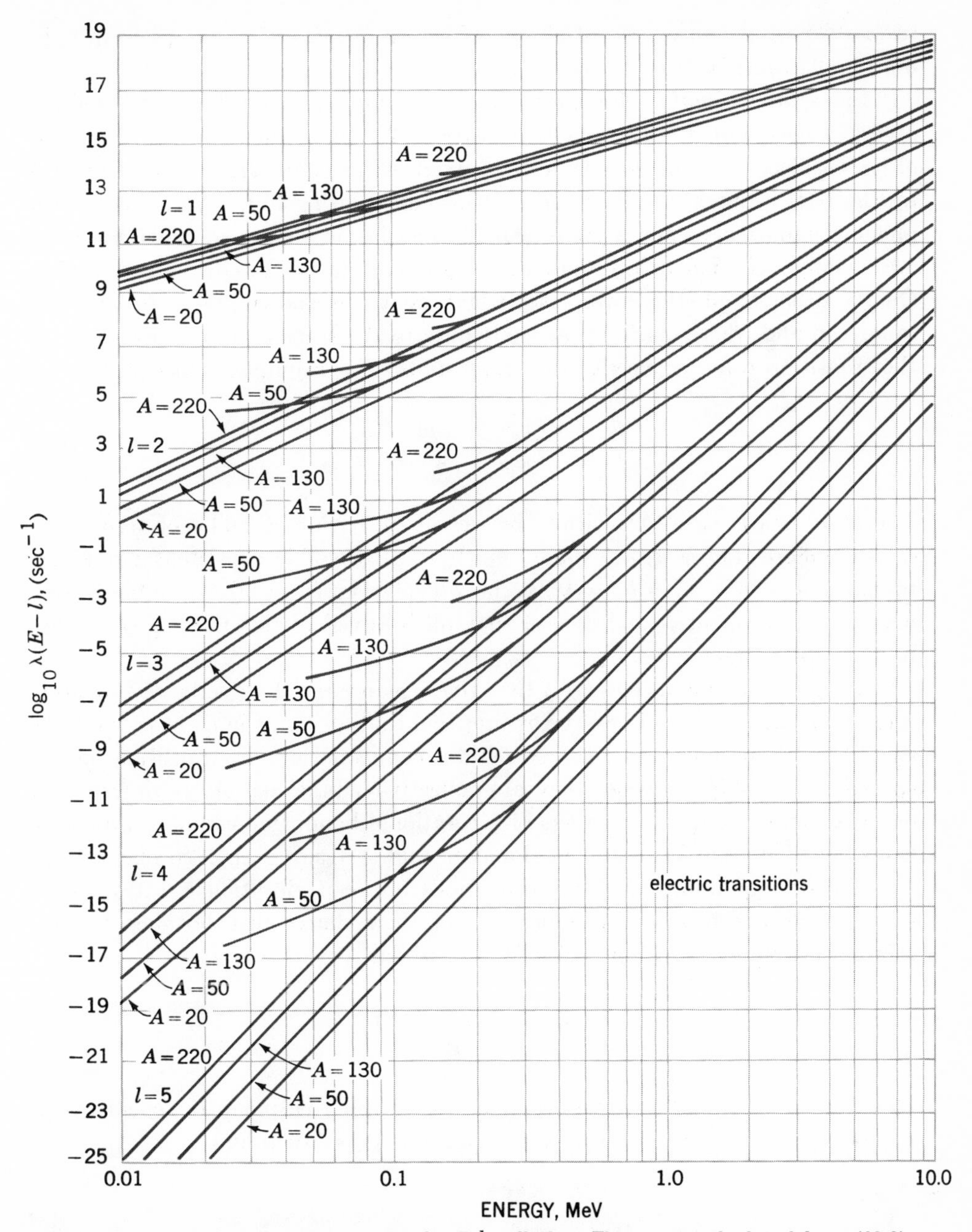

FIGURE 12-1 Single-particle decay rates for E-l radiation. These are calculated from (12-3) and (12-6) with $R = 1.4A^{1/3}$ F. The lower branch is for gamma-ray emission only, and the upper branch includes internal conversion. [*From R. W. Hayward in E. U. Condon and H. Odishaw (eds.), "Handbook of Physics." Copyright* **1967**. *McGraw-Hill Book Company. Used by permission.*]

to (4-19) requires an energy of about 6 MeV in heavy nuclei and more in light nuclei, E1 transitions from these to low-energy states are rather common.

Since wave functions for most states are quite complex, there are generally several terms contributing to any transition. We must then consider the signs of the coefficients of these terms in the wave functions, and sign changes induced by the $H(\text{E-}l)$ operators. In general, terms contributing to Q_l appear with both plus and minus signs and a great deal of cancellation occurs, so transition rates are usually very much less than the single-particle rates shown in Fig. 12-1. On the other hand, there are cases where all terms contribute to Q_l with the same sign and where many terms contribute to the transition in many ways, so the transition rate becomes greater than unity. For example, consider the case

$$\begin{aligned} \psi_i &= a_1(g_{9/2}s_{1/2})_4 + a_2(d_{5/2})_4{}^2 \\ \psi_f &= b_1(g_{9/2}d_{5/2})_2 + b_2(d_{5/2}s_{1/2})_2 \end{aligned} \tag{12-7}$$

There can clearly be E2 transitions between each term of ψ_i and each term of ψ_f. Q_2 is therefore $(a_1b_1 + a_1b_2 + a_2b_1 + a_2b_2)Q_2{}^{SP}$. It would be consistent with (5-10) to have $a_1 = a_2 = b_1 = b_2 = \sqrt{1/2}$, in which case, $Q_2 = 2Q_2{}^{SP}$. If there were many more terms in ψ_i and ψ_f and they all behaved in this way, Q/Q^{SP} could become quite large.

It should be noted, however, that the explanation of these large transition rates depends very heavily on the assumption that all terms in Q_2 have the same sign. This occurs only when the two states connected by the transition have a special relationship, as when they are collective vibrational states differing by one phonon, or adjacent members of a rotational band. Important examples are transitions between the first excited and ground states of even-even nuclei.

These large values of Q_l for transitions between vibrational states can be understood directly from the vibrations of their charge distributions. For Q_2, this can be expressed in terms of β from (7-37) as

$$Q_2 \sim \frac{3}{\sqrt{5\pi}} ZeR^2\beta$$

Due to several factors we have been ignoring in this treatment, the coefficient is reduced from $3/\sqrt{5\pi} \simeq 0.75$ to 0.38. From (12-6), we then find

$$\frac{Q_2}{Q_2{}^{SP}} = \frac{0.38}{3/5} \frac{ZeR^2\beta}{eR^2} \simeq 0.63Z\beta \tag{12-8}$$

In typical cases $\beta \simeq 0.2$ to 0.3, so for most nuclei $Q_2/Q_2{}^{SP}$ is typically between 3 and 10; since transition rates are proportional to Q^2, they are 10 to 100 times the single-particle rate.

A rotational state of a spheroidal nucleus clearly involves an oscillation of electric charge with a frequency equal to twice the rotational frequency. When the rotational state changes, the speed of rotation and hence the rotational frequency changes, whence there is a change in the state of oscillation of electric charge and radiation is emitted. Since the shape of the charge distribution in a rotation as viewed along the axis of rotation oscillates between the shapes shown for a $\lambda = 2$ vibration in Fig. 5-10, the radiation must be E2. As the whole nucleus is involved in the motion, the treatment leading to (12-8) is roughly applicable, so Q_2 is again much larger than the single-particle value.

12-3 Magnetic Multipole Radiation

We noted at the beginning of this chapter that not only can electromagnetic radiation arise from an oscillating electric charge, which sets up an oscillating electric field, but it can equally well arise from an oscillating electric *current* because this sets up an oscillating magnetic field. There are effective electric currents in nuclei due to the spins of nucleons and the orbital motion of protons. In general, the magnetic field due to a distribution of currents in a small source region, as observed at a large distance from the region, can be expressed as an expansion analogous to (1-9) containing magnetic dipole (M1), magnetic quadrupole (M2), etc., terms. Radiation can then be most easily understood as arising from oscillations of these various terms. The formula analogous to (12-3) is

$$\lambda(\text{M-}l) = \frac{2(l+1)}{\hbar l[(2l+1)!!]^2 c^2} \left(\frac{\omega}{c}\right)^{2l+1} A_l{}^2 \qquad \textbf{(12-9)}$$

where A_l is closely related to the amplitude of the magnetic multipole moment oscillations causing the radiation. The operators analogous to $H(\text{E-}l)$ are $H(\text{M-}l)$. The operation $H(\text{M-}l)$ is somewhat complicated,[1] but $H(\text{M-}l)\psi_i$ gives, with amplitudes not much less than unity, all terms in which the orbit of one nucleon is changed such that its total angular momentum is altered vector-addition-wise by l units, its orbital angular momentum is changed vector-addition-wise by $l - 1$ units, and hence its parity is multiplied by $(-1)^{l-1}$. For example, $H(\text{M1})$ can change a $d_{5/2}$ orbit to a $d_{3/2}$ orbit by flipping its spin, or it can convert a $(d_{5/2}d_{3/2})_2$ term to a $(d_{5/2}d_{3/2})_3$ term by changing the m value of one of the orbits, reorienting it in space; $H(\text{M2})$ can change a $p_{1/2}$ orbit to a $d_{3/2}$ or a $d_{5/2}$ orbit; $H(\text{M3})$

[1] A reasonably complete treatment can be found in J. M. Blatt and V. F. Weisskopf, "Theoretical Nuclear Physics," Wiley, New York, 1952.

can change it to a $f_{5/2}$ or $f_{7/2}$ orbit; etc. One can readily check that these changes conform to the stated rules.

The single-particle value of A_1 for an M1 transition is of the order of the magnetic moment change when a spin is flipped or an orbit is reoriented in space. From Sec. 1-6 we estimate that this is of the order of a few times $e\hbar/M_p$. The single-particle rates for M-l transitions, obtained by inserting estimates of this type (with some added sophistication) into (12-9) are shown in Fig. 12-2. They are called *Moszkowski transition rates.* In principle, there should be a difference between single-particle rates for neutrons and protons because of their different magnetic moments and because there is a contribution for protons from their orbital motion, but these differences (typically a factor of ~1.5) are well within the general uncertainty of the whole approach, so the same curves are used for both.

The ratio of λ(E1) to λ(M1) when both proceed at single-particle rates is, from (12-3) and (12-9) with the crude estimates given for Q_1 and A_1,

$$\frac{\lambda(\mathrm{E1})}{\lambda(\mathrm{M1})} \simeq \frac{\sim\frac{1}{2}(eR)^2}{\sim 5(e\hbar/M_p)^2/c^2} = \frac{1}{10}\left(\frac{M_pRc}{\hbar}\right)^2$$

If we take R as the radius of a medium mass nucleus, this is ~1,000. From a comparison between Figs. 12-1 and 12-2 we see that λ(E-l)/λ(M-l) decreases from this value with increasing l.

12-4 Selection Rules

Since gamma rays emitted in E-l transitions carry off angular momentum l and parity $(-1)^l$, conservation of angular momentum and parity requires that the angular momenta and parities of the initial and final nuclear states be related by

E-l: $$\mathbf{I}_i - \mathbf{I}_f = \mathbf{l} \qquad \Delta\pi = (-1)^l \tag{12-10a}$$

In accordance with the properties listed above for the H(M-l) operators, the selection rules for M-l radiation are

M-l: $$\mathbf{I}_i - \mathbf{I}_f = \mathbf{l} \qquad \Delta\pi = (-1)^{l-1} \tag{12-10b}$$

Where the selection rules allow more than one type of transition, the one with the lowest l ordinarily predominates because of its higher rate. However, when E2 transitions have greater than single-particle rates in transitions between collective states, as discussed in connection with (12-8), they often compete favorably with M-1 transitions. For example, the $2^+ \rightarrow 2^+$ transition between the two-phonon and one-phonon collective states in Fig. 5-11 can go by either M1 or E2 or by a combination of both.

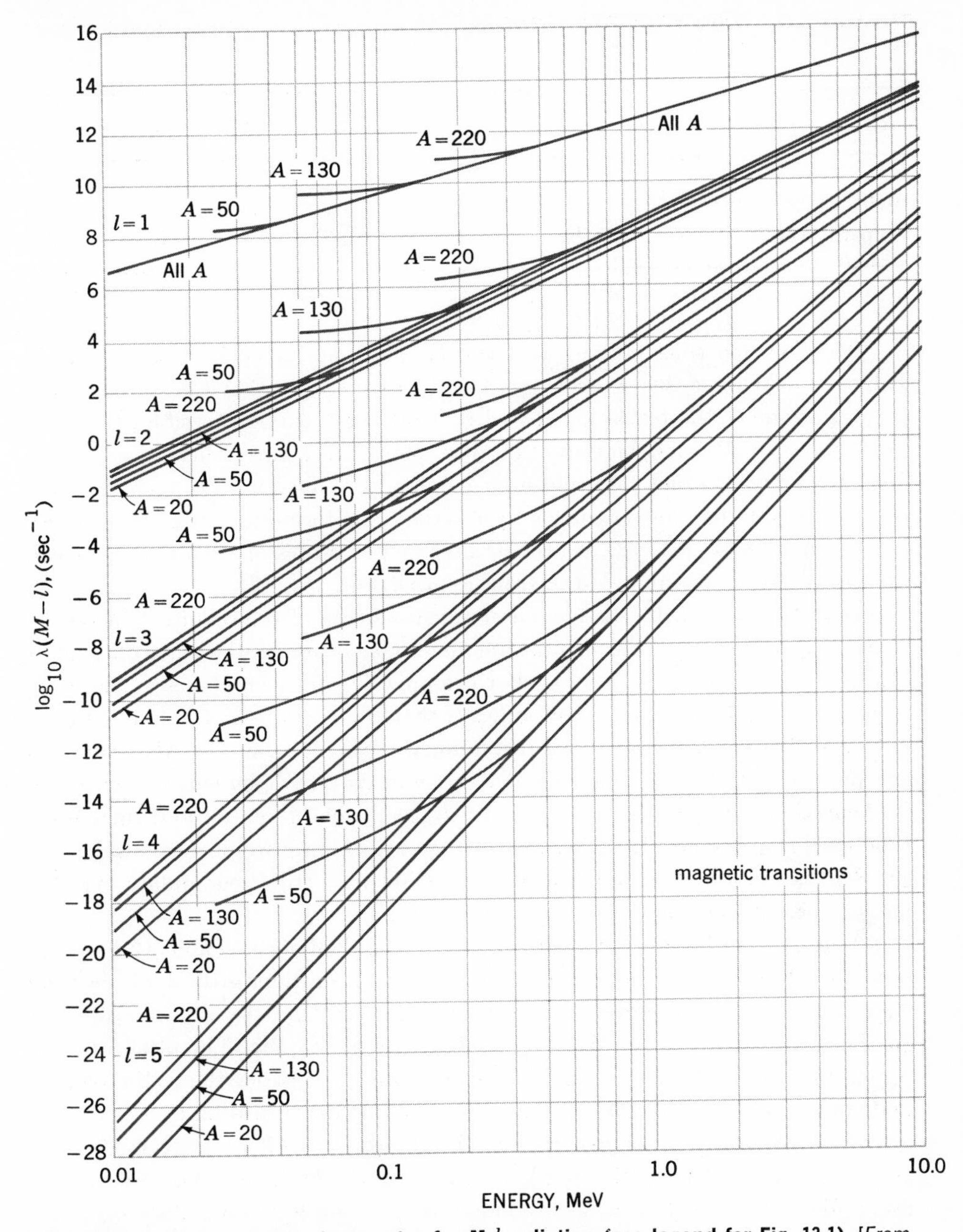

FIGURE 12-2 Single-particle decay rates for M-l radiation (see legend for Fig. 12-1). *[From R. W. Hayward in E. U. Condon and H. Odishaw (eds), "Handbook of Physics." Copyright* **1967.** *McGraw-Hill Book Company. Used by permission.]*

TABLE 12-1: SELECTION RULES FOR GAMMA-RAY TRANSITIONS Transitions in parentheses are not possible if either I_i or I_f is zero

$\Delta I = \lvert I_i - I_f \rvert$	0	1	2	3	4	5
Parity change	(E1)	E1	M2	E3	M4	E5
No parity change	(M1)(E2)	M1(E2)	E2	M3	E4	M5

The selection rules (12-10) with this corollary are given in tabular form in Table 12-1. When either I_i or I_f is zero, (12-10) requires that l be equal to the other, so some of the transitions listed in Table 12-1 are not possible; these are denoted by parentheses. An important example of this type is where $I_i = I_f = 0$. Such transitions cannot take place by gamma-ray emission, although they are possible by internal conversion or by internal pair formation. Especially interesting examples of this type are the 0^+ first excited states of O^{16} and Ca^{40}, which can decay (to the 0^+ ground states) in no other way.

In (12-10) and Table 12-1, no account was taken of the requirement on the H(M-l) operation that the orbital angular momentum of the nucleon cannot change by more than $l - 1$ units. For example, if a nucleon changes from a $d_{3/2}$ to an $s_{1/2}$ orbit, Table 12-1 would indicate that the transition is M1, but this is not in conformity with the above requirement. In such a case, an M1 transition can proceed only through higher-order terms in H(M1) or by small mixtures of configurations such as $[(d_{3/2})_2{}^2 s_{1/2}]_{3/2}$ and $[(d_{3/2})_0{}^2 s_{1/2}]_{1/2}$ in the initial and final wave functions, respectively. In either case, the transition rate would be considerably smaller than in a normal M1 transition, but M1 may still predominate over the alternative E2 transition.

When I^π are known for both the initial and final state, Table 12-1 can be used to determine the transition type and hence to estimate the half-life from Fig. 12-1 or 12-2. As examples, it is clear that $2^+ \rightarrow 0^+$ goes by E2, $3^- \rightarrow 2^+$ goes by E1, $\frac{9}{2}^+ \rightarrow \frac{1}{2}^-$ goes by M4, $\frac{11}{2}^- \rightarrow \frac{1}{2}^+$ goes by E5, etc. The other obvious application of Table 12-1 is where I^π is not known for either the initial or final state but is known for the other. A determination of the transition type may then give the unknown I^π. For example, an E2 transition to a 0^+ state must come from a 2^+ state, E1 transitions to both a 2^+ and a 4^+ state must come from a 3^- state, M4 transitions to a $\frac{1}{2}^-$ state must come from a $\frac{9}{2}^+$ state, etc.

Because of the utility of this method, it is often important to determine experimentally the E-l or M-l of a transition. One method for doing this is by measuring the half-life in a delayed-coincidence experiment (see Sec. 9-7) if it is short or by counting the gamma rays emitted per second as a function of time if it is long. By comparing the measured half-life with the estimates from Figs. 12-1 and 12-2, one can sometimes decide on the transition type. This method is rather dangerous, however, since decay rates often deviate from the single-particle value by several orders of magnitude. A more reliable method is described in the next section.

12-5 Angular-correlation Studies

One very useful method for determining the multipolarities of gamma-ray transitions is by measuring the angular correlation between successively emitted gamma rays. It is illustrated and explained in Fig. 12-3.

The principle behind this method can be understood in terms of what electrical engineers call *antenna patterns*. A dipole antenna, which in its simplest form is a straight rod like the ones used for automobile radios or walkie-talkies, radiates (or receives) power in proportion to $\sin^2 \theta$, where θ is the angle relative to the antenna rod. No power is radiated (or received) in the direction parallel to the rod, a conclusion which can easily be verified with two walkie-talkies. When a photon comes off in a nuclear transition carrying off angular momentum l with z component m, it can be shown that (l,m) determine the antenna pattern. For example, in dipole radiation the $\sin^2 \theta$ antenna pattern mentioned above is characteristic of $l = 1, m = 0$; both $l = 1, m = +1$ and $l = 1, m = -1$ have the antenna pattern $\frac{1}{2}(1 + \cos^2 \theta)$. It should be noted that when all three m states are present in equal amounts, as when radiation is emitted from a group of randomly oriented nuclei, the sum of the three antenna patterns is isotropic and each gives an equal amount of radiation, a result that is physically obvious. It should also be noted that we are still free to choose the z axis in the nuclear problem at our discretion.

Now let us consider the situation illustrated in Fig. 12-4*a*, where a nucleus makes a transition from a 0^+ to a 1^- to a 0^+ state by successive E1 gamma-ray emissions. Since we are free to choose the z axis, we choose it to be along the direction of emission of the first gamma ray γ_1. With this choice, there is no $m = 0$ radiation in γ_1 since the antenna pattern for $m = 0$ ($\sin^2 \theta$) gives no intensity in the z direction. From conservation of angular momentum, this means that the $m = 0$ component of the 1^- state is not populated in the transition, as shown in Fig. 12-4*b*. In the emission of the second gamma ray, angular-momentum conservation then requires that γ_2 come off with $m = 1$ or $m = -1$, both of which have a $\frac{1}{2}(1 + \cos^2 \theta)$ antenna pattern relative to the z axis, which is the direction of γ_1. In summary, if the transitions are $0^+ \rightarrow 1^- \rightarrow 0^+$, the angular correlation of γ_2 relative to γ_1 is given by $1 + \cos^2 \theta$. These angular correlations are very specific to the angular momenta of the three states and the multipolarities of the transitions, so by careful experimental measurements of the

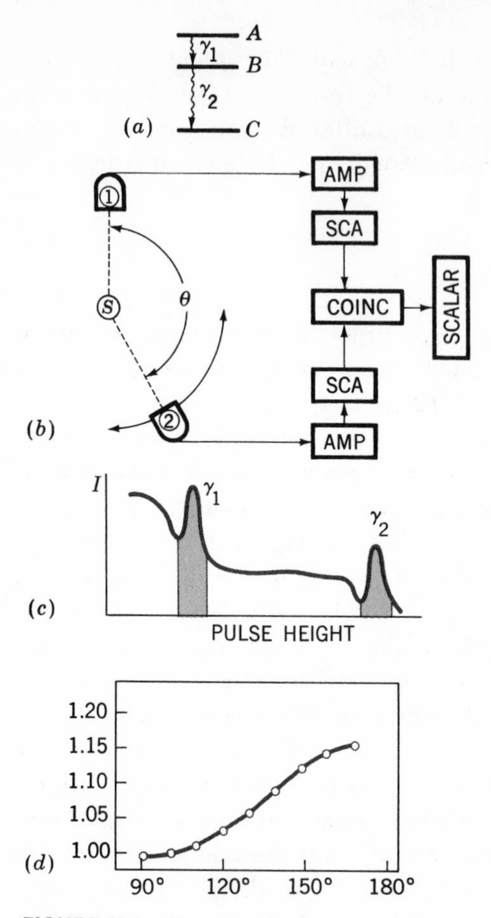

FIGURE 12-3 **For legend see page 301.**

correlations the angular momenta of the three states can be determined. In fact, where there is a mixture of M1 and E2 in the radiation, the angular correlation determines how much of each is present.

While these experiments do a good job of determining multipolarities, they do not determine whether transitions are electric or magnetic, since antenna patterns are the same for both types. This can be understood in terms of radio antennas by comparing the fields produced by current flowing in a rod (an electric dipole antenna) and by current flowing in a circular loop which surrounds the rod (a magnetic dipole antenna). The electric and magnetic fields produced by these two are everywhere proportional but perpendicular to one another; i.e., they have opposite polarizations. This suggests the standard method for distinguishing between electric and magnetic radiations in angular correlation experiments, namely, by a measurement of the polarization. One of the gamma

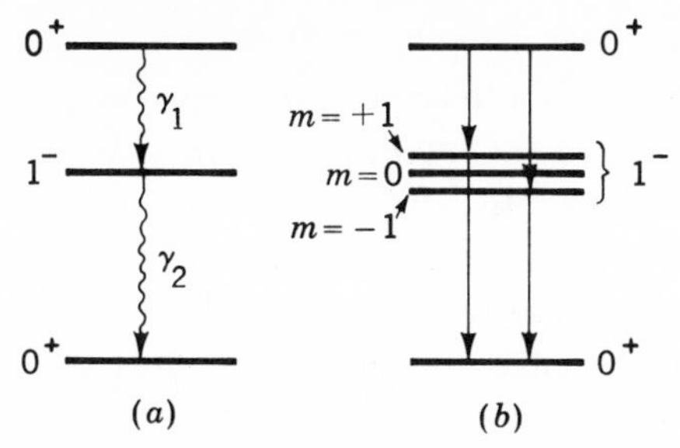

FIGURE 12-4 Successive gamma-ray transitions in the example discussed in the text. In the diagram at the right, the three components of the 1^- state are shown separated even though they are at the same energy. The $m = 0$ component is not populated in the decay if the z axis is chosen to be along the direction of emission of γ_1.

rays is allowed to undergo a Compton scattering; the cross section for that process is sensitive to the angle between the direction of scattering and the polarization vector, so the polarization is easily found and determines whether the radiation is electric or magnetic.

The theory of angular correlations is very well developed, and many different types of angular correlation measurements have given useful results. Correlations between successively emitted beta and gamma rays, between successively emitted alpha particles and gamma rays, between the first and third members of a three-gamma-ray cascade, and between the angular momentum vectors of aligned nuclei and gamma rays emitted from them all have important applications. In all these cases, internal conversion electrons can be substituted for the gamma rays. By applying a magnetic field, the orientation of the intermediate state, for example, the 1^- state in Fig. 12-4, can be made to precess, which leads to a

---

FIGURE 12-3 A typical γ-γ angular-correlation experiment. *(a)* Two successive gamma-ray transitions between nuclear states A, B, and C. *(b)* Experimental arrangement: gamma rays are emitted from the source S and detected by a fixed detector (1) and a movable detector (2). Pulses from the detectors (usually scintillators) are amplified (in AMP) and passed into single-channel analyzers (SCA), instruments which produce an output pulse only when the input pulse heights are between preset limits. These settings are shown by the shaded portions in a typical pulse-height spectrum in *(c)*; the upper SCA is set to detect γ_1 and the lower is set to detect γ_2. Outputs of the two SCA are fed into a coincidence analyzer $(COINC)$, which produces an output pulse when two inputs arrive at the same time, indicating they are from the same decay. These output pulses are counted in a scalar; counts are taken for various angles θ between the two detectors. A typical result is shown in *(d)*. It can be proven that these angular correlations are symmetric about 90°, so data are usually taken only between 90 and 180°.

rotation of the angular correlation pattern; by measuring this rotation as a function of the time delay between the two emissions in a delayed-coincidence experiment,[1] one can determine the magnetic moment of the intermediate state. When the source is in a solid (or a viscous liquid), magnetic and nonuniform electric fields in the crystal acting on the magnetic dipole moment and the electric quadrupole moment of the intermediate state cause random precessions which reduce the magnitude of the variations in the angular correlation, an effect that can be used to investigate these crystalline fields. This is a subject of interest in solid-state physics, but it is a major headache in nuclear experiments.

12-6 Isomerism

Gamma-ray transitions with readily observable half-lives are called *isomers*. While this term is sometimes applied to half-lives as short as 10^{-8} s, the most spectacular cases are where the half-life is of the order of seconds, days, or years. They aroused great interest in the early days of nuclear physics because of the peculiar situation of having two nuclei with the same atomic number and atomic weight. Another interesting aspect of these long-lived isomers is that since their half-lives are comparable with those for beta decay, they can often also decay by that process, and sometimes beta decay is the predominant decay mode.

From Figs. 12-1 and 12-2 we can see that isomerism occurs only for transitions with $l \gtrsim 3$. It is therefore not likely to occur for highly excited states since there are many states of various angular momenta available at lower energy, so $l = 1$ or $l = 2$ transitions are certain to be available. Isomerism is also not likely to occur in even-even nuclei since the vibrational or rotational states provide $I = 0, 2, 4, 6$, etc., states at relatively low energy and in that order, so decays with $l = 1$ or 2 are almost always available. In odd-odd nuclei, as we found in Sec. 6-1, states of rather different angular momenta occur at low energy in a way that is difficult to predict. If the ground state and first excited states happen to have very different angular momenta, the latter will be an isomer. For example, in Sb^{116}, which was discussed in connection with Fig. 6-9, the ground states is 1^+, mostly from $(d_{3/2})_\nu(d_{5/2})_\pi$, and the first excited state is 8^-, mostly from $(h_{11/2})_\nu(d_{5/2})_\pi$. A gamma-ray transition between these would be E7 which would have a half-life in the billions of years. Both the 1^+ and the 8^- states therefore decay by beta decay, the former to low-I and the latter to high-I states of Sn^{116}.

The most widespread and easily understandable type of isomerism occurs in odd-A nuclei. The distribution of isomers in these nuclei as a function of N,

[1] This can be done by introducing various cable lengths in the line from detector 2 in Fig. 12-3*b*.

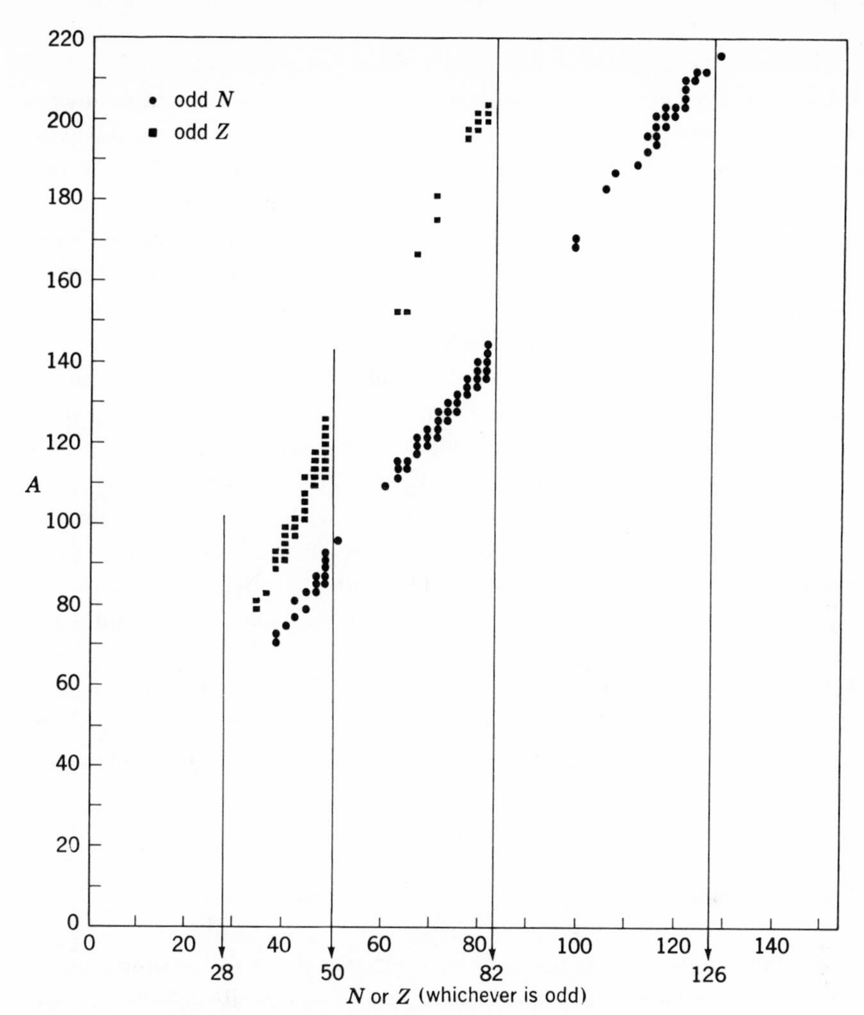

FIGURE 12-5 Distribution of long-lived isomers in odd-A nuclei. [*From M. Goldhaber and R. D. Hill, Rev. Mod. Phys.*, **24: 179 (1952).**]

Z, and A is shown in Fig. 12-5. To understand it, we need only recall that the low-energy states of odd-A nuclei are essentially the SQP states. First let us consider the SQP states in the $\mathfrak{N} = 5$ neutron shell whose energies are shown in Fig. 6-4. Near the beginning of the shell, all higher states can decay to lower states with $l = 1$ or 2 transitions, so no isomerism can occur. There is a short region where the lowest two states are $s_{1/2}$ and $g_{7/2}$, so M3 transitions are expected; this is the situation in Sn^{113}. However, once the $h_{11/2}$ SQP state falls below the $g_{7/2}$, isomerism is expected in every nucleus. In all cases, the $h_{11/2}$ decays to the $d_{3/2}$ SQP state or vice versa by an M4 transition. Some examples of this type can be seen in Fig. 6-6. It is evident from the chart of the nuclides as well as from Fig. 12-5 that there are isomers known for practically every even Z, odd N nucleus with N between 67 and 81. When the $\mathfrak{N} = 5$ shell is filling with protons, the orbit energies are more like those shown for $A = 207$ in Fig. 5-14. The $h_{11/2}$ and $d_{5/2}$ SQP states are close together, so usually there is a $5/2^+$ state for the former to decay into. This is an E3 transition, which often has a half-life short enough to be missed if it is not carefully searched for. Nevertheless there are many odd-Z, even-N nuclei in this region with isomers due to $h_{11/2} \rightarrow d_{5/2}$ (for example, Au^{197}) or $h_{11/2} \rightarrow d_{3/2}$ (for example, Tl^{207}) decays.

From the structure of the $\mathfrak{N} = 4$ shell in Fig. 4-5 it is clear that when the $g_{9/2}$ SQP state falls below the $f_{5/2}$, isomerism will occur. The two lowest-energy states are then $g_{9/2}$ and $p_{1/2}$, which are connected by M4 transitions. Thus, practically all even-Z, odd-N nuclei between $N = 39$ and 49 have isomers, as do practically all even-N, odd-Z nuclei with Z between 39 and 49. In the $\mathfrak{N} = 6$ shell we expect isomerism when the $i_{13/2}$ SQP state falls below the $f_{7/2}$. It is found in nearly all even-Z, odd-N nuclei with N between 115 and 125. Below $N \simeq 115$, the situation is complicated by the occurrence of spheroidal nuclei, so Fig. 4-5 is not applicable there.

All known isomers in odd-A nuclei are shown in Fig. 12-5. It is noteworthy that, with the exception of a few in spheroidal nuclei which were not considered here, essentially all these isomers can be easily understood and could have been predicted.

12-7 Internal Conversion

As was pointed out in Sec. 8-2, transitions between two states of the same nucleus can be induced by the electric and magnetic interactions with its orbital electrons in internal-conversion processes, as well as by gamma-ray emissions. The effects on the nucleus are the same as in the latter case, so here again we have $H(\text{E-}l)$ and $H(\text{M-}l)$ operators. Since the operators are the same and the initial and final nuclear states are the same, Q_l^2 and A_l^2 must be the same for gamma-ray emission and internal-conversion transitions between two states. All other factors determining transition rates can be calculated in a straightforward way, as in (12-3) or (12-9), so the ratio of internal conversion to gamma-ray emission rates $\lambda_{ic}/\lambda_\gamma$ is accurately calculable. This ratio is called the *internal-conversion coefficient* α, and it can be calculated separately for the emission of each orbital electron. Thus α_K, $\alpha_{L_{\text{I}}}$, $\alpha_{L_{\text{II}}}$, $\alpha_{L_{\text{III}}}$, refer to the ratio to the gamma-ray emission rate of the rates for the emission of $K(1s_{1/2})$, $L_{\text{I}}(2s_{1/2})$, $L_{\text{II}}(2p_{1/2})$, and $L_{\text{III}}(2p_{3/2})$ orbital electrons respectively, and α is equal to the sum of these plus all others.

In general, it is not possible to obtain closed-form expressions for the α's, but to see roughly what they involve we introduce an approximate expression for α_K which is valid for transition energies $\hbar\omega$ small compared with the electron rest energy M_ec^2 but large compared with the electron binding energy B_e:

$$\alpha_K(\text{E-}l) \simeq \frac{l}{l+1} Z^3 \left(\frac{1}{137}\right)^4 \left(\frac{2M_ec^2}{\hbar\omega}\right)^{l+5/2} \qquad B_e \ll \hbar\omega \ll M_ec^2$$

The Z^3 dependence is due to the fact that the K electron spends more time near the nucleus as Z increases because of the decrease in the radius of the Bohr orbit. The increase in α with decreasing $\hbar\omega$ is due to the very rapid increase of λ_γ with increasing $\hbar\omega$, as is evident from (12-3). The fact that radiation from an antenna increases rapidly with frequency is familiar from the fact that 60-cycle current flowing in great abundance through wires in every house does not radiate, whereas the design of microwave circuitry is plagued at every turn by the tendency of power to be radiated away.

The results of accurate relativistic calculations of α_K for nuclei of $Z = 30$, 60, and 90 are shown in Fig. 12-6. We see there that the rapid increase in α_K with increasing Z, with decreasing transition energy, and with increasing l is valid at all energies. In addition there is a strong tendency in heavy nuclei for α to be larger for magnetic radiation than for electric radiation. Internal-conversion coefficients for L-shell electrons in $Z = 90$ nuclei are shown in Fig. 12-7. It is seen there that $\alpha_L \ll \alpha_K$ at high energies (and this is even more so in lighter nuclei) but α_L increases more rapidly with decreasing energy. At low energies, L conversion becomes dominant, and K conversion even begins to fall off as the transition energy approaches the binding energy of the K electron.

The upper branches of the curves of Figs. 12-1 and 12-2 show the total single-particle decay rate for emission of gamma rays plus internal-conversion electrons. The total internal-conversion coefficients are the ratio of the difference between the two branches and the lower branch of these curves. They have no direct connection with single-particle rates, and they are, in fact, applicable for any transition.

The energy of the emitted electrons is equal to the transition energy minus the electron binding energy. In heavy nuclei, the binding-energy difference between the three types of L electrons is sufficient to allow conversion electrons from them to be resolved by magnetic spectrographs. An example of such data is shown in Fig. 12-8.

From Figs. 12-6 and 12-7 we can recognize the availability of three methods for determining the E-l or M-l of a transition: (1) by measuring the ratio of conversion electrons to gamma rays, (2) by measuring the ratio of K to L conversion electrons, and (3) by measuring the ratio of $L_{\text{I}}/L_{\text{II}}/L_{\text{III}}$ conversion electrons. These measurements are not difficult to make, and data for many transitions can often be obtained simultaneously, as in Fig. 12-8. The determination of

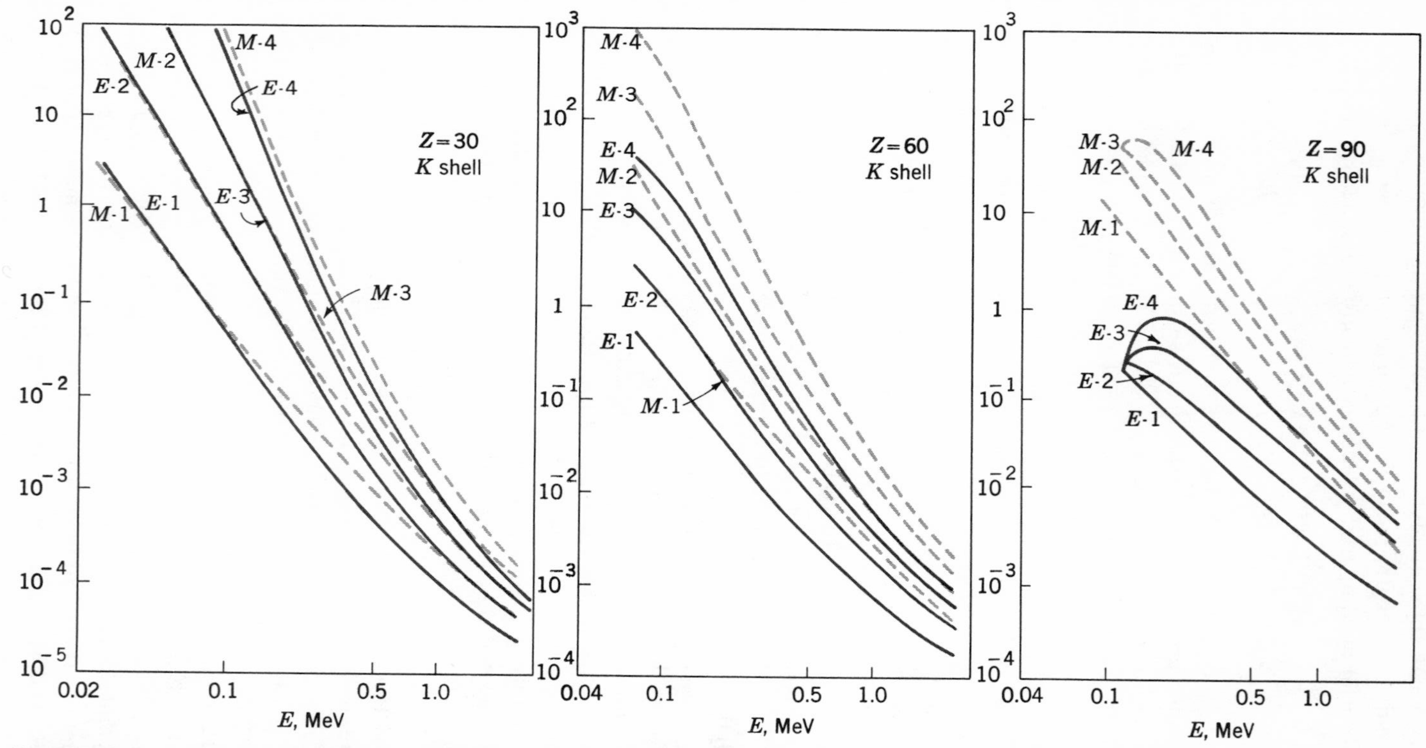

FIGURE 12-6 α_K, **internal-conversion coefficients for** K **electrons, for various E-**l **and M-**l **transitions in nuclei of atomic numbers 30, 60, and 90.** *(From C. M. Lederer et al., "Table of Isotopes," Wiley, New York,* **1966;** *by permission.)*

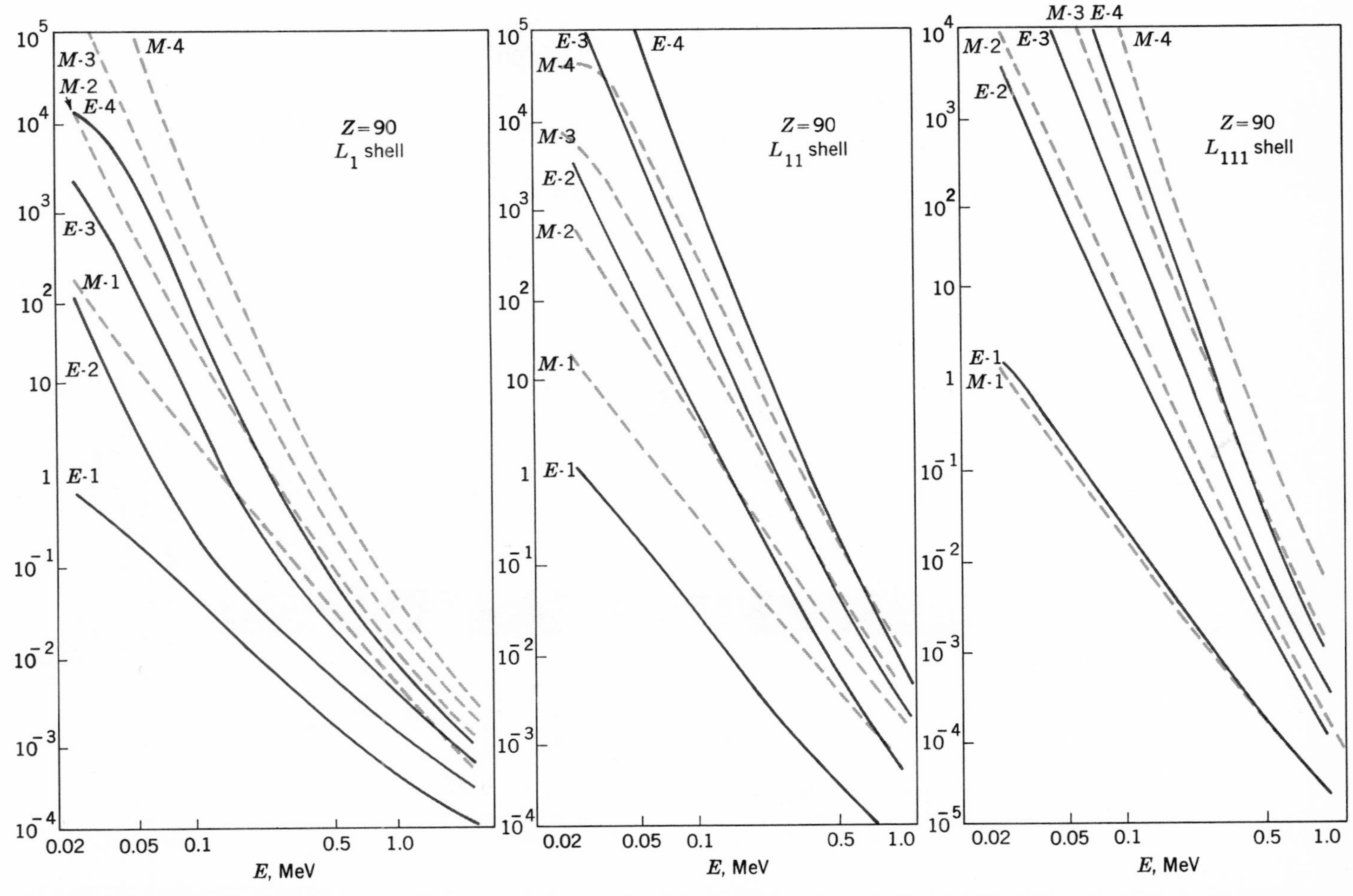

FIGURE 12-7 Internal-conversion coefficients for L_{I}, L_{II}, and L_{III} electrons for various E-l and M-l transitions in nuclei of atomic number 90. *(From C. M. Lederer et al., "Table of Isotopes," Wiley, New York,* **1966;** *by permission.)*

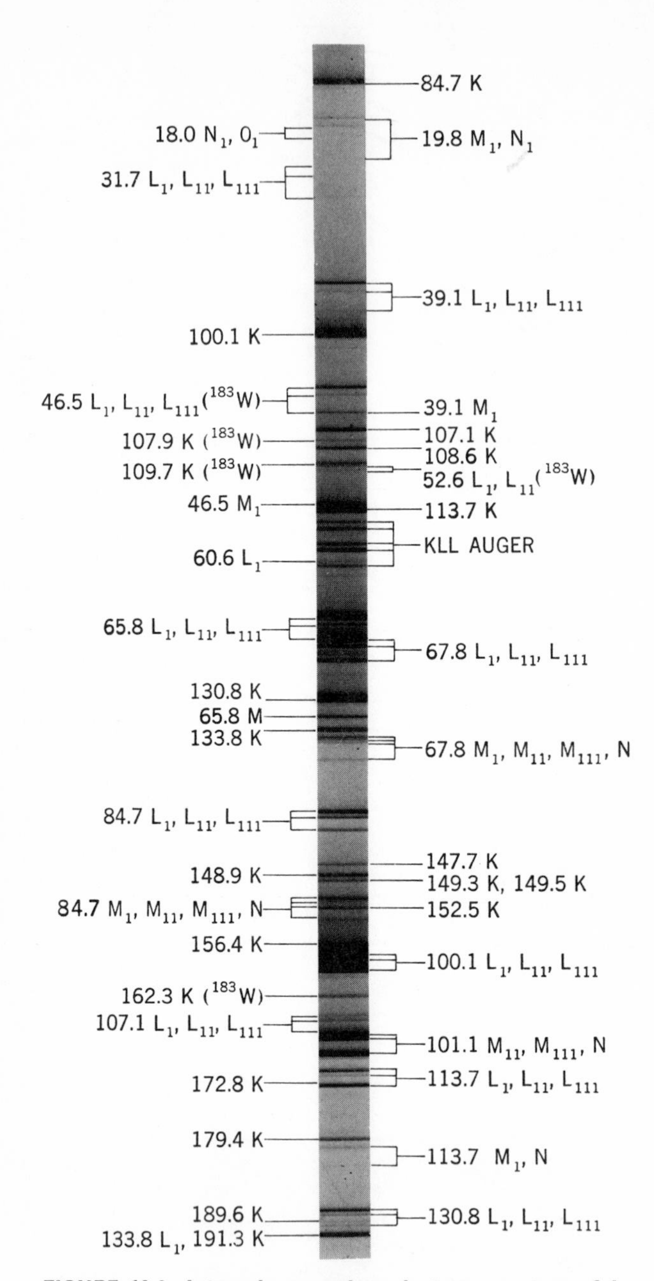

FIGURE 12-8 Internal-conversion electron spectrum following the decay of Re^{182} to W^{182}, recorded on a photographic film on the focal plane of a 180° magnetic spectrograph. *(Courtesy of B. Harmatz, Oak Ridge National Laboratory.)*

transition type from the measurements is usually straightforward by use of curves like those in Figs. 12-6 and 12-7. All three of these methods have been widely used, and, in heavy nuclei at least, internal-conversion experiments have determined far more I^π assignments than any other method.

Problems

12-1 Estimate the transition rate for the gamma-ray decay from the $s_{1/2}$ state (excitation = 0.9 MeV) to the ground ($d_{5/2}$) state of O^{17}. Justify the method used.

12-2 The first excited state (the one-phonon, 2^+ vibrational state) of Cd^{114} has an excitation energy of 0.56 MeV. Estimate the transition rate between it and the ground state.

12-3 Estimate the ratio of E1 to M1 single-particle transition rates in the emission of light by atoms.

12-4 What type of gamma-ray transition is likely to be predominant if the initial and final nuclei are as given below?

(*a*) $3^+ \rightarrow 0^+$
(*b*) $\frac{3}{2}^- \rightarrow \frac{7}{2}^-$
(*c*) $\frac{7}{2}^- \rightarrow \frac{3}{2}^+$
(*d*) $\frac{7}{2}^- \rightarrow \frac{1}{2}^+$
(*e*) $0^+ \rightarrow 0^+$
(*f*) $2^+ \rightarrow 2^+$
(*g*) $1^+ \rightarrow 0^+$
(*h*) $\frac{1}{2}^+ \rightarrow \frac{1}{2}^-$

12-5 The energy levels for the single-hole nucleus Pb^{207} are as shown in Fig. 6-2. What is the principal decay mode of each state? Estimate its half-life. In which of the transitions is internal conversion most important? Are any of these states isomers?

12-6 If nuclei did not become spheroidal as the $\mathfrak{N} = 7$ shell of Fig. 4-5 fills with neutrons, what types of isomerism would be expected in odd-N nuclei?

12-7 Suggest at least three nuclei which probably have isomers but for which none are listed in the chart of nuclides. Consider the problems in finding them.

12-8 A nucleus with $Z = 90$ decays by emission of a 0.4-MeV gamma ray from a $\frac{7}{2}^-$ to a $\frac{1}{2}^-$ state. What is the ratio of internal conversion to gamma-ray emission? What is the K/L ratio? What is the $L_{\mathrm{I}}/L_{\mathrm{II}}/L_{\mathrm{III}}$ ratio? What would these ratios be if the initial state were $\frac{7}{2}^+$?

12-9 What is the internal conversion coefficient for a 0.2-MeV E5 transition in a nucleus with $A = 65$?

Further Reading

See General References, following the Appendix.

Ajzenberg-Selove, F.: "Nuclear Spectroscopy," Academic, New York, 1960.

Hyde, E. K., I. Perlman, and G. T. Seaborg, "The Nuclear Properties of the Heavy Elements," Prentice-Hall, Englewood Cliffs, N.J., 1964.

Mayer, M. G., and J. H. D. Jensen: "Elementary Theory of Nuclear Shell Structure," Wiley, New York, 1955.

Nuclear Data, periodical published by Academic Press, New York.

Segre, E.: "Experimental Nuclear Physics," Wiley, New York, 1953.

Siegbahn, K.: "Alpha, Beta, and Gamma Ray Spectroscopy," North-Holland, Amsterdam, 1965.

Chapter 13

Nuclear Reactions: Compound-nucleus Reactions

There are many types of nuclear reactions, but they can be divided into two more or less distinct classes known as *compound-nucleus reactions* and *direct reactions*. In the former, the incident particle is captured to form a nucleus in a highly excited state, called the *compound nucleus*, which then decays by the processes discussed in the last three chapters. In direct reactions, the incident nucleon interacts with the nucleus as it passes, without forming an intermediate state. Perhaps the simplest quantitative difference between the two reaction mechanisms is the time of interaction. In direct reactions, this is of the order of the transit time of the incident particle over a nuclear diameter, which, from an elementary calculation, is about 3×10^{-22} s; note that the fastest decay times encountered in the last three chapters was orders of magnitude longer than this. Compound-nucleus reactions form the subject matter of this chapter, and direct reactions will be treated in Chap. 14.

13-1 Qualitative Description of Compound-nucleus Reactions—Classical Treatment

We begin our discussion of compound-nucleus reactions with a completely classical treatment, withholding the introduction of quantum effects until Sec. 13-2. We consider a nucleus to be a conglomeration of particles held together by attractive forces acting between them. This is the liquid-drop model introduced in Sec. 7-2. An energetic nucleon approaching this nucleus from the outside is accelerated as it enters by the attractive forces of all the nucleons, and shortly after entering it will probably have a collision with one of them. The way in which the energy is shared in such a collision, as any billiard player knows, depends on the angle of impact, but also on an average the energy of the incident nucleon is shared equally between it and the struck nucleon. Each of these has an excellent chance of having a collision with another nucleon, causing the energy to be shared about equally by four nucleons. All these have further collisions, and so the process continues until the energy is shared among a great many nucleons; in this situation it is called a compound nucleus. The directions of motion of nucleons are changed in collisions, and when a nucleon gets near the edge of the nucleus, it experiences an inward pull from the nuclear forces exerted on it by all the other nucleons, i.e., the shell-theory potential, which can change its direction rather drastically. As a result of these many direction changes, the velocities of nucleons in a compound nucleus are isotropically distributed.

Every once in a while in the course of these collisions a nucleon gets enough energy to come out of the nucleus. If it reaches the surface with enough radial velocity to overcome the attractive forces of the other nucleons, it is emitted. Most nucleons are emitted with relatively low energy compared to that with which the incident nucleon entered; it is most unlikely that nearly *all* the available energy should be concentrated on a single nucleon during this very complex collision process.

To make our discussion more quantitative, let us treat the problem from the standpoint of kinetic theory. The velocities of the nucleons inside the nucleus have a Maxwell distribution, so the number with various kinetic energies E_K is given by

$$N(E_K) \propto \sqrt{E_K}\, e^{-E_K/kt}$$

where k is Boltzmann's constant and T is the "temperature" of the system. The nucleons emitted are in the high-energy tail, so we can ignore the coefficient of the exponential. On coming out of the nucleus, they lose an energy V (the depth of the shell-theory potential) so their kinetic energy outside E is $E_K - V$. If all particles were emitted, we would have

$$\begin{aligned} N(E) &\propto e^{-(E+V)/kt} \\ &\propto e^{-E/kt} \end{aligned}$$

where the second step follows because V is a constant. However, some particles do not have enough energy, or enough of it directed in the r direction, to come out, so this must be modified to

$$N(E) = f(E)e^{-E/kt} \qquad \textbf{(13-1)}$$

where $f(E)$ represents the energy dependence of factors affecting the possibility of emission. A nucleon with large E has a large E_K and hence a better probability for the r component of its velocity to be sufficient to "climb the side of the potential well," so $f(E)$ increases with increasing E_K. This increase is roughly linear (see Prob. 13-1), and (13-1) therefore becomes

$$N(E) \approx\!\!\propto Ee^{-E/kt} \qquad \text{neutrons} \qquad \textbf{(13-2)}$$

which is roughly a Maxwell distribution.

This result is correct only for neutrons. Protons, after getting outside the nucleus, are accelerated by the coulomb forces and thereby receive an additional energy $\mathcal{B}_c$, which is what we have been calling the maximum height of the coulomb barrier. In this classical situation, then, their energy distribution is zero up to $E = \mathcal{B}_c$ and thereafter follows (13-2) with $N(E)$ changed to $N(E + \mathcal{B}_c)$. However, we shall see that this conclusion is considerably modified by quantum effects. Another aspect that will be so modified is the fact that in this picture there is virtually no probability for an alpha particle or any complex nucleus to be emitted. This would be equivalent to the evaporation of ice crystals, i.e., an ordered array, from a water droplet, which we know does not occur.

Now let us consider the question of the angular distribution of emitted nucleons. Since we said that the velocities of the nucleons are isotropically distributed, we might hastily conclude that the angular distribution of the emitted nucleons is isotropic. This conclusion is approximately correct, but it is not quite so. When the incident particle enters, it transfers its angular momentum to the nucleus giving it an overall spin (in the classical sense). The axis of this spin can be different for different incident particles, but in all cases it is perpendicular to the incident direction. When a system is spinning, centrifugal force makes it easier for particles to come off in a direction perpendicular to the axis of spin, so fewer particles come off in the direction of the spin axis. Since the spin axis is perpendicular to the direction of the incident particles, fewer particles are emitted in these directions. The angular distribution of emitted particles therefore has a minimum at 90° to the incident beam and is consequently peaked in the forward and backward directions. Since the spin axis is always at 90°, the angular distribution is symmetric about that angle.

13-2 Qualitative Description of Compound-nucleus Reactions—Quantum Treatment

Since a nucleus is not a classical system, there are many aspects of the discussion of Sec. 13-1 which are not valid. Let us therefore start over and consider the problem in the light of what we know about the nucleus as a quantum system. As an energetic nucleon approaches, it is in some angular-momentum state, characterized by j, l, and m quantum numbers, relative to the center of the nucleus. Although its velocity is greatly increased by the attractive forces of the other nucleons (i.e., the shell-theory potential) as it enters the nucleus, its angular momentum is not changed since these forces act only in the r direction.

When the nucleon enters the nucleus, it enters one of the allowed orbits appropriate to its energy and l, j, m quantum numbers. To be definite, let us say the nucleus is as shown in Fig. 13-1, where $\mathfrak{N} < 5$ orbits are filled, the $\mathfrak{N} = 5$ orbits are partly filled, the $\mathfrak{N} = 6$ orbits are empty but still bound ($E < 0$), and the

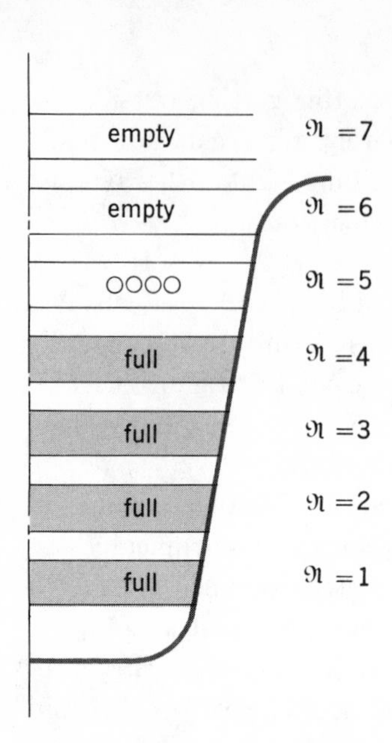

FIGURE 13-1 Status of the target nucleus in example considered in text.

$\mathfrak{N} > 6$ orbits are unbound ($E > 0$); and let us say that the incoming nucleon enters an $\mathfrak{N} = 7$ orbit. In this situation, the rules given in Sec. 5-1 allow a great many collisions. For example, the $\mathfrak{N} = 7$ nucleon can collide with an $\mathfrak{N} = 5$ nucleon leaving the two in $\mathfrak{N} = 6$ orbits, or in other $\mathfrak{N} = 5$ and $\mathfrak{N} = 7$ orbits; or it can collide with an $\mathfrak{N} = 4$ nucleon, leaving itself and its collision partner in $\mathfrak{N} = 5$ and $\mathfrak{N} = 6$ orbits; or it can collide with an $\mathfrak{N} = 3$ nucleon, leaving both in $\mathfrak{N} = 5$ orbits. Once there are nucleons in $\mathfrak{N} = 6$ orbits, they can collide with $\mathfrak{N} = 5$ nucleons, leaving the two in other $\mathfrak{N} = 6$ and $\mathfrak{N} = 5$ orbits; or they can collide with $\mathfrak{N} = 4$ nucleons, leaving both in $\mathfrak{N} = 5$ orbits. Once there are holes in $\mathfrak{N} = 4$ orbits, there can be collisions between $\mathfrak{N} = 4$ and $\mathfrak{N} = 5$ nucleons leading to other $\mathfrak{N} = 4$ and $\mathfrak{N} = 5$ orbits; or there can be collisions between $\mathfrak{N} = 3$ and $\mathfrak{N} = 5$ nucleons leaving both in $\mathfrak{N} = 4$ orbits. In all these collisions, angular momentum and parity must be conserved, but there are always several orbits in each shell which can be reached without violation of these rules.

Thus, while the complete sharing of available energy among all the nucleons in the nucleus envisioned in the classical case is not possible in quantum physics (for example, $\mathfrak{N} = 1$ and $\mathfrak{N} = 2$ nucleons are not excited and no more than one $\mathfrak{N} = 3$ or two $\mathfrak{N} = 4$ nucleons can be excited at any one time), the excitation energy is shared among a goodly number of nucleons. A listing of the orbital configurations that can be reached by these collisions would be quite long. The wave function for the *state* of the compound nucleus includes a term from each of these configurations, in each case with the I^π of the system which is just the total I^π of the original nucleus plus the entering nucleon. Only a small fraction of these terms allow a nucleon to be emitted, namely, those in which one nucleon is in an $\mathfrak{N} = 7$ orbit. Thus it is usually only after a large number of collisions that a nucleon is emitted.

In this discussion, no account has been taken of the fact that after the nucleon is emitted, the residual nucleus must be left in some definite state. In order to make this more explicitly obvious, we may express the wave function of the compound-nucleus state described above as a sum of terms each of which is the wave function for some state of the residual nucleus, i.e., the nucleus with one less nucleon, times the wave function for a single nucleon. This is completely equivalent to the above method of expressing it as a sum of terms each of which is an orbital configuration that can be reached by collisions from the initial configuration, since the various states of the residual nucleus are themselves sums of terms each of which is one of these orbital configurations less a single nucleon. This revised way of writing the wave function of the compound nucleus is just the one used in Sec. 10-2, and indeed the whole process of the decay of the compound nucleus is just that described in Chap. 10. We shall build on it in later sections to derive the energy distribution of emitted particles, and we shall find that where there is enough energy to apply a statistical treatment, the result (13-2) is approximately correct for neutrons.

For protons, however, the classical treatment fails rather badly because it envisions a thoroughgoing exchange of energy by collisions inside the nucleus resulting in neutrons and protons having the same kinetic-energy distribution. In the ground states of stable nuclei, however, we know that there is a difference between the kinetic energies of neutrons and protons arising from the condition for stability against beta decay. This requires equal total energies, whereas the potential energy is larger, i.e., less negative, for protons because of the coulomb potential. Moreover, this difference between neutron and proton kinetic energies cannot be altered by collisions because neutrons are prevented from losing energy by the Pauli exclusion principle; all lower-energy neutron orbits are occupied. The situation where neutron kinetic energies are higher than those of protons persists to a large extent in the compound nucleus, and, as a consequence, protons have a lesser chance of getting out. This may be seen from (10-22), which shows that the difference between neutron and proton emission with a given energy is that proton emission is suppressed by a coulomb barrier penetrability.

Another failure of the classical description in the last section is in the conclusion that alpha particles and other complex nuclei are essentially never emitted. The reason for this in the classical case is that the number of degrees of freedom (i.e., the number of possible patterns of motion, consisting of the positions and velocity vectors of all the particles) available to the residual system increases very rapidly with the number of particles in the system, whence a residual drop with *four* less molecules has far fewer degrees of freedom than a drop with *one* less molecule, so the latter occurs with much higher probability. In quantum systems, however, the number of degrees of freedom is just the number of states, and these are greatly limited by the Pauli exclusion principle. For example, if the system were enough larger to allow one whole extra shell filled with nucleons to be present, only the collisions described near the beginning of this section with each $\mathfrak{N}$ increased by 1 would be allowed, so the number of orbital configurations would be about the same.[1] From this we see that there are not necessarily many more states in a nucleus with one less nucleon than the compound nucleus than in a nucleus with four less nucleons than the compound nucleus. Hence there is no large preference for nucleon emission over alpha-particle emission (or the emission of other complex nuclei) so long as the energies available for the two emissions are not very different. This argument may be recognized as one from statistical physics, and the difference between the classical and quantum systems can be summarized by the perhaps familiar fact that the former obeys *Maxwell-Boltzmann statistics* while the latter is governed by *Fermi-Dirac statistics.*

13-3 Elastic Scattering and Reaction Cross Sections

Now that we have a general understanding of what goes on in a nuclear reaction, let us consider the process in more detail. As a nucleon approaches a nucleus, it encounters a nuclear force, which we have represented by the shell-theory potential. The interaction of a beam of particles with a potential well is a problem easily handled with a quantum-mechanical technique known as *scattering theory*, so one might hope to use this theory to treat the interaction of a nucleon with a nucleus. However, an interaction with a potential well cannot change the energy of a particle or remove it from the beam; it can only deflect it as, for example, in Rutherford scattering or in the nucleon-nucleon scattering processes discussed in Chap. 3. But we know from our qualitative discussions that there is a good chance for an incident nucleon to lose its energy inside the nucleus by collisions, in which case it is at least temporarily absorbed and is consequently lost from the incident beam.

This behavior can be represented by use of a complex potential in what is known as the *optical model.* To understand this, we need only look at the wave function for a traveling wave given in Sec. 10-1

$$\psi = e^{ikx} \qquad k = \left[\frac{2M}{\hbar^2}(E - V)\right]^{1/2}$$

[1] Actually there would be somewhat more because the number of orbits in a shell increases with increasing $\mathfrak{N}$.

If we let the potential be complex by replacing V by $V + iW$, k becomes complex and can be written $k = k_R + ik_I$, whence

$$\psi = e^{ik_R x}e^{-k_I x} \tag{13-3}$$

This is the wave function for a traveling wave whose amplitude is decreasing as it advances; it therefore represents a stream of particles some of which are being absorbed.

With the use of this complex potential, the interaction of an incident particle with a nucleus can be treated by scattering theory. This theory is beyond the scope of this book,[1] but let us see what we might expect to happen. We might think that the cross section for the incident nucleon to be absorbed, called the *reaction cross section* σ_R, could be as large as the actual area within which the absorption is strong, or about πR_0^2. This is roughly correct, but for neutrons scattering theory gives rather

$$\sigma_R \lesssim \pi(R_0 + \lambda\!\!\!^{-})^2 \qquad \text{neutrons} \tag{13-4}$$

This can be understood if we consider the fact that a nucleon is not a point but by virtue of its wave nature is somewhat spread out in space over a radius of the order of $\lambda\!\!\!^{-}$, its wavelength divided by 2π.* There is therefore a large probability for a collision if the nucleon comes within a distance $R_0 + \lambda\!\!\!^{-}$ of the center of the nucleus, which leads to (13-4).

If the incident particle is a proton or some other charged particle, (13-4) must be multiplied by a coulomb barrier-penetration probability. In addition there is a correction for the fact that a charged particle is deflected away by the coulomb forces as it approaches a nucleus and hence will not come within a distance $R_0 + \lambda\!\!\!^{-}$ of the nucleus unless its original line of flight passes somewhat closer than this to the nuclear center.

We know from our studies of wave motion that when a wavefront passes an absorbing obstacle, diffraction occurs. The beam represented by the wavefront is

[1] An excellent and simple treatment of quantum-mechanical scattering theory is given in J. M. Blatt and V. F. Weisskopf, "Theoretical Nuclear Physics," Wiley, New York, 1952.

* A beam of nucleons is treated as a plane wave, but a plane wave is built up of mutually interfering spherical waves, so the direction of motion of any individual nucleon is "uncertain" by some reasonable fraction of a radian. Hence, the uncertainty in the transverse momentum Δp is of the order of its momentum p. From the uncertainty principle, then,

$$\Delta x = \frac{\hbar}{\Delta p} \simeq \frac{\hbar}{p} = \lambda\!\!\!^{-}$$

This is the amount by which its position in a direction transverse to the beam is spread out by its wave nature.

deflected, and the intensity goes through successive maxima and minima as a function of deflection angle, corresponding to directions in which constructive and destructive interference occurs. The familiar formula for the positions of maxima in small-angle diffraction from slits is easily generalized to

$$\theta \propto \frac{\lambda\!\!\!^{-}}{R} \tag{13-5}$$

where R is the radius of the obstacle. A beam of nucleons is a wavefront, so when a portion of it is absorbed out by a nucleus, we may expect a diffraction pattern of scattered nucleons characterized by maxima and minima at angular positions governed by (13-5). This feature is clearly seen in Figs. 13-2 and 13-3, which show measurements of angular distributions of neutrons and protons from elastic scattering. Note that the angle of a given maximum shifts uniformly with the size of the nucleus because of the R dependence in (13-5).

While the simplest features of the curves in Fig. 13-2 and 13-3 can be explained as a diffraction pattern, there are many details that are sensitive to the size and shape of the potential which causes the scattering. For any choice of potential, the angular distributions can be calculated from scattering theory, so the potential can be determined by adjusting it to give the experimental angular distributions. It was in this way that determinations were made of the parameters in (4-2) and that the potential (4-7) was developed, including the coefficient in (4-3). In addition determinations of W come out of this work; they will be presented in the next section. It may be difficult to understand how so many parameters can be determined simultaneously by fitting data, but it should be realized that high-precision measurements are available at a large number of angles for various bombarding energies on various target nuclei. There is truly a great deal of information in these measurements.

Once the parameters of the optical-model potential have been determined, the reaction cross section σ_R can be computed. For energies above 1 or 2 MeV, the results are close to the maximum values from (13-4) for neutrons and to the product of these and coulomb barrier penetrabilities for charged particles. Some of these results for protons and alpha particles are shown in Fig. 13-4.

In scattering theory, nucleons approaching a target nucleus with different l values are treated separately, and the reaction cross section for each l, σ_{R-l}, is calculated. Since they will be important in future discussions, let us estimate them. If the incident nucleon approaches with its momentum along a line whose closest approach to the center of the target nucleus is r, its classical angular momentum is rp. The value of r which gives angular momentum l, r_l, is then

$$r_l p = \sqrt{l(l+1)}\,\hbar \simeq (l + \tfrac{1}{2})\hbar \tag{13-6}$$

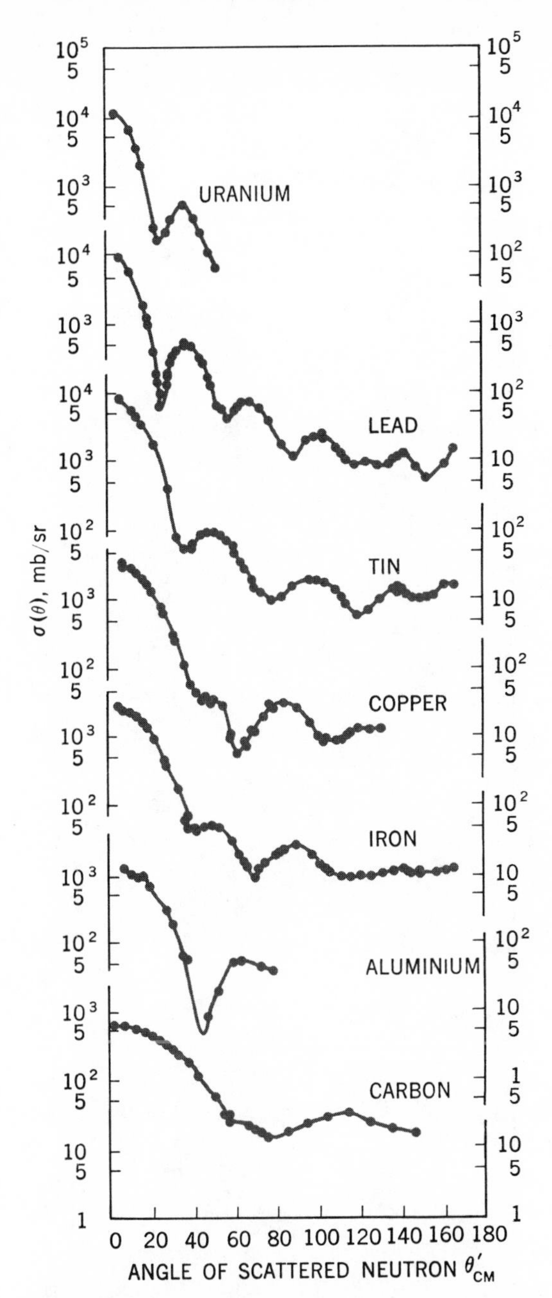

FIGURE 13-2 Angular distributions of ~14.5-MeV neutrons elastically scattered from various nuclei. Dashed lines trace the shifting of peaks due to changing nuclear sizes. [*From J. H. Coon et al., Phys. Rev.,* **111: 250 (1958).**]

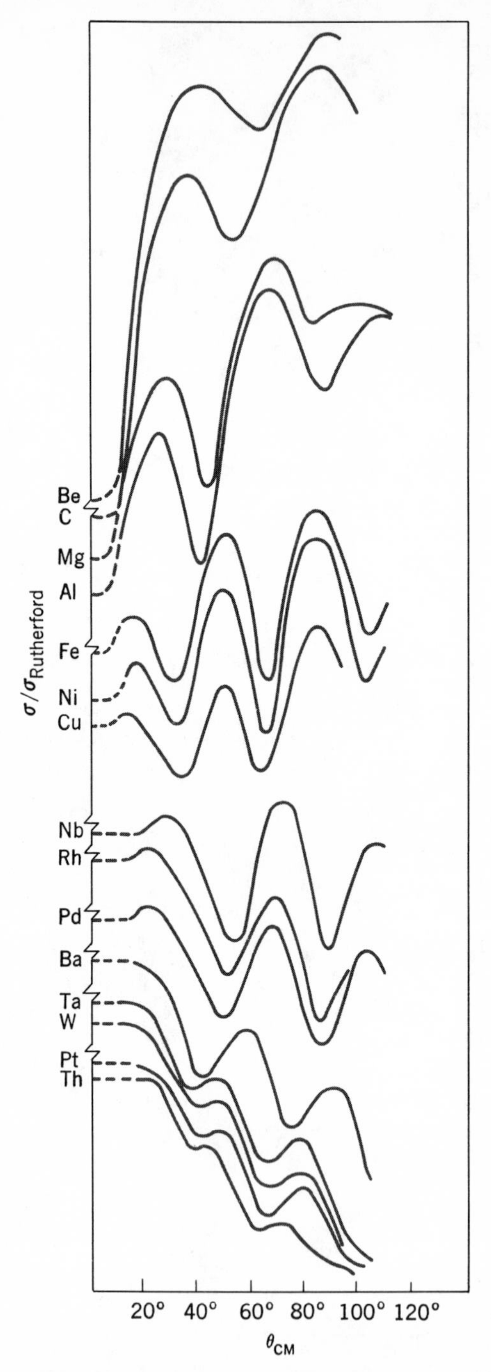

FIGURE 13-3 Angular distribution of 22-MeV protons elastically scattered from various nuclei. Ordinate is the ratio of the measured cross section to that expected from pure Rutherford scattering (due to electric forces only). Dashed lines show the shifting of peaks due to the changing nuclear size. [*From Phys. Rev.*, **93: 282 (1954).**]

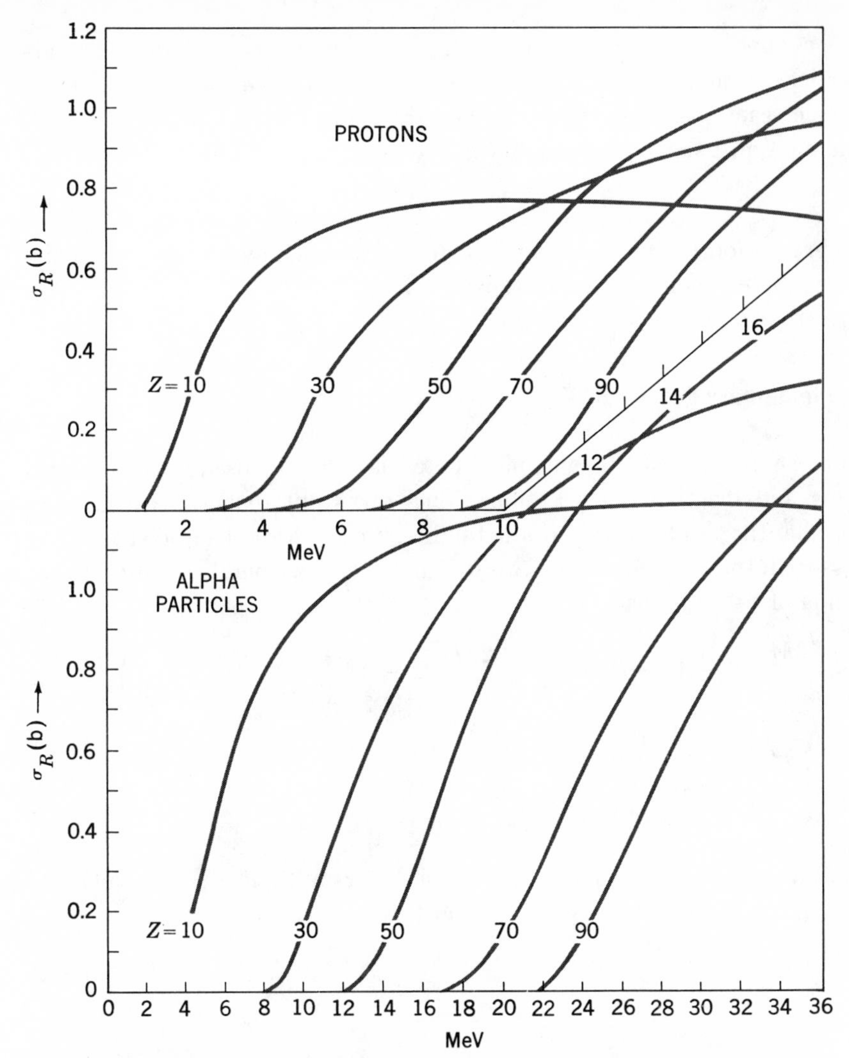

FIGURE 13-4 The reaction cross section σ_R for protons *(upper)* and alpha particles *(lower)* as calculated from optical-model potentials with parameters adjusted to fit the elastic-scattering data.

whence

$$r_l \simeq \left(l + \frac{1}{2}\right)\frac{\hbar}{p} = (l + \tfrac{1}{2})\lambda\!\!\!^{-} \tag{13-7}$$

This implies that a nucleon with an r value between $l\lambda\!\!\!^{-}$ and $(l + 1)\lambda\!\!\!^{-}$ has angular momentum l. The area enclosed between these r values then gives an estimate of σ_{R-l} as

$$\sigma_{R-l} \gtrsim \pi(l + 1)^2\lambda\!\!\!^{-2} - \pi l^2\lambda\!\!\!^{-2}$$
$$\gtrsim (2l + 1)\pi\lambda\!\!\!^{-2}$$

Clearly this cannot be valid for all l or the cross section would be infinite. In particular, particles with large l do not come very close to the nucleus and therefore cannot induce reactions. This effect can be taken into account by including an angular-momentum barrier penetrability $\mathcal{P}_l$, to give

$$\sigma_{R-l} \gtrsim (2l+1)\pi\lambda\!\!\!^{-2}\mathcal{P}_l \tag{13-8}$$

This is the expression we were seeking. When it is summed over l, the total reaction cross section σ_R comes out in agreement with (13-4).[1]

13-4 The Imaginary Potential W

In Sec. 10-3, we introduced the imaginary potential W to be used in calculations of scattering and reaction cross sections. This turns out to be an exceedingly interesting quantity, so let us discuss it further. In all cases of interest here, W is much less than the real potential V, so we can express k from its definition near the beginning of Sec. 13-3 as

$$\begin{aligned} k &= \left[\frac{2M}{\hbar^2}(E - V - iW)\right]^{1/2} = \left[\frac{2M}{\hbar^2}(E - V)\right]^{1/2}\left(1 - i\frac{W}{E-V}\right)^{1/2} \\ &\simeq \left(\frac{2M}{\hbar^2}\tfrac{1}{2}Mv^2\right)^{1/2}\left(1 - \frac{i}{2}\frac{W}{\tfrac{1}{2}Mv^2}\right) \\ &= \frac{Mv}{\hbar}\left(1 - \frac{iW}{Mv^2}\right) = \frac{Mv}{\hbar} - \frac{iW}{\hbar v} \end{aligned}$$

From (13-3) we see that the mean free path F for an incident particle to be removed from the beam is $1/k_I$, whence from the above expression

$$F = \frac{1}{k_I} = \frac{\hbar v}{W}$$

[1] To show this, we note that $\mathcal{P}_l$ changes from near unity to near zero at the value of l for which $r_l = R_0$, the largest radius within which there is a strong interaction. From (13-7), this gives for l_m, the l value where the change occurs,

$$l_m = \frac{R_0}{\lambda\!\!\!^-} - \frac{1}{2} \tag{13-9}$$

From (13-8), then

$$\sigma_R = \sum_{l=0}^{l_m} \sigma_{R-l} \lesssim \pi\lambda\!\!\!^{-2} \sum_{0}^{l_m} (2l+1) = \pi\lambda\!\!\!^{-2}(l_m+1)^2$$

whence we find

$$\sigma_R \lesssim \pi\lambda\!\!\!^{-2}\left(\frac{R_0}{\lambda\!\!\!^-} + \frac{1}{2}\right)^2 = \pi(R_0 + \tfrac{1}{2}\lambda\!\!\!^-)^2$$

Considering the crudeness of the approximations used, this is in agreement with (13-4).

The only way a particle can be removed from the incident beam is by having a collision with one of the nucleons in the nucleus, so F is the mean free path between collisions for an energetic nucleon in the nucleus. The average time between collisions Δt_c is the mean free path divided by the velocity, or

$$\Delta t_c = \frac{\hbar}{W} \tag{13-10}$$

This gives a simple physical interpretation of W: it is a measure of the frequency with which collisions occur.

This interpretation is useful in understanding the r dependence of the imaginary potential. At first one might think that W should be proportional to the nucleon density and hence have the same r dependence as V in (4-1). However, we saw in Sec. 13-2 that collisions can occur much more easily with the least bound nucleons in the nucleus; in the example used there, no collisions could occur with $\mathfrak{N} = 1$ or $\mathfrak{N} = 2$ nucleons and very few with $\mathfrak{N} = 3$ nucleons; most collisions were with nucleons in the $\mathfrak{N} = 4$ and $\mathfrak{N} = 5$ shell. But most nucleons in these shells have high orbital angular momentum—of the 32 orbits with $\mathfrak{N} = 5$, 12 have $l = 5$ and 8 have $l = 4$—and in Fig. 2-4 we found that large-l wave functions are concentrated near the nuclear surface. Moreover, due to the sloping side of the potential well, wavelengths for nucleons in orbits near the top of the well become long at large r, so nucleons in high-n, low-l orbits spend a large fraction of their time at large radii; this is essentially the reason why higher-energy electron shells in atoms have a larger average radius than low-energy shells. Due to both of these effects, an energetic nucleon is more likely to have collisions near the surface than near the center of the nucleus. In view of the interpretation of W from (13-10), the imaginary potential is therefore deepest near the nuclear surface and thus has a rather different r dependence than the real potential V. In our treatment here, however, we shall avoid this complication.

Determinations of W obtained from analyses of elastic-scattering experiments with scattering theory are shown in Fig. 13-5. In view of its interpretation in (13-10) and the similarity between that equation and (5-4), we might at first expect W to be about 1 MeV in conformity with (5-6). However, (5-6) refers to collisions between nucleons in the lowest unfilled shell, where only a limited number of collisions is possible, whereas from the discussion in Sec. 13-2 we know that the number of allowable collisions is greatly increased at the higher excitation energies occurring in nuclear reactions. It therefore seems reasonable for W to be an increasing function of the energy above that available to nucleons in the ground state, $E + S_n \simeq E + 8$ MeV. We have therefore used $E + S_n$ as the abscissa in Fig. 13-5. When the energy of the incident nucleon is zero, the value of the abscissa is 8 MeV.

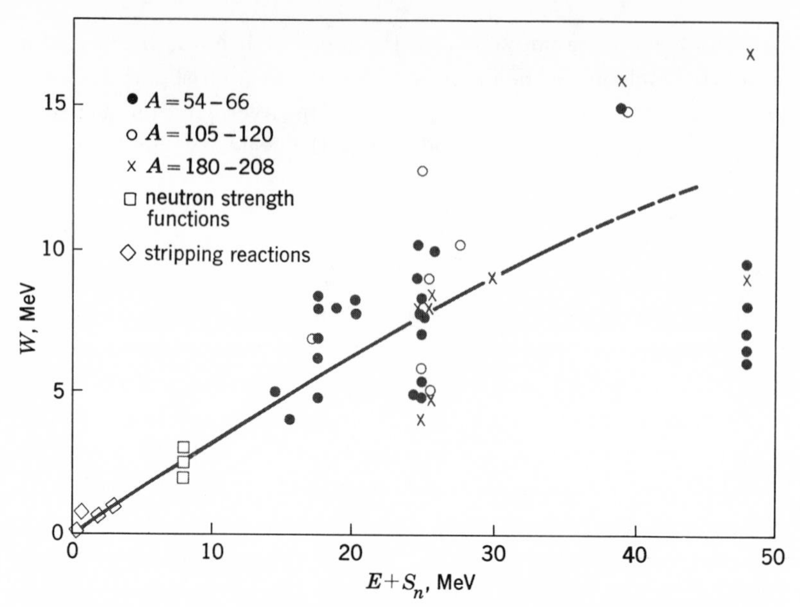

FIGURE 13-5 Determinations of the depth of the imaginary potential W. Abscissa is $E + S_n$, the energy of the incident nucleon plus its separation energy. Values were obtained from various analyses of elastic-scattering data in which the imaginary potential was taken to be of the same shape as the real potential. Points for $E + S_n = 8$ MeV were obtained from analyses of the neutron strength function (see Fig. 13-10), and points near the origin are typical of those obtained from stripping-reaction studies (see Fig. 14-6).

It is interesting to note that (5-6) was the condition for nucleons to make one complete circuit of their orbit between collisions, as in Fig. 4-1*a*. The fact that W becomes considerably larger than 1 MeV at high excitation energies therefore implies that nucleons traverse only a small fraction of an orbit between collisions, so the situation approaches that shown in Fig. 4-1*b*.

Since $\hbar/W$ is the average time between collisions, from the uncertainty principle we may estimate that W is the amount by which energy conservation may be violated in these collisions [see (5-6) and the accompanying discussion]. A given state may therefore include configurations whose energies E_c differ from the energy of that state by as much as W. Or, conversely, a given configuration may be found in states within the energy range $E_c - W$ to $E_c + W$, a range of $2W$. For example, in Sec. 10-2 we discussed the configuration (10-11) and estimated its energy to be about $E_c = 1$ MeV, which corresponds to $E + S_n \simeq 9$ MeV, whence, from Fig. 13-5, $W \simeq 2.5$ MeV. We may therefore expect this configuration to be found with appreciable amplitude in states with energies between -1.5 and $+3.5$ MeV, a spread of 5 MeV. Consequences of this will be discussed in Sec. 13-7.

13-5 Resonances in Nuclear Reactions

When an incident particle enters a nucleus to form a compound nucleus, a quantum *state* of the system must be formed. Since states have definite energies, one might think that a compound nucleus could be formed only if the incident particle had just the right energy to form one of those states. If that were the case, the cross section for any nuclear reaction would behave as shown in Fig. 13-6*a*. However, this is not quite true because the energies of compound nucleus states have finite widths.

A simple understanding of these widths can be derived from the uncertainty principle,

$$\Delta E \, \Delta t \simeq \hbar$$

This says that if a state has a lifetime Δt, it is impossible to know its energy with any better precision that $\Delta E \simeq \hbar/\Delta t$. A state with a finite lifetime therefore has a

FIGURE 13-6 The occurrence of resonances, showing *(a)* the situation if compound-nucleus states had a sharp energy, and *(b)* and *(c)* the effects of finite lifetimes; in *(b)* $\Gamma \ll D$ and in *(c)*, $\Gamma \gg D$.

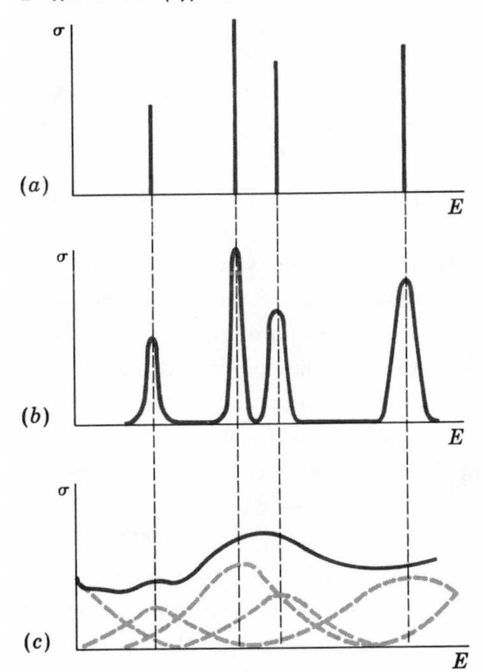

width Γ equal to $\hbar/\Delta t$. Since the lifetime is the inverse of the probability per unit time for decay λ, this leads to

$$\Gamma = \hbar\lambda \tag{13-11}$$

If the widths are less than the energy difference between successive states D, the cross section as a function of incident energy looks something like Fig. 13-6*b*; if the widths are much greater than the energy difference between successive states, that is, $\Gamma \gg D$, the cross section becomes a smoothly varying function of energy, as shown in Fig. 13-6*c*.

When a nucleus can decay in different ways a partial width Γ_i is defined for each type of decay i as

$$\Gamma_i = \hbar\lambda_i \tag{13-11a}$$

where λ_i is the probability per second for the decay i. Since the total probability per unit time for decay λ is

$$\lambda = \sum_i \lambda_i$$

it follows that

$$\Gamma = \sum_i \Gamma_i \tag{13-12}$$

It is quite common to speak of decay rates in terms of widths since they are so simply related by (13-11). For example, the partial width for $l = 0$ neutron emission Γ_n is readily calculated from (10-16) and (10-15) to be

$$\begin{aligned} \Gamma_n &= \hbar[E(\text{MeV})]^{1/2}\left(\frac{A}{57}\right)^{-1/3} \times 3 \times 10^{21} \times \theta^2 \\ &= 1.8 \text{ MeV} \times \theta^2 \times [E(\text{MeV})]^{1/2}\left(\frac{A}{57}\right)^{-1/3} \end{aligned} \tag{13-13}$$

We now use this to investigate the relationship between Γ and D to see whether the true situation is as in Fig. 13-6*b* or *c*.

Consider a reaction induced by a neutron in the energy region of 0.1 MeV on an Fe^{56} nucleus. From (13-8) and the calculations of angular-momentum barrier penetrability in Sec. 10-3, it is clear that a neutron of this energy nearly always enters the nucleus with $l = 0$, so the states of the compound nucleus are $\frac{1}{2}^+$. This energy is not sufficient to allow the compound nucleus to decay by nucleon emission in any way but back to the ground state of Fe^{56} by neutron emission, so that is the dominant mode of decay, whence from (13-12), $\Gamma \simeq \Gamma_n$. We therefore can evaluate Γ by use of (13-13). The reduced width θ^2 in that equation is a measure of the amount of the configuration (10-11), $Fe^{56}(\text{GS}) + n(3s_{1/2})$, in the wave function for the compound-nucleus state. We found in Sec. 13-4 that this configuration is present in $\frac{1}{2}^+$ states over an energy range of about 5 MeV. If the average energy spacing between these states is $\bar{D}$, the number of these states is about 5 MeV/$\bar{D}$. In view of (10-17), the average value of θ^2, $\overline{\theta^2}$, is the reciprocal of this number, or

$$\overline{\theta^2} \simeq \frac{\bar{D}}{5 \text{ MeV}} \tag{13-14}$$

Substituting this into (13-13), we find for average values

$$\bar{\Gamma} = \bar{\Gamma}_n \simeq 0.36[E(\text{MeV})]^{1/2}\bar{D} \tag{13-15}$$

From (13-15) we see that for the situation under consideration, $E \simeq 0.1$ MeV, Γ is typically about $D/10$, whence the curve of cross section vs. bombarding energy is like that in Fig. 13-6*b*. It is characterized by *resonances*, energies near which the cross section is sharply peaked. Moreover, (13-15) indicates that resonances become narrower at lower energies and broader at higher energies. It suggests that Γ remains less than D up to several MeV of bombarding energy; however, as the energy is raised, new processes become possible: $l > 0$ states can be excited, neutron emission can go to excited states, protons can be emitted in (n,p) reactions, etc. The first of these increases the number of states excited and hence decreases $\bar{D}$, and the latter two increase the decay rate and thus increase Γ. Consequently, at energies above about 2 MeV in heavy nuclei, Γ becomes greater than D so the σ-vs.-E curve becomes like that in Fig. 13-6*c*. That energy region will be treated in Sec. 13-9, and in Sec. 13-6 we shall study the lower-energy *resonance region*, where $\bar{\Gamma} \ll \bar{D}$.

Before closing this section, we should note that (13-13) gives a simple relation between θ^2 and Γ_n. The latter was used early in the history of nuclear physics and was called the *width* for obvious reasons. When the theory for this width was later developed in terms of (13-13), θ^2 came to be called the *reduced width* since it contains the factors affecting the width derived from nuclear-structure information.[1]

13-6 Nuclear Reactions in the Resonance Region

In the resonance region, we generally must consider more than one, but not more than a few, types of decay of the compound nucleus. For each of these there is a decay rate and hence a partial width related to it by (13-11*a*). For example, in neutron-induced reactions the compound nucleus might be able to decay to the ground state by neutron emission (decay rate λ_n, partial width $\Gamma_n = \hbar\lambda_n$), to

[1] Actually it is more conventional to define the reduced width as $(R_0/2)\,\theta^2$. Two other quantities are also sometimes called the reduced width. One is γ^2, defined by $\gamma^2 \approx (\hbar^2/2MR_0)\,\theta^2$, and another is Γ_{n_0}, which will be defined in (13-20).

the first excited state by neutron emission ($\lambda_{n'},\Gamma_{n'}$), by proton emission (λ_p,Γ_p), by alpha-particle emission ($\lambda_\alpha,\Gamma_\alpha$), by gamma-ray emission ($\lambda_\gamma,\Gamma_\gamma$), etc.[1] If we define the total width for reactions Γ_R as

$$\Gamma_R = \Gamma_{n'} + \Gamma_p + \Gamma_\alpha + \Gamma_\gamma + \cdots$$

from (13-12), the total width is

$$\Gamma = \Gamma_n + \Gamma_R \tag{13-15a}$$

Our first problem is to calculate the cross sections for elastic scattering $\sigma(n,n)$ and for reactions $\sigma(n,R)$. To begin, let us return to our example of an $s_{1/2}$ neutron entering an Fe^{56} nucleus to form a compound nucleus of Fe^{57}. When the neutron first enters, the configuration must be $[Fe^{56}(GS) + n(s_{1/2})]$, so only compound nucleus states having this configuration in their wave function can be excited, and the probability for a state to be excited is proportional to the fraction of the time it spends in this configuration. But this is just the reduced width θ^2, which, from (13-13), is proportional to $\Gamma_n/E^{1/2}$, so the cross section for formation of any individual state of the compound nucleus is proportional to $\Gamma_n/E^{1/2}$ for that state, or

$$\sigma(n,R) \propto \frac{\Gamma_n}{E^{1/2}}$$

From the discussion of Sec. 13-5, we know that the cross section behaves as in Fig. 13-6*b*, being characterized by resonances of width Γ. In analogy with the theory of *line shape* in the emission of light[2] the shape of these resonances is given by

$$\sigma(n,R) \propto [(E - E_r)^2 + \tfrac{1}{4}\Gamma^2]^{-1}$$

where E_r is the energy at the center of the resonance. This expression has a maximum at $E = E_r$ and falls to half that value at $E - E_r = \pm\Gamma/2$, which are the properties we expect here.

From (13-8) we know that the maximum cross section for an $l = 0$ neutron to induce a reaction is $\pi\lambda\!\!\!^{-2}$, so it seems reasonable to expect

$$\sigma(n,R) \propto \pi\lambda\!\!\!^{-2}$$

[1] As in the case of any other excited state, a compound nucleus can decay by gamma-ray emission. The probability per unit time for this decay is just the sum of λ_γ from (12-3) and (12-9) for transitions to all lower-energy excited states, and this decay rate gives a Γ_γ through (13-11*a*).

[2] J. D. Jackson, "Classical Electrodynamics," p. 601, Wiley, New York, 1962; L. I. Schiff, "Quantum Mechanics," p. 256, McGraw-Hill, New York, 1949.

In addition, $\sigma(n,R)$ must be proportional to the probability for the compound nucleus to decay by emission of a reaction product rather than by scattering, whence

$$\sigma(n,R) \propto \frac{\lambda_R}{\lambda_R + \lambda_n} = \frac{\Gamma_R}{\Gamma_R + \Gamma_n} = \frac{\Gamma_R}{\Gamma}$$

Collecting all these terms and inserting a proportionality factor f gives

$$\sigma(n,R) = \pi \bar{\lambda}^2 \frac{\Gamma_n \Gamma_R}{(E - E_r)^2 + \frac{1}{4}\Gamma^2} \frac{f}{\Gamma E^{1/2}}$$

From this formula we see that the maximum possible cross section at a given energy occurs when $E_r = E$, $\Gamma_n = \Gamma_R = \frac{1}{2}\Gamma$, in which case

$$\sigma(n,R) = \pi \bar{\lambda}^2 \frac{f}{\Gamma E^{1/2}}$$

But we know from (13-8) that the maximum possible cross section is $\pi\bar{\lambda}^2$, so f must be equal to $\Gamma E^{1/2}$. We therefore obtain

$$\sigma(n,R) = \pi \bar{\lambda}^2 \frac{\Gamma_n \Gamma_R}{(E - E_R)^2 + \frac{1}{4}\Gamma^2} \qquad \textbf{(13-16)}$$

If a reaction takes place, the probability that it will involve emission of some particular particle x, and thus be an (n,x) reaction, is just Γ_x/Γ_R. Using this factor in (13-16) gives

$$\sigma(n,x) = \pi \bar{\lambda}^2 \frac{\Gamma_n \Gamma_x}{(E - E_r)^2 + \frac{1}{4}\Gamma^2} \qquad \textbf{(13-17)}$$

This is known as the *Breit-Wigner formula;* it can be derived more rigorously,[1] and it has played an important part in the development of nuclear physics.

Since the relative probability for decay of the compound nucleus by scattering and by a reaction is Γ_n/Γ_R, multiplying (13-16) by this factor gives the scattering cross section $\sigma(n,n)$ as

$$\sigma(n,n) = \pi \bar{\lambda}^2 \frac{\Gamma_n^2}{(E - E_r)^2 + \frac{1}{4}\Gamma^2} \qquad \textbf{(13-18)}$$

Actually this gives only the *resonance scattering,* the scattering via formation of a compound nucleus. In addition, there is scattering, i.e., a deflection of the incident neutron, by the nuclear force as represented by the shell-theory potential, called *potential scattering.* This occurs at all energies and without formation of a compound nucleus, and its cross section therefore is not characterized by resonances.

[1] See footnote 1, page 317.

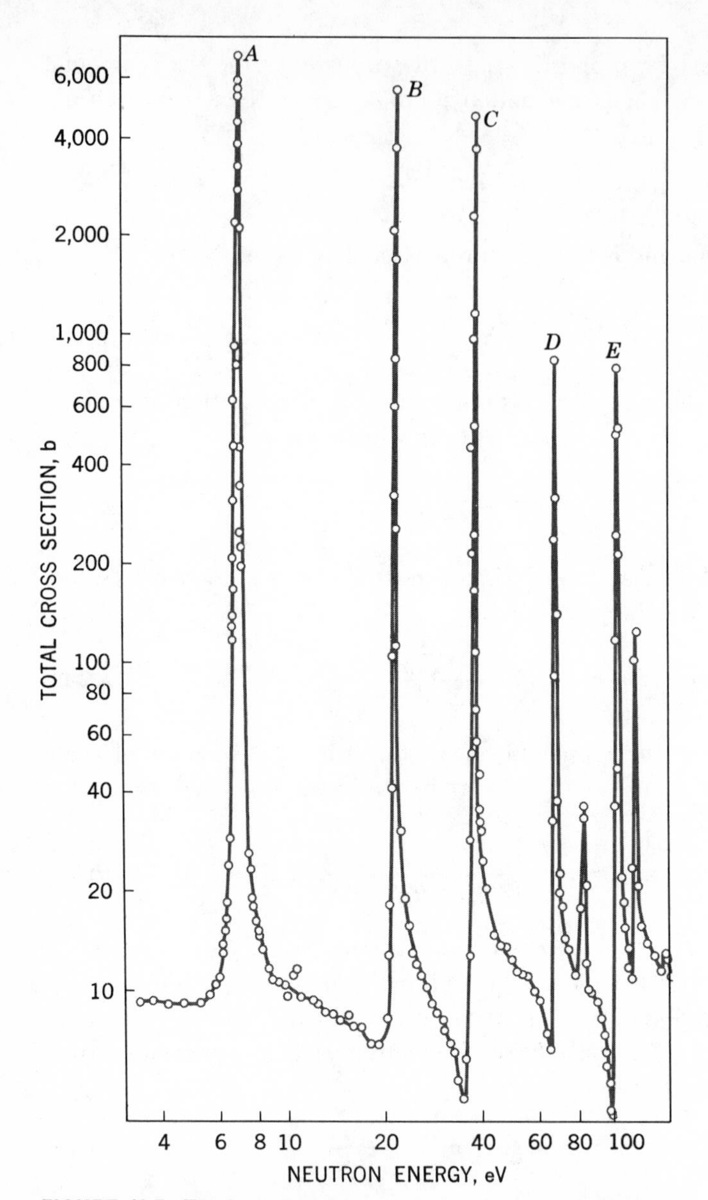

FIGURE 13-7 Total neutron cross section vs. energy for U^{238}. Letters above the peaks identify them for the discussion in the text. *(From Brookhaven Natl. Lab. Rept. BNL* **325.***)*

These two types of scattering are not simply additive, but, as in other wave phenomena, interference effects occur. Examples of this are peaks C, D, and E in Fig. 13-7; the potential and resonance scattering there interfere destructively on the low-energy side of the resonance and constructively on the high-energy side. This pattern is characteristic of $l = 0$ resonances. Actually, Fig. 13-7 gives measurements of the total cross section,

$$\sigma_T = \sigma(n,n) + \sigma(n,R)$$

This can be measured rather easily by an attenuation experiment as explained in connection with (8-17). Peaks C, D, and E are due to states of the compound nucleus for which $\Gamma_n \gg \Gamma_R$, so $\sigma_T \simeq \sigma(n,n)$; peak A is due to one for which $\Gamma_R \gg \Gamma_n$, whence $\sigma_T \simeq \sigma(n,R)$, and it is governed by (13-17). Between the resonances in Fig. 13-7 we see the effects of potential scattering only.

At higher energies, compound nuclei can be formed by capture of neutrons with $l > 0$, in which case the maximum reaction cross section from (13-8) is $(2l + 1)\pi\lambda\!\!\!^{-2}$; hence (13-17) contains an additional factor $2l + 1$. If the reaction is induced by a proton or alpha particle, the Γ_n in (13-17) is replaced by Γ_p or Γ_α, respectively. Note that the effect of the coulomb barrier on impeding the

FIGURE 13-8 Cross section vs. energy for protons on Ca^{48} leading to *(a)* elastic scattering at 165°, and *(b)* (p,n) reactions leaving the final nucleus, Sc^{48}, in its ground state. Note the interference between resonance and potential scattering in *(a)*, whereas the reaction cross sections in *(b)* are given essentially by (13-17). Solid curves are theoretical fits to the data. [*From P. Wilhjelm et al., Phys. Rev.,* **177: 1553 (1969).**]

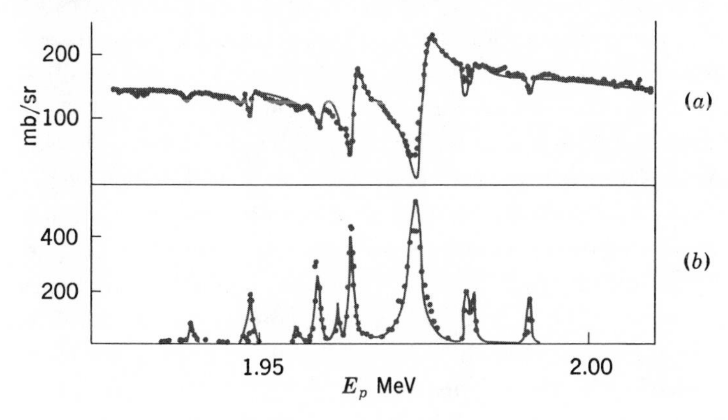

capture of charged particles such as protons is automatically taken into account since $\Gamma_p = \hbar\lambda_p$ from (13-11*a*), and λ_p contains the barrier-penetration probability from (10-22). An example of a resonance in (*p,p*) and (*p,n*) reactions is shown in Fig. 13-8. Here again there is interference between the potential and resonance scattering in (*p,p*), but (*p,n*), being a reaction rather than a scattering, is represented by a simple resonance curve as given by (13-17).

13-7 Strength Functions

In the derivation of (13-15), which gives a relationship between the average values of Γ_n and D, an estimate was made of the amount of the configuration (10-11), $[Fe^{56}(GS) + n(s_{1/2})]$, in the states of the compound nucleus excited by low-energy neutrons; in this section we shall consider this problem in more detail. In Sec. 10-2 it was pointed out that in the absence of configuration mixing (i.e., collisions) all this configuration would be in a single state whose energy, the *configuration energy* defined in Sec. 5-3, corresponds to a state of the compound nucleus excited by about 1-MeV neutrons (we shall refer to this as a state at 1 MeV energy); this state would then have $\theta^2 = 1$, and all others would have $\theta^2 = 0$. As a result of configuration mixing, this configuration is spread over a large number of states, over an energy range estimated in Sec. 13-4 to be about $2W \simeq 5$ MeV, centered at 1 MeV. To be more specific, let us define $\Sigma'\theta^2$ as the sum of the θ^2 values for all $\frac{1}{2}^+$ states over energy intervals much smaller than 5 MeV but large enough to even out fluctuations in θ^2 from one state to another. We then expect $\Sigma'\theta^2$ to behave as shown in Fig. 13-9*a*. As explained in connection with (5-15), the probability for a state to have a given configuration decreases as its distance from the configuration energy increases because its having that configuration involves an increasing violation of energy conservation. Since $\bar{D}$ is proportional to the inverse of the number of states involved in a $\Sigma'\theta^2$, and $\overline{\theta^2}$ is $\Sigma'\theta^2$ divided by this number, $\Sigma'\theta^2$ is proportional to $\overline{\theta^2}/\bar{D}$. The plot in Fig. 13-9*a* is therefore a plot of $\overline{\theta^2}/\bar{D}$.

Now let us consider the question of how this curve differs for different nuclei. According to Fig. 4-5, the $3s_{1/2}$ neutron orbit is in the $\mathfrak{N} = 5$ shell, which begins to fill when a nucleus has more than 50 neutrons. It is therefore not unreasonable if we estimate that in the 51-neutron nucleus Zr^{91}, the $3s_{1/2}$ orbit is bound by about 6 MeV, or in terms of our usage in this section, its configuration energy is -6 MeV. The $4s_{1/2}$ orbit is in the $\mathfrak{N} = 7$ shell, so in accordance with (4-19) its energy is about 15 MeV higher, or $+9$ MeV. The distribution of $\overline{\theta^2}/\bar{D}$ for these states is therefore as shown in Fig. 13-9*b*. Note that the widths of these curves vary with excitation energy in accordance with Fig. 13-5, which gives $2W \simeq 10$ MeV for a state with $E + S_n = 9 + 8$ MeV $= 17$ MeV.

It is obvious from Fig. 13-9 that $\overline{\theta^2}/\bar{D}$ in the region near $E = 0$ is very differ-

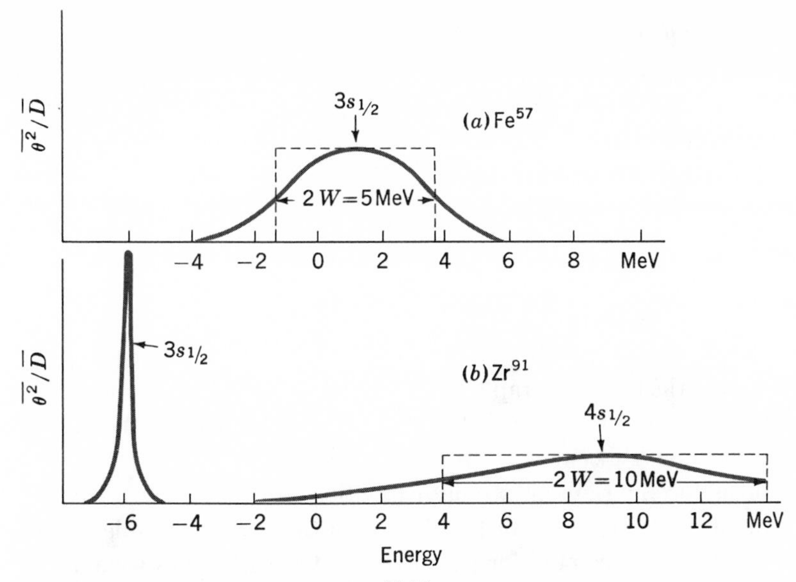

FIGURE 13-9 $\Sigma'\theta^2$, **or equivalently** $\overline{\theta^2}/\bar{D}$, **for** $\frac{1}{2}^+$ **states in** *(a)* **Fe^{57} and** *(b)* **Zr^{91}. Dashed lines show a rectangular approximation which is used to estimate values of the ordinates.**

ent for Fe^{57} and Zr^{91}. If we employ the rectangular approximation shown in Fig. 13-9*a*, the number of states is 5 MeV/$\bar{D}$, whence, from (10-17),

$$\overline{\theta_{Fe}{}^2} \simeq \frac{\bar{D}}{5\text{ MeV}} \tag{13-19a}$$

For Zr at $E \simeq 9$ MeV, a similar approximation gives $\overline{\theta_{Zr}{}^2} \simeq \bar{D}/10$ MeV; but at $E \simeq 0$, the curve in Fig. 13-9*b* is down from this value by a factor of about 5, whence, in the latter energy region

$$\overline{\theta_{Zr}{}^2} \simeq \frac{\bar{D}}{50\text{ MeV}} \tag{13-19b}$$

In general, when one of the $s_{1/2}$ states of Fig. 4-5 comes at an energy near $E = 0$, we have a large $\overline{\theta^2}/\bar{D}$; this happens for the $3s_{1/2}$ orbit at $A \simeq 50$ and for the $4s_{1/2}$ orbit at $A \simeq 155$. For A about halfway between these, as in Zr, $\overline{\theta^2}/\bar{D}$ is a minimum.

A more conventional notation in these matters is obtained by defining a new quantity Γ_{n_0} as

$$\Gamma_{n_0} = \Gamma_n[E(\text{eV})]^{-1/2} \tag{13-20}$$

where E is the energy of the state. From (13-12), then

$$\theta^2 \simeq \frac{1{,}000\Gamma_{n_0}}{1.8\text{ MeV}}\left(\frac{A}{57}\right)^{1/3} \tag{13-21}$$

The *neutron strength function* (SF) is defined as

$$\mathrm{SF} = \frac{\bar{\Gamma}_{n_0}}{\bar{D}} \tag{13-22}$$

$$\mathrm{SF} = 0.018 \text{ MeV} \frac{\overline{\theta^2}}{\bar{D}} \left(\frac{A}{57}\right)^{-1/3}$$

where the second expression uses (13-21). Inserting (13-19) into it gives

$$\mathrm{SF} \simeq \begin{cases} 3.6 \times 10^{-4} & \text{for Fe} \\ 0.3 \times 10^{-4} & \text{for Zr} \end{cases}$$

From the discussion in the last paragraph, we expect these to be a maximum and a minimum in an oscillation of the strength function with increasing A, and we expect another maximum to occur near $A = 155$.

Experimental determinations of strength functions are made by analyzing experimental data of the type shown in Fig. 13-7 with theoretical formulas based on (13-17) and (13-18), as corrected for potential scattering, to obtain values of Γ_n for each resonance. Γ_{n_0} is then calculated from (13-20). The spacing D between resonances comes directly from the data as the distance of a resonance from its nearest neighbors. By making determinations of this type for a number of resonances, the average values of Γ_{n_0} and D are obtained, so the strength function can be determined from (13-22). Experimental determinations of SF obtained in this way are shown in Fig. 13-10. The large error bars are due to statistical uncertainties arising from the fact that they are determined from a relatively few resonances. In spite of this, the predicted behavior is clearly evident. The peak near $A = 155$ is broadened due to the fact that nuclei in that region are non-spherical. Its detailed behavior has been explained by calculations which take this into account.[1]

It is interesting to point out that the data in Fig. 13-10 are sensitive to W. For example, if W were very large, we can see from Fig. 13-9 that all configurations would be evenly spread out over energy and there would be no variations with A (this is sometimes called the *black-nucleus approximation*). Theoretical fits to the data in Fig. 13-10 therefore give determinations of W. These are plotted in Fig. 13-5 at $E + S_n = 8$ MeV, which corresponds to $E = 0$. The different points there are from slightly different theoretical analyses.

From (13-17) and (13-18) we see that the average cross section for exciting states is proportional to $\bar{\Gamma}_n$, which is proportional to $\bar{\Gamma}_{n_0}$; the number of states is proportional to $1/\bar{D}$, whence the average cross section, averaged over an energy range large enough to include many states, is proportional to the strength func-

[1] D. M. Chase, L. Wilets, and A. R. Edmonds, *Phys. Rev.*, **110:** 1091 (1958).

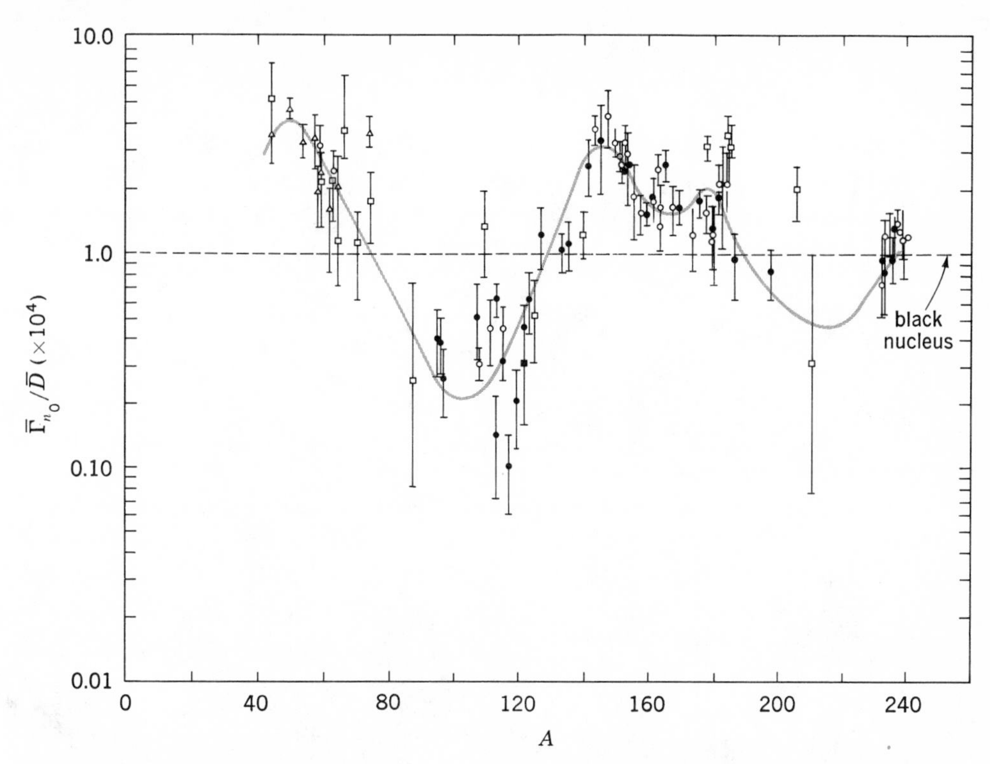

FIGURE 13-10 Strength function for $l = 0$ neutrons with low energies for various nuclei. *(From E. Vogt, Brookhaven Natl. Lab. Rept. BNL* **331.)**

tion. From Fig. 13-10 we therefore expect the average cross section for low-energy neutrons to exhibit maxima and minima as a function of target mass. This may be seen in Fig. 13-11*a*, which shows total neutron cross sections averaged over many resonances as a function of E and A. The peaks in that figure near $E = 0$ for $A \simeq 50$ and $A \simeq 155$ are clearly evident. On the other hand, the low-energy cross section for $A \simeq 90$ is not particularly low; this is because $l = 1$ neutron capture has a maximum in that region (that is where the $2p_{3/2}$ and $2p_{1/2}$ orbits of Fig. 4-5 are near zero energy) and the angular-momentum barrier penetrability becomes appreciable above about 0.1 MeV. All these effects, including those of higher l, are taken into account in the calculations shown in Fig. 13-11*b*, and it may be seen that they explain the general behavior of the measured cross sections. As is evident from the figure, cross sections become very slowly varying with both E and A at higher neutron energies. This is because strength function peaks for different l values occur in different regions and because they become very broad as can be seen by comparing parts (*a*) and (*b*) of Fig. 13-9 or by noting the increase of W with increasing energy in Fig. 13-5.

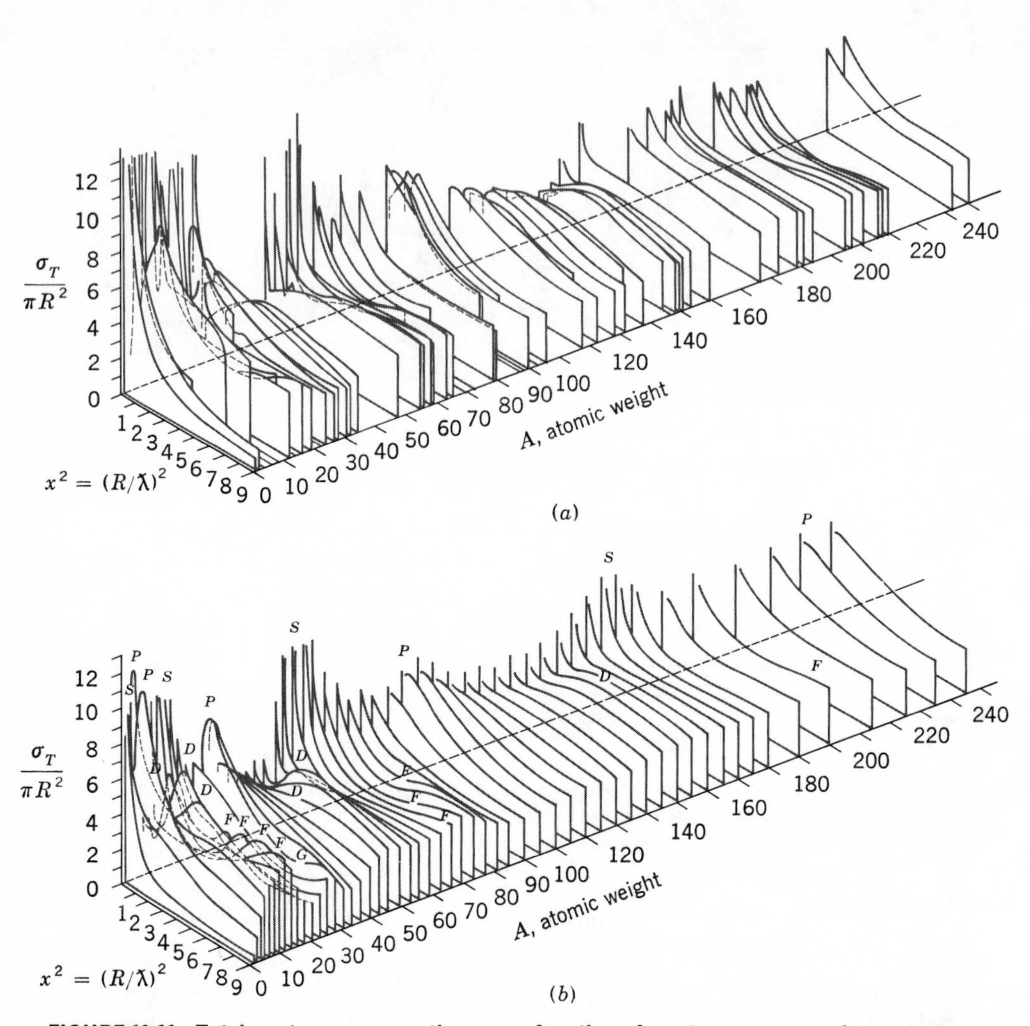

FIGURE 13-11 Total neutron cross sections as a function of neutron energy and target mass number: ***(a)*** **experimental data and** ***(b)*** **results of calculations.** S, P, D, **etc., above peaks indicates they are from resonances for** $l = 0, 1, 2$, **etc. neutrons.** [*From H. Feshbach, C. E. Porter, and V. F. Weisskopf, Phys. Rev.,* **96: 448 (1954).**]

13-8 Nuclear Reactions Induced by Low-energy Neutrons

We shall see in Sec. 13-9 that low-energy neutrons are the principal particle emitted in nuclear reactions, and if these neutrons have further reactions, the principal products are lower-energy neutrons. When the energy is too low for such reactions, the predominant process is elastic scattering. While this leaves the struck nucleus in its original state, it recoils as the neutron bounces off, carrying away some kinetic energy; neutrons thereby lose energy even in elastic collisions so long as their kinetic energy is higher than that of the nuclei with which they collide. By repeated collisions of this type, they eventually come into thermal equilibrium with their surroundings, which means that they just as often gain as lose energy in elastic collisions. In this situation they are referred to as *thermal neutrons.* Their average energy is kT, where T is the temperature of the surrounding material and k is the Boltzmann constant. For $T = 300°K$, $kT = 0.025$ eV.

In accordance with (13-17), cross sections for nuclear reactions are proportional to $\lambdabar^2\Gamma_n$; $\lambdabar^2$ is proportional to $1/E$, and from (13-13), Γ_n is proportional to $E^{1/2}$, whence cross sections are proportional to $E^{-1/2}$. They can therefore be a thousand times larger at 1 eV than at 1 MeV. Since neutrons spend by far the largest amount of their time at low energies, and since reaction cross sections are especially large there, we might expect reactions induced by low-energy neutrons to be of great practical importance. This is indeed the case, and we therefore devote this section to a discussion of them.

Low-energy neutrons can induce (n,γ) reactions[1] on almost any nucleus and can induce (n,p), (n,α), and (n,f) reactions on a few nuclei in which these reactions are energetically favorable. This last group includes some reactions which are very important from a practical point of view:

1. $N^{14}(n,p)C^{14}$. As a result of this reaction induced by neutrons from cosmic rays, nitrogen in the earth's atmosphere is converted to C^{14}, which beta-decays with a half-life of 5,700 years. The C^{14} mixes with ordinary carbon and is taken up by plants as CO_2, and perhaps the plants are eaten by animals. In any case, when life ceases, no more carbon is taken up and the C^{14} decays. The ratio of C^{14} to ordinary carbon therefore decreases with a half-life of 5,700 years, so by measuring this ratio, one can determine the age of wood, leather, and other objects derived from plant or animal life. This so-called *radiocarbon dating* has been extremely important in archeology and related fields.
2. $Li^6(n,\alpha)H^3$. This reaction allows H^3, known as tritium, to be produced cheaply and abundantly. Tritium is the principal ingredient in hydrogen bombs and may some day be useful in thermonuclear reactors for controlled production of energy (see Sec. 15-5).
3. $B^{10}(n,\alpha)Li^7$. This reaction serves as a basis for the most commonly used neutron detectors (see Sec. 9-6). It is also used in shielding against neutrons because boron (in the form of borax) is relatively cheap.
4. $U^{235}(n,f)$. This reaction is the key to all nuclear energy. Not only is it the principal reaction in almost all nuclear reactors and in many atomic bombs, but

[1] An (n,γ) reaction is a neutron-induced reaction in which *only* gamma rays are emitted.

it is used in the production of all other fissionable materials which form the basis for these devices. While many nuclei are now known which undergo (n,f) reactions with low energy neutrons, the only one found in nature is U^{235}, which is a 0.7 percent abundant isotope of ordinary uranium.

Aside from these, the most interesting reactions induced by low-energy neutrons are (n,γ), often called *neutron capture*. In view of the discussion of Sec. 8-4, it may seem surprising that gamma-ray emission can compete with neutron emission, so let us investigate this competition. The excitation energy of a compound nucleus formed by capture of a low-energy neutron is about equal to the neutron-separation energy, or about 8 MeV. From Fig. 12-1, the single-particle decay rate for an E1 transition of this energy is about 10^{15} s^{-1}, but E1 decays in this energy region are typically about a thousand times slower than the single-particle rate. On the other hand, there is a large number of states that can be reached by gamma-ray emission—all the states below 8 MeV excitation energy. As a result of these considerations, it turns out that the total decay rate is about 3×10^{14} s^{-1}, which corresponds via (13-11*a*) to

$$\Gamma_\gamma \simeq 0.15 \text{ eV} \tag{13-23}$$

Since this represents the sum of individual decay rates to a great many individual states, (13-23) is not subject to wide fluctuations from one resonance to another; the transition rate to any one state may vary, but when so many transitions are involved, these variations average out. In fact Γ_γ seldom varies from (13-23) by much more than a factor of 2 for all nuclei with $A > 60$.

From (13-20) and Fig. 13-10, the average width for neutron emission is roughly

$$\bar{\Gamma}_n \simeq 10^{-4}\bar{D}[E(\text{eV})]^{1/2} \tag{13-24}$$

Estimates of $\bar{D}$ can be obtained from (5-37) or from experimental measurements like those in Fig. 13-7. As examples, in odd-Z nuclei, $\bar{D}$ for states of a given I^π at the excitation energy reached by capture of low-energy neutrons is about 2,000, 40, and 10 eV for $A = 60$, 110, and 180, respectively. In order to find the energy at which $\bar{\Gamma}_n = \Gamma_\gamma$, we equate the right sides of (13-23) and (13-24) with these values of $\bar{D}$ inserted. The results are $E = 0.8$ eV, 1.4 keV, and 22 keV, respectively. For neutron energies above these, $\Gamma_n > \Gamma_\gamma$, so scattering predominates, but below these energies, (n,γ) reactions are most likely to result when a neutron is captured to form a compound nucleus. Of course elastic scattering by the potential-scattering mechanism (this accounts for the region between the resonances in Fig. 13-7) takes place at all energies, so neutrons can still be thermalized even in materials consisting of heavy elements.

From (13-23) and (13-24) we see that in any energy region, Γ_γ/Γ_n varies inversely as $\bar{D}$, so nuclei with large $\bar{D}$ have small (n,γ) cross sections. From the discussion of Chap. 5 we know that closed-shell nuclei have fewer states and consequently larger $\bar{D}$ than neighboring nuclei, whence they have smaller (n,γ) cross sections. This effect is enhanced by the fact the nearest resonance to any randomly chosen energy is usually further away in these nuclei because of the larger $\bar{D}$. We shall see examples of this along with an important application in Fig. 15-8.

Since neutrons spend most of their lifetime at thermal energies, the (n,γ) cross section for thermal neutrons, σ_{th}, is an important property of a material. Its values are therefore included on the chart of nuclides. When (13-20) is substituted into the Breit-Wigner formula (13-17), the latter becomes

$$\sigma = \pi \lambda\!\!\!^{-2}[E(\text{eV})]^{1/2} \frac{\Gamma_{n_0}\Gamma_\gamma}{(E - E_R)^2 + \frac{1}{4}\Gamma^2}$$

Inserting the values of $\lambda\!\!\!^{-}$ and E at thermal energy, 0.025 eV, this becomes

$$\sigma_{th} = 4 \times 10^6 \text{ barns} \frac{\Gamma_{n_0}\Gamma_\gamma}{(E - E_R)^2 + \frac{1}{4}\Gamma_\gamma^2} \tag{13-25}$$

where E_R is the energy of the closest resonance. In order to estimate σ_{th} for a typical situation, we may set $\Gamma_{n_0} \approx 10^{-4}\bar{D}$ from Fig. 13-10, and we may estimate that the lowest-energy resonance is at $\bar{D}/2$, whence $E - E_R = \bar{D}/2$. Inserting these and taking Γ_γ from (13-23), we find

$$\sigma_{th} \approx \frac{240 \text{ eV}}{\bar{D}} \text{ barns} \qquad \text{typical} \tag{13-25a}$$

While this estimate may be typical, it is subject to wide variations depending on the location of the lowest-energy resonance. If there should happen to be a resonance at 0.025 eV, $E - E_R = 0$, whence (13-25) gives

$$\sigma_{th} = 10^4\bar{D} \text{ barns/eV} \qquad \text{resonance at 0.025 eV}$$

For example, with $A \simeq 110$, where $\bar{D} \simeq 40$ eV, this gives $\sigma_{th} = 4 \times 10^5$ barns, whereas a typical value from (13-25*a*) would be 6 barns. This fantastic difference, a factor of nearly 10^5, arises from the fact that an energy level of a nucleus at about 8 MeV excitation is shifted in energy by only 20 eV, or only 2.5 parts in 10^6! For all practical purposes we can therefore say that the occurrence of large thermal neutron cross sections is essentially random among heavy nuclei. There is certainly no hope of ever understanding the energies of nuclear states to that accuracy.

The most common material with a large thermal-neutron capture cross section is cadmium since the isotope Cd^{113} has $\sigma_{th} = 20{,}000$ barns. Cadmium is therefore used to absorb thermal neutrons in reactors and in many experimental situations. There are many nuclei with larger σ_{th} (for example, σ_{th} for Xe^{135} is 3×10^6 barns) but they are not as readily available.

From (13-25a) we see that typical thermal-neutron capture cross sections vary inversely as $\bar{D}$, whence σ_{th} are small in light nuclei. This is basically because the nearest resonance is likely to be very far away, making $(E - E_R)^2$ in the denominator of (13-25) large. As examples of this effect, for H_2, Be^9, C^{12}, and O^{16}, σ_{th} is 5×10^{-4}, 9×10^{-3}, 3×10^{-3}, and 2×10^{-4} barn, respectively. Beryllium, carbon, and heavy water are therefore useful when it is desirable to avoid neutron capture. An application of this will be discussed in Sec. 15-6.

13-9 Compound-nucleus Reactions—Statistical Region

In Sec. 13-5, we found that when the energy of the incident particle is a few MeV or more, resonance widths become greater than the spacing between resonances, so the cross section as a function of energy becomes smooth, as in Fig. 13-6c. When there are a great many resonances contributing to the reaction cross section at any energy, this cross section builds up to near the maximum value given by (13-4) or Fig. 13-4. Once the compound nucleus is formed, it can decay in a great many ways, by emission of neutrons, protons, etc., each leaving the final nucleus not only in its ground state but in a great many excited states; in fact, since there are so many excited states available, the chance of the final nucleus being left in its ground state is very small. In this region, the best method for achieving an understanding is through a statistical approach; that is the subject of this section.

From (10-22), the decay rate of a compound nucleus to some final state i is given by

$$\lambda_i = \lambda_0 \theta_i^2 \mathcal{P}_i$$

If the decay can go to many states, the total decay rate is obtained by summing this expression over i. Let us say that the compound nucleus has an excitation energy such that if nucleon emission leaves the residual nucleus in its ground state, the energy of the emitted nucleon is E_0; in this section we shall assume that E_0 is several MeV. Nucleon emission can also lead to excited states of the residual nucleus, giving a nucleon energy E if the state has an excitation energy $E_0 - E$. Since this is a statistical treatment, we can use the *density-of-states* concept defined in connection with (5-34). In the present situation $\omega(E_0 - E)$ is the number of states per unit energy in the residual nucleus at excitation energy $E_0 - E$. We also define $\overline{\theta^2}$ and $\bar{\mathcal{P}}$ as the average reduced width and barrier penetrability for transitions to these states. The probability per unit time for compound-nucleus decays in which nucleons are emitted with energies between E and $E + dE$, $\lambda'(E)\,dE$, is then obtained by summing λ_i over all states in the energy range dE. The number of such states is $\omega(E_0 - E)\,dE$, whence

$$\lambda'(E)\,dE = \lambda_0 \overline{\theta^2} \bar{\mathcal{P}} \omega(E_0 - E)\,dE$$

From (10-15), λ_0 is proportional to $E^{1/2}$. It is usually assumed that $\overline{\theta^2}$ is independent of E, although this may be open to some question. For neutrons the factor $\bar{\mathcal{P}}$ represents penetration through angular-momentum barriers. From (13-9) we see that the largest l value for which $\mathcal{P}$ is near unity increases approximately as $E^{1/2}$; for closely related reasons, the effective value of $\bar{\mathcal{P}}$ increases about as $E^{1/2}$. Thus we have for neutrons (subscript n)

$$\lambda_n'(E)\,dE \simeq C'E\omega(E_{0n} - E)\,dE \qquad \text{neutrons} \tag{13-26a}$$

where C' is the product of all the energy-independent factors. For protons, $\bar{\mathcal{P}}$ includes a coulomb barrier-penetration factor $\mathcal{P}_p$, whence

$$\lambda_p'(E)\,dE \simeq C''E\mathcal{P}_p(E)\omega(E_{0p} - E)\,dE \qquad \text{protons} \tag{13-26b}$$

The other type of nucleon decay which is ordinarily energetically favorable is alpha-particle emission, for which λ' is given by (13-26*b*) with p's changed to α's. A more elaborate theoretical development (presented at the end of this section) gives, for the emission of any particle x,

$$\lambda_x'(E)\,dE = Cm_xE\sigma_{Rx}(E)\omega(E_{0x} - E)\,dE \tag{13-26}$$

where m_x is the mass and σ_{Rx} is the reaction cross section σ_R for particle x and C is another energy-independent factor. From (13-4) it is seen that σ_R has only slight energy dependence for neutrons, and it was pointed out in connection with that equation that for charged particles, σ_R differs from it principally by a coulomb barrier penetrability, whence (13-26) is essentially the same as (13-26*a*) and (13-26*b*).

From (5-34) we see that ω increases very rapidly with increasing excitation energy, so that factor tends to make $\lambda'(E)$ increase rapidly with decreasing E. The other terms, however, cause $\lambda'(E)$ to decrease at small E and go to zero at $E = 0$. The curves of $\lambda'(E)$ vs. E are therefore as shown in Fig. 13-12. They are drawn there for the same E_0, although for any given compound-nucleus decay E_0 differs for the three emitted particles.

These curves can be related to familiar ideas if we express $\omega(E_0 - E)$ in terms of S from (5-35) and expand S about its value at E_0 as

$$S(E_0 - E) = S(E_0) + \left(\frac{\partial S}{\partial E}\right)_{E_0}(-E) + \cdots$$

If we apply the second of (5-35) and then set the exponentials of both sides of the first of (5-35) equal to each other, we find

$$\omega(E_0 - E) = \omega(E_0)e^{-E/kT(E_0)}$$

Inserting this in (13-26) then gives

$$\lambda_x'(E)\,dE = C'''\sigma_{Rx}Ee^{-E/kT}\,dE \tag{13-27}$$

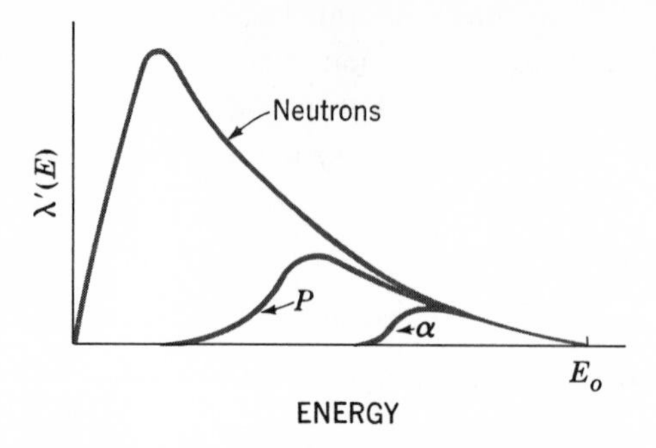

FIGURE 13-12 Energy spectra of neutrons, protons, and alpha particles emitted from a compound nucleus whose excitation energy is in the statistical region. These are plots of (13-26) for $E_{0n} = E_{0p} = E_{0\alpha}$.

For neutrons, σ_R is approximately constant, so (13-27) is just the result (13-2) obtained from the classical treatment. The energy distributions of emitted neutrons follow roughly a Maxwell distribution, and those for protons and alpha particles are a Maxwell distribution cut at low energies by a coulomb barrier-penetration probability. This is qualitatively evident from Fig. 13-12.

Since measurements of energy distributions of particles emitted in nuclear reactions are readily carried out, they can be fitted to (13-26) to obtain information on the level density ω. It is in this way that values for the parameters in (5-34) and (5-34*a*) were obtained.

The total decay rate for neutron emission λ_n is the integral under the curve in Fig. 13-12, or

$$\lambda_n(E_{0n}) = \int_0^{E_{0n}} \lambda_n'(E)\, dE \tag{13-28}$$

and there are similar expressions for protons and alpha particles; curves for $\lambda(E_0)$ are shown in Fig. 13-13.

The total decay rate λ_T is

$$\lambda_T = \lambda_n + \lambda_p + \lambda_\alpha \tag{13-29}$$

and the total probabilities for the compound nucleus to decay by neutron, proton, and alpha-particle emission are then λ_n/λ_T, λ_p/λ_T, and λ_α/λ_T, respectively. These quantities are readily obtainable from Fig. 13-13. If E_0 is not much different for the three particles, these probabilities are just proportional to the areas under the three curves in Fig. 13-12. It is obvious from that figure that neutron emission predominates. The extent of this predominance increases with increasing atomic

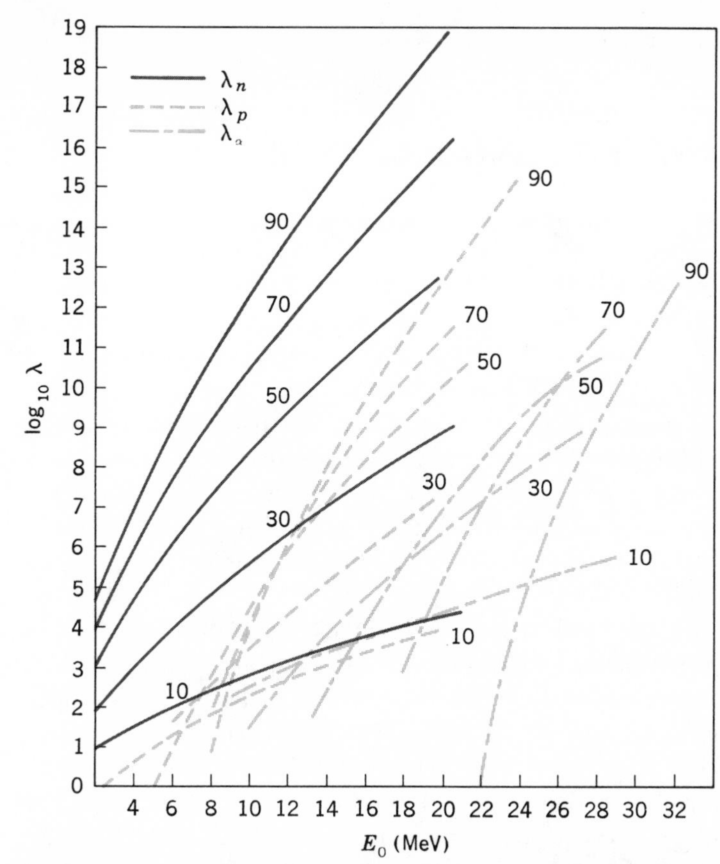

FIGURE 13-13 Plots of $\lambda_i(E_{0i})$ where i is a neutron, proton, and alpha particle, for various nuclei. These curves were calculated by G. R. Rao from (13-28) by use of (13-26); in using (13-26), σ was taken from Fig. 13-4 and (13-4), and ω was taken from a refined version of (5-34*a*) given by A. Gilbert and A. G. C. Cameron, *Can. J. Phys.*, 43: 1446 (1965). If the final nucleus is even-even or odd-A, before using these curves one should reduce E_0 by 2Δ or Δ, respectively, in view of the discussion following (5-34*a*); values of Δ are given in the above reference.

number as the coulomb barrier cuts out an increasing amount of the area under the curves for protons and alpha particles.

The cross section for a specific nuclear reaction consists of the product of two parts: the cross section for formation of a compound nucleus, as given by (13-4) and Fig. 13-4, and the probability for the compound nucleus to decay by emission of the particle of interest, as given in the previous paragraph. For example, the cross section for an (n,p) reaction is

$$\sigma(n,p) = \pi(R_0 + \lambda\!\!\!^-)^2 \frac{\lambda_p}{\lambda_T}$$

and the cross section for a (p,α) reaction is

$$\sigma(p,\alpha) = \sigma_{Rp} \frac{\lambda_\alpha}{\lambda_T}$$

where σ_{Rp} is from Fig. 13-4. For an (α,n) reaction,

$$\sigma(\alpha,n) = \sigma_{R\alpha} \frac{\lambda_n}{\lambda_T} \simeq \sigma_{R\alpha}$$

where the last step is justified by the fact that $\lambda_n \gg \lambda_p \gg \lambda_\alpha$, whence from (13-29), $\lambda_T \simeq \lambda_n$.

If the incident particle has a high energy, it often happens that after the first particle is emitted there is still sufficient excitation energy for further nucleon emission. The residual nucleus after emission of the first particle may then be considered as a second compound nucleus, and all the above considerations apply to its decay. Thus we get $(p,\alpha n)$, $(n,2n)$, $(\alpha,2p)$, etc., reactions. Since the decay of the second compound nucleus depends on its excitation energy, which in turn varies with the energy of the first emitted particle, the calculation of cross sections for these reactions requires a double integration.

When the energy of the incident particle is sufficiently high, three, four, or more nucleons can be emitted by a succession of these compound-nucleus decays. When a large number of particles is emitted, we have the spallation process mentioned in Sec. 8-6. For reasons already stated, most emitted particles are neutrons, especially in the case of heavy nuclei. If the incident particle is a proton, for example, the predominant reaction changes from (p,n) to $(p,2n)$, to $(p,3n)$, to $(p,4n)$, etc., as the energy of the incident proton is increased. An example of this is shown in Fig. 13-14.

In the remainder of this section, we give a derivation for (13-26). It will be presented on a level for advanced students and may be omitted without loss of continuity. We consider the transition

$$C(E_0 + S_x) \rightarrow X(E_0 - E) + x + E \tag{13-30}$$

from a compound state C at excitation energy $E_0 + S_x$ (where S_x is the separation energy of particle x) to a final nucleus X at excitation energy $E_0 - E$, plus particle x emitted with energy E. From (11-1) we can express the rate for this transition proceeding to the right with x having an energy between E and $E + dE$, $\lambda'_{\rightarrow}(E)\,dE$, as

$$\lambda'_{\rightarrow}(E)\,dE = \frac{2\pi}{\hbar} |M|^2 \omega_X(E_0 - E) \frac{dn_f}{dE}\,dE \tag{13-31}$$

However, (13-30) can also proceed to the left, and from (11-1) the transition rate $\lambda_{\leftarrow}$ for this is

$$\lambda_{\leftarrow} = \frac{2\pi}{\hbar} |M|^2 \omega_c(E_0 + S_x) \tag{13-32}$$

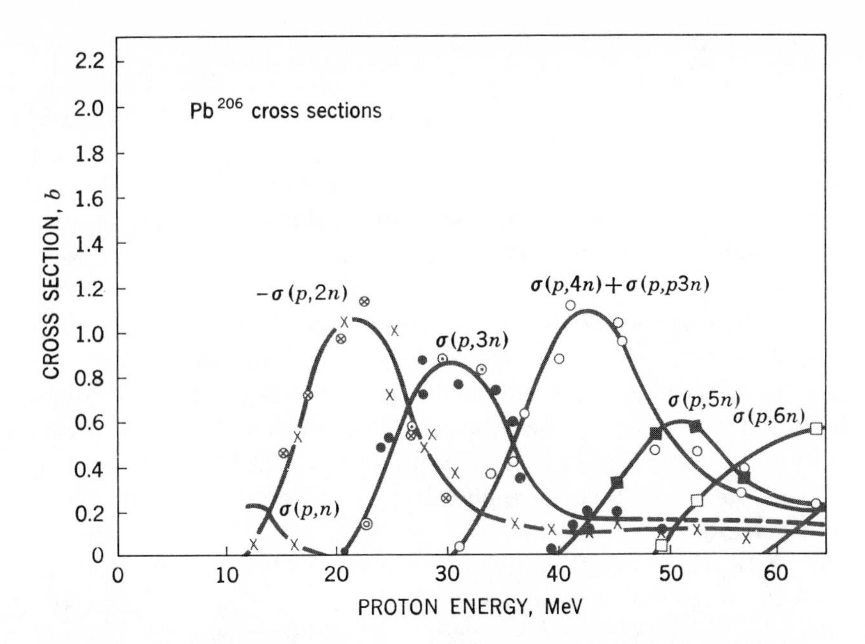

FIGURE 13-14 Cross sections for (p,xn) reactions on Pb^{206} as a function of proton energy. [*From R. E. Bell and H. M. Skarsgard, Can. J. Phys.,* **34: 745 (1956);** *reproduced by permission of the National Research Council of Canada.*]

It is important to note that if H in (11-2) is a hermitian operator, as will always be the case in physically meaningful situations, the $|M|^2$ in (13-31) and (13-32) are identical. There is another way to express the transition rate of (13-30) to the left, namely, in terms of its cross section σ_{Rx}, as

$$\lambda_{\leftarrow} \propto v\sigma_{Rx}(E)$$
$$\propto \left(\frac{E}{m_x}\right)^{1/2} \sigma_{Rx}(E)$$

Equating this to (13-32) and retaining only terms which depend on E gives

$$|M|^2 \propto \left(\frac{E}{m_x}\right)^{1/2} \sigma_x(E)$$

From the discussion leading to (11-4)

$$\frac{dn_f}{dE}\, dE \propto p^2\, dp \propto m_x^{3/2} E^{1/2}\, dE$$

Inserting these into (13-31) then gives

$$\lambda'_{\rightarrow}(E)\, dE \propto m_x E \sigma_{Rx}(E) \omega_X(E_0 - E)\, dE$$

This is the same as (13-26) except that E_0 is there called E_{0x}.

13-10 Nuclear Reactions Induced by Gamma Rays

When a gamma ray is incident upon a nucleus, transitions can occur which are just the inverse of those discussed in connection with gamma decay in Chap. 12 and the cross section for them to give a transition between two states is proportional to the Q_l^2 and A_l^2 discussed there. For example, if the gamma-ray energy is just equal to the energy of the transition between the ground and first excited state of an even-even nucleus (plus the correct additional energy needed to give the nucleus the kinetic energy needed to absorb the momentum of the gamma ray), this transition can take place. Its cross section is relatively large because the transition was shown in connection with (12-8) to have a large Q_2 value. For such a reaction to occur the gamma-ray energy must be correct to within the width of the state it excites; but from Fig. 12-1 and (13-11), we see that this is only a small fraction of 1 eV. Thus, while there are applications of this type of reaction (especially *resonance fluorescence*, which is a method for measuring gamma-ray-emission half-lives, and the *Mossbauer effect*, which is an important technique for measuring the hyperfine structure in the emission and is widely used in solid-state physics), the energy-matching requirements are so severe that these reactions are not generally encountered.

At higher energies, where the widths of energy levels become larger than the spacings between them, that is, $\Gamma > D$, gamma rays of any energy can excite a nucleus. For reasons given in Sec. 12-1, cross sections are by far the largest for E1 transitions, and these are "made to order" to excite the collective vibration shown in Fig. 5-10c; Q_1 may be shown to be very large for such a transition by arguments analogous to those used in obtaining (12-8).

In the simplest situation, that vibration would be in a single 1^- nuclear state, which we have said is at about 14 MeV excitation energy in heavy nuclei. However, there are clearly large numbers of 1^- states in the same energy region, and these mix configurations with the vibrational state, whence a great many 1^- states in this energy region contain a part of the vibrational state. They will then be excited by gamma rays with a cross section which is proportional to the size of that part, i.e., to the fraction of its time the nucleus, when in that state, spends in the collective vibration. The gamma-ray cross section as a function of its energy therefore behaves as shown in Fig. 13-15. Its entire character is dominated by the collective vibration. It may be noted from Fig. 13-15 that the energy of the 1^- vibration shifts to higher energy as A decreases.

Once a nucleus is excited into one of these states, it behaves like any compound nucleus with an excitation of about 14 MeV. The most probable decay, according to Sec. 13-9, is by neutron emission to give a (γ,n) reaction. If the energy is high enough for the residual nucleus following a (γ,n) reaction to emit another neutron, the reaction becomes $(\gamma,2n)$; etc. Of course (γ,p), (γ,α), (γ,f), etc., reactions can also occur.

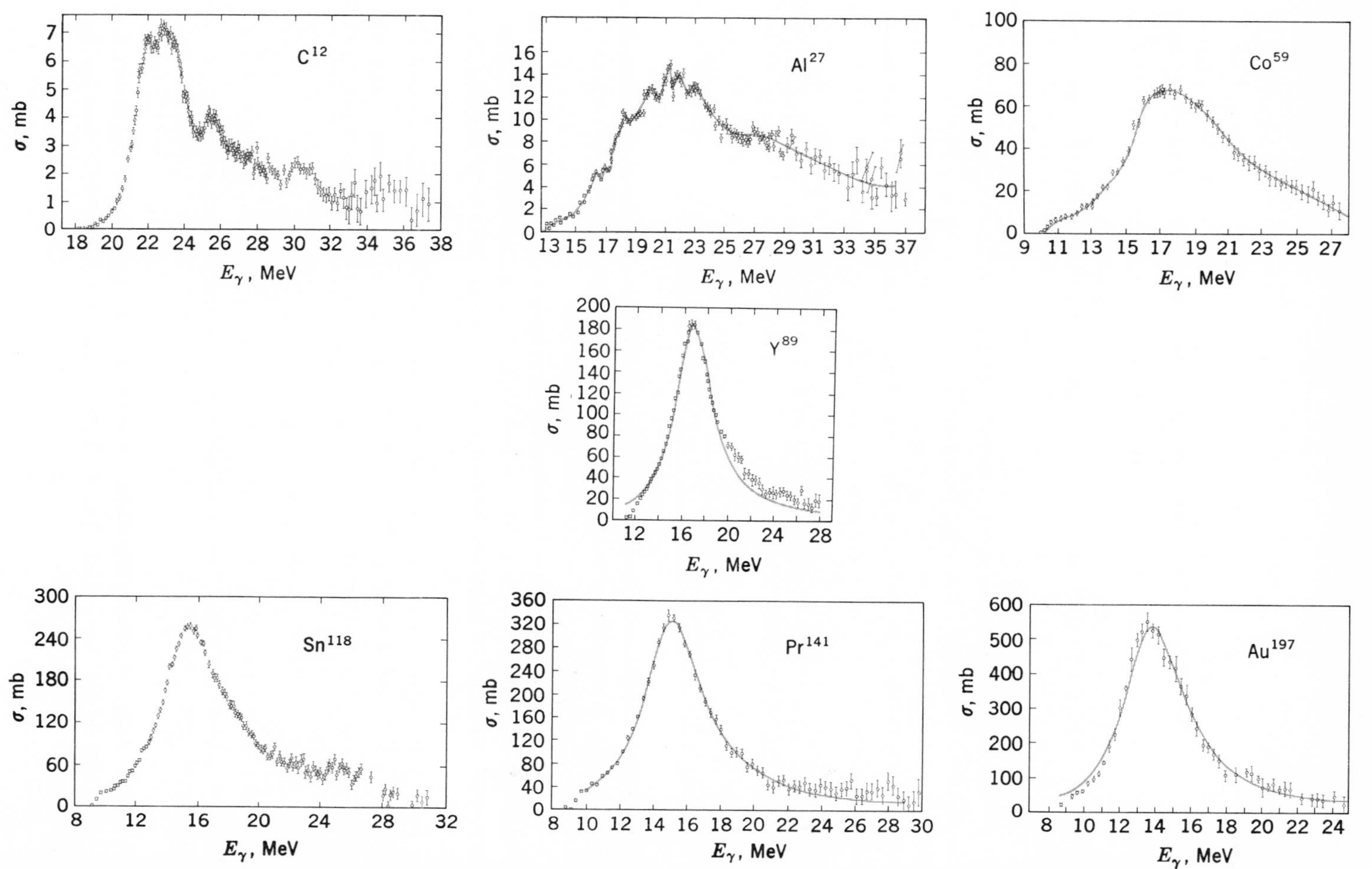

FIGURE 13-15 Cross section for reactions induced by gamma rays vs. their energy E_γ in various nuclei. Actually these data include only reactions in which neutrons are emitted, for example, (γ,n), $(\gamma,2n)$, (γ,pn), but these account for nearly all the reactions in most cases. Note that: these cross sections are dominated by the resonance due to the E1 vibration of Fig. 5-10c. [From S. C. Fultz, R. L. Bramblett, J. T. Caldwell, N. A. Kerr, N. E. Hansen, C. P. Jupiter, B. L. Berman, R. R. Harvey, and M. A. Kelly, *Phys. Rev.*, 127: 1273 (1962); 128: 2345 (1962);143: 790 (1966); 148 1198 (1966); 162: 1098 (1967); 186, 1255 (1969).]

Problems

13-1 Assume that inside the nucleus a neutron has ϵ_r ($= p_r^2/2M$) distributed according to $2^{6-\epsilon_r}$, and similarly for the two orthogonal components ϵ_θ and ϵ_ϕ (note that this is a Maxwell distribution), and that in coming out of the nucleus the component in the r direction is reduced by three units so that the energy on emergence is $E = (\epsilon_r - 3) + \epsilon_\theta + \epsilon_\phi$. By sampling all combinations of integer ϵ_r, ϵ_θ, ϵ_ϕ weighted with their probability from the above distribution function, calculate the probability of various values of E and compare with (13-1) to determine $f(E)$.

13-2 If the incident neutron enters an $\mathfrak{N} = 8$ orbit in Fig. 13-1, what collisions are allowed? Can $\mathfrak{N} = 1$ nucleons be excited?

13-3 What is the cross section for (p,n) reactions induced by 8-MeV protons on a Sn nucleus?

13-4 Plot the reaction cross section for neutrons vs. their energy from 1 to 20 MeV on Pb, Sn, Cu, and Al.

13-5 If 10-MeV neutrons induce nuclear reactions on Cu^{63}, what is the largest l value with which they have a good chance of entering the nucleus?

13-6 What is the mean free path inside a nucleus for nucleons of 1, 10, and 30 MeV? Compare these with the radius of an average nucleus.

13-7 A nucleus has a neutron resonance at 80 eV and no other resonances nearby. For the 80-eV resonance, $\Gamma_n = 5$ eV, $\Gamma_\gamma = 1$ eV, $\Gamma_\alpha = 3$ eV, and all other Γ's are negligible. What are the cross sections for (n,γ) and (n,α) reactions at 85 eV?

13-8 The ratio of C^{14} to C^{12} in a wooden object is one-tenth as large as in recently cut wood. How old is the object?

13-9 How thick a sheet of natural cadmium is needed to attenuate an incident flux of thermal neutrons by a factor of 100?

13-10 If D is taken from Sec. 5-11 with the excitation energy obtained from Table A-3, find Γ_{n_0} for neutrons incident on Sn^{116}. What is the typical width for a state at 50 eV?

13-11 An (n,p) reaction induced by 15-MeV neutrons on Sn is endothermic by 2 MeV; that is, $E_{0n} = 15$ MeV, $E_{0p} = 13$ MeV. Estimate the cross section for this reaction.

Further Reading

See General References, following the Appendix.

Ajzenberg-Selove, F.: "Nuclear Spectroscopy," Academic, New York, 1960.
Endt, P. M., and M. Demeur: "Nuclear Reactions," Interscience, New York, 1959.
Hughes, D. J.: "Pile Neutron Research," Addison-Wesley, Cambridge, Mass., 1953.
Nuclear Data, periodical published by Academic Press, New York.
Segre, E.: "Experimental Nuclear Physics," Wiley, New York, 1953.
Siegbahn, K.: "Alpha, Beta, and Gamma Ray Spectroscopy," North-Holland, Amsterdam, 1965.

Chapter 14
Nuclear Reactions: Direct Reactions

The compound-nucleus process discussed in the last chapter is the dominant one in nuclear reactions induced by low-energy particles, and, as such, it is extremely important in practical applications. However, the information it has given us about the nucleus itself is minimal. In fact, it was responsible for a "bum steer" that held back the development of nuclear-structure theory for more than a decade. When resonances were discovered and understood in the mid-1930s, they were correctly interpreted to mean that collisions take place before a nucleon can traverse more than a small fraction of a turn around an orbit (see Sec. 13-4). It was assumed that this conclusion was equally applicable to nuclei in their ground states and hence that the situation was like that shown in Fig. 4-1*b* rather than Fig. 4-1*a*. All attempts to use a shell theory were therefore abandoned in about 1936, and they were not resumed until 1949. One can easily imagine the difficulty we would have had in Chap. 5 if the three 0^+ states in O^{18} were mixed into thousands of nuclear states spread out over an energy range of 5 or 10 MeV—not to mention the fact that there would have been no way of identifying the orbits or determining their energies. The error was in not realizing that the energy range W over which mixing occurs decreases with decreasing energy, as we have seen in Fig. 13-5; the concept of W was not, of course, available at that time.

On this basis, one might think that nuclear reactions become even less useful for shedding light on nuclear structure as the energy is increased since W becomes even larger. However, this is most definitely not the case. As bombarding energies are increased, *direct reactions* become increasingly important, and, as we shall see, they have played a vital role in revealing the structure of nuclei. How this comes about is the subject of this chapter.

14-1 Mechanisms in Direct Reactions

Let us begin by describing the principal mechanisms in direct reactions. As illustrated by simple examples, they are:

1. An incident nucleon enters the nucleus, has a collision in which it loses some of its energy, and then comes off.
2. An incident nucleon enters the nucleus, has a collision in which the struck nucleon takes most of the energy, and the latter comes off.
3. An incident particle comes close enough for the force it exerts on several nucleons in the nucleus to set the nucleus into vibration or, if it is a nonspherical nucleus, into rotation. The incident particle is deflected by the interaction, but it continues on its way.
4. A proton enters the nucleus, exchanges its charge with one of the neutrons in the nucleus by exchanging mesons, and comes off as a neutron.
5. The incident nucleon comes close enough to pick up a nucleon from the nuclear surface, and the two come off together as a deuteron.
6. An incident deuteron comes close enough to a nucleus for one of the two nucleons of which it is composed to enter the nucleus, while the other is deflected but continues on its way.

All these examples have in common that the time of interaction is of the order of the time it takes the incident particle to traverse the nuclear diameter, which is typically about 3×10^{-22} s. The fact that this time is much shorter than the times required for compound-nucleus reactions is the basic distinction between the two processes. Another common feature of the above examples is that there is no mechanism for greatly altering the direction of the momentum carried in by the incident particle. In the compound nucleus, the direction is changed by repeated collisions and, more importantly, by reflections back into the nucleus from the nuclear surface, so angular distributions of emitted particles are basically isotropic. In direct reactions there are few collisions, and there are no reflections, so the direction of the incident momentum is very influential throughout, whence the direction of emission of emitted particles is highly correlated with it. In more common parlance, the angular distribution of emitted particles is peaked in the forward direction.

The probabilities for examples 1 and 2 are much less than one might at first think. When a 20-MeV nucleon enters a nucleus, its kinetic energy increases to about 70 MeV ($E = 20$ MeV and $V \simeq -50$ MeV, so $E - V = 70$ MeV; it is accelerated by the average force exerted on it by the nucleons in the nucleus as represented by the shell-theory potential). The nucleons with which it is likely to have collisions in accordance with the discussions of Sec. 13-2 have kinetic energies between about 30 and 40 MeV, so in an average collision the nucleon will lose about 17 MeV, leaving both it and its collision partner only about 3 MeV of energy in excess of that needed to escape. From (10-14), nucleons of this energy require about 2×10^{-22} s to escape due to reflections at the nuclear surface if they happen to be neutrons with $l = 0$. If, as is more likely, they have angular momentum (i.e., their velocity has an appreciable component in the θ direction) or are protons, there are barriers to penetrate so it takes much longer. If we take $W \simeq 3$ MeV from Fig. 13-5, the average time between collisions, from (13-10), is 2×10^{-22} s, so even for $l = 0$ neutrons there is a good probability for further collisions before emission. If a second collision occurs, there is not ordinarily enough energy left for escape and a compound nucleus is formed. If the first collision leaves one of the nucleons with more than 3 MeV of energy, the time for escape, from (10-14), is decreased (as $E^{-1/2}$) but W increases with increasing energy so collisions become more probable, whence the chance of a nucleon escaping without further collisions does not increase rapidly with increasing energy. At high enough energy, of course, reflection at the nuclear surface becomes negligible and barriers are easily penetrated, so direct reactions like those in examples 1 and 2 become more important than simple compound-nucleus formation above about 50 MeV. In the 10 to 30-MeV range of incident energy they do play a role, but they contribute nucleons principally to the low-energy portion of the energy spectrum of emitted particles. This is the region to which compound nucleus reactions contribute so heavily, as can be seen from Fig. 13-12, so contributions from direct reactions can only be vaguely distinguished by their generally forward-peaked angular distribution.

This is in sharp contrast to the situation in reactions of the types represented by examples 3 to 6 at the beginning of this section. In them, the disturbance to the nucleus is minimal, whence there is a strong tendency for it to be left in a state of low excitation. Since these are the states whose structure we have the best chance of understanding, a great deal of interest and effort has been devoted to using these direct reactions for quantitative studies of the structure of low-energy states of nuclei. Much of the discussion of Chaps. 4 to 6 has been developed and tied firmly to experimental reality through these studies.

Example 3, direct-reaction inelastic scattering, is clearly a good technique for studying vibrational and rotational states of nuclei, and it has been widely used for that purpose. Example 4, called *charge-exchange scattering*, provides a method for studying isobaric analog states and related phenomena. Examples 5 and 6 are *transfer reactions;* as an incident particle passes a nucleus, nucleons are transferred from one to the other. Example 5 is known as a *pickup reaction* for obvious reasons, and example 6 is called *stripping* since the nucleons are stripped off the incident particle by the nucleus. There are many other types of stripping and pickup reactions than the ones given in the examples. Among one-nucleon stripping reactions that have been studied are not only (d,p) and (d,n), as described in the example, but (t,d), $(\mathrm{He},{}^3d)$, (α,He^3), (α,t), $(\mathrm{C}^{13},\mathrm{C}^{12})$, etc. Many experiments have been done with two-nucleon stripping reactions like (t,p), (t,n), (He_3,p), (α,d), (Li^6,α), $(\mathrm{O}^{16},\mathrm{N}^{14})$, etc., three-nucleon stripping reactions like (α,p), four-nucleon stripping reactions like (Li^6,d) and (Li^7,t), five-nucleon stripping like

(Li^7,d), etc. In all these cases, the inverse pickup reactions have also been studied. In the following sections we shall see how some of these direct reactions are used to obtain nuclear structure information.

14-2 Angular Distributions of Particles Emitted in Direct Reactions

One feature of all direct reactions is the relationship between the angular momentum transferred in the reaction and the angular distribution of the emitted particle. To understand this, consider the momentum vector diagram in Fig. 14-1. The momentum of the incident particle p_i must be equal to the vector sum of the momentum of the emitted particle p_e plus the momentum transferred to the target nucleus p_t. The trigonometric solution for p_t is

$$p_t^2 = p_i^2 + p_e^2 - 2p_ip_e \cos \theta$$

If we let $p_i = p$, $p_e = p - \delta$, this becomes

$$p_t^2 = 2p^2(1 - \cos \theta)\left(1 - \frac{\delta}{p}\right) + \delta^2$$

$$= p^2\left[\theta^2\left(1 - \frac{\delta}{p}\right) + \left(\frac{\delta}{p}\right)^2\right]$$

where the second step uses the small-angle expansion for cos θ with only the first term retained. Solving this for θ^2 gives

$$\theta^2 = \frac{(p_t/p)^2 - (\delta/p)^2}{1 - \delta/p} \tag{14-1}$$

FIGURE 14-1 Momentum vector diagram showing the transferred momentum p_t as the vector difference between the momenta of the incident and emitted particles p_i and p_e, respectively.

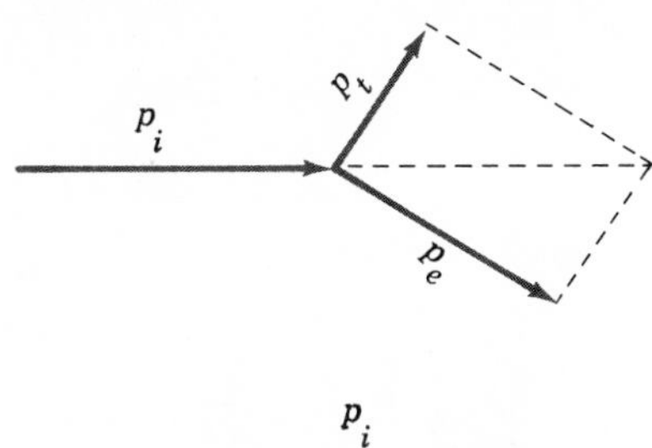

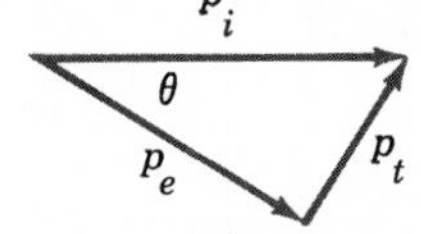

The orbital angular momentum transferred, $\hbar\sqrt{l_t(l_t+1)}$, is classically $\mathbf{r} \times \mathbf{p}_t$, which must be equal to or less than $R'p_t$, where R' is the radius at which most of the reactions take place. Hence the minimum value of p_t is

$$p_t \geq \frac{\hbar\sqrt{l_t(l_t+1)}}{R'} \tag{14-2}$$

Substituting (14-2) into (14-1) gives

$$\theta^2 \geq \frac{(\lambda\!\!\!^{-}/R')^2 l_t(l_t+1) - (\delta/p)^2}{1 - \delta/p} \tag{14-3}$$

where we have used the familiar relation $\lambda\!\!\!^{-} = \hbar/p$.

We have noted previously that the angular distribution for emitted particles in direct reactions tends to be concentrated in the forward direction, as in the curve labeled $l_t = 0$ in Fig. 14-2*a*. If this is the angular distribution when $l_t = 0$, (14-3) requires that the angular distribution for $l_t = 1$ differ from it in that the

FIGURE 14-2 Elementary derivation of angular distributions in direct reactions. The basic angular distribution is forward-peaked as the one labeled $l_t = 0$ in the upper drawing. For $l_t = 1, 2, \ldots$, increasing portions of the low-angle part are eliminated to satisfy angular-momentum conservation. Due to wave effects, the angular distributions in the upper drawing become like those in the lower drawing.

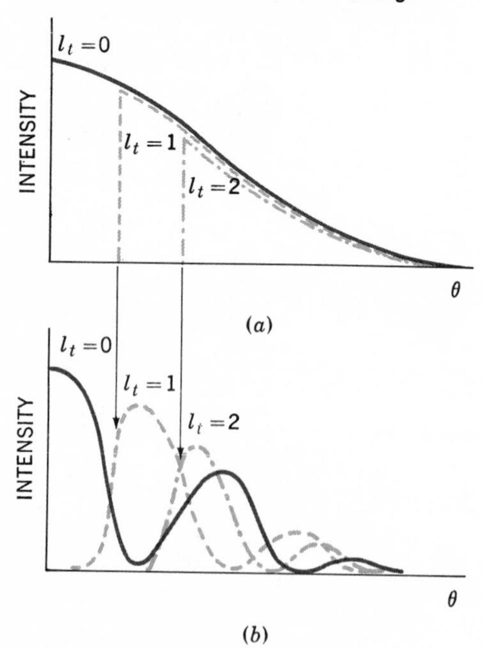

small-angle portion is removed, so to a first approximation we get an angular distribution like that labeled $l_t = 1$. For $l_t = 2$, (14-3) requires that a still more extensive portion be removed at small angles, as in the curve labeled $l_t = 2$; and similarly for larger l_t.

Since these angular distributions result from a superposition of waves emanating from all parts of the nuclear surface, we expect interference effects leading to a diffractionlike pattern of intensity variations. These variations are superimposed on the basic angular distributions of Fig. 14-2*a* to give distributions somewhat like those shown in Fig. 14-2*b*. The first maximum in the angular distribution should be at an angle somewhat larger than θ given by (14-3).

The angular distributions in Fig. 14-2*b* can be put on a semiquantitative basis by use of what is known as the plane-wave Born approximation (PWBA). We shall show that this gives an angular distribution $I(\theta)$ as

$$I(\theta) \propto [j_{l_t}(qR')]^2 \tag{14-4}$$

where j_{l_t} is the spherical Bessel function of order l_t, values of which are readily available in tables,[1] and $q = p_t/\hbar$, which is a function of θ through (14-1). The values of qR' at the first maximum of the expression in (14-4), C_l, are listed in Table 14-1. The angle at which they occur can be calculated by setting p_t in (14-1) equal to $\hbar q = \hbar C_l/R'$, which gives

$$\theta^2 = \frac{C_l{}^2(\lambda\!\!\!^{-}/R')^2 - (\delta/p)^2}{1 - \delta/p} \tag{14-5}$$

By comparing this with (14-3) we see that (14-5) is compatible with the conclusion of the last paragraph if C_l is "somewhat larger" than $[l_t(l_t + 1)]^{1/2}$. They are compared in Table 14-1; we see there that the angle of the first maximum is typically about 30 percent larger than given by (14-3), which is well within expectations.

[1] See, for example, M. Abramovitz and I. A. Stegun, "Handbook of Mathematical Functions," GPO, Washington, 1967.

TABLE 14-1: VALUES OF qR' FOR FIRST MAXIMUM IN (14-4), C_l, COMPARED WITH $[l_t(l_t + 1)]^{1/2}$

l_t	0	1	2	3	4	5
C_l	0	2.10	3.35	4.50	5.65	6.75
$[l_t(l_t + 1)]^{1/2}$	0	1.41	2.45	3.46	4.47	5.48
Ratio		1.49	1.37	1.30	1.26	1.24

In using (14-4) it is customary to choose R' to fit the data; a single choice is required to fit data for all l_t, and it is not permitted to deviate too far from the nuclear radius.

While the PWBA calculation is usually sufficiently accurate to give the location of the first and perhaps the second maximum, it does not give a very accurate fit to the angular distribution and fails completely in predicting absolute cross sections. The difficulty is that in PWBA it is assumed that the only interaction is that causing the reaction. This ignores the other effects of the nuclear and coulomb potentials, which, when a particle approaches close enough to have a reaction, lead to scattering and perhaps absorption of the particle, as described in Sec. 13-3. This complication is taken into account in the distorted-wave Born approximation (DWBA) by treating the incident and emitted particles as moving under the influence of the optical-model potential. A DWBA calculation involves extensive computation which can only be done with computers, but the result is a detailed prediction of the cross section as a function of angle without use of an arbitrary radius parameter like R' in (14-4).

As an illustration of angular distributions in direct reactions, Fig. 14-3 shows angular distributions of protons from 15-MeV-deuteron-induced (d,p) reactions on Ni^{58} leading to various states of Ni^{59}; the energies of the outgoing protons are about 21 MeV ($\delta/p \simeq 0.16$). Application of (14-5) with $R' = 8.1$ F then predicts that the first maxima should fall at 0, 13, 19, 30, and 40° for $l_t = 0$, 1, 2, 3, and 4, respectively. From Fig. 14-3 we see that these predictions are in quite good agreement with the data. The lines in the figure are the results of the DWBA calculations. They give reasonably detailed and accurate fits at least at forward angles. From the DWBA calculations it can be determined that most of the reactions actually do occur in the general region of $r = 8.1$ F,[1] although there are contributions from a wide range of radii.

In addition to giving angular distributions, the DWBA calculations give absolute cross sections, σ_{DWBA}, under the assumption that the nuclear-structure change in the reaction follows some simple model. For example, in a (d,p) reaction the model is that the transferred neutron enters one of the orbits of Fig. 4-5 without otherwise disturbing the nucleus. In general, then, the cross section can be written

$$\sigma = \sigma_{DWBA}\mathcal{S} \tag{14-6}$$

where $\mathcal{S}$ is a factor describing the degree to which the model used in the calculation correctly describes the nuclear structure changes. It will be discussed extensively in the following sections.

[1] It may be noted that this is "outside" of the nuclear potential well, but the wave functions of loosely bound nucleons extend well outside of the well; an extreme example of this was shown in Fig. 3-1.

In the remainder of this section we give a derivation of (14-4), the angular distribution as calculated with the PWBA, and some further comments on DWBA calculations. It will be understandable only by advanced students, and it may be omitted without loss of continuity. If we assume that the incident (i) and emitted (e) particles have no other interactions with the nucleus, they may be represented by plane waves before and after the reaction whence the total initial (I) and final (F) wave functions are

$$\begin{aligned} \Psi_I &= \exp(i\mathbf{k}_i \cdot \mathbf{r}_i)\psi_I \\ \Psi_F &= \exp(i\mathbf{k}_e \cdot \mathbf{r}_e)\psi_F \end{aligned} \tag{14-7}$$

where ψ_I and ψ_F are the internal wave functions for the initial (target plus particle i) and final (residual nucleus plus particle e) nuclei, respectively. From

FIGURE 14-3 **Angular distributions of protons from 15-MeV-deuteron-induced (d,p) reactions on Ni^{58} leading to various states of Ni^{59} with different orbital-angular-momentum transfers, l. Curves are results of DWBA calculations. Data for $l = 4$ are from a 12-MeV-deuteron-induced reaction.** [*From Phys. Rev.,* **126: 698 (1962); 133, B995 (1964).**]

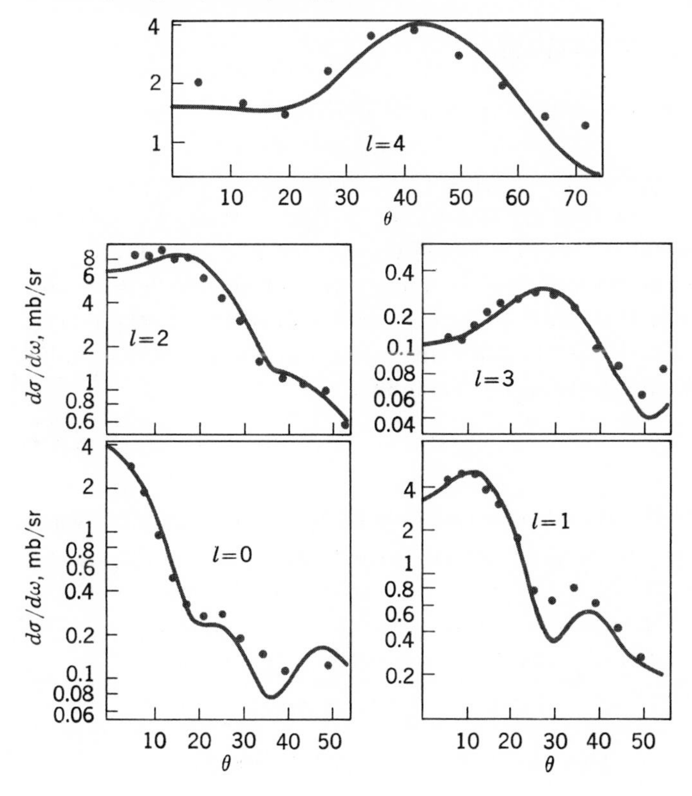

(11-1) the transition rate is proportional to the square of the matrix element, which we shall call T_{IF}, times dn_f/dE. From the treatment in Sec. 11-1,

$$dn_f \propto p_e^2\, dp_e \propto p_e\, dE$$

where E is the energy of the emitted particle, whence $dn_f/dE \propto p_e$, which is not a function of θ. The angular distribution is therefore determined by the matrix element, which, from (11-2), is

$$T_{IF} = \int \Psi_F^* V \Psi_I\, d\tau \tag{14-8}$$

We take V to be a zero-range potential, i.e., one acting only if i and e are at the same point, and to simplify the calculation we assume that it acts only at a single radius R'; thus

$$V = V_0 \delta(\mathbf{r}_i - \mathbf{r}_e)\delta(r_i - R') \tag{14-9}$$

Inserting (14-7) and (14-9) into (14-8) gives

$$T_{IF} \propto V_0 \int \exp\,[i(\mathbf{q}\cdot\mathbf{R}')]\psi_F^*\psi_I\, d\tau \tag{14-10}$$

where

$$\mathbf{q} = \mathbf{k}_i - \mathbf{k}_e = \frac{\mathbf{p}_t}{\hbar}$$

We can use the mathematical expansion

$$\exp\,[i(\mathbf{q}\cdot\mathbf{R}')] = \sum_l i^l \sqrt{4\pi(2l+1)}\, j_l(qR')Y_{l,0}(\alpha,\phi)$$

where, to avoid confusion, we use α to represent the angular variable in the nucleus usually represented by θ. In this expansion, the various terms represent the angular-momentum components l in the plane wave associated with linear momentum $\hbar q = p_t$; these l are therefore l_t. If conservation of angular momentum allows only a single l_t, only that term in the expansion contributes to the transition. When it is inserted into (14-10) and the integrations over α and the variables in ψ_I and ψ_F are carried out, the result is therefore

$$T_{IF} \propto j_{l_t}(qR')$$

which leads to (14-4).

In a DWBA calculation, the plane waves in (14-7) are replaced by the wave functions of particles moving in an optical-model potential (see Sec. 13-3), and the factor $\delta(r_i - R')$ in (14-9) is not used. This latter change means that the interaction is not confined to $r = R'$, and the final integration must then include an integration over r. This last complication is sometimes included in PWBA calculations under simplified conditions.

In some DWBA calculations, the $\delta(\mathbf{r}_i - \mathbf{r}_e)$ in (14-9) is replaced by some finite range of interaction, but this usually does not alter the results substantially.

In more extensive calculations, the velocity dependence in the interaction V (which arises from the velocity dependence of the nucleon-nucleon force) is taken into account. This is referred to as use of *nonlocal interactions*, and it makes an appreciable difference in the results in some cases.

Since DWBA calculations require the use of optical-model potentials for all particles involved in nuclear reactions, including deuterons, alpha particles, etc., optical-model potentials have been developed for them by measuring elastic scattering of these particles from nuclei and going through the fitting procedures discussed in Sec. 13-3. All in all, the use of DWBA calculations has developed into an extensive and highly refined technology.

14-3 Nuclear-structure Studies with One-nucleon Transfer Reactions

A (d,p) reaction is essentially a fancy way of inserting a neutron into a nucleus, so l_t is just the orbital angular momentum with which the neutron is inserted. Its total angular momentum is j_t, the vector sum of l_t and its spin. If it enters the nucleus Ni^{58}, the configuration upon its entry is

$$Ni^{58}(GS) + n(l_t,j_t)$$

For example, if $l = 1$, $j = 3/2$ (hereafter we drop the subscript t), the configuration is

$$Ni^{58}(GS) + n(p_{3/2}) \tag{14-10a}$$

As the ground state of Ni^{58} is 0^+, this configuration gives a term with I^π of $3/2^-$, so the state of Ni^{59} formed must be $3/2^-$. Clearly that state must contain the configuration (14-10a) in its wave function, and furthermore the probability for any particular $3/2^-$ state of Ni^{59} to be formed in this reaction is proportional to the fraction of its time it spends in the configuration (14-10a). To be more explicit, if the wave function for some $3/2^-$ state i, $\psi_{3/2}(i)$, is written

$$\psi_{3/2}(i) = \theta_i\psi[Ni^{58}(GS)]\psi(p_{3/2}) + \cdots$$

the cross section for exciting the state i must be proportional to θ_i^2; hence, from (14-6), $\mathcal{S}$ must be proportional to θ_i^2. If $\theta_i^2 = 1$, we have just the model used in calculating σ_{DWBA}, as explained before (14-6), so $\mathcal{S} = 1$. In general, then, $\mathcal{S}$ must be *equal* to θ_i^2. Even though we see that it is identical with the reduced width, in direct-reaction applications $\mathcal{S}$ is called the *spectroscopic factor*. We append the j value of the orbit it refers to as a subscript. In the case we have been discussing, then, $\mathcal{S}_{3/2}(i) = \theta_i^2$, and we define $\mathcal{S}_j(i)$ for (d,p) on an even-even target as[1]

$$\psi_A(i) = [\mathcal{S}_j(i)]^{1/2}\psi_{A-1}(GS)\psi(j) + \cdots \qquad (d,p) \tag{14-11}$$

[1] The accurate definition of $\mathcal{S}$ is the overlap integral as for θ^2 in Sec. 10-5.

All these considerations are, of course, equally applicable if the neutron is inserted with any other I^π. If it enters the nucleus with $l = 3$, $j = \frac{5}{2}$, the configuration must be

$$\text{Ni}^{58}(\text{GS}) + n(f_{5/2}) \tag{14-10b}$$

and the cross section for exciting any $\frac{5}{2}^-$ state i of Ni^{59} in the reaction is given by (14-6) with $\mathcal{S} = \mathcal{S}_{5/2}(i)$, which is the square of the amplitude of the term (14-10*b*) in the wave function for the state i.

An experimental study with a (d,p) reaction, as illustrated with the example of Ni^{58}, proceeds as follows. A thin target of Ni^{58} is bombarded with monoenergetic deuterons, and the energy distribution of the protons emitted at various angles relative to the incident beam is measured. A typical measurement is shown in Fig. 14-4. The various peaks correspond to reactions in which Ni^{59} is left in its various states; the highest-energy proton peak corresponds to cases where Ni^{59} is left in its ground state, the next highest energy peak corresponds to cases where it is left in its first excited state, etc. Peaks are labeled by the excitation energy of the corresponding states in Ni^{59}. From the size of each peak at various angles, angular distributions are obtained like those shown in Fig. 14-3, and, as described

FIGURE 14-4 Energy distribution of protons from 15-MeV-deuteron-induced $\text{Ni}^{58}(d,p)\text{Ni}^{59}$ reactions observed at $\theta = 9°$ with a magnetic spectrograph and photographic-plate detection. Numbers attached to peaks are excitation energies (in MeV) of corresponding states in Ni^{59} and orbital-angular-momentum transfers l in parentheses. [*From Phys. Rev.*, **126: 698 (1962).**]

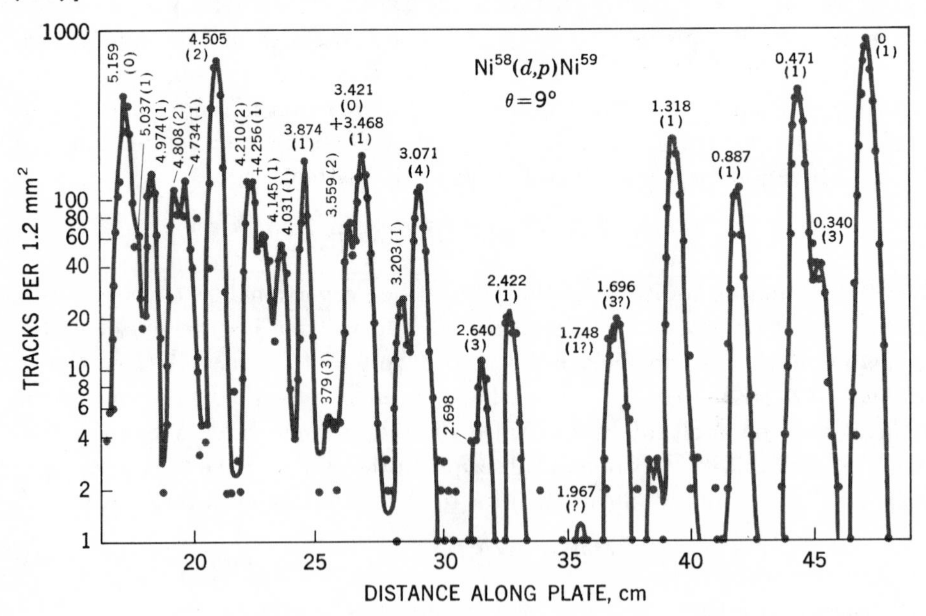

previously, these lead to determinations of the l value of the orbit the neutron entered. These are the digits in parentheses in Fig. 14-4. It is then generally easy to surmise which of the orbits of Fig. 4-5 it corresponds to. In Ni^{58} there are 30 neutrons, whence all orbits up through $\mathfrak{N} = 3A$ in Fig. 4-5 are filled and there are two neutrons in $\mathfrak{N} = 4$ orbits. The lowest-energy states excited will therefore be ones in which the neutron enters $\mathfrak{N} = 4$ orbits; in cases where $l = 3$, it is the $1f_{5/2}$ orbit; for $l = 4$ it is the $1g_{9/2}$ orbit; and for $l = 1$ it is the $2p_{3/2}$ or $2p_{1/2}$ orbit. At somewhat higher excitation energies one finds the states in which the neutron enters the lower-energy $\mathfrak{N} = 5$ orbits; in $l = 2$ cases it is probably the $2d_{5/2}$ orbit, and for $l = 0$ it is the $3s_{1/2}$ orbit. In situations where there are two orbits with the same l value, as for $2p_{3/2}$ and $2p_{1/2}$ here, distinguishing between the two is sometimes difficult, but there are several methods available for doing so. The simplest in principle is to take advantage of the fact that the accurate calculation of angular distributions by the DWBA method is sensitive to the detailed shape of the nuclear potential, and, as we know from Fig. 4-4, this shape is somewhat different for $p_{3/2}(S\|l)$ than for $p_{1/2}(S\nparallel l)$ neutrons. As a consequence, there are differences between the angular distributions, especially at back angles. An example of this difference is shown in Fig. 14-5. Clearly not only l but also j values can be determined from these angular distributions.

Once the j value has been assigned for each state of Ni^{59}, a comparison of the cross section for exciting it with the DWBA calculation allows its $\mathcal{S}_j$ value to be determined from (14-6). Let us consider what may be expected for $\mathcal{S}$ in some simple cases. If the target is even-even, and if the state excited is a pure SQP state, we see from (6-1) and (14-11) that

$$\mathcal{S}_j(\text{SQP}) = 1 - V_j^2$$

For example we found in our discussion of Fig. 6-5 that the lowest three states of Sn^{117} are nearly pure SQP states, so the $Sn^{116}(d,p)$ reaction leading to these states gives $\mathcal{S}$ values corresponding to $(1 - V_{1/2}^2)$, $(1 - V_{3/2}^2)$, and $(1 - V_{11/2}^2)$. If the SQP state is mixed among several nuclear states, each is excited with an $\mathcal{S}$ value of $(1 - V_j^2)$ times the fraction of the SQP state it contains. That was the situation in Pd shown in Fig. 6-7 and holds also for the Ni^{59} case under consideration here. In such situations, the fractions appearing in all nuclear states must add up to unity, whence the sum of the $\mathcal{S}$ values for these states must be $1 - V_j^2$, or

$$\sum_i \mathcal{S}_j(i) = 1 - V_j^2 \qquad (d,p),\ \text{even-even target} \tag{14-12}$$

The sum rule (14-12) must be valid separately for each neutron orbit j. This gives a method for determining V_j^2 and thereby allows curves like those in Fig. 5-6 to be checked experimentally. In Ni^{58} the results show $1 - V_j^2$ to be about 0.69, 0.91,

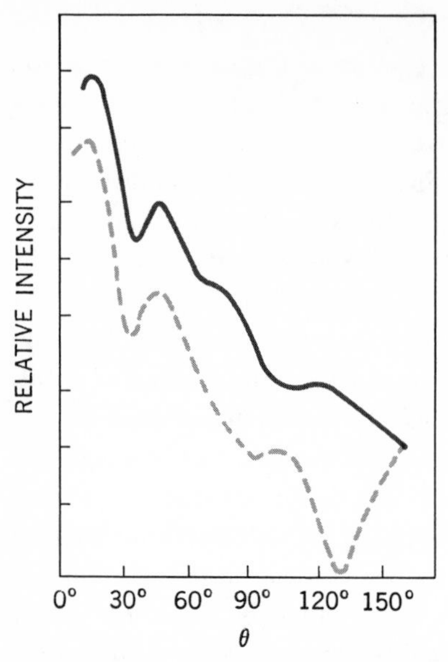

FIGURE 14-5 Angular distributions of protons from 10 MeV-deuteron-induced $Fe^{54}(d,p)$ reactions leading to the ground state (solid line) and 0.41-MeV state (dashed line) of Fe^{55}. Both are $l = 1$ transitions, but the former state is $\frac{3}{2}^-$ and the latter is $\frac{1}{2}^-$. Note the difference between the two at back angles due to their different values of total-angular-momentum transfer j. [*From L. L. Lee and J. P. Schiffer, Phys. Rev. Letters,* **12: 108 (1964).**]

0.96, and 0.99 for the $2p_{3/2}$, $1f_{5/2}$, $2p_{1/2}$, and $1g_{9/2}$ states, respectively.[1] Since each state can contain $2j + 1$ neutrons, this means that, on an average, the ground state of Ni^{58} contains about 1.24 $2p_{3/2}$ neutrons, 0.58 $1f_{5/2}$ neutron, 0.08 $2p_{1/2}$ neutron, and 0.10 $1g_{9/2}$ neutron. Since we know from Sec. 5-3 that at any one time there are two neutrons in the same orbit, they must spend 62 percent of their time in $2p_{3/2}$ orbits, and 29, 4, and 5 percent of their time in $1f_{5/2}$, $2p_{1/2}$, and $1g_{9/2}$ orbits, respectively.

The fractional contribution of each nuclear state to the sum in (14-12) is a measure of the fraction of the SQP state it contains. This was the method used in determining the numbers given for this in Figs. 6-6 and 6-7. When these fractions, or just the $\mathcal{S}$ values for each state to which they are proportional, are plotted at the energy of the nuclear state as in Fig. 14-6, we obtain a graphic description of

[1] *Phys. Rev.*, **126:** 698 (1962); **133:** B995 (1964).

how individual SQP states are distributed among nuclear states. From the discussion in Sec. 13-4, the widths of these distributions are $2W$; these data therefore give rough determinations of W which are included in Fig. 13-5.

The "centers of gravity" of the distributions in Fig. 14-6 give the energies of the SQP states which we called E_j in Sec. 6-1. These experimental determinations are used for checking curves like those in Figs. 6-3 and 6-4. From the discussion in connection with Figs. 6-1 and 6-2, we know that in single-particle nuclei, the E_j are just the orbit energies of Fig. 4-5. This method has therefore been used to put Fig. 4-5 on a quantitative basis. We shall return to this matter shortly.

Pickup reactions like (p,d) give similar but complementary information. An experimental study of the $Ni^{60}(p,d)Ni^{59}$ reaction would proceed in the same way as was described above for the (d,p) reaction, except that the value of $\mathcal{S}$ in this case gives the fraction of the time each state in Ni^{59} has the configuration $Ni^{60}(GS)$ plus a hole; e.g., if the state is $\frac{3}{2}^-$, its $\mathcal{S}$ value gives the fraction of the time its configuration is

$$Ni^{60}(GS) + n(p_{3/2})^{-1}$$

The formal definition of $\mathcal{S}$ in a (p,d) reaction on an even-even target is

$$\psi_A(i) = \frac{1}{2j+1}\,[\mathcal{S}_j(i)]^{1/2}\psi_{A+1}(GS)\psi(j)^{-1} + \cdots \qquad (p,d) \tag{14-13}$$

FIGURE 14-6 Typical distributions of nuclear states containing parts of a given SQP state. Labels give the SQP state and the nucleus in each case. Abscissas are excitation energies of the states, and heights of the lines are the $\mathcal{S}$ values in (d,p) reactions. Circles with crosses indicate "centers of gravity" obtained by weighting each state with its $\mathcal{S}$ value; they determine E_j. Bars show the width $2W$ from the curve in Fig. 13-5. They apparently underestimate the width for the Pd data. [*From Phys. Rev.*, **126: 698 (1962); 131: 2184 (1963); 161: 1257 (1967).**]

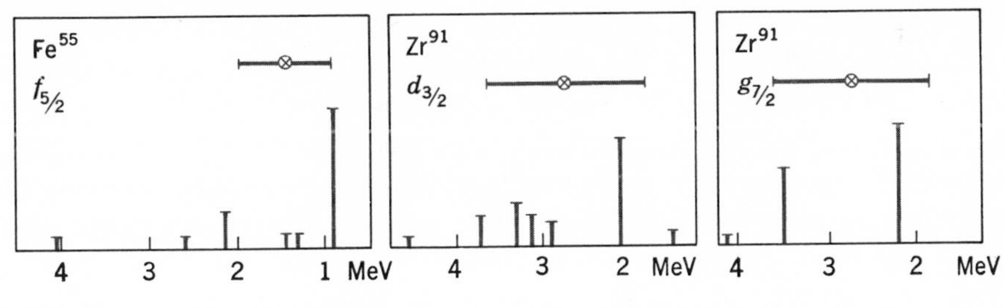

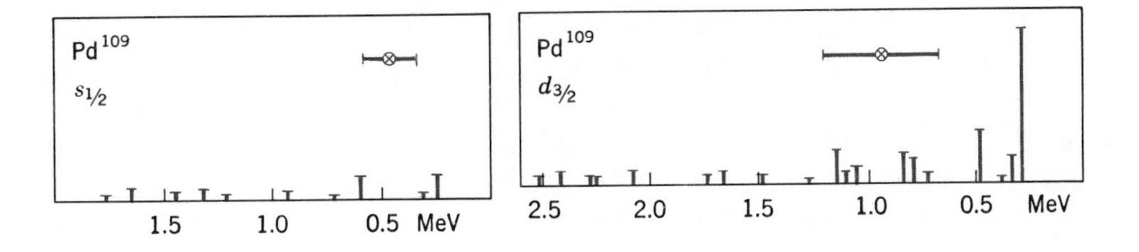

If the state is a pure SQP state, from (6-1) and this definition we find

$$\mathcal{S}_j(\text{SQP}) = (2j + 1)V_j^2$$

If the SQP state is mixed among several nuclear states, the $\mathcal{S}_j$ value for each is $(2j + 1)V_j^2$ times the fraction of the SQP state it contains, whence a sum over all these states gives

$$\sum_i \mathcal{S}_j(i) = (2j + 1)V_j^2 \qquad (p,d),\text{ even-even target} \tag{14-14}$$

The $\mathcal{S}$ values for each state can be determined from measurements of cross sections with (14-6), and from them the values of V_j^2 can be determined from (14-14). Similarly, all the other applications of (d,p) reactions discussed above, such as determinations of E_j and of W, can be carried out equally well with (p,d) reactions. In fact, when V_j^2 is smaller than $\frac{1}{2}$, it is more accurate to use pickup reactions for these purposes.

If a given nuclear state contains a fraction f of the SQP state, from the discussion preceding (14-12) and (14-14) the ratio of its $\mathcal{S}$ values in (d,p) and (p,d) reactions is

$$\frac{\mathcal{S}(d,p)}{\mathcal{S}(p,d)} = \frac{f(1 - V_j^2)}{f(2j + 1)V_j^2} = \frac{1 - V_j^2}{(2j + 1)V_j^2} \tag{14-15}$$

This serves as the basis for another method of determining j values for states. For example, of the two types of $l = 1$ states excited in Ni, $p_{3/2}$ and $p_{1/2}$, the former is lower in energy and will therefore be fuller, whence it will have a considerably smaller value for the ratio $(1 - V_j^2)/V_j^2$. In a (d,p) study, a measurement of (p,d) at a single well-chosen angle is usually sufficient to determine the j value from (14-15) since the l values are known from the (d,p) work. This method is therefore much easier than the one given in connection with Fig. 14-5 for determining j values.

From the discussion of Sec. 6-1 we know that in single-hole nuclei the E_j are the orbit energies of Fig. 4-5, so measurements of E_j in these by pickup reactions on closed-shell nuclei have also been used to put Fig. 4-5 on a quantitative basis. A comparison of these results with those from (d,p) reactions gives information on the variation of these orbit energies with mass number A. For example, the $\mathfrak{N} = 3$ orbit energies can be determined for $A = 17$ from $O^{16}(d,p)$ and for $A = 39$ from $Ca^{40}(p,d)$, whence their variation in this region can be surmised. Results of this type for all neutron single-particle and single-hole nuclei are summarized in Fig. 14-7. A brief explanation of the behavior of the measured orbit energies is given in the legend for that figure.

The discussion up to this point has centered on applications of stripping and pickup reactions in which the target nucleus is even-even. There are also applications for experiments on odd-A targets which give information on odd-odd and even-even nuclei. An example of the latter is a (d,p) reaction on Pd^{105} lead-

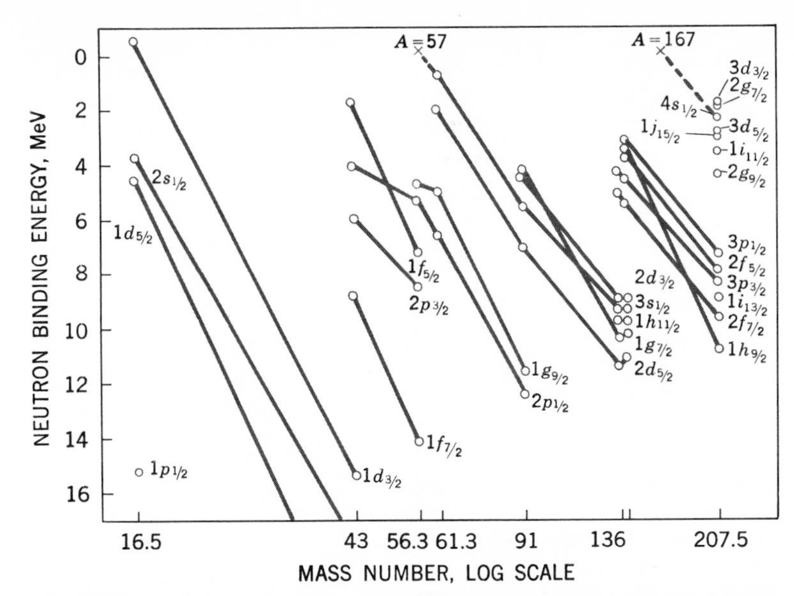

FIGURE 14-7 Energies of neutron orbits in various single-particle and single-hole nuclei. Small corrections for symmetry energy have been applied to correct the data to the line of beta stability. The slopes are due to the fact that nuclear radii increase with increasing A, causing the wavelength of a given orbit to increase and its energy to decrease in accordance with (2-1). Extrapolating the data for $s_{1/2}$ states to zero energy predicts locations of the maxima in the strength functions of Fig. 13-10; these are shown by crosses. Note the faster decrease in energies of high-angular-momentum orbits as discussed in connection with Fig. 5-14. Energy differences between major shells, obtained from this figure, gave the determination of M^* discussed after (4-18). [*From Am. J. Phys.*, **33: 1011 (1965).**]

ing to the 2^+ vibrational state of Pd^{106}. The angular distribution of the protons can be decomposed into a sum of $l = 0$, $l = 2$, and $l = 4$ parts, and $\mathcal{S}$ values can be obtained for each. These give the amount of $d_{5/2}/2s_{1/2}$, $(d_{5/2})^2$ plus $d_{5/2}d_{3/2}$, and $d_{5/2}/2g_{7/2}$ configurations, respectively, in the vibrational state. Note that all these configurations can give 2^+ terms.

An example of results from stripping reactions on an odd-proton nucleus is given in Fig. 14-8. The reaction here is $Y^{89}(d,p)Y^{90}$, and it is compared with $Zr^{90}(d,p)Zr^{91}$. The ground state of Y^{89} differs from that of Zr^{90} in having a $p_{1/2}$ proton hole, so we might expect the results of the two reactions to be similar except that for each state excited in Zr^{91} there should be two in Y^{90}, corresponding to the two possible couplings with the $p_{1/2}$ proton hole. For example, the ground state excited in the $Zr^{90}(d,p)$ reaction is the $d_{5/2}$ single-particle state. In the $Y^{89}(d,p)$ reaction, this is $(p_{1/2})_\pi^{-1}(d_{5/2})_\nu$, which can couple to give 2^- and 3^- states. The cross section for exciting these is proportional to the number of ways each

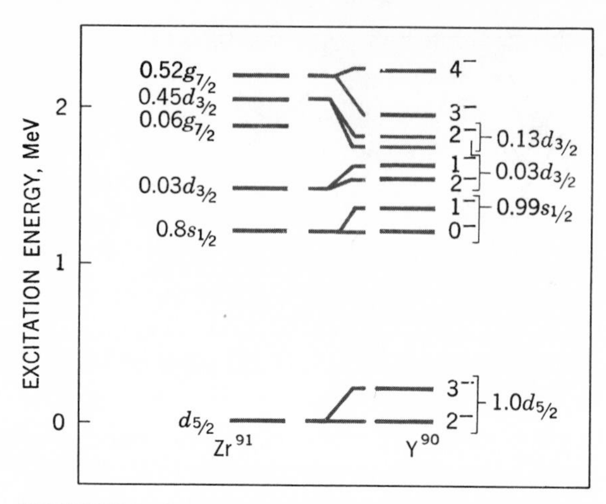

FIGURE 14-8 States excited in Zr^{91} and Y^{90} by (d,p) reactions on Zr^{90} and Y^{89}, respectively. Attached figures are the fraction of the SQP state contained in the nuclear state; for example, $0.8s_{1/2}$ means that the nuclear state contains 80 percent of the $s_{1/2}$ SQP state. For each state excited in Zr^{91} there are two in Y^{90} corresponding to the two angular-momentum values obtained by coupling with the $p_{1/2}$ hole. The two states are excited in proportion to $2I + 1$, and this allows a determination of I. [*From A. I. Hamburger and E. W. Hamburger, Phys. Lett.,* **4: 223 (1963).]**

can be formed, which is proportional to $2I + 1$, corresponding to the number of possible values of M_I. Measurements of cross sections therefore distinguish between the two states.

Other single-nucleon transfer reactions have been used for the same purposes as (d,p) and (p,d), and they have advantages in particular cases; (d,t) reactions are more favorable energetically then (p,d) and are often used for neutron pickup. We see from (14-5) that the first maximum in the angular distribution for a large l_t can be moved to small angles by choosing a reaction with a large δ, such as (α,He_3). This allows states of high l to be more easily excited and more easily studied. All the examples used here have involved the transfer of a single neutron, but corresponding proton-transfer experiments have been done using (d,n), (He^3,d), and (α,t) stripping reactions and their inverses as pickup reactions.

14-4 Nuclear-structure Studies with Isobaric Analog States

Since a (d,p) reaction is essentially a fancy way of inserting a neutron into a nucleus, we might ask at this point why so much information about nuclear structure has been obtained from these reactions whereas so little on this subject has been learned from neutron-induced reactions. The reduced widths θ^2 from the latter are completely equivalent to the spectroscopic factors $\mathcal{S}$ from the former, and the l value with which the neutron enters can be obtained from the shape of the resonance in elastic scattering (see Fig. 13-7, where the shapes of peaks C, D, and E indicate that the reactions are induced by $l = 0$ neutrons) just as it can be determined from the angular distribution in stripping reactions. The difference is principally in the energy regions available for excitation by the two reactions. Neutron-induced reactions can excite only states with excitation energies above 8 MeV, and in this region each single-particle state is mixed into thousands of nuclear states; in fact there are so many states that experimental energy resolutions are not sufficient to resolve them except in a narrow energy range. And even with unlimited experimental resolutions, the energy range is badly limited by the fact that widths of resonances become wider than the spacings between them ($\Gamma > D$) above about 2 MeV in heavy nuclei (see Sec. 13-5). Stripping reactions, on the other hand, can be used to study the low-excitation-energy region where nuclear structure is much simpler and better understood and the energy range available is much wider.

However, there are compound-nucleus reactions in which the energy-range handicap is not applicable, namely, in the excitation of isobaric analog states with protons. To understand this, let us consider the example of $A = 117$ nuclei shown in Fig. 14-9. A (d,p) reaction on Sn^{116} is exothermic; the neutron enters with negative energy (the excess energy is carried off by the proton) as it excites, let us say, the $\frac{3}{2}^+$ state of Sn^{117} labeled A in the figure. If the wave function for this state is written

$$\begin{aligned}\psi_A = {} & \theta_{AG}\psi[\mathrm{Sn}^{116}(\mathrm{GS})]\psi(d_{3/2})_\nu \\ & + \theta_{AH}\psi[\mathrm{Sn}^{116}(H)][C_1\psi(s_{1/2}) + C_2(d_{3/2}) + C_3(d_{5/2}) + C_4(g_{7/2})]_\nu \\ & + \theta_{AK}\psi[\mathrm{Sn}^{116}(K)]\psi(d_{3/2})_\nu + \cdots\end{aligned} \tag{14-16}$$

where GS, H, K are the states of Sn^{116} so labeled in the figure, the spectroscopic factor is

$$\mathcal{S}_{3/2}(A) = \theta_{AG}^2$$

In a proton-induced compound-nucleus reaction on Sn^{116}, the isobaric analog of state A, labeled A', is excited. From the method of generating isobaric analog states explained in Fig. 6-11, we see that the first term in (14-16) leads to $2T$ terms in the wave function for A', one for each orbit occupied by a neutron and not by a proton. One of these is exactly the same as the first term in (14-16) except that $(d_{3/2})_\nu$ is changed to $(d_{3/2})_\pi$. Hence, the reduced width $\theta_{A'G}^2$ for excitations of state A' is

$$\theta_{A'G}^2 = \frac{1}{2T}\,\theta_{AG}^2 = \frac{1}{2T_0 + 1}\,\theta_{AG}^2 = \frac{1}{2T_0 + 1}\,\mathcal{S}(A) \tag{14-17}$$

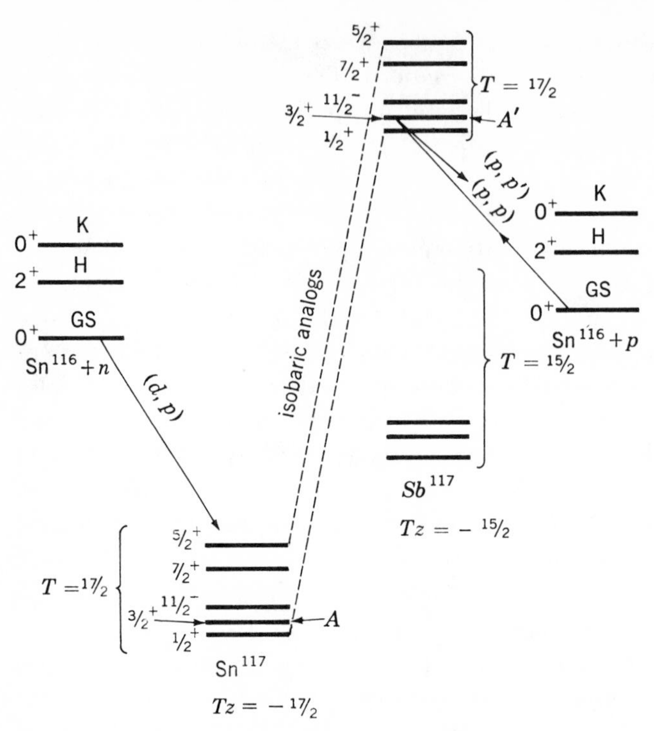

FIGURE 14-9 Energy levels in $A = 117$ nuclei.

where T_0 is the T value for the target nucleus [in this case, Sn^{116}(GS)], which is one-half unit smaller than T for the state in (14-16). The relation (14-17) is, of course, valid for any state.

Since A is a low-energy excited state, it is likely to contain an appreciable fraction of the $d_{3/2}$ SQP state (from Fig. 6-6, in this case it contains virtually all of it), whence from (14-12) $\mathcal{S}(A)$ is an appreciable fraction of $1 - V_j^2$, which is typically $\lesssim 0.2$. From (14-17), then, $\theta_{A'G}{}^2$ is at least of the order of 0.01. While this may seem to be small, it is much larger than θ^2 for other states in the same energy region, as can be understood from the discussion of Sec. 13-5. The resonance due to excitation of the isobaric analog state is therefore easily seen in the experimental data. An example is shown in Fig. 14-10. As we see from that figure, the shape of the resonance in elastic scattering varies with angle. This is due to the change in the interference between resonance and potential scattering, and the resulting resonance shapes are different for each proton orbital angular momentum l. Measurements of these shapes therefore give determinations of l just as angular distributions give it in (d,p) reactions. In summary, then, measurements of elastic proton scattering through resonances due to isobaric analog states give the same information as (d,p) reactions and can be used for the same purpose.

However, in addition one can measure inelastic scattering through these resonances, as shown at the right in Fig. 14-10. Consider, as an example, the (p,p') reaction leading to the second excited state of Sn^{116} labeled K in Fig. 14-9. From (13-17) the cross section for this reaction is proportional to $\Gamma_p \Gamma_p'$, which is proportional to $\theta_{A'G}{}^2 \theta_{A'K}{}^2$. Since $\theta_{A'G}{}^2$ is known from the elastic scattering, these measurements give $\theta_{A'K}{}^2$, which gives $\theta_{AK}{}^2$ in (14-16) by relationships analogous to (14-17). This is the information that could be obtained from (d,p) reactions only by bombarding a target of Sn^{116} in the excited state K, but of course such an experiment is impossible.

Inelastic-scattering studies of this type can be extended to other excited states of Sn^{116}; when the final state is not 0^+, as for the first excited state labeled H in Fig. 14-9, $C_1\theta_{AH}$, $C_2\theta_{AH}$, $C_3\theta_{AH}$, and $C_4\theta_{AH}$ in (14-16) can in principle be determined with the help of angular distributions of the inelastically scattered protons. In principle, then, the entire wave function in (14-16) can be determined experimentally. In practice, however, angular distributions are not very different for different l, so terms like those in the second line of (14-16) are usually difficult to determine. This difficulty applies to all non-0^+ states, so the information obtained is mostly qualitative or semiquantitative. In a few cases, however, detailed quantitative wave functions have been worked out. For example, in a study of this type with a Ba^{138} target, the second excited state of Ba^{139} for which I^π is $\frac{3}{2}^-$ was found[1] to have the wave function

$$\begin{aligned}\psi(\tfrac{3}{2}^-) = \sqrt{0.3}\,\psi[\mathrm{Ba}^{138}(\mathrm{GS})]\,\psi(p_{3/2})_\nu + \sqrt{0.7}\,\psi[\mathrm{Ba}^{138}(2^+)]\psi(f_{7/2})_\nu \\ + \sqrt{0.0}\,\psi[\mathrm{Ba}^{138}(2^+)]\psi(p_{3/2})_\nu + \sqrt{0.0}\,\psi[\mathrm{Ba}^{138}(2^+)]\psi(p_{1/2})_\nu\end{aligned}$$

14-5 Multinucleon Transfer Reactions

After this diversion to a compound-nucleus reaction, let us return to the subject of direct reactions and consider multinucleon transfer reactions. The simplest reactions of this type are those in which two identical nucleons are transferred, as in (t,p), (p,t), or (He_3,n). Investigations of the structure of the triton reveal that the two neutrons in it are paired, with $S = 0$ well over 90 percent of the time. Hence, when the two neutrons are transferred in a (t,p) reaction, effectively a 0^+ particle is being transferred. If the target nucleus is even-even (and consequently 0^+), the I^π of the final state must then be given entirely by the transferred orbital angular momentum l. We therefore have simply

$$\begin{aligned} I &= l \\ \pi &= (-1)^l \end{aligned} \qquad (t,p),\ \text{even-even} \tag{14-18}$$

[1] S. A. A. Zaidi, P. Von Brentano, D. Rieck, and J. P. Wurm, *Phys. Lett.*, **19:** 45 (1965).

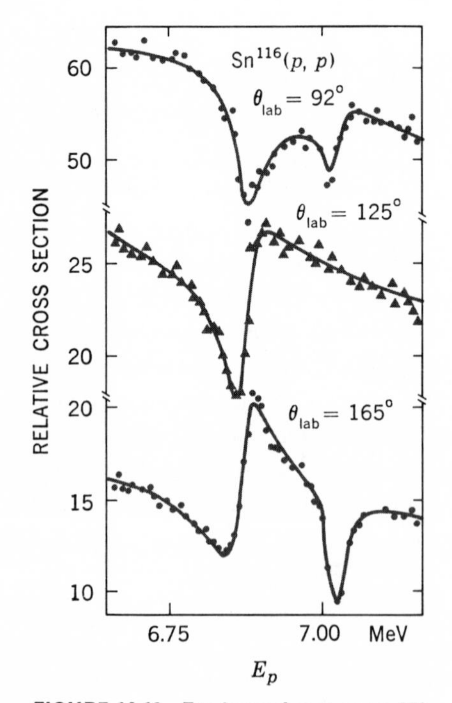

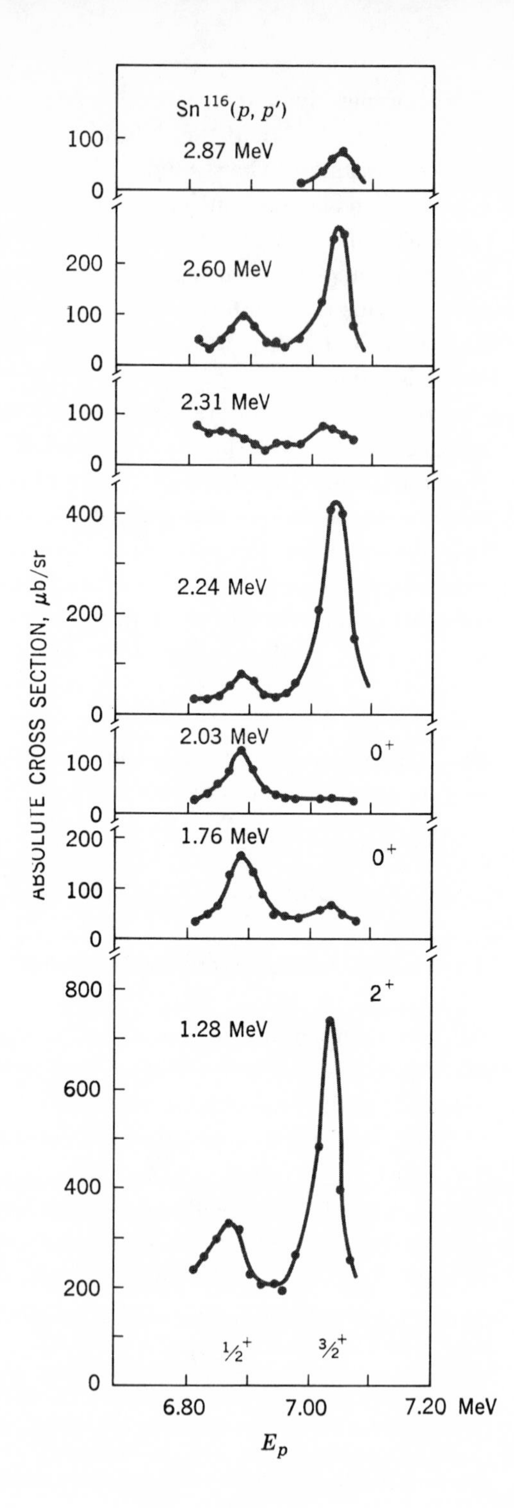

FIGURE 14-10 For legend see page 371.

This is simpler than the situation in single nucleon transfer, where, due to the spin of the nucleon, $I = l \pm \frac{1}{2}$. This was the cause of the difficulty in determining j ($= I$) discussed in Sec. 14-3.

Angular distributions from a typical study with (t,p) reactions are shown in Fig. 14-11. The locations of the first maxima as predicted by (14-4) are shown at the top, and it is seen that there is good agreement for the $l = 0$, 2, 3, and 4 cases where l is known through (14-18) from the known I^π of the states excited. Since there has been considerable experience with agreements of this nature, the location of the first maximum is widely used to determine the I^π of nuclear states. For example, the two states excited in the transitions whose angular distributions are unlabeled in Fig. 14-11 were assigned 1^- and 5^- (or possibly 6^+).

The simplest transitions for the calculations of spectroscopic factors $\mathcal{S}$ are those between ground states of even-even nuclei, as, for example, in $Sn^{118}(p,t)$-Sn^{116}(GS). The problem here is very similar to that in alpha-particle emission except that only a pair of neutrons is removed instead of pairs of both neutrons and protons, as in the latter case. The result here is therefore analogous to (10-37) as

$$\mathcal{S}(p,t) \propto \left[\sum_j (j + \tfrac{1}{2}) V_j (1 - V_j^2)^{1/2}\right]^2 \qquad \text{GS} \rightarrow \text{GS} \tag{14-19}$$

where the j are all the neutron orbits involved. Here again we have a coherence effect, causing transitions between ground states to be very strong, much stronger than if only a single orbit were available for the neutron pair to be transferred into. As explained in connection with (10-36), this occurs only when the states involved have a special relationship with one another, as for transitions from a ground state to another ground state, to a collective 2^+ or 3^- state, or to a so-called *pairing-vibration state*, one which differs from the ground state by a special combination of excited pairs. All these states are strongly excited in (p,t) and (t,p) reactions, but essentially all other states are rather weakly excited because the expression for $\mathcal{S}$ in those transitions includes terms with both plus and minus signs so a lot of cancellation occurs.

Equation (14-6), which is valid for single-nucleon transfer reactions, implies that the angular distributions are independent of nuclear-structure effects. The processes involved in two-nucleon transfers, however, are sufficiently more complicated to make the simplification of (14-6) much less valid. Aside from the

---

FIGURE 14-10 Cross section vs. energy for proton-induced reactions on Sn^{116}. States at 6.87 and 7.03 MeV are isobaric analogs of the ground and first excited states of Sn^{117} (see Fig. 14-9). Note that the shape of the resonance varies with angle in elastic scattering; for $l = 2$ the resonance disappears at 125°. For (p,p'), the shape of the resonance is given by (13-17). The various curves are for transitions to the various states of Sn^{116} whose excitation energies and I^π are attached. The 1.28- and 1.76-MeV states are those labeled H and K in Fig. 14-9. [(p,p) *data from P. Richard et al., Phys. Rev.,* **145: 971 (1966);** (p,p') *data from Phys. Rev.,* **161: 1208 (1967)].**

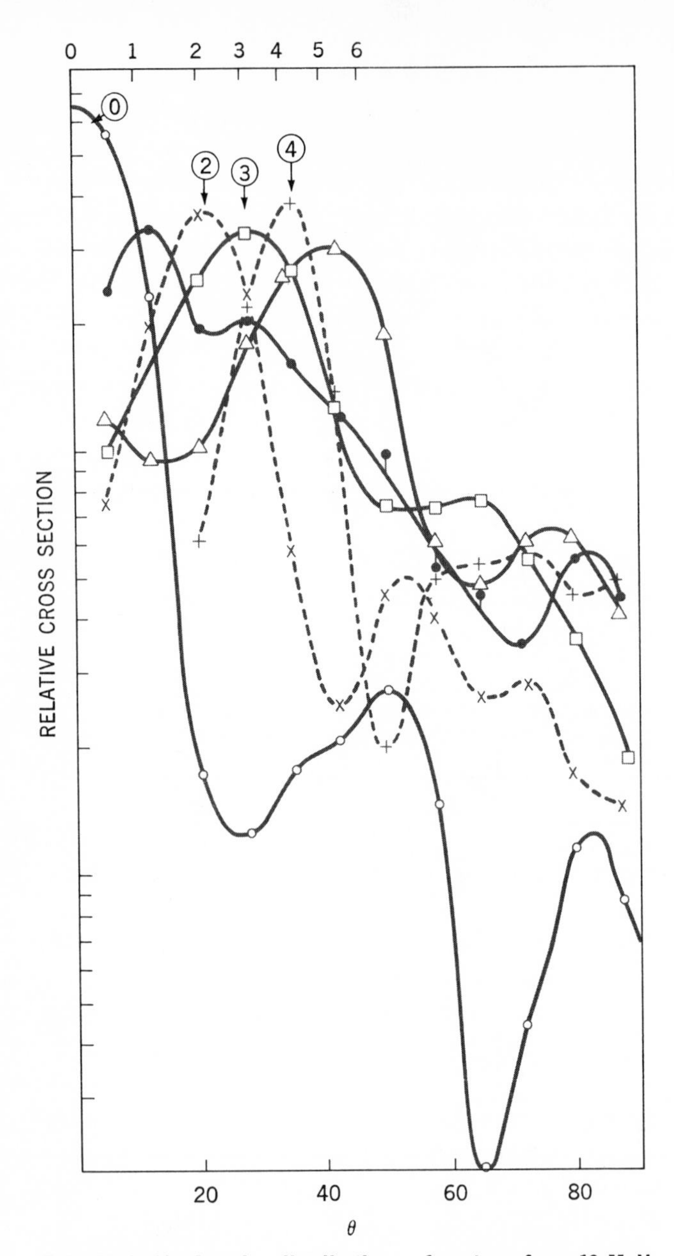

FIGURE 14-11 Angular distributions of protons from 13-MeV-triton-induced (t,p) reactions on Fe^{54} and Fe^{56} leading to selected states of Fe^{56} and Fe^{58}, respectively. Encircled figures are l values in cases where these are known. Shown at the top are angular positions predicted by (14-4) for the first maximum in the angular distribution for each l. [*From Phys. Rev.*, **157, 1033 (1967); 146: 748 (1966).**]

position of the first maximum, the angular distribution is appreciably sensitive to the details of nuclear structure. In principle this can be an advantage; more information can be obtained about nuclear structure here than in single-nucleon transfer. However, the process of analyzing the data to obtain this information is very slow and difficult, and as yet only a few transitions are well understood.

According to (14-18), only 0^+, 1^-, 2^+, 3^-, etc., states (these are called *natural-parity states*) can be excited in (t,p) and (p,t) reactions, but this is not the case in (He^3,p) or (p,He^3), where the total spin of the transferred neutron and proton may be either 0 or 1. We therefore have, for an even-even target nucleus,

$$\begin{aligned} I &= l, l \pm 1 \\ \pi &= (-1)^l \end{aligned} \qquad (\text{He},p), \text{ even-even} \tag{14-20}$$

There is therefore an uncertainty in I assignments made with this reaction.

FIGURE 14-12 Angular distribution of alpha particles from the 22-MeV-proton-induced (p,α) reaction leading to a state with $Q = 1.08$ MeV. The curves are from various DWBA calculations for angular-momentum transfer $l = 1, j = \frac{1}{2}$. [*From C. B. Fulmer and J. B. Ball, Phys. Rev.*, **140, B330 (1965).]**

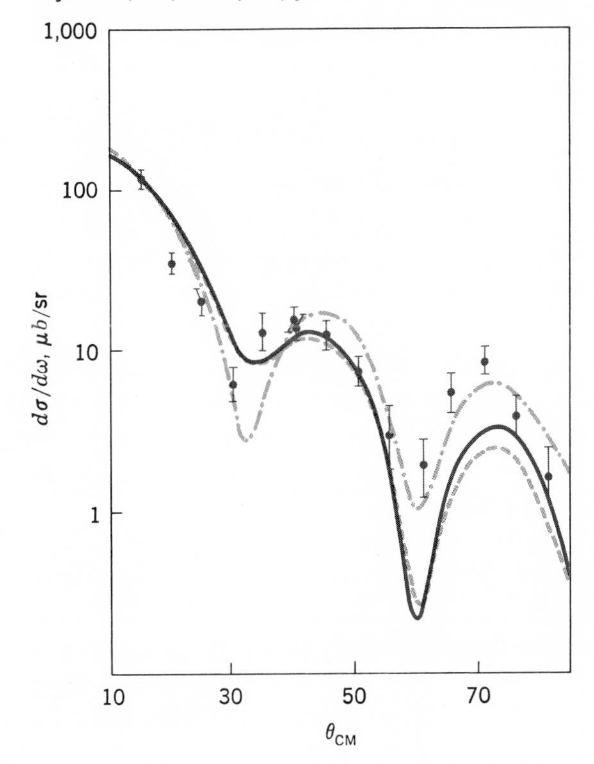

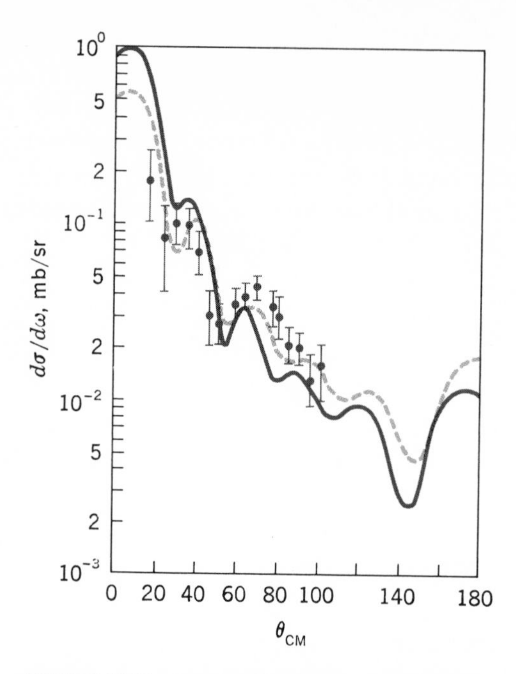

FIGURE 14-13 Angular distribution of Li^6 nuclei from F^{19} (d,Li^6) N^{15}(GS) induced by 15-MeV deuterons. Curves are DWBA calculations for $l = 1$. [*From W. W. Daehnick and L. J. Denes, Phys. Rev.,* **136: B1325 (1964).**]

The situation is somewhat more favorable in (α,d) or (d,α) reactions since the transferred neutron and proton must be in a $S = 1$ state to cancel the $S = 1$ of the deuteron when they add to give $S = 0$ for the alpha particle. For an even-even target nucleus, then **I** is the vector sum of l and 1.

$$\begin{aligned} \mathbf{I} &= \mathbf{l} + \mathbf{1} \\ \pi &= (-1)^l \end{aligned} \qquad (\alpha,d), \text{ even-even} \tag{14-21}$$

We see immediately that a 0^+ state cannot be excited, so a state excited by $l = 0$ in (He^3,p) and not excited in (α,d) is almost surely 0^+. If $l = 0$ is found in an (α,d) reaction, the final state must be 1^+. If an angular distribution shows a combination of $l = 1$ and $l = 3$ transfers, the state excited must be 2^-. If only $l = 1$ is observed, it could be 1^- or 2^-, but it is more probably 1^- because if it were 2^-, a mixture with $l = 3$ would be likely. A similar situation occurs for a pure $l = 2$ angular distribution. From (14-21), it could be 1^+, 2^+, or 3^+, but if it were 1^+, there would probably be a mixture of $l = 0$, and if it were 3^+, there would probably be a mixture of $l = 4$, so it is most probably 2^+. When mixtures are found, the assignment is, of course, definite. For example, an $l = 2$ plus $l = 4$ mixture definitely indicates 3^+.

Examples of three- and four-nucleon transfer reactions are shown in Figs. 14-12 and 14-13. As the number of transferred nucleons increases, angular distributions become increasingly sensitive to nuclear-structure details, but it is still possible to determine l values. These give the parity and a range of possible I values for the final state.

14-6 Other Types of Direct Reactions

In all cases where nucleon transfer is possible, it is the dominant process in direct reactions, but there are reactions in which it is not possible, as in inelastic scattering and (p,n) reactions. These are the cases described in examples 3 and 4 near the beginning of Sec. 14-1.

In inelastic scattering of alpha particles, (α,α'), on an even-even target, both participants are 0^+, whence conservations of angular momentum and parity require

$$\begin{aligned} I &= l \\ \pi &= (-1)^l \end{aligned} \qquad (\alpha,\alpha'),\ \text{even-even} \tag{14-22}$$

A special simplification occurs in these reactions due to the fact that the interactions take place almost exclusively at the nuclear surface; alpha particles have little chance of retaining their identity once inside the nucleus. Hence the approximation leading to (14-4) assumes validity, and the locations of maxima are determined by the maxima of $[j_l(qR')]^2$. Moreover, $ƛ = \hbar/\sqrt{2ME}$ is especially small for alpha particles, so at energies like 30 to 40 MeV, where these experiments are usually done, $ƛ/R'$ is typically about 0.04. Using this in (14-5), we find that the first maximum even for large l occurs below about 20°, whence over most of the angular range we may use the large qR' approximation for $[j_l(qR')]^2$. The angular distribution is therefore

$$I(\theta) \propto [j_l(qR')]^2 \simeq (qR')^{-2} \sin^2\left(qR' - \frac{l\pi}{2}\right) \tag{14-23}$$

We see that the oscillations with angle are in the same phase for all odd l, and these are 180° out of phase with the oscillations for all even l. It can be shown that the angular oscillations in elastic scattering are in phase with those for odd l. An example of this simple behavior is shown in Fig. 14-14. As a result, (α,α') reactions are excellent for determining parities but not of much use for determining I.

Inelastic proton scattering is somewhat more complicated; since the reaction is not confined to the nuclear surface, angular distributions are not sharply varying as in Fig. 14-14. Examples of DWBA fits to (p,p') data are shown in Fig. 14-15.

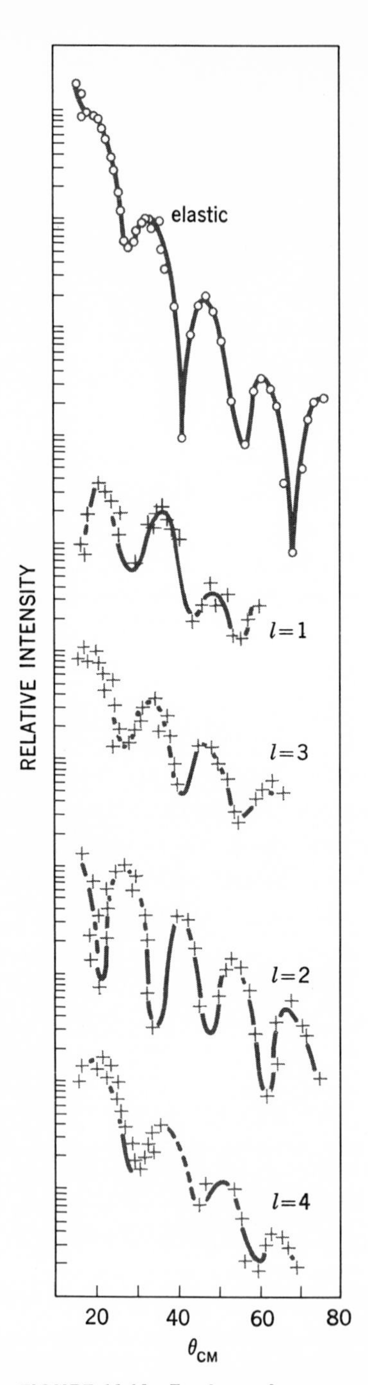

FIGURE 14-14 For legend see page 377.

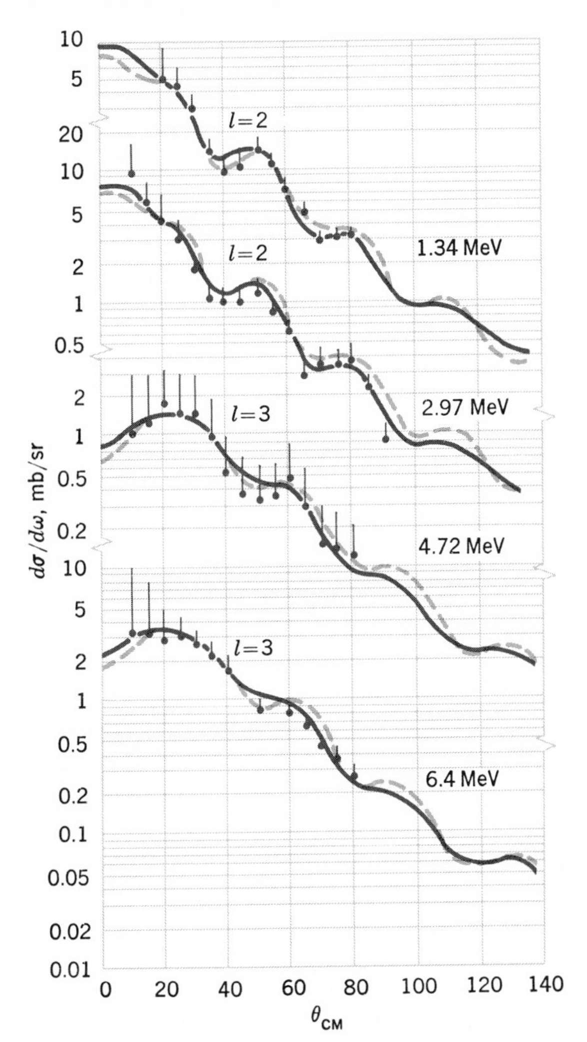

FIGURE 14-15 **Angular distribution of protons inelastically scattered from Ni^{60} leading to states whose excitation energies are attached. Curves are DWBA calculations for the l values shown. The incident proton energy was 40 MeV.** [*From M. Fricke and G. R. Satchler, Phys. Rev.*, **139: B567 (1965).**]

FIGURE 14-14 **Angular distributions from 31-MeV-alpha-particle-induced elastic scattering and (α,α') reactions on $Ca^{40,42}$. Note that the elastic $l = 1$, and $l = 3$ angular distributions are in phase while the $l = 2$ and $l = 4$ angular distributions are 180° out of phase with them.** *(From A. M. Bernstein, in M. Baranger and E. Vogt (eds.), "Advances in Physics,"* **vol. III,** *Plenum Press; by permission.)*

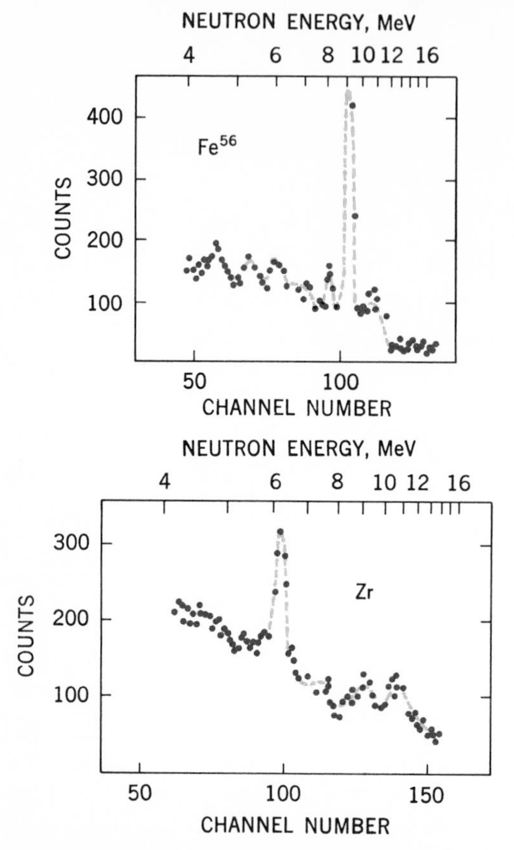

FIGURE 14-16 Energy distribution of neutrons from 18-MeV-proton-induced (p,n) reactions. Peaks are due to excitation of isobaric analog states. [*From J. D. Anderson, C. Wong, and J. W. McClure, Phys. Rev.,* **129: 2718 (1963).**]

The tendency of the oscillations in the angular distributions for even and odd l to be 180° out of phase is still evident.

The dominant process in inelastic scattering is excitation of collective vibrations and rotations. In the former case, (14-22) requires $l = \lambda$, where λ is the vibration mode as defined in Sec. 5-9. The two most strongly excited states in any reaction are almost invariably the $\lambda = 2$ (2^+) and $\lambda = 3$ (3^-) collective vibrations. If a vibrational model is used in the DWBA calculations, the measured cross section can be interpreted to give the amplitude of the vibration, i.e., the maximum value of β from Sec. 6-7 during the course of the vibration. As we shall see in the next section, there is an easier way to study the $\lambda = 2$ vibration, but the discovery of the $\lambda = 3$ vibration, its location in various nuclei, and measurements of its amplitude were all done with inelastic-scattering reactions.

In studies of rotational states of even-even spheroidal nuclei by inelastic scattering, the angular distribution can be analyzed to determine the shape of the nucleus in its ground state. In particular, if the shapes of these nuclei are described by giving the distance from the center to the surface $r(\theta)$ as

$$r(\theta) = r_0[1 + \beta_2 P_{20}(\theta) + \beta_4 P_{40}(\theta)]$$

where $P_{\lambda\mu}$ are the associated Legendre polynomials given in Table 2-2, the values of β_2 and β_4 can be determined. Note that odd values of λ are not included in the expression for $r(\theta)$; if they were of appreciable magnitude, the angular momentum of successive rotational states would be 0^+, 1^-, 2^+, 3^-, . . . rather than 0^+, 2^+, 4^+, 6^+, . . . as we know them to be from Fig. 6-17.

Some attempts have been made to study transitions between single-particle

FIGURE 14-17 Coulomb-energy differences in energies of isobaric analog states. Circles are from data of the type shown in Figs. 6-12*b* and 6-13, and squares are from (p,n) reactions of the type shown in Fig. 14-16. Data of this type are also available from the energies of the resonances in Fig. 14-10 and similar studies. *(From J. D. Anderson, Argonne Natl. Lab. Rept. 6878, 1964.)*

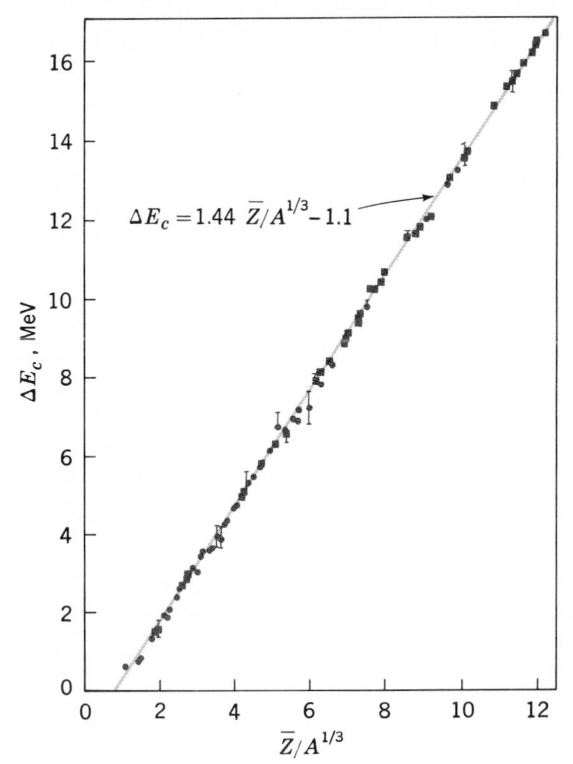

states with direct-reaction inelastic scattering, but the angular distributions obtained from DWBA calculations with this model do not fit the data and the values of $\mathcal{S}$ obtained are too large. The mechanism in these reactions so strongly favors collective motions that even in transitions between predominantly single-particle states the small amounts of collective configurations in the wave functions have important effects.

The dominant process in (p,n) reactions is the excitation of isobaric analog states. Examples of this are shown in Fig. 14-16. The angular distributions can be fit with DWBA calculations. Measurements of this type give the energies of the isobaric analog states, and these determine the coulomb energy by which isobaric analog states are shifted, as in Fig. 6-12*b*. A plot of the results of these studies is shown in Fig. 14-17.

14-7 Coulomb Excitation

When an electrically charged particle passes near a nucleus on a path described in classical treatments of Rutherford scattering, the nucleus experiences a changing electric field similar to what it would experience if it were struck by a gamma ray. Unlike the latter situation, however, the time variation of the electric field is not purely sinusoidal, so it may be Fourier analyzed into a sum of contributions from a range of frequencies extending uniformly from zero up to roughly ω_c, given by

$$\omega_c = \frac{v_i}{a_c}$$

where v_i is the velocity of the incident particle and a_c is one-half the distance of closest approach in 180° Rutherford scattering,

$$a_c = \frac{Z_1 Z_2 e^2}{M v_i^2}$$

A nuclear state with energy $\hbar\omega$ less than $\hbar\omega_c$ can therefore be excited in a manner similar to its excitation by a gamma ray of energy $\hbar\omega$, as discussed in Sec. 13-10. This process is called *coulomb excitation*. Another way to explain it is to use the picture we developed in Sec. 3-9, where electric forces are thought of as arising from a constant exchange of photons just as the nuclear forces arise from meson exchanges. When one of these photons is absorbed, exciting the nucleus to a higher-energy state, we have coulomb excitation.

Since coulomb excitation is effectively a reaction induced by gamma rays, a state of $I = l$, $\pi = (-1)^l$ must be effectively excited by $E - l$ gamma rays and the cross section is proportional to Q_l^2. In accordance with the discussion connected with (12-8), Q_l^2 is by far the largest for transitions between the ground states and the one-phonon vibrational states of spherical nuclei or the lowest-energy ($I = 2$) rotational states of spheroidal nuclei. These states are therefore by far the most strongly excited. In some cases where the interaction is particularly strong, as when the incident particle has a large electric charge, more than one excitation can be given to the same nucleus, raising it to successively higher vibrational or rotational states. This is called *multiple coulomb excitation*. From the discussion connected with (12-8) we see that transitions between adjacent states in vibrational and rotational bands have large Q_l.

The theory of coulomb excitation is rather involved and specialized, so we shall not treat it in detail here.[1] However, the fact that the process is so simple and well understood has made it extremely useful in a wide range of studies of collective states. It has been used to locate the 2^+ collective states in virtually all nuclei and to determine Q_2 in transitions between them and the ground states. These may be used directly to determine the half-lives of these states through (12-3) and indirectly to determine the amplitudes of the vibration. The high-energy states of the rotational bands shown in Fig. 6-17 were found in multiple-coulomb-excitation experiments. The measurements of the static electric quadrupole moments of collective states mentioned at the end of Sec. 7-5 were obtained from second-order differences between coulomb excitation by two types of particles.

Problems

14-1 Estimate the angles of the first maxima in the angular distributions for various l from (14-5) for 40 MeV-proton-induced (p,p') reactions on nickel. Compare the results with Fig. 14-15.

14-2 Do the calculation of Prob. 14-1 for 40-MeV-alpha-particle-induced (α,α') reactions.

14-3 With (14-4), calculate the angle expected for the second maxima in the $l = 0$ and $l = 1$ angular distributions of Fig. 14-3.

14-4 If the $d_{5/2}$ state is half full in Pd^{108}, from Fig. 6-7 calculate $\mathcal{S}$ for the $Pd^{108}(d,p)Pd^{109}$(GS) reaction.

14-5 Plot the data for $d_{5/2}$ in Sn^{119} from Fig. 6-6 as in Fig. 14-6. Locate the center of gravity E_j and estimate a value of W. Plot the latter in Fig. 13-5. Also plot the values of W for $s_{1/2}$, $d_{3/2}$, and $h_{11/2}$ in Fig. 13-5.

[1] A detailed treatment will be found in K. Alder et al., *Rev. Mod. Phys.*, **28**: 432 (1956).

14-6 Calculate the slopes expected for the lines in Fig. 14-7 connecting the $3s_{1/2}$ states in the approximation that the potential is a square well with radius given by R in (4-2) and assuming $\psi = 0$ at $r = R$. Compare with the slope in Fig. 14-7.

14-7 If the $d_{5/2}$ and $d_{3/2}$ states are three-fourths and one-fourth full, respectively, in Sn^{116}, how sensitive is (14-15) for distinguishing between $3/2^+$ and $5/2^+$ states?

14-8 From Fig. 14-10 discuss qualitatively the wave function for the lowest-energy $3/2^+$ state in Sn^{117}. Compare this with the information in Fig. 6-6, which was obtained from (d,p) reactions.

14-9 What is the l value in the reaction $Sn^{117}(t,p)Sn^{119}$(GS)? In the reaction $Sn^{119}(t,p)Sn^{121}$(GS)? Figure 6-6 may be useful here.

14-10 In a (p,n) reaction induced by 20-MeV protons, what is the energy of the peak in the energy distribution of neutrons from Sn^{117}?

14-11 Calculate ΔE_c for Sn^{117} from the location of the ground-state isobaric-analog-state resonance in Fig. 14-10 and masses in Table A-3. Plot this point in Fig. 14-17.

14-12 What is the minimum energy proton needed to excite the lowest-energy 2^+ states in Cd^{114} and Pt^{192} (see Fig. 5-11) by coulomb excitation? At what proton energies do nuclear reactions become important in these nuclei (see Fig. 13-4)?

Further Reading

See General References following the Appendix.

Ajzenberg-Selove, F.: "Nuclear Spectroscopy," Academic, New York, 1960.

Butler, S. T.: "Nuclear Stripping Reactions," Wiley, New York, 1957.

Endt, P. M., and M. Demeur: "Nuclear Reactions," Interscience, New York, 1959.

Kikuchi, K., and M. Kawai: "Nuclear Matter and Nuclear Reactions," North-Holland, Amsterdam, 1968.

Mayer, M. G., and J. H. D. Jensen: "Elementary Theory of Nuclear Shell Structure," Wiley, New York, 1955.

Nuclear Data, periodical published by Academic Press, New York.

Siegbahn, K.: "Alpha, Beta, and Gamma Ray Spectroscopy," North-Holland, Amsterdam, 1965.

Chapter 15

Applications of Nuclear Physics

In this chapter we consider some of the ways in which the nucleus affects our lives and the world around us. Its most important role is an indirect one as the center of the atom; its electric charge determines the number of electrons it attracts to form atoms, which in turn determines all chemical behavior and physical properties of materials. However, there are several areas in which the structure of the nucleus, its decay properties, and the reactions it undergoes have a direct and controlling influence. These include the production of energy in stars, the origin of the elements (*nucleosynthesis*), the development of energy sources to power our technological civilization, and a wide variety of uses for radioactivity. These form the subject of this chapter.

15-1 Applications of Radioactivity

The earliest uses of radioactivity were for dating, taking advantage of the decay laws derived in Sec. 8-5. Ages of rocks, and hence the age of the earth, were determined from the ratio of uranium to lead (into which uranium eventually decays) in them. Numerous other dating techniques have been developed for other age ranges, down to thousands of years as in the case of radiocarbon dating (see Sec. 13-8) and even to a few years as in the determination of the age of water accumulations by their content of tritium (half-life = 12 years).

A broad group of applications takes advantage of the fantastic sensitivity with which radioactivity can be detected; the decay of a single atom can be registered by an instrument at a remote location outside a container wall, whereas other methods of detection, e.g., chemical, require the presence of billions of atoms and generally involve disturbance of the material. Making use of this high sensitivity, elements can be *tagged* by mixing in a small amount of a radioactive isotope of the same element and then followed through various chemical, biological, and physical processes. This *tracer technique* has proved to be an important research tool in various phases of medicine, chemistry, engineering, agriculture, metallurgy, hydrology, oceanography, ecology, zoology, etc., and has found practical application in medical diagnostics, criminology, leak detection, fuel transport, and many other technological problems.

Let us cite a few examples of the use of tracers. The wear on piston rings is rapidly and sensitively measured by incorporating radioactive iron isotopes into them and measuring the radioactivity accumulating in the lubricating oil. This has been done as a function of automobile speed and engine warm-up time and for different lubricating oils and additives, and much useful information has been obtained on what oils and additives are best for various situations. Analogous studies have been done on automobile tire wear; some P^{32} was incorporated into the outer tread, and the decrease in count rate was measured with a radiation detector mounted under the fender. In this way studies were made of the effects on tire wear of load, speed, road type, etc. By incorporating a radioactive tracer into the lubricating oil, the degree of oil burning in diesel engines is checked by measuring the radioactivity in the exhaust gases. Different types of oil filters are tested by mixing an antimony compound tagged with Sb^{124} into the oil and checking the activity in the oil after it has passed through the filter.

The mercury used in fluorescent lamps constitutes a health hazard in plants manufacturing them; a small amount of Hg^{197} is mixed with the mercury, and the amount of mercury vapor in the air is thereby continually monitored by passing air samples over a Geiger counter. Similar methods are used for keeping track of other dangerous air contaminants like chlorine (with Cl^{36}) and ammonia (with H^3). The efficiency of washing machines is checked by soaking rags in oil containing P^{32} and measuring the radioactivity after washing. The same method was used by a testing service to compare the efficiency of various laundry detergents.

An important agricultural problem is getting pesticides into the leaves of trees to protect them from being eaten by insect larvae. By labeling the pesticide with S^{35} and measuring the activity inside leaves at later times, it was found that some methods of application were preferable to others. For example, application as a miscible oil is preferable to use of a powder dissolved in water, and introducing the pesticide through the roots concentrates it in the young leaves; exposure to light was found to increase the speed of assimilation. Studies of the spread of radioactively labeled fertilizer through soil have shown that the migration rate is much higher than had been expected; this led to the idea of spreading fertilizer for cereals in lines rather than by scattering it uniformly, and thereby saved almost half of the cost. The habits of insect pests have been studied by feeding them radioactive materials and following their movements with radiation detectors, thereby developing means of destroying them.

Perhaps the most important applications of tracers are in medical diagnostics. Iodine taken into the body concentrates strongly in the thyroid gland, and an overactive thyroid holds it much longer than a normal thyroid does. If a patient

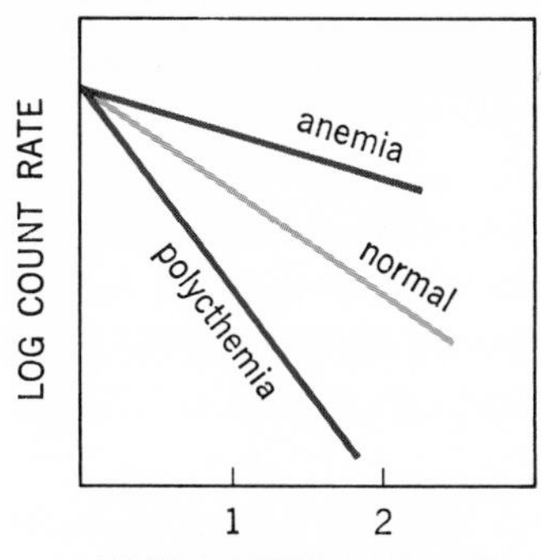

FIGURE 15-1 Count rate from blood samples as a function of time after injection of a radioactive iron isotope. The three curves are for patients with anemia and polycythemia and for a normal patient. *(From Dorothy Parker, "Isotopes in Action,"* **The Macmillan Company, 1963;** *by permission.)*

drinks a glass of water containing I^{131} (as NaI), the count rate in a detector placed near his neck 24 hr later immediately determines whether his thyroid is normal, overactive, or underactive. A similar technique is used to diagnose anemia; as shown in Fig. 15-1, an anemic patient retains iron in the blood longer than a normal patient so by injecting a radioactive iron isotope and later withdrawing blood and measuring its activity, anemia or its converse, polycythemia, can be detected. By placing a detector inside a thick lead shield which gamma rays can enter easily only through a long, narrow opening, the direction from which gamma rays are coming can be determined. This device is moved back and forth to scan an area of a patient's body to determine the location from which gamma rays are being emitted. For example, this method is used to determine the size and shape of the thyroid gland when I^{131} is concentrated in it. A more important application is to locate brain tumors; albumin (a substance in the blood), arsenic, and other substances tend to concentrate in brain tumors, so by labeling one of these with radioactive isotopes, the size and location of a brain tumor can be accurately determined, allowing surgery for its removal to be carried out with precision. The flow of blood in a patient can be studied by injecting Na^{24} as a tracer and detecting its radiations at various parts of the body; this method is used to locate blockages and to diagnose conditions characterized by a sluggish flow. Pernicious anemia is caused by the failure of the body to absorb vitamin B_{12}; if the vitamin is tagged with Co^{58} or Co^{60}, its path in passing through the body can be followed with a radiation detector to locate the trouble.

Another use for the high detection sensitivity of radioactive atoms is in *activation analysis*, which allows one to determine the quantity of a trace element in a substance. The substance is bombarded with particles, e.g., neutrons, which can induce nuclear reactions in the element under study, and then the radioactive products of those reactions are detected. Radiochemical separations to remove activities induced in the principal constituents of the material are often useful in this work. In this way, concentrations of a few parts per billion of an element can be accurately measured.

Another group of applications takes advantage of the availability of radioactive sources and of detectors to study properties of materials through their interaction with radiations. These include measuring thicknesses of everything from the thinnest metal foils to thick coal seams by the fraction of the gamma rays or electrons from a beta-ray emitter they stop, or by the amount by which they reduce the energy of alpha particles. They include measurements of densities of materials by similar methods, the determination of the levels to which nontransparent containers are filled, and prospecting for oil by studying the scattering of neutrons and gamma rays from soil structures. They include inspection for imperfections inside materials by noting the variations in their absorption of gamma rays; this was formerly done exclusively with x-rays but *gamma radiography* has the advantages of being a much smaller, simpler, cheaper, and more mobile operation.

Let us consider some specific examples of these applications. Thickness gauges based on measuring the attenuation of nuclear radiations are used routinely in the manufacture of paper, rubber, plastics, and metal sheets. They often incorporate methods for correcting deviations from the proper thickness in the rolling apparatus. If the count rate in the thickness gauge gets too high, indicating that the sheet is too thin, a signal is sent to the rollers which automatically increases the gap between them. In many cases, an entire process consisting of several passes through different types of rollers is completely and automatically controlled by instruments of this type located between each rolling stage. By use of a very low energy beta-ray emitter, very thin films can be measured and controlled. The thickness of ink on printing rollers is kept between 3 and 7 μm in this way by use of Ni^{63}. When the ink gets too thick, the radiation is more strongly absorbed, causing a decrease in count rate, and this sends a signal which reduces the ink supply; conversely, an increase in count rate causes the ink supply to be increased.

Ketchup manufacturers store their product in large vats, but the properties of ketchup are such that it is not easy to tell the level to which the vat is filled. By placing a radioactive source and a detector at the same level on opposite sides of the vat and moving them up and down together, one observes a drastic change in count rate when the level of the ketchup is passed since it absorbs gamma rays much more strongly than does air. A similar scheme is used in determining the height to which grain storage bins are filled.

Radiations from radioactive nuclei are used directly in many industrial processes, taking advantage of their ability to knock electrons out of their location in atoms, molecules, and crystals and thereby to induce chemical, physical, or biological changes. For example, they induce polymerization in some chemicals, increase the hardness or toughness of many materials, change the color of diamonds, produce ozone, and disperse static electricity by ionizing air and thereby making it electrically conducting. When polyethylene is irradiated with gamma rays, its flow temperature can be increased from 120 to over 200°C while its tensile strength is increased by 60 percent. The former improvement allows it to go through sterilization procedures necessary in medical applications. By irradiating rubber tires, their wear life can be doubled while improving their smoothness in running. Air friction on airplane wings causes charge to accumulate, which sometimes encourages icing; this problem can be overcome by mounting radioactive sources on the edge of the wings. The effective heights of lightning rods are substantially increased by mounting small radioactive sources on their top points to dissipate static charges that accumulate there. The accuracy of high-precision balances may be affected by charge accumulations; at least one manufacturer fits his balances with a small radioactive ring to dissipate these charges.

By changing molecules, radiation produces genetic mutations. Taking advantage of this, geneticists have bred new types of insects in a much shorter time than would be possible with normal mutation rates. On the other hand, excessive mutations pose a threat to human beings since the vast majority of mutations introduce undesirable characteristics in offspring. It was this problem that led to the international agreement to end testing of nuclear bombs. Enough radiation causes sexual sterility; by inducing sterility in a large number of males, populations of harmful insects have been vastly decreased. This method is especially effective for species in which females mate only once.

If enough molecules are changed, a cell will die, and if enough cells are killed, an organ will cease to function, causing death. Large doses of radiation can therefore be fatal to human beings, which is the reason for using *fallout shelters* after a nuclear bomb attack. On the other hand, cancerous cells are more susceptible to damage from radiation than normal cells, so by proper administration of radiation, cancer can sometimes be controlled or cured. By ingesting I^{131}, which, as noted previously, concentrates in the thyroid, an enlarged thyroid can be reduced to normal size. This was previously accomplished by major surgery. The use of radiation to kill microorganisms has found application in the preservation of food as a substitute for refrigeration—irradiated potatoes have been kept 18 months without deterioration or sprouting—and in the sterilization of surgical bandages and instruments. Disposable needles for innoculations, for example, are ordinarily sterilized by exposure to radioactive sources.

The energy released in radioactivity has been used to produce electric power, either directly (but in very small quantities) by using the electric charge emitted, or by absorbing the energy as heat, which is then converted to electric power. Power sources exceeding 100 W have been produced with the latter method; they are useful, for example, in unmanned Arctic meteorological stations and in unmanned orbiting satellites. The energy from radioactivity can be converted directly into light by exposing luminescent materials to it, as is done in luminous watch dials. Lamps that can be seen a quarter mile away on a dark night have been produced in this way.

Radioactivity is used in mineral prospecting, not only for uranium and thorium, whose radiations can be detected directly, but for beryllium, whose nucleus has the lowest energy threshold for (γ,n) reactions, whence it is the only material in which neutrons can be produced by irradiation with gamma rays from many radioactive sources.

All in all, radioactivity is playing a very important role in our technology. It has saved agriculture and industry tens of billions of dollars and has saved thousands of lives through its medical applications. And its use is still in a rapidly growing stage.

15-2 Energy Production and Thermonuclear Reactions

Historically, energy has been produced on earth by molecular reactions—better known as chemical reactions—such as the burning of wood, coal, gas, and petroleum products and the utilization of food in the bodies of animals. However, it was realized long ago that chemical reactions could not be the source of the energy radiated by stars such as our sun; if the sun were made entirely of coal and oxygen, it could not produce energy at its present rate for more than about a thousand years, whereas it is well known that it has already done so for several billion years. The only other reaction processes that produce energy are nuclear reactions, and since nuclear energies are typically a million times larger than atomic energies—MeV vs. eV—it was apparent that nuclear reactions must be responsible for energy production in stars.

In order to produce large quantities of energy, it is necessary to have reactions between matter in bulk, as in air-fuel mixtures; but in bulk matter nuclei are kept far apart by their mutual coulomb repulsion. This problem has an analog in chemical reactions since there is an electrical repulsion between two *molecules* as they come together. In order to overcome this repulsion, it is necessary for the interacting particles to have kinetic energies of the order of the maximum potential energy arising from the repulsion, which we have referred to as the height of the barrier. This is typically a few electron volts or less in the case of molecules. By virtue of their thermal motion, all particles have some kinetic energy, as can be calculated from the Maxwell distribution for velocities

$$p(v)\,dv \propto v^2 \exp\left(\frac{-\frac{1}{2}Mv^2}{kT}\right) dv \tag{15-1}$$

where $p(v)\,dv$ is the probability for a velocity between v and $v + dv$ and k is the Boltzmann constant,

$$k = \frac{1 \text{ eV}}{1.16 \times 10^4 \text{ °K}} \tag{15-2}$$

At room temperature and at furnace temperatures, kT is about 0.025 and 0.1 eV, respectively, whence there is a reasonable chance for reactions to occur between molecules in the high-energy tail of the Maxwell distribution by taking advantage of the quantum-mechanical barrier penetration. We therefore have chemical reactions between matter in bulk, especially at elevated temperatures.

For nuclear reactions, on the other hand, the problem is much more serious. In the typical case calculated near the end of Sec. 10-4, the barrier height was 9.6 MeV, and the barrier-penetration probability even for a 1-MeV proton was 10^{-13}. That was for a proton interacting with a medium-mass nucleus, but even for two protons to approach within the distance at which the nuclear attractive force is equal to the coulomb repulsive force, which works out to be about 7 F, the coulomb barrier height is 0.2 MeV, and the barrier penetrability from (10-27) is about 10^{-3} at 0.01 MeV and 10^{-13} at 0.001 MeV (1,000 eV). There can clearly be no nuclear reactions at earthly temperatures. On the other hand, temperatures near the center of stars and in atomic bomb explosions are sufficiently high to allow these so-called *thermonuclear* reactions to take place, and there is some hope of achieving these temperatures in a controlled manner on earth. Before discussing these matters, let us treat thermonuclear reactions in more detail.

Consider a homogeneous mixture of two gases containing n_1 and n_2 particles per unit volume of the two types moving randomly with speed v, and let us say that a particle of gas 1 can have a reaction with a particle of gas 2 with a cross section σ. From the definition of cross section, the probability per unit distance travelled for a given particle of gas 1 to have a reaction is $n_2\sigma$. The distance it travels per unit time is v, whence its probability per unit time for a reaction is $n_2\sigma v$. Since there are n_1 particles per unit volume like the one we are considering, the total reaction rate per unit volume $\mathcal{R}$ is

$$\mathcal{R} = n_1 n_2 \sigma v \tag{15-3}$$

In accordance with the Breit-Wigner formula (13-17), the cross section for a (p,q) reaction, where p and q are any two particles, is

$$\sigma = \pi \lambda\!\!\!^{-2} \frac{\Gamma_p \Gamma_q}{(E - E_r)^2 + \frac{1}{4}\Gamma^2} \tag{15-4}$$

If p is a charged particle, we see from (13-11a) and (10-22) that Γ_p includes a coulomb barrier penetrability $\mathcal{P}$. If we assume as we did for thermal neutrons in Sec. 13-8 that the lowest-energy resonance is at a much higher energy than the region we are considering (in very light nuclei, spacings between resonances are typically tens or hundreds of keV), the energy variation in (15-4) at very low energies is dominated by $\lambda\!\!\!^{-2}$, which is proportional to $1/E$, and by $\mathcal{P}$, whence

$$\sigma \simeq \frac{S_0 \mathcal{P}}{E} \tag{15-5}$$

where S_0, which includes all the other factors in (15-4), is relatively energy-independent. In the expression (10-27a) for the penetrability, we note that at very low energies $x \rightarrow 0$ whence $f(x) \rightarrow \pi/2$, so $\mathcal{P}$ becomes

$$\mathcal{P} = \exp\,(-\alpha E^{-1/2}) \tag{15-6}$$

where

$$\begin{aligned} \alpha &= 0.63 \frac{\pi}{2} zZ \left(\frac{M}{M_p}\right)^{1/2} (\mathrm{MeV})^{1/2} \\ &= 31.3 Z_1 Z_2 \left(\frac{M_1 M_2}{M_1 + M_2}\right)^{1/2} (\mathrm{keV})^{1/2} \end{aligned}$$

In the second line we have changed to a more symmetric notation between the two particles, introduced M_1 and M_2 as the masses of particles 1 and 2 in units of the proton mass, replaced M by the reduced mass as is necessary if both particles are about equally massive, and changed units from MeV to keV. If we express (15-1) in terms of energy, it becomes

$$p(E)\, dE \propto E^{1/2}\, dE \exp\left(-\frac{E}{kT}\right) \tag{15-1a}$$

$\mathcal{R}(E)\, dE$, the number of reactions per unit time in the energy interval E to $E + dE$, is found by multiplying (15-1a) by (15-3), which gives, after using (15-5) and (15-6),

$$\mathcal{R}(E) \propto S_0 E \exp - \left(\alpha E^{-1/2} + \frac{E}{kT}\right) \tag{15-7}$$

The makeup of the expression for the reaction rate, (15-7), can be understood with the help of Fig. 15-2. Its energy dependence is determined by the

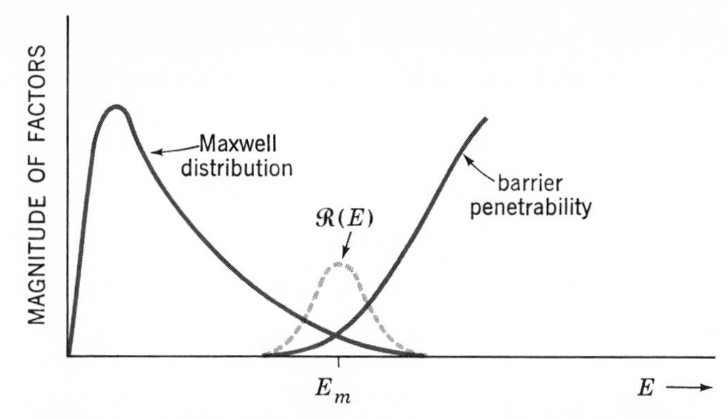

FIGURE 15-2 The principal factors in (15-7), the Maxwell distribution and the barrier penetrability, and their product $\mathcal{R}$ (E), which is large only in the region of E_m, the effective thermal energy.

product of two very strongly energy-dependent factors, the Maxwell distribution (15-1a), which decreases exponentially with increasing E, and the barrier penetrability (15-6), which increases sharply with increasing E. As we see in Fig. 15-2, the product of these has a maximum at some energy E_m and most reactions occur in this energy region. E_m is therefore called the *effective thermal energy*.

Since the energy dependence of $\mathcal{R}(E)$ is nearly all contained in the exponent of (15-7), E_m can be determined by setting the derivative of that exponent equal to zero, which gives

$$E_m = (\tfrac{1}{2}\alpha kT)^{2/3} \tag{15-8}$$

It is typically several times kT. For protons colliding with protons, for example, $\alpha = 22(\text{keV})^{1/2}$, whence $E_m = 5$ keV for $kT = 1$ keV, and $E_m = 23$ keV for $kT = 10$ keV. The temperature dependence of $\mathcal{R}$ can be roughly estimated by inserting E_m for E in the exponent of (15-7), which easily gives

$$\mathcal{R} \propto \exp\,[-2.0\alpha^{2/3}(kT)^{-1/3}] \tag{15-9}$$

This has the general behavior of the accurate result which is obtained by integrating (15-7) over energy. That result, including constant factors that we have not carried, is

$$\mathcal{R} = 7.20 \times 10^{-19}\, n_1 n_2 \left(\frac{Z_1 Z_2 M_1 M_2}{M_1 + M_2}\right)^{-1} S_0 \tau^2 e^{-\tau} \tag{15-10}$$

where

$$\tau = 2.37\alpha^{2/3}(kT)^{-1/3} \tag{15-11}$$

and $\mathcal{R}$ is in reactions per cm³-s if n_1 and n_2 are in particles per cubic centimeter and S_0 is in keV-barns. It is conventional to define T_6 as the temperature in units of 10^6 °K, in which notation (15-11) and (15-8) are

$$\begin{aligned} \tau &= 42.48 \left(\frac{Z_1{}^2 Z_2{}^2 M_1 M_2}{M_1 + M_2}\right)^{1/3} T_6{}^{-1/3} \\ E_m &= 1.220 \left(\frac{Z_1{}^2 Z_2{}^2 M_1 M_2}{M_1 + M_2}\right)^{1/3} T_6{}^{2/3} \qquad \text{keV} \end{aligned} \tag{15-12}$$

The mean lifetime τ_1, of particle 1 due to these reactions is clearly

$$\tau_1 = \frac{n_1}{\mathcal{R}\nu_1} \tag{15-13}$$

where ν_1 is the number of them consumed in a reaction.

All of the nuclear-physics information in this treatment is contained in the factor S_0, which is defined by (15-4) and (15-5). It depends on Γ_q, on that part of Γ_p not determined by the barrier penetration ($\theta_p{}^2$), and on the energy of the lowest-energy resonance. This is similar to the situation for thermal neutrons, as explained after (13-25*d*), and the result is qualitatively the same: S_0 varies by large factors for different reactions in a more or less random fashion. The only meaningful procedure in most cases is therefore to determine S_0 experimentally by measurements of σ vs. E and application of (15-5). Since $\mathcal{P}$, and therefore the cross section, becomes extremely small in the region near E_m, these experiments ordinarily cannot be extended to that low an energy, but S_0 is usually essentially independent of energy, so determinations from measurements at higher energies are often reliable. In reactions in which there is a resonance in the thermal-energy region, clearly this procedure is not valid; S_0, as defined by (15-5), then varies rapidly with energy, and in the absence of measurements, its values can only be crudely estimated from indirect evidence. In all cases, of course, it is highly desirable to extend measurements of σ vs. E to as low an energy as is possible, and a great deal of effort has been expended by nuclear physicists in this direction.

15-3 Energy Production in Stars

Stars are formed by a collection of *interstellar dust*, contracting together under the influence of their mutual gravitational attraction. As the collection becomes more concentrated, forces increase (due to the $1/r^2$ gravitational-force law), causing velocities to increase, and the density rapidly builds up to the point where collisions occur, so we can speak of a temperature.[1] The process continues,

[1] A minor event in the contraction of interstellar dust to form a star is the formation of planets, which proves to be the easiest way for the angular momentum to be disposed of.

with the density and temperature increasing rapidly until they reach a point where thermonuclear reactions can occur in the central region of the star. The energy generated in these reactions increases with increasing density and temperature in accordance with (15-10) until the outward pressure of radiant energy balances the inward pressure of gravity, at which point the star is in equilibrium. Calculations based on energy-transport theory indicate that at the center of stars like our sun, the density is about 100 g/cm³ and the temperature is about 12 million degrees ($kT \simeq 1$ keV).

About 99.9 percent of all nuclei in the universe are hydrogen or helium (in about a 10:1 ratio), and for reasons we shall explain in Sec. 15-4, helium cannot undergo nuclear reactions in the conditions under consideration. The principal material of interest is therefore hydrogen, or rather protons, since atoms are ionized at these temperatures. We found in Chap. 3 that the nucleus consisting of two protons has no bound states, so when two protons interact, they do not stay together and no energy is released. However, in the short time they are together, we can say that we have a nucleus of He^2, and this nucleus can beta-decay to H^2, the deuteron. The reaction can be written

$$p + p \rightarrow d + e^+ + \nu + 0.42 \text{ MeV} \tag{15-14}$$

Since the time required for beta decay is many orders of magnitude longer than the time during which the two protons interact, such a reaction is fantastically improbable, but this is still the only reaction that can occur between two protons. Let us estimate S_0 for it.

From (15-5), S_0 is σE at energies above the coulomb barrier, so we calculate it at about 0.5 MeV. We know from Chap. 3 that there is a resonance in this region, the $S = 0$ state of He^2 shown in Fig. 6-9. Since the time of interaction is very short, Γ_p is very large, so $\Gamma_p \gg E - E_r$ and $\Gamma \simeq \Gamma_p$, whence (15-4) reduces to

$$\sigma \simeq 4\pi\lambdabar^2 \frac{\Gamma_\beta}{\Gamma_p} \tag{15-15}$$

where Γ_β is the partial width for decay into the products on the right side of (15-14). From (13-11*a*), Γ_β/Γ_p is just the inverse of the ratio of the times required for proton emission and beta decay. The former is the time of interaction of the two protons, which is several fermi divided by the velocity as reduced by the coulomb force, or about 6×10^{-22} s. The beta decay (15-14) is a superallowed Gamow-Teller transition since only a spin-flip is required in addition to changing a proton into a neutron. Its $fT_{1/2}$ value is therefore about 3,000, and from Fig. 9-2, $f \simeq 0.15$, so $T_{1/2}$ is about 2×10^4 s. On this basis $\Gamma_\beta/\Gamma_p \simeq (6 \times 10^{-22})/(2 \times 10^4) \simeq 3 \times 10^{-26}$. Since $\lambdabar^2 = 400/E$ keV-barns, (15-15) gives for S_0

$$\begin{aligned} S_0 = \sigma E &\simeq 4\pi \times 400 \text{ keV-barns} \times 3 \times 10^{-26} \\ &\simeq 1.5 \times 10^{-22} \text{ keV-barn} \end{aligned}$$

Note that this value is rather insensitive to the energy assumed in the calculation. Our treatment here was very crude, but an elaborate and accurate treatment is possible and gives

$$S_0 \simeq 3.0 \times 10^{-22} \text{ keV-barn}$$

If this value is inserted into (15-10) along with the values for n_1 $(= n_2)$ corresponding to a density of 100 g/cm^3 and the result is used in (15-13), the time required for the hydrogen in the sun to be consumed is

$$\tau_{\text{H}} \simeq 10 \times 10^9 \text{ years}$$

Fortunately for us, the sun is only about 5×10^9 years old, so it can be estimated that it will last for about another 5 billion years.

Once deuterium is formed by reaction (15-14), it very rapidly undergoes the reaction

$$\text{H}^2 + p \rightarrow \text{He}^3 + \gamma$$

for which $S_0 = 7.8 \times 10^{-5}$ keV-barn; the mean lifetime of deuterium in the sun, as calculated from (15-13) and (15-10) is about 2 s. The most probable fate of the He^3 is to undergo the reaction

$$\text{He}^3 + \text{He}^3 \rightarrow \text{He}^4 + p + p$$

for which $S_0 \simeq 5 \times 10^3$. As a result of these reactions, four protons are converted into a He^4 nucleus, releasing an energy equal to the binding energy of He^4 less twice the neutron-proton mass difference, or about 26.7 MeV. This comes out as kinetic energy of the particles emitted in the reactions, and all but a few tenths MeV carried by the neutrinos is quickly dissipated in collisions and appears as heat.

The process we have been describing is called the *proton-proton* or *p-p cycle*. Since the sun contains many elements other than hydrogen, there are many other reactions taking place. For example, the so called *carbon cycle*

$$\begin{array}{ll}
\text{C}^{12} + p \rightarrow \text{N}^{13} + \gamma & S_0 = 1.2 \text{ keV-barns} \\
\text{N}^{13} \rightarrow \text{C}^{13} + e^+ + \nu & T_{1/2} = 10 \text{ min} \\
\text{C}^{13} + p \rightarrow \text{N}^{14} + \gamma & S_0 = 6 \text{ keV-barns} \\
\text{N}^{14} + p \rightarrow \text{O}^{15} + \gamma & S_0 = 2.8 \text{ keV-barns} \\
\text{O}^{15} \rightarrow \text{N}^{15} + e^+ + \nu & T_{1/2} = 2 \text{ min} \\
\text{N}^{15} + p \rightarrow \text{C}^{12} + \alpha & S_0 = 1.1 \times 10^5 \text{ keV-barns}
\end{array}$$

also converts four protons into a He^4 nucleus, giving the same energy release. We see that the carbon nucleus acts only as a catalyst, since each one consumed at the beginning of the chain is replaced by a new one produced at the end.

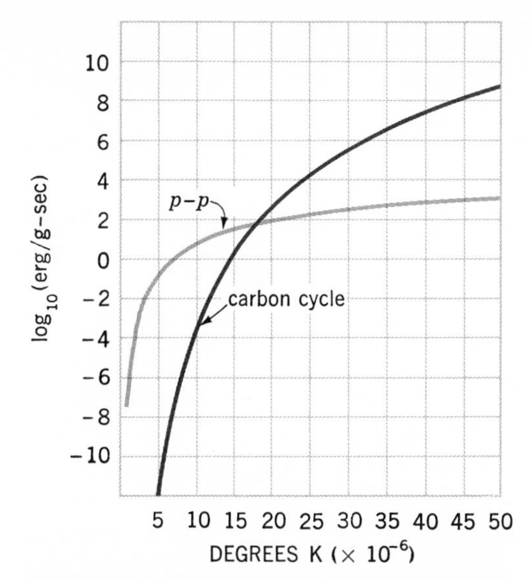

FIGURE 15-3 Rate of energy production from the proton-proton cycle and from the carbon cycle as a function of temperature. Densities are those in the sun, 100 g/cm^3 for protons and 1 g/cm^3 for C^{12}. [*From E. M. Burbidge et al., Rev. Mod. Phys.,* **29: 547 (1957).**]

The rate of energy production from (15-10) at various temperatures when the density of protons is 100 g/cm^3 and the density of C^{12} is 1 g/cm^3, their densities near the center of the sun, is shown in Fig. 15-3. Note that the curve for the carbon cycle increases much more rapidly with temperature than does that for the *p-p* cycle. This can be understood from (15-10) and (15-12), which shows that the reaction rates for the two processes go approximately as

$$\mathcal{R} = \begin{cases} \exp \dfrac{T^{1/3}}{33} & \text{for } p + p \\ \exp \dfrac{T^{1/3}}{151} & \text{for } p + \mathrm{N}^{14} \end{cases}$$

[note from (15-10) and the above values of S_0 that $p + \mathrm{N}^{14}$ is the slowest reaction in the carbon cycle] the difference being due to the great differences in the coulomb barriers.

We see from Fig. 15-3 that for temperatures above 18 million degrees, the carbon cycle becomes the dominant energy producer. Such temperatures occur in the central region of stars more massive than the sun by about 50 percent or more.

15-4 The Origin of Complex Nuclei

The product of hydrogen burning is He^4, the alpha particle, so we must now consider what reactions it can initiate. A collision between a proton and an alpha particle leads to Li^5, but, as we found in Sec. 8-3, Li^5 has no bound states and these interactions therefore lead only to scattering. When two alpha particles collide, we get Be^8, but, as we found in Sec. 8-3, Be^8 is unstable, decaying into two alpha particles. However, in this decay a compound-nucleus state is formed, and when it decays, there is a barrier to be penetrated. Since the energy release is quite low, the half-life is relatively long, about 3×10^{-16} s. If, during this time, a third alpha particle strikes the Be^8 nucleus, an excited state of C^{12} (7.65 MeV) is formed, but this state generally decays by alpha emission back to Be^8. However, here again the energy is low, so the half-life is relatively long, about 5×10^{-17} s. This 7.65-Mev state, it turns out, is a two-phonon vibrational state which can decay to the 4.41 MeV one-phonon vibrational state by an E2 gamma-ray transition at about 10 times the single-particle rate. From Fig. 12-1 we see that this transition has a half-life of about 5×10^{-14} s, whence from Sec. 8-4, the ratio of alpha-particle to gamma-ray emission is about 1,000. If a gamma ray is emitted, alpha decay is no longer energetically possible, so the decay is by another gamma-ray emission to the ground state of C^{12}. Hence we have a reaction of three alpha particles to form a C^{12} nucleus. It is only by this chain of unlikely events that there can be reactions between He^4 nuclei.

This chain of events is so improbable that it practically never occurs under conditions like those in the sun. Therefore, when the hydrogen of such a star is burned out, there are no further nuclear reactions to produce outward radiation pressure, so gravitational contraction once more begins and continues until the density and temperature of the central region reach the point where helium burning, i.e., reactions between alpha particles, can take place. This is at a density of about 10^5 g/cm^3 (1 ton/in.^{3}) and a temperature of 10^8 °K. During this contraction of the stellar interior, the star undergoes profound changes: its outer surface expands by a large factor, and it becomes what is known as a *red giant*. When this happens to our sun in about 5 billion years, the temperature on the earth will increase enormously, bringing to an end life as we know it.

Once C^{12} is formed, other reactions become possible, like

$$C^{12}(\alpha,\gamma)O^{16}$$
$$O^{16}(\alpha,\gamma)Ne^{20}$$
$$Ne^{20}(\alpha,\gamma)Mg^{24}$$

The rates of these reactions from (15-13) and (15-10) are shown as a function of temperature in Fig. 15-4. We see there that they are already appreciable under

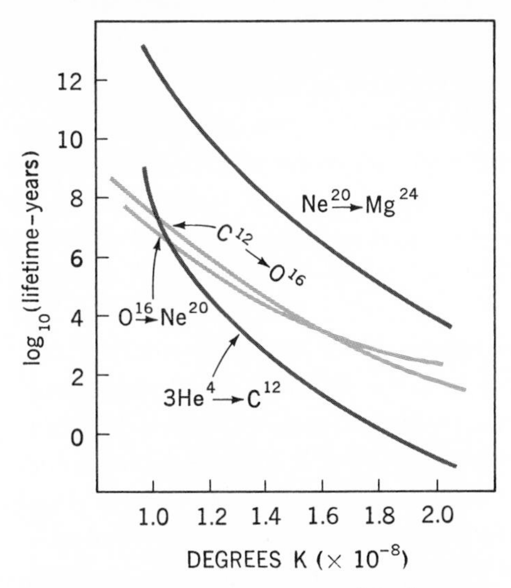

FIGURE 15-4 Lifetime in years (the inverse of the rate of destruction) for various nuclei due to (α,γ) reactions as a function of temperature at a density of 10^5 g/cm^3 of He^4. [*From E. M. Burbidge et al., Rev. Mod. Phys.,* **29: 547 (1957).**]

the conditions we are considering, so O^{16}, Ne^{20}, and some Mg^{24} as well as C^{12} are produced during the helium-burning stage.

After the helium is consumed, further gravitational contraction with its consequent heating takes place, until reactions between C^{12} nuclei (at 7×10^8 °K), between O^{16} nuclei (at 1.0×10^9 °K), and between other pairs of C, O, Ne, and Mg nuclei become frequent. For example, the principal reactions between two C^{12} nuclei are

$$\begin{aligned} C^{12} + C^{12} \rightarrow\ & Ne^{20} + \alpha \\ & Na^{23} + p \\ & Mg^{23} + n \\ & Mg^{24} + \gamma \end{aligned}$$

In all these reactions, gamma rays are released, and they, as well as the neutrons, protons, and alpha particles released, have further reactions which continually create and destroy nuclei. The least likely nucleus to be destroyed by gamma-ray-induced reactions is Si^{28} because (γ,n), (γ,p), and (γ,α) reactions on that nucleus have the largest negative Q values. In addition, the cross section for forming heavier nuclei by capture of protons and alpha particles is not large in so heavy a nucleus. As a result, when the temperature reaches about 2×10^9 °K, a large fraction of all nuclei are Si^{28}, and the nuclear reaction rate slows down.

What happens beyond this point is somewhat less certain, but a widely accepted theory is that this slowdown brings on further gravitational contraction leading to a further temperature increase until, at a temperature of about 4×10^9 °K the gamma rays from *blackbody radiation* become sufficiently energetic to induce at a relatively slow rate the reaction

$$Si^{28}(\gamma,\alpha)Mg^{24} \tag{15-16}$$

The alpha particles released in this reaction induce reactions on other Si^{28} nuclei, leading to the formation of heavier nuclei and the emission of neutrons and protons. The neutrons, protons, and alpha particles induce further reactions, causing a buildup of heavier and heavier nuclei until a quasi-equilibrium is reached among the product nuclei and these particles in which the rate of their formation becomes equal to the rate of their destruction, while the Si^{28} is slowly consumed by (15-16). It happens that neutron separation energies are generally higher than those for protons among stable nuclei in this mass region, so there are many more protons than neutrons in the quasi equilibrium. Proton-induced reactions therefore occur most frequently, and they lead to proton-rich product nuclei. The most abundant heavy nucleus in the quasi-equilibrium is Ni^{56}, which is an especially tightly bound proton-rich nucleus because it has closed shells (28) of both neutrons and protons. The Ni^{56} is eventually converted to Fe^{56} by successive beta decays, so the latter is the most abundant heavy nucleus found in nature. The results of a detailed calculation of the abundances of nuclei produced in Si^{28} burning are shown in Fig. 15-5, where they are compared with natural abundances in the solar system. It is seen that these calculations explain the occurrence of the most abundant nuclei between silicon and iron.

The conditions under which the calculations shown in Fig. 15-5 give the correct results are $T = 4.4 \times 10^9$ °K and a density of 10^8 g/cm^3 (1,000 tons/in.^{3}). If the density were much lower, for example, Fe^{54} would be more abundant than Fe^{56}. Under these conditions, the entire Si^{28} burning process occurs in a fraction of a second, so it is part of a supernova explosion.

Whether by the route described or by some other, it is apparently fairly common for stars eventually to become supernovas. In these explosions enormous temperatures and pressures develop, allowing all sorts of nuclear reactions that cannot otherwise occur. Two of these will be described later in this section. In some supernovas, all the outer layers of the star are blown off while the central core collapses. In this central core, consisting mostly of nuclei in the iron region, no further exothermal nuclear reactions can occur—note from Fig. 7-1 that the binding energy per nucleon is a maximum there, and in fact that is basically why these nuclei are the most abundant—so there is no outward radiation pressure. Gravitational contraction therefore proceeds unimpeded, leading to higher

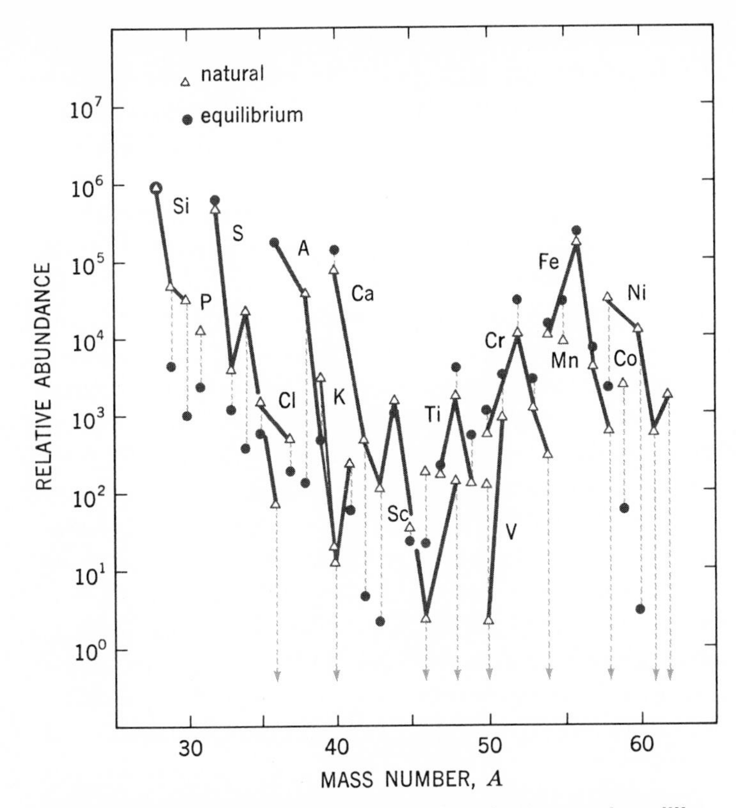

FIGURE 15-5 Abundances of various nuclei in the quasi equilibrium accompanying burning of Si^{28} as calculated for a density of 10^8 g/cm^3 and a temperature of 4.4×10^9 °K when 65 percent of the Si^{28} has been consumed. Triangles are the relative abundances found in nature. [*From D. Bodansky, D. D. Clayton, and W. A. Fowler, Phys. Rev. Lett.*, **20: 161 (1968).**]

and higher densities and pressures. Temperatures are sufficiently high to allow electron-capture processes (8-3) to take place as

$$e^- + (Z, A) \rightleftharpoons (Z\text{-}1, A) + \nu \tag{15-17}$$

even though they are endothermal, and an equilibrium is established between the two sides of (15-17). When the density exceeds about 10^9 g/cm^3, the pressure due to the electrons becomes so large that this concentration equilibrium is shifted in favor of the right side of (15-7) because an electron is absorbed out when that equation goes to the right. This shift is greatly enhanced by a quantum-mechanical effect known as degeneracy: if the whole star is considered to be a single potential well for the n electrons in it, these electrons are restricted to allowed orbits and energies, as explained in Chap. 2, and in accordance with the Pauli exclusion principle, the n lowest energy of these orbits are occupied (neglecting excitation). From (2-5) we see that the energy of the nth orbit varies inversely as square of the size of the star (L, the width of the well), so when the star becomes very small, its energy may be several MeV. This means that all electron orbits with energies less than several MeV are occupied. It is then impossible for (15-17) to proceed to the left if the energy of the emitted electron is only a few MeV. Hence (15-17) goes only to the right until all nuclei (Z,A) are replaced by $(Z - 1, A)$.

In this way, nuclei become more and more neutron-rich until, as explained in Sec. 8-3, the neutron separation energy becomes less than zero so neutrons are emitted. The residual nuclei continue to capture electrons as in (15-17) and therefore continue to emit neutrons until, when the density reaches 10^{12} g/cm^3, the core of the star consists only of neutrons. It is called a *neutron star*, and it continues to contract under gravity until nuclear densities are reached and exceeded. It typically reaches a density of about 10^{15} g/cm^3, whence the diameter of the entire star is reduced to about 10 km. In this fantastic shrinkage process, angular momentum must be conserved, which means that the angular velocity must increase inversely as the square of the diameter. In the final condition, the star makes one revolution in a small fraction of a second. This causes variations in the emitted light with this frequency, and these variations have been observed. Such stars are called *pulsars*.

Neutrons are produced in various stages of stellar evolution, and their principal reactions, as explained in Sec. 13-8, are (n,γ) and elastic and inelastic scattering. As scattering leads back to the original nucleus, only (n,γ) is of consequence in forming new elements. These reactions increase A by one unit, whence they slowly but surely build up heavier and heavier nuclei. The number of nuclei with A nucleons, $n(A)$, is increased by (n,γ) reactions on nuclei with $A - 1$ nucleons at a rate proportional to $n(A - 1)$ and to the cross section for those reactions, $\sigma(A - 1)$; it is simultaneously decreased by (n,γ) reactions leading to nuclei with $A + 1$ nucleons which occur at a rate proportional to $n(A)\sigma(A)$. Due to these two effects

$$\frac{dn(A)}{dt} \propto [n(A-1)\sigma(A-1) - n(A)\sigma(A)]$$

When this process reaches equilibrium, $dn(A)/dt = 0$, whence

$$n(A-1)\sigma(A-1) = n(A)\sigma(A)$$

Since a similar equation relates A and $A + 1$, $A + 1$ and $A + 2$, etc., we have

$$n(A)\sigma(A) = \text{const} \tag{15-18}$$

through the chain of successive neutron captures.

There are two types of neutron-capture chains, the *s* (for slow) process, in which the time between successive captures is long enough for the product to beta-decay to a stable nucleus, and the *r* (for rapid) process, in which this time is too short for all but very fast beta decays. The *s* process is typical of those in red giant stars, and the *r* process occurs in supernova explosions. In the *s* process, the nuclei that capture neutrons are stable, so their cross sections can be measured, allowing an experimental check on (15-18). Two such checks are shown in Figs. 15-6 and 15-7. In Fig. 15-6, a comparison is made between the abundances of elements in the solar system and the inverse of the (n,γ) cross sections for neutrons of about 1 MeV energy (this is the average energy of neutrons emitted in fission, so intense sources are available). We see there the peaks in $1/\sigma(n,\gamma)$ for closed-shell nuclei explained in Sec. 13-8, and we also see that these explain the anomalously high abundances of these nuclei. Actually, 1 MeV is considerably higher than the neutron energies in red giant stars, so measurements have been made at about 25 keV average neutron energy, and the results are shown in Fig. 15-7. We see there, and also in Fig. 15-6, that the product

FIGURE 15-6 Natural abundances for nuclei made in the *s* process of successive neutron capture (slow chain). Data are shown separately for odd-A and even-A nuclei. Solid curve is $1/\sigma(n,\gamma)$ from cross sections measured with fission ($\sim$1-MeV) neutrons. Note the similarities between the curves, and the peaks in both for closed-shell nuclei. [*From E. M. Burbidge et al., Rev. Mod. Phys.*, **29: 547 (1957).**]

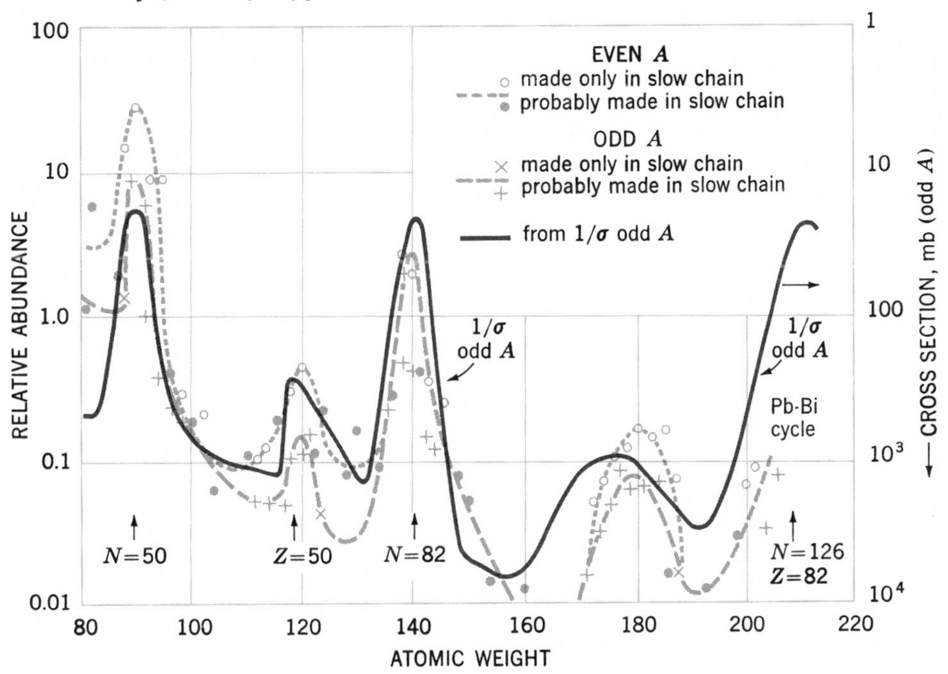

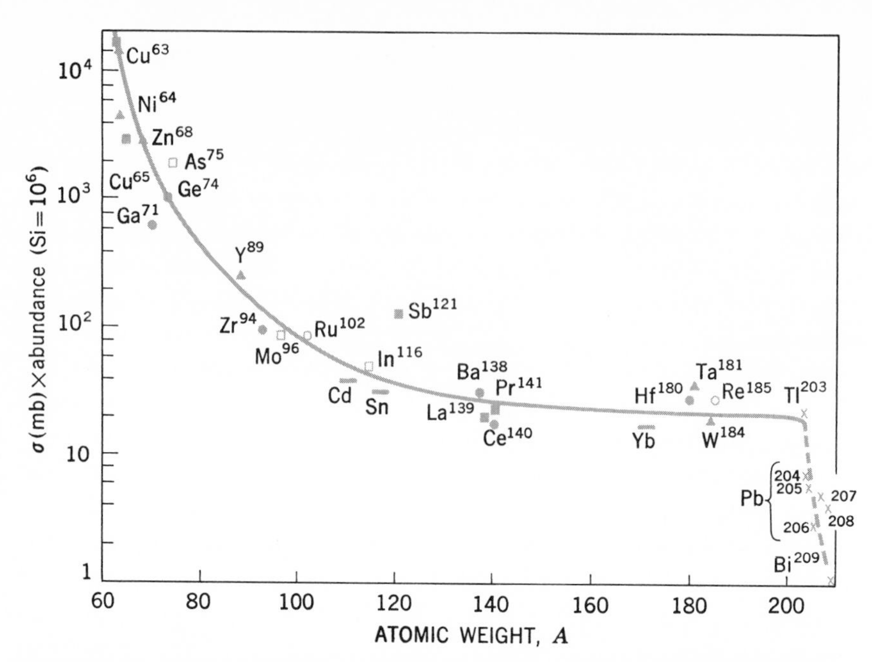

FIGURE 15-7 Product of natural abundances and (n,γ) cross sections as measured with ~25-keV neutrons for various nuclei. (Bars over symbols indicate the average value for natural isotopic mixtures.) The fact that all points lie on a smooth curve decreasing monotonically with increasing A is strong support for the thesis that these nuclei were produced by the s process of neutron capture. The fact that the curve is not constant for all A indicates that the process is most often interrupted before the heavy nuclei can be formed. [*From D. D. Clayton et al., Ann. Phys.*, **12: 331 (1961); by permission.**]

$n(A)\sigma(A)$ is constant over a wide range of A values but not for all nuclei formed by the s process. This indicates that in most cases the process was cut off before the heavier nuclei could be formed. Nuclei heavier than Bi^{209} cannot be formed by the s process since the beta-stable nuclei in that region undergo alpha decay with short half-lives (see, for example, the $A = 211$ nuclei). Thorium and uranium can therefore be formed only by the r process.

Some proton-rich heavy nuclei that are found in nature could not be formed by successive neutron capture. For example, we can see from the chart of nuclides that neutron captures by either the s or r process cannot produce any of the tin isotopes with A less than 116. These were produced by (p,γ) reactions; since the coulomb barrier in such a reaction is quite high, they can take place only at very high temperatures, as in supernova explosions.

We have sketched only some of the most important aspects of nucleosynthesis. In addition there are reactions occurring outside the central regions of stars, where temperatures are cooler, and even near the stellar surface in localized explosions. Mixing of the gases is important in many cases. All in all, there is a fairly good understanding of how virtually all nuclei were formed and reasonable explanations for their abundances.

For various reasons, stars sometimes shoot some of their material out into space, where it becomes interstellar dust. This happens to some extent in the red giant stage and most spectacularly in supernovas. This interstellar dust may later be collected to form new stars. The fact that our solar system contains all elements indicates that it is made of material which was blown out of stars, and there is evidence that it has gone through this star–interstellar dust–star cycle more than once.

15-5 Thermonuclear Reactions on the Earth

Our civilization consumes energy at an enormous and rapidly increasing rate, and it is clear that our supplies of conventional fuel are limited. It is estimated that as much fuel will be consumed between the years 1950 and 2000 as in all previous history and that several times this much will be consumed in the first half of the twenty-first century. At this rate our supply of conventional fuels cannot last more than another century or two. The only solution to the problem this poses is to develop methods of using nuclear reactions to produce energy. The two known methods of doing this will be discussed in this section and the next.

Since energy is produced so copiously in stars, we might expect to use the same methods here, but it is clear that this is hopeless. A density of 100 g/cm^3 has never been attained or even approached on earth, let alone at a temperature of 10 million degrees, and a fuel supply requiring 10 billion years to burn would not produce enough energy per unit mass to be practical. Fortunately, however, there are more favorable reactions than (15-14). The most favorable are the reactions between two deuterons ($d - d$) and the reactions between a deuteron and a triton ($d - t$), which can be written

$$\left.\begin{aligned} H^2 + H^2 &\rightarrow He^3 + n + 3.25\ \text{MeV} \\ &\rightarrow H^3 + p + 4.0\ \text{MeV} \end{aligned}\right\} d - d \quad \text{or} \qquad H^2 + H^3 \rightarrow He^3 + n + 17.6\ \text{MeV} \qquad d - t \tag{15-19}$$

The values of $\overline{\sigma v}$ for these reactions at various temperatures are shown in Fig. 15-8. The rather large values for the $d - t$ reaction are due to a resonance near 100 keV. Reaction rates can be calculated by inserting these values in (15-3).

To produce thermonuclear energy, we must create a gas mixture of H^2 and H^3 at thermonuclear temperatures. Atoms would be completely ionized at these temperatures, so the gas would be a *plasma* of nuclei and electrons. If the temperature of the plasma is to be maintained by the energy produced in the reactions, the reaction rate must be large enough to overcome the energy radiated out of the

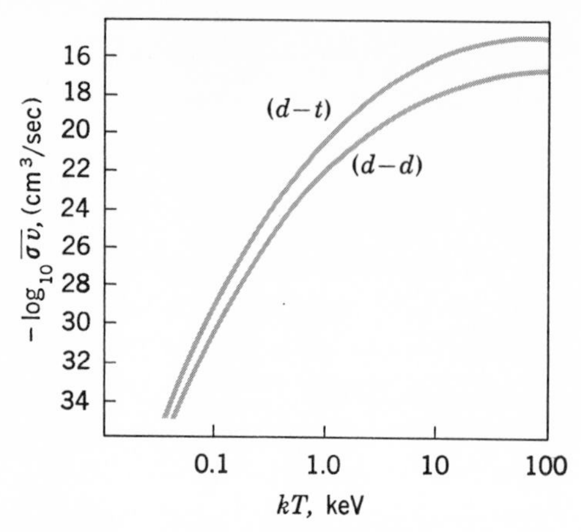

FIGURE 15-8 The average value of σv, the product of cross section at a given velocity times that velocity, averaged over the Maxwell distribution of velocities, for the $d - d$ and $d - t$ reactions as a function of temperature. These values can be used in (15-3) to calculate the reaction rates.

system by bremsstrahlung in electron-ion collisions. This requires temperatures of at least 10^8 deg ($kT \lesssim 10$ keV). In order to produce interesting amounts of power at this temperature, the density should be at least of the order of 10^{15} particles/cm^3. The mean lifetime of a nucleus under these conditions, as calculated from (15-13), is about 10 s for $d - t$ and 1,000 s for $d - d$.

If a system of this type could be constructed, fuel sources would be no problem. Deuterium is present as 150 parts per million in normal hydrogen, and hydrogen, of course, is the principal constituent of water. We would then be "burning the seas." In fact the energy produced by seawater burned in this way is 10 times that produced by an equal amount of gasoline burned in the usual chemical reaction.

The principal difficulty in such a system is containment. No material container could withstand such temperatures, and if it could, collisions with the wall material would lead to fatal energy losses. The obvious solution is then to contain the plasma with magnetic fields. It is not difficult to devise magnetic-field configurations that will contain charged particles indefinitely, but when the density of the plasma gets large, instabilities develop leading to losses of material. The longest containment times achieved to date are of the order of milliseconds, and even this has not been attained at the temperature-density conditions of interest.

If and when the containment problem is solved, there are many others, but all seem to be readily solvable. The initial heating of the plasma can be accomplished by electromagnetic induction or by bombarding with intense beams of high-energy particles. There seems to be no great difficulty in the injection of fuel or in the removal of energy.

The difficulties in containment are not present in bombs. Thermonuclear temperatures at densities approaching those in normal matter ($\sim 10^{21}$ particles/cm^3) are achieved in a limited volume and for a very short time by atomic bombs. From (15-13) and (15-10), the time required varies inversely as the density, so the 10-s lifetime given above for 10^{15} particles/cm^3 is reduced to the microsecond region. From (15-19), we see that the total energy produced by $d - t$ reactions is 17.6 MeV/5 amu which is 40×10^9 kcal/lb. Since the energy release in the explosion of TNT is 450 kcal/lb, 1 lb of a deuterium-tritium mixture burning by $d - t$ reactions gives the same energy release as 40 kilotons (90 million pounds) of TNT.

15-6 Fission as a Source of Energy

In Sec. 15-2 it was stated that nuclear reactions cannot occur at earthly temperatures because nuclei are kept apart by their mutual electrical repulsion. However there is one obvious exception to this rule, namely, reactions induced by neutrons. In fact, as we have seen in Sec. 13-8, cross sections for neutron-induced reactions become larger as the energy decreases. The difficulty, however, is that neutrons do not occur naturally in large quantities, so in order to have reactions between matter in bulk, we must have reactions in which more neutrons are produced than consumed. Reactions such as $(n,2n)$ have this property, but they are highly endothermal (by the separation energy of a neutron). The only exothermal reaction that produces more neutrons than it consumes is fission.

Successive stages in the fission process are illustrated in Fig. 15-9; we see there that when the two fragments break apart, their shape is very nonspherical. Their surface area is then a few percent larger than in a spherical shape; this requires a surface energy which, from (7-8) and (7-23), is about 15 MeV. When the nuclei are pulled into a spherical shape by the surface tension, this energy is available as excitation energy, so we have compound nuclei with about 15 MeV of excitation. As we know from Sec. 13-9, by far the most probable way for them to decay is by neutron emission,[1] and if their excitation energy is sufficient, they may emit more than one neutron. In the thermal-neutron-induced fission of U^{235}, an average of 2.5 neutrons are emitted (from both fragments combined).

[1] Note that the neutron/proton ratio in fission fragments is the same as in the original nucleus, e.g., uranium; so judged by stable nuclei in their mass region, they are very neutron-rich. Hence separation energies are lower for neutrons than for protons, and E_{0n} is larger than E_{0p} in (13-28). This enhances the ordinarily strong preference for neutron emission in heavy nuclei.

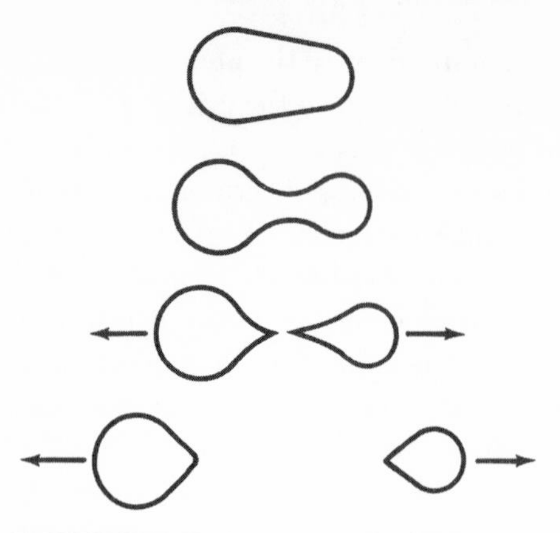

FIGURE 15-9 Various stages in the fission process, progressing from top to bottom. This demonstrates that fission-product nuclei have a large surface energy due to their distorted shape just after fission occurs. This energy is eventually converted into ordinary excitation energy of the nuclei, so neutron emission is energetically possible.

In order to produce energy from fission it is therefore only necessary to assemble a large enough quantity of U^{235} so that at least 40 percent of the neutrons produce a fission reaction before escaping out the side. The minimum size of such an assembly is called the *critical mass*.

One difficulty in this, as can be seen qualitatively from Fig. 13-12 or calculated from (13-26) with (5-34*a*), is that the neutrons emitted in nuclear reactions have an average energy of about 1 MeV whereas the fission cross section is much more favorable at lower energies. The critical mass can therefore be substantially reduced by the introduction of *moderators* to degrade the neutron energy. A neutron loses more energy on an average in an elastic collision with a light nucleus than with a heavy nucleus, whence moderators should be low-A materials. However, it is also important that all materials used in fission *reactors* have low neutron-capture cross sections since they would otherwise drain off neutrons. As mentioned at the end of Sec. 13-8, carbon and heavy water meet these requirements and are therefore good moderators.

The requirement of low neutron-capture cross sections must be carefully considered in the choice of materials used in reactor construction. Unusual elements such as zirconium and beryllium have therefore found wide usage. Control of a reactor, on the other hand, is accomplished with a material of high neutron cross section, such as cadmium. When the reaction rate becomes too high, a cadmium rod is pushed in to absorb neutrons, and conversely the reaction rate can be increased by withdrawing the cadmium rod.

About 80 percent of the energy released in fission comes off as kinetic energy of the fragments, and nearly all the rest is in the energies of the neutrons and gamma rays emitted as the fragments give up their excitation energy and in the electrons and gamma rays emitted in the subsequent beta-decay processes. This energy is rapidly lost and turned into heat in the surrounding material, and the heat is removed from the reactor by a circulating coolant. The energy released is about 190 MeV/235 amu, which is about 8×10^9 kcal/lb. This may be compared with 3600 kcal/lb from the burning of coal.

The fuel we have been discussing, U^{235}, is rather rare in nature, occurring only as 0.7 percent of natural uranium. However, it can be used to produce two other fissionable materials, U^{233} and Pu^{239}, by neutron capture and subsequent beta decay in Th^{232} and U^{238}, respectively. It is therefore possible to produce more fissionable material than is consumed in a reactor. Such a reactor is called a *breeder*. Of the 2.5 neutrons released per fission, one must replace the neutron which induced the fission and one must be used to replace the nucleus which underwent fission, so less than 20 percent of the neutrons can be wasted by being captured in other materials or by escaping out the sides of the reactor.

Uranium is found in small quantities in granite, which is one of the principal rock materials. If the rich deposits of uranium were used up, it would be economically feasible to recover it from granite; the mining operation would be of the same magnitude as present coal-mining operations and would provide sufficient energy for billions of years. We would then be powering our civilization by "burning the rocks." It is not now clear whether burning the rocks or burning the seas as discussed in the last section will be more feasible, but it seems certain that we must do one or the other.

There are other uses for reactors than production of power. The neutrons they produce can be used in a wide variety of research activities, or they can be used for making radioactive isotopes by neutron-induced reactions. Indeed reactors are the principal source of the radioactive materials whose applications were discussed in Sec. 15-1. An atomic bomb is basically a fission reactor with a mass very much larger than the critical one. Instead of limiting the reaction rate with control rods, steps are taken to accentuate it. From the energy equivalents given above, we may easily deduce that each pound of fuel undergoing fission releases the explosive energy of 9 kilotons of TNT.

Problems

15-1 How can the Pb formed from the decay of uranium be distinguished from naturally occurring Pb?

15-2 In many medical diagnostic applications it is desirable to use short-half-life activities to minimize radiation exposure to the patient. Suggest a suitable carbon isotope and a method for producing it.

15-3 A chemical engineering problem requires a tracer of manganese with a half-life between 1 h and 1 day. Suggest an isotope and a method of producing it.

15-4 What weight of Po^{210} would be needed to produce 100 W of power?

15-5 Which is better for treatment of deep cancers, Cs^{137} or Co^{60}?

15-6 Calculate the reaction rate of the $p - p$ reaction at a density of 100 g/cm^3 for $T_6 = 3$, 10, and 30. For each case, determine the mean lifetime of a proton and the ratio E_m/kT.

15-7 If the nuclear force were slightly stronger so that He^2 was stable against nucleon emission and the cross section for the reaction $p + p \rightarrow He_2 + \gamma$ was $0.1\pi\lambda\!\!\!^{-2}$ (a reasonably typical value), how would the lifetime of the sun be affected?

15-8 If the sun contained 10 times as much carbon as it actually does, how would its energy production rate be affected?

15-9 If Li^5 were stable, discuss briefly how the evolution of stars would be changed. Consider the same question if Be^8 were stable.

15-10 If a star like our sun (mass $= 2 \times 10^{30}$ kg, diameter $=$ 865,000 miles) were to shrink to a diameter of 10 km, what would be the energy of the lowest unoccupied orbits for electrons? Ignoring variations of density with radius (a very poor approximation), what would be its period of rotation (present period $=$ 25 days)?

15-11 If an Sn^{122} nucleus undergoes two successive neutron captures, what nucleus is most probably eventually formed if the time between captures is 10 s? If it is 100 years?

15-12 Determine whether the following isotopes are made in the *s* process or the *r* process: In^{115}, Sn^{119}, Sn^{124}, Te^{126}.

15-13 By what process is Cd^{106} made?

15-14 What happens to the deuterium contained in the material from which a star is formed?

15-15 What is the minimum weight of a 100-megaton hydrogen bomb?

15-16 A 100-MW reactor consumes half of its fuel in 3 years. How much U^{235} does it contain?

Further Reading

Aller, L. H.: "Astrophysics," Ronald Press, New York, 1954.

Annual Reviews of Nuclear Science, Annual Reviews Inc., Stanford, Calif., issued annually.

Arya, A. P.: "Fundamentals of Nuclear Physics," Allyn and Bacon, Boston, 1966.

Belcham, R. F. K.: "A Guide to Nuclear Energy," Philosophical Library, New York, 1958.

Bishop, A. S.: "Project Sherwood: The U.S. Program in Controlled Fusion," Addison-Wesley, Reading, Mass., 1958.

Cohen, B. L.: "The Heart of the Atom," Doubleday, Garden City, N.J., 1967.

Duncan, J. F., and G. B. Cook: "Isotopes in Chemistry," Clarendon Press, Oxford, 1968.

Enge, H. A.: "Introduction to Nuclear Physics," Addison-Wesley, Reading, Mass., 1966.

Glasstone, S., and M. C. Edlund: "The Elements of Nuclear Reactor Theory," Van Nostrand, New York, 1952.

Goodman, C. D.: "The Science and Engineering of Nuclear Power," Addison-Wesley, Reading, Mass., 1947.

Kamen, M. D.: "Isotopic Tracers in Biology," Academic, New York, 1957.

Littler, D. J., and J. F. Raffle: "An Introduction to Reactor Physics," McGraw-Hill, New York, 1955.

Liverhant, S. E.: "Elementary Introduction to Reactor Physics," Wiley, New York, 1960.

Murray, R. L.: "Introduction to Nuclear Engineering," Prentice-Hall, Englewood Cliffs, N.J., 1961.

Parker, D.: "Isotopes in Action," Macmillan, New York, 1963.

Piraux, J.: "Radioisotopes and Their Industrial Applications," Charles C Thomas, Springfield, Ill., 1964.

Rose, D. J., and M. Clark, Jr.: "Plasmas and Controlled Fusion," Technology Press, Cambridge, Mass., 1961.

Sacks, J.: "Isotopic Tracers in Biochemistry and Physiology," McGraw-Hill, New York, 1953.

Salmon, A.: "The Nuclear Reactor," Wiley, New York, 1964.

Simon, A. I.: "An Introduction to Thermonuclear Research," Pergamon, New York, 1959.

Smith, H. E.: "Isotopic Power Sources," U.S. Atomic Energy Commission Report, Washington, 1961.
Soodak, H., and E. C. Campbell: "Elementary Pile Theory," Wiley, New York, 1950.
U.S. Atomic Energy Commission: Annual Report, GPO, Washington.
Weinberg, A. M., and E. P. Wigner: "The Physical Theory of Neutron Chain Reactors," University of Chicago Press, Chicago, 1958.
Wolf, G.: "Isotopes in Biology," Academic, New York, 1964.

Appendix

TABLE A-1: USEFUL CONSTANTS

Speed of light in vacuum	$c = 2.9979 \times 10^{8}$ m/s
Elementary charge	$e = 1.60210 \times 10^{-19}$ C
Avogadro's number	$N_A = 6.02252 \times 10^{26}$ molecules/kg mole
Electron rest mass	$M_e = 9.1091 \times 10^{-31}$ kg
	$= 511.01$ keV
Proton rest mass	$M_p = 1.6725 \times 10^{-27}$ kg
	$= 938.26$ MeV
Neutron rest mass	$M_n = 1.6748 \times 10^{-27}$ kg
	$= 939.55$ MeV
Planck's constant	$h = 6.6256 \times 10^{-34}$ J-s
	$\hbar = 1.05450 \times 10^{-34}$ J-s
	$= 6.5820 \times 10^{-22}$ MeV-s
Nuclear magneton	$e\hbar/M_p = 5.0505 \times 10^{-27}$ J-m²/Wb
Boltzmann constant	$k = 1.3805 \times 10^{-23}$ J/°K
Electron volt	$\mathrm{eV} = 1.6021 \times 10^{-19}$ J/eV

TABLE A-2: TOTAL ANGULAR MOMENTUM I, MAGNETIC DIPOLE MOMENT μ (IN UNITS OF NUCLEAR MAGNETONS, $e\hbar/M_p$), AND ELECTRIC QUADRUPOLE MOMENT (IN UNITS OF $e \times 10^{-24}$ CM^2) FOR VARIOUS NUCLEI† [ ] indicates I is not directly measured. Abbreviations are s = seconds, m = minutes, h = hours, d = days, y = years, ky = 10^3 years, My = 10^6 years, Gy = 10^9 years, Ty = 10^{12} years

Nucleus	*T½ or level (keV)*	*I*	*μ*	*Q*
${}_{0}n^{1}{}_{1}$	12 m	1/2	−1.9131	
${}_{1}\bar{H}^{1}{}_{0}$			−1.8	
${}_{1}H^{1}{}_{0}$		1/2	+2.79278	
${}_{1}H^{2}{}_{1}$		1	+0.85742	+0.0028
${}_{1}H^{3}{}_{2}$	12 y	1/2	+2.9789	
${}_{2}He^{3}{}_{1}$		1/2	−2.1276	
${}_{2}He^{3+}{}_{1}$		1/2		
${}_{2}He^{4}{}_{2}$		0		
${}_{2}He^{6}{}_{4}$	0.8 s	0		
${}_{3}Li^{6}{}_{3}$		1	+0.82202	−0.0008
${}_{3}Li^{7}{}_{4}$		3/2	+3.2564	−0.04
${}_{3}Li^{8}{}_{5}$	0.8 s	[2]	+1.6532	
${}_{4}Be^{9}{}_{5}$		3/2	−1.1776	+0.05
${}_{5}B^{10}{}_{5}$		3	+1.8007	+0.08
${}_{5}B^{11}{}_{6}$		3/2	+2.6885	+0.04
${}_{5}B^{12}{}_{7}$	0.020 s	[1]	+1.002	
${}_{6}C^{11}{}_{5}$	21 m	3/2	±1.03	±0.031
${}_{6}C^{12}{}_{6}$		0		
${}_{6}C^{13}{}_{7}$		1/2	+0.7024	
${}_{6}C^{14}{}_{8}$	5.6 ky	0		
${}_{7}N^{12}{}_{5}$	0.012 s	[1]	±0.46	
${}_{7}N^{13}{}_{6}$	10 m	1/2	±0.3221	
${}_{7}N^{14}{}_{7}$		1	+0.4036	+0.01
${}_{7}N^{15}{}_{8}$		1/2	−0.2831	
${}_{8}O^{15}{}_{7}$	2.1 m	1/2	±0.7189	
${}_{8}O^{16}{}_{8}$		0		
${}_{8}O^{17}{}_{9}$		5/2	−1.8937	−0.026
${}_{8}O^{18}{}_{10}$		0		
${}_{9}F^{17}{}_{8}$	66 s	[5/2]	±4.722	
${}_{9}F^{19}{}_{10}$		1/2	+2.6288	
${}_{9}F^{20}{}_{11}$	11 s	[2]	+2.094	
${}_{10}Ne^{19}{}_{9}$	18 s	1/2	−1.887	
${}_{10}Ne^{20}{}_{10}$		0		
${}_{10}Ne^{21}{}_{11}$		3/2	−0.6618	+0.09
${}_{10}Ne^{22}{}_{12}$		0		
${}_{10}Ne^{23}{}_{13}$	38 s	[5/2]	−1.08	
${}_{11}Na^{21}{}_{10}$	23 s	3/2	+2.386	
${}_{11}Na^{22}{}_{11}$	2.6 y	3	+1.746	
${}_{11}Na^{23}{}_{12}$		3/2	+2.2175	+0.14
${}_{11}Na^{24}{}_{13}$	15 h	4	+1.690	
${}_{12}Mg^{24}{}_{12}$		0		
${}_{12}Mg^{25}{}_{13}$		5/2	−0.8551	+0.22
${}_{12}Mg^{26}{}_{14}$		0		
${}_{13}Al^{27}{}_{14}$		5/2	+3.6414	+0.15
${}_{14}Si^{28}{}_{14}$		0		
${}_{14}Si^{29}{}_{15}$		1/2	−0.5553	
${}_{14}Si^{30}{}_{16}$		0		
${}_{15}P^{30}{}_{15}$	2.6 m	1		
${}_{15}P^{31}{}_{16}$		1/2	+1.1317	
${}_{15}P^{32}{}_{17}$	14 d	1	−0.2523	
${}_{16}S^{32}{}_{16}$		0		
${}_{16}S^{33}{}_{17}$		3/2	+0.6433	−0.055
${}_{16}S^{34}{}_{18}$		0		
${}_{16}S^{35}{}_{19}$	87 d	3/2	+1.00 or −1.07	+0.04
${}_{16}S^{36}{}_{20}$		0		
${}_{17}Cl^{35}{}_{18}$		3/2	+0.82183	−0.079
${}_{17}Cl^{36}{}_{19}$	0.3 My	2	+1.285	−0.017
${}_{17}Cl^{37}{}_{20}$		3/2	+0.68411	−0.062
${}_{18}Ar^{35}{}_{17}$	1.8 s	[3/2]	+0.63	
${}_{18}Ar^{36}{}_{18}$		0		
${}_{18}Ar^{37}{}_{19}$	34 d	3/2	+0.95	
${}_{18}Ar^{38}{}_{20}$		0		
${}_{18}Ar^{39}{}_{21}$	265 y	7/2	−1.3	
${}_{18}Ar^{40}{}_{22}$		0		
${}_{19}K^{37}{}_{18}$	1.2 s	3/2	+0.204	
${}_{19}K^{38}{}_{19}$	7.7 m	3	+1.374	
${}_{19}K^{39}{}_{20}$		3/2	+0.3914	+0.055
${}_{19}K^{40}{}_{21}$	1.3 Gy	4	−1.298	−0.07
${}_{19}K^{41}{}_{22}$		3/2	+0.2149	+0.067
${}_{19}K^{42}{}_{23}$	12 h	2	−1.141	
${}_{19}K^{43}{}_{24}$	22 h	3/2	±0.163	
${}_{19}K^{45}{}_{26}$	20 m	3/2	±0.173	
${}_{20}Ca^{40}{}_{20}$		0		
${}_{20}Ca^{41}{}_{21}$	110 ky	7/2	−1.595	
${}_{20}Ca^{43}{}_{23}$		7/2	−1.317	
${}_{21}Sc^{43}{}_{22}$	3.9 h	7/2	+4.62	−0.26
${}_{21}Sc^{44}{}_{23}$	3.9 h	2	+2.56	+0.10
${}_{21}Sc^{44m}{}_{23}$	2.4 d	6	+3.88	−0.19
${}_{21}Sc^{45}{}_{24}$		7/2	+4.7564	−0.22
${}_{21}Sc^{46}{}_{25}$	84 d	4	+3.03	+0.12
${}_{21}Sc^{47}{}_{26}$	3.4 d	7/2	+5.34	−0.22
${}_{21}Sc^{48}{}_{27}$	1.8 d	6		
${}_{22}Ti^{45}{}_{23}$	3.1 h	7/2	±0.095	≈±0.02
${}_{22}Ti^{47}{}_{25}$		5/2	−0.7883	+0.29
${}_{22}Ti^{49}{}_{27}$		7/2	−1.1039	+0.24
${}_{23}V^{47}{}_{24}$	31 m	3/2		

† From *Nucl. Data*, **5:** 443 (1969) by permission of Academic Press, New York.

TABLE A-2: (Continued)

Nucleus	*T½ or level (keV)*	*I*	*μ*	*Q*
${}_{23}V^{48}{}_{25}$	16 d	4		
${}_{23}V^{49}{}_{26}$	330 d	7/2	±4.5	
${}_{23}V^{50}{}_{27}$	10^{15} y	6	+3.3470	±0.06
${}_{23}V^{51}{}_{28}$		7/2	+5.149	−0.05
${}_{24}Cr^{49}{}_{25}$	42 m	5/2	±0.48	
${}_{24}Cr^{51}{}_{27}$	28 d	7/2	±0.94	
${}_{24}Cr^{53}{}_{29}$		3/2	−0.4744	±0.03
${}_{25}Mn^{51}{}_{26}$	45 m	5/2	±3.57	
${}_{25}Mn^{52}{}_{27}$	5.7 d	6	±3.05	
${}_{25}Mn^{52m}{}_{27}$	21 m	2	±0.0076	
${}_{25}Mn^{53}{}_{28}$	2 My	7/2	±5.01	
${}_{25}Mn^{54}{}_{29}$	290 d	3	±3.28	
${}_{25}Mn^{55}{}_{30}$		5/2	+3.444	+0.4
${}_{25}Mn^{56}{}_{31}$	2.6 h	3	+3.218	
${}_{26}Fe^{57}{}_{31}$		1/2	+0.0902	
${}_{26}Fe^{57m}{}_{31}$	14.4	[3/2]	−0.1547	+0.2
${}_{26}Fe^{57m}{}_{31}$	136	[5/2]	±1.0	
${}_{26}Fe^{57m}{}_{31}$	367	[3/2]	<0.5	
${}_{26}Fe^{59}{}_{33}$	45 d	3/2		
${}_{27}Co^{56}{}_{29}$	77 d	4	±3.83	
${}_{27}Co^{57}{}_{30}$	270 d	7/2	±4.62	
${}_{27}Co^{58}{}_{31}$	71 d	1, 2	+4.03	
${}_{27}Co^{59}{}_{32}$		7/2	+4.62	+0.4
${}_{27}Co^{60}{}_{33}$	5.3 y	5	+3.78	
${}_{27}Co^{60m}{}_{33}$	10.5 m	2		
${}_{28}Ni^{61}{}_{33}$		3/2	−0.7487	+0.16
${}_{28}Ni^{61m}{}_{33}$	68	[5/2]	+0.42	
${}_{29}Cu^{60}{}_{31}$	24 m	2	+1.22	
${}_{29}Cu^{61}{}_{32}$	3.3 h	3/2	+2.13	
${}_{29}Cu^{62}{}_{33}$	9.9 m	1	−0.38	
${}_{29}Cu^{63}{}_{34}$		3/2	+2.223	−0.180
${}_{29}Cu^{64}{}_{35}$	13 h	1	−0.216	
${}_{29}Cu^{65}{}_{36}$		3/2	+2.382	−0.195
${}_{29}Cu^{66}{}_{37}$	5.2 m	1	±0.283	
${}_{30}Zn^{63}{}_{33}$	38 m	3/2	−0.282	+0.31
${}_{30}Zn^{64}{}_{34}$		0		
${}_{30}Zn^{65}{}_{35}$	245 d	5/2	+0.769	−0.026
${}_{30}Zn^{66}{}_{36}$		0		
${}_{30}Zn^{67}{}_{37}$		5/2	+0.8754	+0.17
${}_{30}Zn^{68}{}_{38}$		0		
${}_{31}Ga^{66}{}_{35}$	9.5 h	0		
${}_{31}Ga^{67}{}_{36}$	78 h	3/2	+1.850	+0.22
${}_{31}Ga^{68}{}_{37}$	68 m	1	±0.0117	±0.03
${}_{31}Ga^{69}{}_{38}$		3/2	+2.016	+0.19
${}_{31}Ga^{70}{}_{39}$	21 m	1		
${}_{31}Ga^{71}{}_{40}$		3/2	+2.562	+0.12
${}_{31}Ga^{72}{}_{41}$	14 h	3	−0.1322	+0.59

Nucleus	*T½ or level (keV)*	*I*	*μ*	*Q*
${}_{32}Ge^{70}{}_{38}$		0		
${}_{32}Ge^{71}{}_{39}$	11 d	1/2	+0.546	
${}_{32}Ge^{72}{}_{40}$		0		
${}_{32}Ge^{73}{}_{41}$		9/2	−0.8792	−0.28
${}_{32}Ge^{74}{}_{42}$		0		
${}_{32}Ge^{75}{}_{43}$	82 m	1/2	±0.51	
${}_{32}Ge^{76}{}_{44}$		0		
${}_{33}As^{70}{}_{37}$	52 m	4		
${}_{33}As^{72}{}_{39}$	26 h	2		
${}_{33}As^{75}{}_{42}$		3/2	+1.439	+0.29
${}_{33}As^{76}{}_{43}$	26 h	2	−0.905	±7.8
${}_{34}Se^{74}{}_{40}$		0		
${}_{34}Se^{75}{}_{41}$	120 d	5/2		+1.0
${}_{34}Se^{76}{}_{42}$		0		
${}_{34}Se^{77}{}_{43}$		1/2	+0.534	
${}_{34}Se^{78}{}_{44}$		0		
${}_{34}Se^{79}{}_{45}$	60 ky	7/2	−1.02	+0.8
${}_{34}Se^{80}{}_{46}$		0		
${}_{34}Se^{82}{}_{48}$		0		
${}_{35}Br^{76}{}_{41}$	17 h	1	±0.548	±0.25
${}_{35}Br^{77}{}_{42}$	58 h	3/2		
${}_{35}Br^{79}{}_{44}$		3/2	+2.106	+0.31
${}_{35}Br^{80}{}_{45}$	18 m	1	±0.514	±0.18
${}_{35}Br^{80m}{}_{45}$	4.5 h	5	+1.317	+0.71
${}_{35}Br^{81}{}_{46}$		3/2	+2.270	+0.26
${}_{35}Br^{82}{}_{47}$	36 h	5	±1.626	±0.70
${}_{36}Kr^{82}{}_{46}$		0		
${}_{36}Kr^{83}{}_{47}$		9/2	−0.970	+0.26
${}_{36}Kr^{83m}{}_{47}$	9.3	[7/2]		+0.44
${}_{36}Kr^{84}{}_{48}$		0		
${}_{36}Kr^{85}{}_{49}$	11 y	9/2	±1.005	+0.43
${}_{36}Kr^{86}{}_{50}$		0		
${}_{37}Rb^{81}{}_{44}$	4.7 h	3/2	+2.05	
${}_{37}Rb^{81m}{}_{44}$	32 m	9/2		
${}_{37}Rb^{82m}{}_{45}$	6.3 h	5	+1.643	
${}_{37}Rb^{83}{}_{46}$	83 d	5/2	+1.4	
${}_{37}Rb^{84}{}_{47}$	33 d	2	−1.32	
${}_{37}Rb^{85}{}_{48}$		5/2	+1.3524	+0.26
${}_{37}Rb^{86}{}_{49}$	19 d	2	−1.691	
${}_{37}Rb^{87}{}_{50}$	47 Gy	3/2	+2.7500	+0.12
${}_{37}Rb^{88}{}_{51}$	18 m	2	±0.51	
${}_{38}Sr^{86}{}_{48}$		0		
${}_{38}Sr^{87}{}_{49}$		9/2	−1.093	+0.3
${}_{38}Sr^{88}{}_{50}$		0		
${}_{39}Y^{89}{}_{50}$		1/2	−0.1373	
${}_{39}Y^{90}{}_{51}$	64 h	2	−1.63	−0.15
${}_{39}Y^{91}{}_{52}$	58 d	1/2	±0.164	
${}_{40}Zr^{91}{}_{51}$		5/2	−1.303	

TABLE A-2: (Continued)

Nucleus	*T½ or level (keV)*	*I*	*μ*	*Q*
$_{41}\mathrm{Nb}^{93}_{52}$		9/2	+6.167	−0.22
$_{42}\mathrm{Mo}^{92}_{50}$		0		
$_{42}\mathrm{Mo}^{94}_{52}$		0		
$_{42}\mathrm{Mo}^{95}_{53}$		5/2	−0.9133	±0.12
$_{42}\mathrm{Mo}^{96}_{54}$		0		
$_{42}\mathrm{Mo}^{97}_{55}$		5/2	−0.9325	±1.1
$_{42}\mathrm{Mo}^{98}_{56}$		0		
$_{42}\mathrm{Mo}^{100}_{58}$		0		
$_{43}\mathrm{Tc}^{99}_{56}$	210 ky	9/2	+5.68	+0.3
$_{44}\mathrm{Ru}^{99}_{55}$		5/2	−0.63	$\lvert 99m/99 \rvert \geq 3$
$_{44}\mathrm{Ru}^{99m}_{55}$	90	3/2	−0.29	≥0.1
$_{44}\mathrm{Ru}^{101}_{57}$		5/2	−0.69	
$_{45}\mathrm{Rh}^{103}_{58}$		1/2	−0.0883	
$_{46}\mathrm{Pd}^{105}_{59}$		5/2	−0.642	+0.8
$_{47}\mathrm{Ag}^{102}_{55}$	15 m	5		
$_{47}\mathrm{Ag}^{102m}_{55}$	7 m	2	+4	
$_{47}\mathrm{Ag}^{103}_{56}$	59 m	7/2	+4.4	
$_{47}\mathrm{Ag}^{104}_{57}$	1.2 h	5	+4.0	
$_{47}\mathrm{Ag}^{104m}_{57}$	27 m	2	+3.7	
$_{47}\mathrm{Ag}^{105}_{58}$	40 d	1/2	±0.101	
$_{47}\mathrm{Ag}^{106}_{59}$	24 m	1	+2.9	
$_{47}\mathrm{Ag}^{106m}_{59}$	8.3 d	6		
$_{47}\mathrm{Ag}^{107}_{60}$		1/2	−0.1135	
$_{47}\mathrm{Ag}^{108}_{61}$	2.4 m	1	+2.80	
$_{47}\mathrm{Ag}^{109}_{62}$		1/2	−0.1305	
$_{47}\mathrm{Ag}^{109m}_{62}$	40 s	7/2	±4.3	
$_{47}\mathrm{Ag}^{110}_{63}$	24 s	1	+2.85	
$_{47}\mathrm{Ag}^{110m}_{63}$	253 d	6	+3.604	
$_{47}\mathrm{Ag}^{111}_{64}$	7.5 d	1/2	−0.145	
$_{47}\mathrm{Ag}^{112}_{65}$	3.2 h	2	±0.054	
$_{47}\mathrm{Ag}^{113}_{66}$	5.3 h	1/2	±0.159	
$_{48}\mathrm{Cd}^{105}_{57}$	55 m	5/2	−0.74	+0.5
$_{48}\mathrm{Cd}^{107}_{59}$	6.7 h	5/2	−0.6144	+0.8
$_{48}\mathrm{Cd}^{109}_{61}$	470 d	5/2	−0.8270	+0.8
$_{48}\mathrm{Cd}^{110}_{62}$		0		
$_{48}\mathrm{Cd}^{111}_{63}$		1/2	−0.5943	
$_{48}\mathrm{Cd}^{111m}$	49 m	11/2	−1.11	−1.0
$_{48}\mathrm{Cd}^{112}_{64}$		0		
$_{48}\mathrm{Cd}^{113}_{65}$	$>10^{15}$ y	1/2	−0.6217	
$_{48}\mathrm{Cd}^{113m}_{65}$	14 y	11/2	−1.087	−0.8
$_{48}\mathrm{Cd}^{114}_{66}$		0		
$_{48}\mathrm{Cd}^{115}_{67}$	2.3 d	1/2	−0.6477	
$_{48}\mathrm{Cd}^{115m}_{67}$	43 d	11/2	−1.040	−0.6
$_{48}\mathrm{Cd}^{116}_{68}$		0		
$_{49}\mathrm{In}^{109}_{60}$	4.3 h	9/2	+5.53	+0.86
$_{49}\mathrm{In}^{110}_{61}$	66 m	2	+4.36	+0.36
$_{49}\mathrm{In}^{110m}_{61}$	4.9 h	7	+10.4 or −10.7	−0.21 +0.22
$_{49}\mathrm{In}^{111}_{62}$	2.8 d	9/2	+5.53	+0.85
$_{49}\mathrm{In}^{112}_{63}$	14 m	1	+2.81	+0.089
$_{49}\mathrm{In}^{112m}_{63}$	21 m	4		
$_{49}\mathrm{In}^{113}_{64}$		9/2	+5.523	+0.82
$_{49}\mathrm{In}^{113m}_{64}$	1.7 h	1/2	−0.210	
$_{49}\mathrm{In}^{114m}_{65}$	50 d	5	+4.7	
$_{49}\mathrm{In}^{115}_{66}$	600 Ty	9/2	+5.534	+0.83
$_{49}\mathrm{In}^{115m}_{66}$	4.5 h	1/2	−0.244	
$_{49}\mathrm{In}^{116m}_{67}$	54 m	5	+4.3	
$_{49}\mathrm{In}^{117}_{68}$	45 m	9/2		
$_{49}\mathrm{In}^{117m}_{68}$	1.9 h	1/2	−0.2515	
$_{50}\mathrm{Sn}^{113}_{63}$	118 d	1/2	±0.88	
$_{50}\mathrm{Sn}^{115}_{65}$		1/2	−0.918	
$_{50}\mathrm{Sn}^{116}_{66}$		0		
$_{50}\mathrm{Sn}^{117}_{67}$		1/2	−1.000	
$_{50}\mathrm{Sn}^{118}_{68}$		0		
$_{50}\mathrm{Sn}^{119}_{69}$		1/2	−1.046	
$_{50}\mathrm{Sn}^{119m}_{69}$	24	[3/2]	+0.68	−0.08
$_{50}\mathrm{Sn}^{120}_{70}$		0		
$_{50}\mathrm{Sn}^{121}_{71}$	27 h	3/2	±0.70	±0.08
$_{51}\mathrm{Sb}^{115}_{64}$	31 m	5/2	+3.46	−0.27
$_{51}\mathrm{Sb}^{116}_{65}$	15 m	3		
$_{51}\mathrm{Sb}^{117}_{66}$	2.8 h	5/2	+2.67	−0.4
$_{51}\mathrm{Sb}^{118m}_{67}$	3.5 m	1	±2.5	
$_{51}\mathrm{Sb}^{119}_{68}$	38 h	5/2	+3.45	−0.30
$_{51}\mathrm{Sb}^{120}_{69}$	16 m	1	±2.3	
$_{51}\mathrm{Sb}^{121}_{70}$		5/2	+3.359	−0.29
$_{51}\mathrm{Sb}^{121m}_{70}$	37	[7/2]	+2.51	−0.4
$_{51}\mathrm{Sb}^{122}_{71}$	2.8 d	2	−1.90	+0.69
$_{51}\mathrm{Sb}^{123}_{72}$		7/2	+2.547	−0.37
$_{51}\mathrm{Sb}^{124}_{73}$	60 d	3		
$_{51}\mathrm{Sb}^{125}_{74}$	2.7 y	[7/2]	±2.6	
$_{52}\mathrm{Te}^{116}_{64}$	2.5 h	0		
$_{52}\mathrm{Te}^{117}_{65}$	61 m	1/2		
$_{52}\mathrm{Te}^{119}_{67}$	16 h	1/2	±0.25	
$_{52}\mathrm{Te}^{119m}_{67}$	4.5 d	11/2		
$_{52}\mathrm{Te}^{123}_{71}$	>50 Ty	1/2	−0.7359	
$_{52}\mathrm{Te}^{125}_{73}$		1/2	−0.8871	
$_{52}\mathrm{Te}^{125m}_{73}$	35.5	3/2	+0.60	−0.2
$_{52}\mathrm{Te}^{126}_{74}$		0		
$_{52}\mathrm{Te}^{128}_{76}$		0		
$_{52}\mathrm{Te}^{130}_{78}$		0		
$_{53}\mathrm{I}^{123}_{70}$	13 h	5/2		
$_{53}\mathrm{I}^{124}_{71}$	4.0 d	2		
$_{53}\mathrm{I}^{125}_{72}$	60 d	5/2	+3.0	−0.89
$_{53}\mathrm{I}^{126}_{73}$	13 d	2		
$_{53}\mathrm{I}^{127}_{74}$		5/2	+2.808	−0.79
$_{53}\mathrm{I}^{127m}_{74}$	59	[7/2]		−0.71

TABLE A-2: (Continued)

Nucleus	*T½ or level (keV)*	*I*	*μ*	*Q*
$_{53}I^{128}{}_{75}$	25 m	1		
$_{53}I^{129}{}_{76}$	16 My	7/2	+2.617	−0.55
$_{53}I^{129m}{}_{76}$	27	[5/2]	+2.8	−0.68
$_{53}I^{130}{}_{77}$	12 h	5		
$_{53}I^{131}{}_{78}$	8.1 d	7/2	+2.74	−0.40
$_{53}I^{132}{}_{79}$	2.3 h	4	±3.08	±0.08
$_{53}I^{133}{}_{80}$	21 h	7/2	+2.84	−0.26
$_{53}I^{135}{}_{82}$	6.7 h	7/2		
$_{54}Xe^{129}{}_{75}$		1/2	−0.7768	
$_{54}Xe^{129m}{}_{75}$	40	[3/2]		±0.41
$_{54}Xe^{131}{}_{77}$		3/2	+0.6908	−0.12
$_{54}Xe^{132}{}_{78}$		0		
$_{54}Xe^{134}{}_{80}$		0		
$_{54}Xe^{136}{}_{82}$		0		
$_{55}Cs^{125}{}_{70}$	45 m	1/2	+1.41	
$_{55}Cs^{127}{}_{72}$	6.2 h	1/2	+1.46	
$_{55}Cs^{129}{}_{74}$	31 h	1/2	+?1.479	
$_{55}Cs^{130}{}_{75}$	30 m	1	+1.37 or −1.45	
$_{55}Cs^{131}{}_{76}$	10 d	5/2	+3.54	−0.57
$_{55}Cs^{132}{}_{77}$	6.2 d	2	+2.22	+0.46
$_{55}Cs^{133}{}_{78}$		7/2	+2.578	−0.003
$_{55}Cs^{133m}{}_{78}$	81	[5/2]	+3.44	
$_{55}Cs^{134}{}_{79}$	2.2 y	4	+2.990	+0.36
$_{55}Cs^{134m}{}_{79}$	3.1 h	8	+1.096	
$_{55}Cs^{135}{}_{80}$	2 My	7/2	+2.729	+0.044
$_{55}Cs^{136}{}_{81}$	13 d	5	+3.70	
$_{55}Cs^{137}{}_{82}$	30 y	7/2	+2.838	+0.045
$_{55}Cs^{138}{}_{83}$	32 m	3	±0.5	
$_{56}Ba^{134}{}_{78}$		0		
$_{56}Ba^{135}{}_{79}$		3/2	+0.8365	+0.18
$_{56}Ba^{136}{}_{80}$		0		
$_{56}Ba^{137}{}_{81}$		3/2	+0.9357	+0.28
$_{56}Ba^{138}{}_{82}$		0		
$_{57}La^{138}{}_{81}$	0.1 Ty	5	+3.707	±0.8
$_{57}La^{139}{}_{82}$		7/2	+2.778	+0.22
$_{57}La^{140}{}_{83}$	40 h	3		
$_{58}Ce^{141}{}_{83}$	33 d	7/2	±0.9	
$_{58}Ce^{143}{}_{85}$	33 h	3/2		
$_{59}Pr^{141}{}_{82}$		5/2	+4.3	−0.07
$_{59}Pr^{142}{}_{83}$	19 h	2	±0.25	±0.03
$_{59}Pr^{143}{}_{84}$	14 d	7/2		
$_{60}Nd^{141}{}_{81}$	2.4 h	3/2		
$_{60}Nd^{143}{}_{83}$		7/2	−1.08	−0.48
$_{60}Nd^{145}{}_{85}$		7/2	−0.66	−0.25
$_{60}Nd^{147}{}_{87}$	11 d	5/2	±0.59	
$_{60}Nd^{149}{}_{89}$	1.9 h	5/2		

Nucleus	*T½ or level (keV)*	*I*	*μ*	*Q*
$_{61}Pm^{147}{}_{86}$	2.6 y	7/2	+2.7	+0.7
$_{61}Pm^{148}{}_{87}$	5.4 d	1	+2.0	+0.2
$_{61}Pm^{149}{}_{88}$	54 h	7/2		
$_{61}Pm^{151}{}_{90}$	28 h	5/2	±1.6	±1.9
$_{62}Sm^{147}{}_{85}$	0.1 Ty	7/2	−0.813	−0.20
$_{62}Sm^{149}{}_{87}$		7/2	−0.670	+0.058
$_{62}Sm^{149m}{}_{87}$	22	[5/2]	−0.62	+0.4
$_{62}Sm^{152m}{}_{90}$	122	[2]	+0.84	
$_{62}Sm^{153}{}_{91}$	47 h	3/2	−0.022	+1.0
$_{62}Sm^{155}{}_{93}$	24 m	3/2		±0.9
$_{63}Eu^{151}{}_{88}$		5/2	+3.464	+1.1
$_{63}Eu^{151m}{}_{88}$	21.7	7/2	+2.57	+1.6
$_{63}Eu^{152}{}_{89}$	13 y	3	±1.924	±3.0
$_{63}Eu^{152m}{}_{89}$	9.3 h	0		
$_{63}Eu^{153}{}_{90}$		5/2	+1.530	+2.8
$_{63}Eu^{153m}{}_{90}$	97	[5/2]	+3.2 or −0.5	
$_{63}Eu^{153m}{}_{90}$	103	[3/2]	+2.0	
$_{63}Eu^{154}{}_{91}$	16 y	3	±2.000	
$_{64}Gd^{153}{}_{89}$	242 d	3/2		
$_{64}Gd^{155}{}_{91}$		3/2	−0.254	+2.3
$_{64}Gd^{155m}{}_{91}$	86	5/2	+0.9 or −0.5	±0.2
$_{64}Gd^{155m}{}_{91}$	105	3/2	+0.1 or −0.4	±1
$_{64}Gd^{156m}{}_{92}$	89	[2]	±0.79	±1
$_{64}Gd^{157}{}_{93}$		3/2	−0.339	+1.7
$_{64}Gd^{157m}{}_{93}$	64	[5/2]		±2.9
$_{64}Gd^{158m}{}_{94}$	79	[2]	±0.86	±1.1
$_{64}Gd^{159}{}_{95}$	18 h	3/2		
$_{64}Gd^{160m}{}_{96}$	75	[2]		±1.1
$_{65}Tb^{157}{}_{92}$	>30 y	[3/2]	±2.0	
$_{65}Tb^{158}{}_{93}$	150 y	3	±1.74	+2.7
$_{65}Tb^{159}{}_{94}$		3/2	±1.99	+1.3
$_{65}Tb^{159m}{}_{94}$	58	[5/2]	±2	
$_{65}Tb^{160}{}_{95}$	73 d	3	±1.68	+3.0
$_{65}Tb^{161}{}_{96}$	6.9 d	3/2		
$_{66}Dy^{159}{}_{93}$	144 d	3/2		
$_{66}Dy^{160m}{}_{94}$	87	[2]	±0.74	negative
$_{66}Dy^{161}{}_{95}$		5/2	−0.46	+2.3
$_{66}Dy^{161m}{}_{95}$	26	5/2	+0.55	+2.3
$_{66}Dy^{161m}{}_{95}$	74	[3/2]	−0.39	+1.4
$_{66}Dy^{162m}{}_{96}$	81	[2]	+0.74	
$_{66}Dy^{163}{}_{97}$		5/2	+0.64	+2.5
$_{66}Dy^{164m}{}_{98}$	73	[2]	+0.66	−1.9
$_{66}Dy^{165}{}_{99}$	2.3 h	7/2	±0.50	
$_{66}Dy^{166}{}_{100}$	82 h	0		
$_{67}Ho^{161}{}_{94}$	2.5 h	7/2		

TABLE A-2: (Continued)

Nucleus	*T½ or level (keV)*	*I*	*μ*	*Q*
${}_{67}\mathrm{Ho}^{165}{}_{98}$		7/2	+4.12	+3.0
${}_{67}\mathrm{Ho}^{166}{}_{99}$	27 h	0		
${}_{68}\mathrm{Er}^{160}{}_{92}$	29 h	0		
${}_{68}\mathrm{Rr}^{163}{}_{95}$	75 m	5/2	+1.1	+3.9
${}_{68}\mathrm{Er}^{164m}{}_{96}$	92	[2]	±0.71	
${}_{68}\mathrm{Er}^{165}{}_{97}$	10 h	5/2	±0.65	±2.2
${}_{68}\mathrm{Er}^{166m}{}_{98}$	81	[2]	±0.63	−2.0
${}_{68}\mathrm{Er}^{167}{}_{99}$		7/2	−0.564	+2.8
${}_{68}\mathrm{Er}^{168m}{}_{100}$	80	[2]	±0.67	
${}_{68}\mathrm{Er}^{169}{}_{101}$	9.4 d	1/2	+0.513	
${}_{68}\mathrm{Er}^{170m}{}_{102}$	79	[2]	±0.63	±2.1
${}_{68}\mathrm{Er}^{171}{}_{103}$	7.5 h	5/2	±0.70	±2.3
${}_{69}\mathrm{Tm}^{163}{}_{94}$	1.8 h	1/2	±0.08	
${}_{69}\mathrm{Tm}^{165}{}_{96}$	29 h	1/2		
${}_{69}\mathrm{Tm}^{166}{}_{97}$	7.7 h	2	±0.047	±4.5
${}_{69}\mathrm{Tm}^{167}{}_{98}$	9.6 d	1/2		
${}_{69}\mathrm{Tm}^{169}{}_{100}$		1/2	−0.323	
${}_{69}\mathrm{Tm}^{169m}{}_{100}$	8.4	[3/2]	+0.52	−1.3
${}_{69}\mathrm{Tm}^{170}{}_{101}$	127 d	1	±0.246	±0.59
${}_{69}\mathrm{Tm}^{171}{}_{102}$	1.9 y	1/2	±0.229	
${}_{70}\mathrm{Yb}^{170m}{}_{100}$	84	[2]	±0.67	negative
${}_{70}\mathrm{Yb}^{171}{}_{101}$		1/2	+0.4919	
${}_{70}\mathrm{Yb}^{171m}{}_{101}$	67	[3/2]	±0.35	
${}_{70}\mathrm{Yb}^{171m}{}_{101}$	76	[5/2]	+1.01	
${}_{70}\mathrm{Yb}^{172m}{}_{102}$	79	[2]	±0.66	
${}_{70}\mathrm{Yb}^{'73}{}_{103}$		5/2	−0.6776	+3.0
${}_{70}\mathrm{Yb}^{174m}{}_{104}$	76	[2]	±0.67	
${}_{70}\mathrm{Yb}^{176m}{}_{106}$	82	[2]	±0.76	
${}_{71}\mathrm{Lu}^{169}{}_{98}$	1.5 d	7/2		
${}_{71}\mathrm{Lu}^{170}{}_{99}$	2.0 d	0		
${}_{71}\mathrm{Lu}^{171}{}_{100}$	8.3 d	7/2		
${}_{71}\mathrm{Lu}^{175}{}_{104}$		7/2	+2.23	+5.6
${}_{71}\mathrm{Lu}^{176}{}_{105}$	20 Gy	7	+3.18	+8.0
${}_{71}\mathrm{Lu}^{176m}{}_{105}$	3.7 h	1	+0.318	−2.3
${}_{71}\mathrm{Lu}^{177}{}_{106}$	6.8 d	7/2	+2.24	+5.4
${}_{72}\mathrm{Hf}^{177}{}_{105}$		7/2	+0.61	+3
${}_{72}\mathrm{Hf}^{178}{}_{106}$		0		
${}_{72}\mathrm{Hf}^{178m}{}_{106}$	93	[2]		
${}_{72}\mathrm{Hf}^{179}{}_{107}$		9/2	−0.47	+3
${}_{72}\mathrm{Hf}^{180}{}_{108}$		0		
${}_{73}\mathrm{Ta}^{181}{}_{108}$		7/2	+2.36	+4.2
${}_{73}\mathrm{Ta}^{181m}{}_{108}$	6.2	[9/2]	+5.1	+3
${}_{73}\mathrm{Ta}^{183}{}_{110}$	5.0 d	7/2		
${}_{74}\mathrm{W}^{182}{}_{108}$		0		
${}_{74}\mathrm{W}^{182m}{}_{108}$	100	[2]	±0.50	
${}_{74}\mathrm{W}^{183}{}_{109}$		1/2	+0.117	
${}_{74}\mathrm{W}^{183m}{}_{109}$	46	[3/2]	−0.1	
${}_{74}\mathrm{W}^{183m}{}_{109}$	99	[5/2]	±0.7	
${}_{74}\mathrm{W}^{184}{}_{110}$		0		
${}_{74}\mathrm{W}^{184m}{}_{110}$	111	[2]	±0.55	
${}_{74}\mathrm{W}^{185}{}_{111}$	74 d	3/2		
${}_{74}\mathrm{W}^{186}{}_{112}$		0		
${}_{74}\mathrm{W}^{186m}{}_{112}$	122	[2]	±0.65	
${}_{74}\mathrm{W}^{187}{}_{113}$	24 h	3/2		
${}_{75}\mathrm{Re}^{185}{}_{110}$		5/2	+3.172	+2.7
${}_{75}\mathrm{Re}^{186}{}_{111}$	90 h	1	±1.73	~±0.4
${}_{75}\mathrm{Re}^{187}{}_{112}$	60 Gy	5/2	+3.204	+2.6
${}_{75}\mathrm{Re}^{188}{}_{113}$	17 h	1	±1.78	~±0.4
${}_{76}\mathrm{Os}^{186m}{}_{110}$	137	[2]	±0.64	
${}_{76}\mathrm{Os}^{187}{}_{111}$		1/2	+0.0643	
${}_{76}\mathrm{Os}^{188m}{}_{112}$	155	[2]	±0.62	
${}_{76}\mathrm{Os}^{189}{}_{113}$		3/2	+0.6566	+0.8
${}_{77}\mathrm{Ir}^{191}{}_{114}$		3/2	+0.145	+1.3
${}_{77}\mathrm{Ir}^{191m}{}_{114}$	82	[1/2]	+0.546	
${}_{77}\mathrm{Ir}^{192}{}_{115}$	74 d	4		
${}_{77}\mathrm{Ir}^{193}{}_{116}$		3/2	+0.158	+1.2
${}_{77}\mathrm{Ir}^{193m}{}_{116}$	73	1/2		+0.468
${}_{77}\mathrm{Ir}^{194}{}_{117}$	19 h	1		
${}_{78}\mathrm{Pt}^{194}{}_{116}$		0		
${}_{78}\mathrm{Pt}^{195}{}_{117}$		1/2	+0.6060	
${}_{78}\mathrm{Pt}^{195m}{}_{117}$	99	[3/2]	−0.60	
${}_{78}\mathrm{Pt}^{196}{}_{118}$		0		
${}_{78}\mathrm{Pt}^{197}{}_{119}$	20 h	1/2		
${}_{79}\mathrm{Au}^{190}{}_{111}$	40 m	1	±0.066	
${}_{79}\mathrm{Au}^{191}{}_{112}$	3.0 h	3/2	±0.137	
${}_{79}\mathrm{Au}^{192}{}_{113}$	4.1h	1	±0.00785	
${}_{79}\mathrm{Au}^{193}{}_{114}$	18 h	3/2	±0.139	
${}_{79}\mathrm{Au}^{194}{}_{115}$	39 h	1	±0.074	
${}_{79}\mathrm{Au}^{195}{}_{116}$	192 d	3/2	±0.147	
${}_{79}\mathrm{Au}^{196}{}_{117}$	6.2 d	2	+0.58 or −0.62	
${}_{79}\mathrm{Au}^{196m}{}_{117}$	9.7 h	12		
${}_{79}\mathrm{Au}^{197}{}_{118}$		3/2	+0.14486	+0.58
${}_{79}\mathrm{Au}^{197m}$	77	[1/2]	+0.42	
${}_{79}\mathrm{Au}^{198}{}_{119}$	2.7 d	2	+0.590	
${}_{79}\mathrm{Au}^{199}{}_{120}$	3.2 d	3/2	+0.270	
${}_{80}\mathrm{Hg}^{193}{}_{113}$	6 h	3/2	−0.62	−2
${}_{80}\mathrm{Hg}^{193m}{}_{113}$	11 h	13/2	−1.063	−1.2
${}_{80}\mathrm{Hg}^{195}{}_{115}$	9.5 h	1/2	+0.538	
${}_{80}\mathrm{Hg}^{195m}{}_{115}$	40 h	13/2	−1.049	+1.3
${}_{80}\mathrm{Hg}^{197}{}_{117}$	65 h	1/2	+0.524	
${}_{80}\mathrm{Hg}^{197m}{}_{117}$	24 h	13/2	−1.032	+1.5
${}_{80}\mathrm{Hg}^{198}{}_{118}$		0		
${}_{80}\mathrm{Hg}^{199}{}_{119}$		1/2	+0.5027	

TABLE A-2: (Continued)

Nucleus	*T½ or level (keV)*	*I*	μ	*Q*
$_{80}Hg^{200}{}_{120}$		0		
$_{80}Hg^{201}{}_{121}$		3/2	−0.5567	+0.45
$_{80}Hg^{202}{}_{122}$		0		
$_{80}Hg^{203}{}_{123}$	47 d	5/2	+0.84	$\pm \leq 13$
$_{80}Hg^{204}{}_{124}$		0		
$_{81}Tl^{195}{}_{114}$	1.2 h	1/2		
$_{81}Tl^{197}{}_{116}$	2.7 h	1/2	+1.55	
$_{81}Tl^{198}{}_{117}$	5.3 h	2	$\pm <0.002$	
$_{81}Tl^{198m}{}_{117}$	1.8 h	7		
$_{81}Tl^{199}{}_{118}$	7.4 h	1/2	+1.59	
$_{81}Tl^{200}{}_{119}$	26 h	2	$\pm \leq 0.15$	
$_{81}Tl^{201}{}_{120}$	72 h	1/2	+1.60	
$_{81}Tl^{202}{}_{121}$	12 d	2	$\pm \leq 0.15$	
$_{81}Tl^{203}{}_{122}$		1/2	+1.6115	
$_{81}Tl^{204}{}_{123}$	3.9 y	2	±0.089	
$_{81}Tl^{205}{}_{124}$		1/2	+1.6274	
$_{81}Tl^{206}{}_{125}$	4.2 m	0		
$_{82}Pb^{206}{}_{124}$		0		
$_{82}Pb^{207}{}_{125}$		1/2	+0.5895	
$_{82}Pb^{208}{}_{126}$		0		
$_{83}Bi^{199}{}_{116}$	25 m	9/2		
$_{83}Bi^{200}{}_{117}$	35 m	7		
$_{83}Bi^{201}{}_{118}$	1.8 h	9/2		
$_{83}Bi^{202}{}_{119}$	1.6 h	5		
$_{83}Bi^{203}{}_{120}$	12 h	9/2	+4.59	−0.64
$_{83}Bi^{204}{}_{121}$	12 h	6	+4.25	−0.41
$_{83}Bi^{205}{}_{122}$	15 d	9/2	+5.5	
$_{83}Bi^{206}{}_{123}$	6.3 d	6	+4.56	−0.19
$_{83}Bi^{209}{}_{126}$	$>10^{18}$ y	9/2	+4.080	−0.35
$_{83}Bi^{210}{}_{127}$	5 d	1	±0.0442	±0.13
$_{84}Po^{201}{}_{117}$	18 m	3/2		
$_{84}Po^{202}{}_{118}$	51 m	0		
$_{84}Po^{203}{}_{119}$	42 m	5/2		
$_{84}Po^{204}{}_{120}$	3.5 h	0		
$_{84}Po^{205}{}_{121}$	1.8 h	5/2	≈+0.26	+0.17
$_{84}Po^{206}{}_{122}$	8.8 d	0		
$_{84}Po^{207}{}_{123}$	6.0 h	5/2	≈+0.27	+0.28
$_{84}Po^{209}{}_{125}$	103 y	1/2	+0.76	
$_{84}Po^{210}{}_{126}$	138 d	0		
$_{85}At^{211}{}_{126}$	7.2 h	9/2		
$_{89}Ac^{227}{}_{138}$	22 y	3/2	+1.1	+1.7
$_{90}Th^{229}{}_{139}$	7.3 ky	5/2	+0.38	≈4.6
$_{91}Pa^{231}{}_{140}$	34 ky	3/2	±1.98	
$_{91}Pa^{233}{}_{142}$	27 d	3/2	+3.4	−3.0
$_{92}U^{233}{}_{141}$	0.2 My	5/2	+0.54	+3.5
$_{92}U^{235}{}_{143}$	0.7 Gy	7/2	−0.35	+4.1
$_{93}Np^{237}{}_{144}$	2.2 My	5/2	+3.3	Negative
$_{93}Np^{237m}{}_{144}$	60	[5/2]	+1.8	
$_{93}Np^{238}{}_{145}$	2.1 d	2		
$_{93}Np^{239}{}_{146}$	2.3 d	5/2		
$_{94}Pu^{239}{}_{145}$	24 ky	1/2	+0.200	
$_{94}Pu^{241}{}_{147}$	13 y	5/2	−0.73	+5.6
$_{95}Am^{241}{}_{146}$	460 y	5/2	+1.59	+4.9
$_{95}Am^{242}{}_{147}$	16 h	1	±0.382	±2.8
$_{95}Am^{243}{}_{148}$	8 ky	5/2	+1.4	+4.9
$_{96}Cm^{242}{}_{146}$	160 d	0		
$_{97}Bk^{249}{}_{152}$	314 d	7/2		
$_{99}Es^{253}{}_{154}$	20 d	$\geq 7/2$		

TABLE A-3: MASSES OF VARIOUS NUCLEI†

A	EL	MASS EXCESS (KEV)	
1	N	8071.53	0.07
	H	7289.04	0.05
2	H	13135.9	0.1
3	H	14950.00	0.17
	HE	14931.35	0.17
4	H	25920	500
	HE	2424.92	0.25
	LI	25130	300
5	H	33790	800
	HE	11390	50
	LI	11680	50
6	HE	17596.8	3.9
	LI	14087.0	0.7
	BE	18374	5
7	HE	26110	30
	LI	14907.8	0.8
	BE	15769.5	0.8
	B	27940	100
8	HE	31650	120
	LI	20946.6	1.1
	BE	4941.7	0.5
	B	22921.6	1.2
9	LI	24966	5
	BE	11348.1	0.6
	B	12415.4	0.9
	C	28911	6
10	BE	12607.7	0.7
	B	12051.98	0.39
	C	15702.4	1.8
11	BE	20179	15
	B	8667.70	0.28
	C	10650.2	1.2
12	B	13370.0	1.3
	C	0.0	0.0
	N	17343	5
13	B	16562	4
	C	3125.14	0.22
	N	5345.6	0.9
	O	23110	70
14	C	3019.89	0.28
	N	2863.74	0.13
	O	8007.4	0.4
15	C	9873.3	0.9
	N	101.6	0.4
	O	2860	1
16	C	13693	16
	N	5683.4	2.4
	O	-4736.58	0.19
	F	10693	14
17	N	7871	15
	O	-807.4	0.9
	F	1951.7	0.5
	NE	16470	190
18	N	13274	30
	O	-782.50	0.27
	F	872.8	0.9
	NE	5319	5
19	O	3332.2	2.5
	F	-1486.2	0.7
	NE	1752.0	1.1
	NA	12970	70
20	O	3799	8
	F	-15.9	0.8
	NE	-7041.5	0.4
	NA	6840	40
21	F	-46	7
	NE	-5729.9	1.2
	NA	-2183	9
	MG	10620	120
22	F	2828	30
	NE	-8025.0	0.5
	NA	-5183.0	0.7
	MG	-372	15
23	NE	-5150.0	2.8
	NA	-9528.4	1.5
	MG	-5471.9	2.7
	AL	6770	80
24	NE	-5948	10
	NA	-8416.2	1.5
	MG	-13930	1
	AL	-48	7
25	NA	-9356	9
	MG	-13190.8	1.2
	AL	-8911.9	1.4
26	NA	-7510	300
	MG	-16212	1
	AL	-12208.0	1.1
	SI	-7146	8
27	NA	-6580	700
	MG	-14584.1	1.4
	AL	-17194.3	0.9
	SI	-12384.7	1.9
28	MG	-15016.4	2.3
	AL	-16848.2	1.1
	SI	-21490.4	0.8
	P	-7153	8
29	AL	-18212	5
	SI	-21892	1
	P	-16949	8
30	AL	-15890	40
	SI	-24430.5	1.1
	P	-20203.2	2.8
	S	-14064	10
31	SI	-22946.4	1.3
	P	-24438.7	0.8
	S	-18997	11
32	SI	-24090	7
	P	-24303.3	0.8
	S	-26013.2	0.7
	CL	-13262	12
33	P	-26336.1	2.4
	S	-26585.1	1.4
	CL	-21001.8	3.5
34	P	-24830	200
	S	-29928.5	1.5
	CL	-24437.7	1.9
	AR	-18394	12
35	S	-28845.1	0.4
	CL	-29012.52	0.36
	AR	-23048.9	1.7
36	S	-30664.4	1.2
	CL	-29520.7	1.2
	AR	-30229.4	1.2
37	S	-26906	30
	CL	-31760.5	0.3
	AR	-30946.6	0.7
	K	-24797.4	1.8
	CA	-13230	50
38	S	-26863	30
	CL	-29799	4
	AR	-34714	1
	K	-28790	8
	CA	-22022	16
39	CL	-29801	18
	AR	-33239	5
	K	-33803	1
	CA	-27280	7
40	CL	-27540	500
	AR	-35038.3	0.6
	K	-33533	1
	CA	-34844	1
	SC	-20520	7
41	AR	-33066.6	1.4
	K	-35559.2	1.3
	CA	-35134.7	1.3
	SC	-28639	5
42	AR	-34420	40
	K	-35020.8	1.6
	CA	-38537.1	2.2
	SC	-32106.0	3.2
	TI	-25119	6
43	K	-36581	10
	CA	-38398.2	2.2
	SC	-36177.3	2.8
	TI	-29319	10
44	K	-35780	40
	CA	-41462.5	2.4
	SC	-37813	6
	TI	-37544	7
45	K	-36613	11
	CA	-40805.3	2.4
	SC	-41061.9	2.1
	TI	-38998.7	3.8

† This table was prepared by A. H. Wapstra and N. B. Gove in June, 1970, from a preliminary version of the 1970 mass adjustment (to be published) and is reproduced here with their kind permission. Atomic masses in $C^{12} = 12$ scale are calculated by a least-squares fit to all available experimental data. The quantity tabulated is the mass excess, (mass $- A)X_0$, where

$$X_0 = 931481 \pm 5 \text{ keV}$$

followed by the standard error (in keV) of this mass excess.

TABLE A-3: (Continued)

A	EL	MASS EXCESS (KEV)	
46	K	-35425	16
	CA	-43140	5
	SC	-41757.3	2.2
	TI	-44124.3	1.7
	V	-37070.0	2.8
47	K	-35703	9
	CA	-42340	4
	SC	-44327.5	2.4
	TI	-44927.8	1.6
	V	-42011	4
48	CA	-44221	5
	SC	-44494	6
	TI	-48484.1	1.5
	V	-44468.6	3.1
	CR	-42815	18
49	CA	-41291	6
	SC	-46550	5
	TI	-48555.9	1.5
	V	-47954.7	1.8
	CR	-45386	10
50	CA	-39576	9
	SC	-44543	16
	TI	-51434.9	2.9
	V	-49215.5	2.9
	CR	-50253	2
	MN	-42622.7	3.3
51	SC	-43225	21
	TI	-49741	6
	V	-52195.4	1.6
	CR	-51444.1	1.9
	MN	-48238	5
52	TI	-49472	10
	V	-51435.1	1.9
	CR	-55413.6	2.2
	MN	-50704	4
	FE	-48332	13
53	V	-51859	25
	CR	-55282.5	2.2
	MN	-54684.8	2.6
	FE	-50942	15
54	CR	-56931.1	2.2
	MN	-55556	4
	FE	-56251.5	2.8
	CO	-48002	5
55	CR	-55120	6
	MN	-57710.2	2.7
	FE	-57478.5	2.7
	CO	-54012.5	3.4
56	CR	-55265	30
	MN	-56909.9	2.8
	FE	-60606.4	2.5
	CO	-56038	3
	NI	-53908	11
57	MN	-57620	50
	FE	-60181.1	2.5
	CO	-59344.3	3.2
	NI	-56101	8
58	MN	-56050	110
	FE	-62152.7	2.7
	CO	-59846	5
	NI	-60233.0	3.1
	CU	-51666	4
59	FE	-60668.1	3.7
	CO	-62233.9	2.8
	NI	-61160	3
	CU	-56361	11
60	FE	-61433	30
	CO	-61654.1	2.9
	NI	-64477	3
	CU	-58351	5
	ZN	-54191	18
61	FE	-59030	70
	CO	-62919	18
	NI	-64225	3
	CU	-61980.3	3.8
	ZN	-56580	200
62	CO	+61530	40
	NI	-66750.9	3.1
	CU	-62804	5
	ZN	-61114	10
63	CO	-61859	19
	NI	-65516.9	3.2
	CU	-65582.8	3.2
	ZN	-62217	4
64	CO	-60110	500
	NI	-67105.0	3.7
	CU	-65427.5	3.2
	ZN	-66002.7	3.3
	GA	-58931	30
65	NI	-65130	6
	CU	-67261.4	3.5
	ZN	-65910.7	3.4
	GA	-62652	15
66	NI	-66058	30
	CU	-66257.7	3.8
	ZN	-68893.2	3.5
	GA	-63718	5
	GE	-61616	14
67	NI	-63200	300
	CU	-67301	8
	ZN	-67875.4	3.5
	GA	-66874.9	3.9
	GE	-62450	50
68	CU	-65420	60
	ZN	-70003.1	3.4
	GA	-67084	5
69	CU	-65930	70
	ZN	-68413	5
	GA	-69321.4	2.8
	GE	-67095.9	3.7
	AS	-63200	300
70	ZN	-69558.3	3.1
	GA	-68904.5	2.9
	GE	-70558.4	1.6
	AS	-64336	20
71	ZN	-67331	15
	GA	-70137.1	2.5
	GE	-69902.0	2.3
	AS	-67893	7
72	ZN	-68130	7
	GA	-68586.7	2.7
	GE	-72580.4	1.7
	AS	-68229	7
73	GA	-69740	40
	GE	-71293.2	1.8
	AS	-70954	15
	SE	-68214	18
	BR	-63510	500
74	GA	-67920	50
	GE	-73422.7	1.8
	AS	-70858.9	3.3
	SE	-72212	5
75	GA	-68540	200
	GE	-71841	12
	AS	-73028.0	2.3
	SE	-72163.3	2.5
	BR	-69153	20
76	GE	-73212.3	2.3
	AS	-72284.7	2.3
	SE	-75252.9	2.1
77	GE	-71160	40
	AS	-73916	9
	SE	-74599.8	2.1
	BR	-73235.3	3.5
	KR	-70235	30
78	GE	-71780	100
	AS	-72760	100
	SE	-77025	2
	BR	-73452	4
	KR	-74145	5
79	GE	-69390	210
	AS	-73690	50
	SE	-75931	6
	BR	-76072.3	3.3
	KR	-74441	9
	RB	-70920	50
80	AS	-71760	200
	SE	-77755.4	2.4
	BR	-75883	3
	KR	-77894	6
81	AS	-72590	200
	SE	-76385	5
	BR	-77972	5
	KR	-77680	100
	RB	-75420	100
82	SE	-77585	6
	BR	-77502	7
	KR	-80590	5
	RB	-76420	30
83	SE	-75438	32
	BR	-79016	7
	KR	-79985	4
	RB	-78947	32
	SR	-76697	30
84	SE	-75920	70
	BR	-77730	50
	KR	-82433.9	3.2
	RB	-79754	4
	SR	-80638.3	3.5
	Y	-73688	30
85	BR	-78690	100
	KR	-81488	6
	RB	-82164.9	3.4
	SR	-81095	6
	Y	-77835	12
86	BR	-75960	400
	KR	-83259.5	3.8
	RB	-82744.3	3.4
	SR	-84505.4	2.3
	Y	-79232	10
87	KR	-80699	9
	RB	-84589.3	2.5
	SR	-84862.4	2.2
	Y	-82980	7
	ZR	-79480	21
88	KR	-79700	100
	RB	-82600	15
	SR	-87904.2	2.3
	Y	-84286	5
	ZR	-83620	200
89	KR	-76560	33
	RB	-81710	13
	SR	-86195.6	3.8
	Y	-87687.0	2.9
	ZR	-84853	4
	NB	-80980	100
	MO	-75010	320
90	KR	-74740	90
	RB	-79300	90
	SR	-85930.1	3.6
	Y	-86476	3
	ZR	-88768.7	2.6
	NB	-82658	6
	MO	-80171	7

TABLE A-3: (Continued)

A	EL	MASS EXCESS (KEV)	
91	RB	-78000	150
	SR	-83677	11
	Y	-86342	6
	ZR	-87887.0	2.6
	NB	-86625	7
	MO	-82182	29
92	SR	-82920	70
	Y	-84831	20
	ZR	-88454.0	2.5
	NB	-86450	8
	MO	-86805.5	2.8
	TC	-78860	140
93	SR	-79950	70
	Y	-84250	20
	ZR	-87139.8	3.5
	NB	-87203	3
	MO	-86806.6	3.9
	TC	-83621	14
94	SR	-78740	230
	Y	-82260	200
	ZR	-87262.6	2.6
	NB	-86360	3
	MO	-88406.4	2.2
	TC	-84146	6
	RU	-82750	300
95	Y	-81226	20
	ZR	-85656.4	2.7
	NB	-86783.7	1.8
	MO	-87708.7	1.7
	TC	-86008	11
	RU	-83448	11
96	ZR	-85425	4
	NB	-85604	4
	MO	-88791.4	1.7
	TC	-85850	50
	RU	-86071	5
97	ZR	-82931	16
	NB	-85603	16
	MO	-87538.4	1.7
	TC	-87193	9
	RU	-86040	100
	RH	-82550	100
98	ZR	-81272	20
	NB	-83510	100
	MO	-88109.1	1.7
	TC	-86520	200
	RU	-88221	4
	RH	-83250	400
99	MO	-85953.6	1.9
	TC	-87325.9	3.2
	RU	-87618.1	2.8
	RH	-85566	20
	PD	-82161	28
	AG	-76120	100
100	MO	-86183.2	2.9
	TC	-85850	60
	RU	-89220.1	2.8
	RH	-85590	20
101	MO	-83503	19
	TC	-86323	25
	RU	-87953.8	2.6
	RH	-87399	13
	PD	-85409	23
	AG	-80740	80
	CD	-75210	150
102	RU	-89098.4	2.8
	RH	-86776	7
	PD	-87925	9
	AG	-82365	22
103	TC	-84910	100
	RU	-87258	6
	RH	-88013	4
	PD	-87463	27
	AG	-84780	60
104	RU	-88092	5
	RH	-86941	4
	PD	-89411	10
	AG	-85311	32
105	TC	-82530	200
	RU	-85932	6
	RH	-87850	4
	PD	-88415.8	3.5
	AG	-87081	20
	CD	-84280	100
106	RU	-86326	11
	RH	-86365	11
	PD	-89905.4	3.4
	AG	-86931	11
	CD	-87128.3	3.8
	IN	-80388	30
107	RU	-83710	300
	RH	-86860	40
	PD	-88367.5	3.4
	AG	-88402.7	3.3
	CD	-86986	5
	IN	-83500	150
108	RU	-83700	610
	RH	-85020	600
	PD	-89522.8	3.7
	AG	-87600.2	3.5
	CD	-89245.5	3.6
	IN	-84100	80
109	PD	-87601.5	3.6
	AG	-88717.4	3.2
	CD	-88535	4
	IN	-86516	8
110	RH	-82840	500
	PD	-88338	13
	AG	-87452.7	3.2
	CD	-90342.2	1.9
	IN	-86410	50
111	PD	-86020	50
	AG	-88222.2	3.5
	CD	-89250.2	1.9
	IN	-88425	29
	SN	-85917	17
112	PD	-86270	50
	AG	-86570	50
	CD	-90573	2
	IN	-87989	9
	SN	-88647	8
	SB	-81850	50
113	AG	-87033	20
	CD	-89043.1	1.8
	IN	-89340	9
	SN	-88312	13
	SB	-84450	40
114	CD	-90012.4	1.8
	IN	-88581	7
	SN	-90562	7
	SB	-84870	50
115	AG	-84910	100
	CD	-88088	10
	IN	-89538	8
	SN	-90025	6
	SB	-86995	21
	TE	-82450	200
116	CD	-88713.0	2.1
	IN	-88299	12
	SN	-91520.1	3.2
	SB	-87020	40
	TE	-85460	110
117	CD	-86406	13
	IN	-88927	10
	SN	-90390	2
	SB	-88640	40
	TE	-8[illegible]150	50
	I	-80840	110
118	CD	-86702	20
	IN	-87450	300
	SN	-91646	2
	SB	-87951	6
119	CD	-84210	300
	IN	-87712	38
	SN	-90059	2
	SB	-89481	20
	TE	-87187	20
	I	-83990	400
	XE	-79000	420
120	IN	-85490	600
	SN	-91092.4	2.1
	SB	-88412	7
	TE	-89400	13
	I	-84320	100
121	IN	-85820	40
	SN	-89200	3
	SB	-89587.9	2.1
122	IN	-83230	800
	SN	-89933.6	3.1
	SB	-88323.7	2.3
	TE	-90302.2	3.2
	I	-86160	40
123	IN	-83410	50
	SN	-87807	6
	SB	-89217.0	2.4
	TE	-89159.9	2.8
124	SN	-88227.5	3.9
	SB	-87612.3	2.6
	TE	-90512.2	2.6
	I	-87342	20
	XE	-87450	140
125	SN	-85888	6
	SB	-88260	4
	TE	-89025.1	2.8
	I	-88877	3
126	SN	-86012	15
	SB	-86330	150
	TE	-90063.0	2.8
	I	-87912	6
	XE	-89163	7
127	SB	-86706	7
	TE	-88287	4
	I	-88979.4	3.4
	XE	-88315	5
	CS	-86225	21
128	SN	-83400	210
	SB	-84700	150
	TE	-88986.9	3.1
	I	-87733.3	3.6
	XE	-89858.0	1.6
	CS	-85951	25
129	SB	-84589	30
	TE	-87002	8
	I	-88502	6
	XE	-88692	4
130	TE	-87343.4	3.6
	I	-86887	10
	XE	-89878.6	1.6
	CS	-86853	10
	BA	-87294	10
131	TE	-85190	6
	I	-87444.2	3.7
	XE	-88415.0	3.7
	CS	-88060	7
	BA	-86716	18
	LA	-83760	40
132	TE	-85187	22
	I	-85692	16
	XE	-89272	4
	CS	-87173	23
	BA	-88447	9
	LA	-83740	50

TABLE A-3: (Continued)

A	EL	MASS EXCESS (KEV)	
133	TE	-82900	110
	I	-85860	50
	XE	-87656	8
	CS	-88083	8
	BA	-87568	8
134	I	-83970	60
	XE	-88121	4
	CS	-86902	7
	BA	-88961	7
	LA	-85251	26
135	I	-83778	30
	XE	-86498	11
	CS	-87655	9
	BA	-87864	7
	LA	-86820	120
136	I	-79420	100
	XE	-86421	5
	CS	-86354	8
	BA	-88901	7
	LA	-86030	70
	CE	-86450	31
	PR	-81384	36
	ND	-78920	60
137	XE	-82211	21
	CS	-86561	8
	BA	-87734	7
138	BA	-88274	7
	LA	-86493	17
	CE	-87524	15
	PR	-83087	18
139	XE	-75980	220
	CS	-80780	100
	BA	-84926	7
	LA	-87183	12
	CE	-86908	19
	PR	-84796	28
140	CS	-77540	100
	BA	-83238	16
	LA	-84273	12
	CE	-88038	11
	PR	-84650	13
141	BA	-79970	110
	LA	-82965	32
	CE	-85395	11
	PR	-85976	11
	ND	-84171	19
	PM	-80460	40
142	BA	-77770	100
	LA	-79966	13
	CE	-84483	12
	PR	-83748	11
	ND	-85912	11
	PM	-81090	100
	SM	-79040	120
143	LA	-78210	40
	CE	-81589	13
	PR	-83034	11
	ND	-83966	11
	PM	-82901	17
	SM	-79422	32
	EU	-74420	200
144	CE	-80399	12
	PR	-80715	12
	ND	-83712	11
	PM	-81340	40
	SM	-81900	12
	EU	-75573	32
145	CE	-77110	100
	PR	-79595	15
	ND	-81400	11
	PM	-81230	13
	SM	-80592	12
	EU	-77872	19

A	EL	MASS EXCESS (KEV)	
146	CE	-75730	120
	PR	-76810	100
	ND	-80894	11
	PM	-79418	26
	SM	-80944	21
	EU	-77072	23
147	PR	-75430	200
	ND	-78126	11
	PM	-79020	11
	SM	-79245	11
	EU	-77482	15
	GD	-75154	29
148	ND	-77378	12
	PM	-76849	15
	SM	-79314	11
	EU	-76214	32
	GD	-76207	16
	TB	-70590	300
149	PR	-71370	200
	ND	-74374	15
	PM	-76043	12
	SM	-77115	11
	GD	-75072	14
	TB	-71372	20
150	ND	-73659	12
	PM	-73530	80
	SM	-77030	11
	EU	-74716	22
	GD	-75725	21
	TB	-71057	23
151	ND	-70896	13
	PM	-73362	15
	SM	-74550	12
	EU	-74626	12
	GD	-74183	16
	TB	-71553	15
	DY	-68548	29
152	ND	-70123	32
	PM	-71250	500
	SM	-74748	12
	EU	-72890	12
	GD	-74711	13
	TB	-70891	33
	DY	-70058	18
	HO	-63670	300
153	PM	-70740	100
	SM	-72543	12
	EU	-73345	12
	GD	-73104	12
	DY	-69090	16
	HO	-64829	20
154	SM	-72450	12
	EU	-71718	14
	GD	-73696	13
	DY	-70353	29
	HO	-64595	23
155	SM	-70193	12
	EU	-71820	12
	GD	-72066	12
	TB	-71221	21
	DY	-69122	20
156	SM	-69360	18
	EU	-70074	15
	GD	-72526	12
	DY	-70492	18
157	EU	-69462	19
	GD	-70822	12
	TB	-70758	13
	DY	-69395	16
158	EU	-67250	120
	GD	-70681	12
	TB	-69441	15
	DY	-70385	14
	HO	-66408	15

A	EL	MASS EXCESS (KEV)	
159	EU	-65920	50
	GD	-68554	13
	TB	-69505	13
	DY	-69139	13
160	GD	-67935	13
	TB	-67815	13
	DY	-69649	13
	HO	-66729	33
161	GD	-65496	16
	TB	-67446	15
	DY	-68029	13
	HO	-67210	40
	ER	-65162	20
	TM	-61640	100
162	GD	-64290	120
	TB	-65690	70
	DY	-68152	14
	HO	-66002	33
	ER	-66300	18
	TM	-61400	100
163	TB	-64670	50
	DY	-66352	14
	HO	-66343	14
	ER	-65135	15
	TM	-62718	25
164	TB	-62600	100
	DY	-65935	14
	HO	-64954	16
	ER	-65919	15
	TM	-61957	25
165	DY	-63579	14
	HO	-64874	14
	ER	-64503	14
	TM	-62938	33
	YB	-60186	39
166	DY	-62565	15
	HO	-63046	14
	ER	-64905	14
	TM	-61870	18
	YB	-61610	27
167	HO	-62300	24
	ER	-63270	14
	TM	-62516	30
	YB	-60561	22
	LU	-57490	70
168	HO	-60200	100
	ER	-62970	14
	TM	-61260	60
	YB	-61544	18
	LU	-56740	400
169	HO	-58750	100
	ER	-60902	14
	TM	-61246	14
	YB	-60340	18
	LU	-58070	35
170	HO	-56390	300
	ER	-60092	14
	TM	-59769	14
	YB	-60737	14
	LU	-57297	24
171	ER	-57697	14
	TM	-59186	14
	YB	-59283	14
172	ER	-56476	19
	TM	-57365	17
	YB	-59235	14
173	ER	-53410	300
	TM	-56212	33
	YB	-57532	14
	LU	-56842	33

TABLE A-3: (Continued)

A	EL	MASS EXCESS (KEV)	
174	TM	-53870	40
	YB	-56929	14
	LU	-55558	16
	HF	-55756	23
175	TM	-52280	50
	YB	-54678	14
	LU	-55146	14
	HF	-54539	16
176	TM	-49340	70
	YB	-53481	15
	LU	-53367	15
	HF	-54556	15
	TA	-51460	200
177	YB	-50972	17
	LU	-52368	15
	HF	-52865	15
	TA	-51707	15
178	LU	-50170	50
	HF	-52420	15
	TA	-50510	100
	W	-50420	100
	RE	-45760	110
179	LU	-49100	40
	HF	-50448	15
	TA	-50329	21
180	LU	-46460	100
	HF	-49763	15
	TA	-48838	27
	W	-49650	200
	RE	-45850	200
181	HF	-47386	15
	TA	-48410	15
	W	-48223	18
182	HF	-45900	200
	TA	-46402	15
	W	-48206	15
	RE	-45346	25
183	HF	-43218	31
	TA	-45258	18
	W	-46326	15
	RE	-45770	17
184	TA	-42635	34
	W	-45665	15
	OS	-44156	23
	IR	-39440	190
185	TA	-41380	100
	W	-43344	16
	RE	-43773	15
	OS	-42758	15
186	TA	-38570	60
	W	-42474	16
	RE	-41880	16
	OS	-42957	16
	IR	-39126	25
187	W	-39868	16
	RE	-41180	16
	OS	-41182	16
188	W	-38633	16
	RE	-38982	16
	OS	-41100	16
	IR	-38267	19
	PT	-37727	21
189	W	-35440	200
	RE	-37941	25
	OS	-38951	16
190	RE	-35490	200
	OS	-38673	16
	IR	-36620	150
	PT	-37292	25
191	OS	-36362	16
	IR	-36672	16
192	OS	-35849	16
	IR	-34799	16
	PT	-36256	17
	AU	-32742	26
193	OS	-33412	17
	IR	-34544	16
	PT	-34483	16
194	OS	-32442	16
	IR	-32539	16
	PT	-34780	13
	AU	-32271	20
	HG	-32221	28
195	OS	-29780	510
	IR	-31780	100
	PT	-32780	13
	AU	-32551	13
196	IR	-29460	60
	PT	-32630	13
	AU	-31149	12
	HG	-31833	12
197	IR	-28410	200
	PT	-30411	11
	AU	-31158	6
	HG	-30743	21
198	IR	-25500	300
	PT	-29903	21
	AU	-29599	6
	HG	-30972	6
	TL	-27510	80
199	PT	-27403	26
	AU	-29097	6
	HG	-29550	6
200	AU	-27310	100
	HG	-29507	6
	TL	-27053	7
201	PT	-23500	110
	AU	-26160	100
	HG	-27664	6
	TL	-27250	60
202	AU	-23850	200
	HG	-27347	6
	TL	-26110	21
	PB	-26060	36
203	HG	-25267	6
	TL	-25758	6
	PB	-24776	13
	BI	-21590	50
204	AU	-20190	300
	HG	-24687	6
	TL	-24343	6
	PB	-25106	6
205	HG	-22160	100
	TL	-23812	6
	PB	-23769	6
	BI	-21065	10
206	HG	-20937	21
	TL	-22244	7
	PB	-23777	6
	BI	-20127	26
	PO	-18309	37
207	TL	-21014	8
	PB	-22446	6
	BI	-20041	10
	PO	-17132	12
	AT	-13290	50
208	TL	-16750	7
	PB	-21743	6
	BI	-18875	13
	PO	-17462	6
209	TL	-13629	16
	PB	-17609	8
	BI	-18257	7
	PO	-16367	7
	AT	-12882	10
210	TL	-9225	14
	PB	-14721	6
	BI	-14784	6
	PO	-15945	6
	AT	-12068	26
	RN	-9724	37
211	PB	-10464	10
	BI	-11840	8
	PO	-12430	6
	AT	-11637	10
	RN	-8742	13
	FR	-4210	50
212	PB	-7545	8
	BI	-8118	7
	PO	-10364	6
	AT	-8624	15
	RN	-8647	8
213	BI	-5223	13
	PO	-6646	9
	AT	-6450	200
	RN	-5697	11
	FR	-3554	11
214	PB	-150	11
	BI	-1185	14
	PO	-4461	6
	AT	-3410	21
	FR	-1055	27
	RA	-27	37
215	BI	1720	90
	PO	-515	10
	AT	-1255	22
	RN	-1171	21
	FR	377	32
	RA	2546	14
	AC	5970	50
216	PO	1785	9
	AT	2259	10
	RN	261	12
	RA	3258	31
217	AT	4401	13
	RN	3666	10
	FR	4430	280
	RA	5884	23
218	PO	8387	11
	AT	8115	14
	RN	5231	11
	FR	7012	23
219	AT	10540	80
	RN	8854	10
	FR	8613	29
	RA	9393	29
220	RN	10614	9
	FR	11482	12
	RA	10278	15
221	FR	13282	13
	RA	12974	11
	AC	14600	350
	TH	16940	30
222	RN	16398	10
	FR	16361	23
	RA	14333	12
	AC	16568	25

TABLE A-3: (Continued)

A	EL	MASS EXCESS (KEV)	
223	FR	18404	10
	RA	17255	10
	AC	17818	30
	TH	19281	35
224	RA	18826	9
	AC	20222	14
	TH	20007	18
225	RA	22032	11
	AC	21641	13
	TH	22319	12
	PA	24410	350
226	RA	23690	10
	AC	24323	11
	TH	23209	11
	PA	25978	27
227	RA	27198	22
	AC	25868	10
	TH	25824	10
	PA	26824	30
228	RA	28958	11
	AC	28904	12
	TH	26768	9
	PA	28874	15
	U	29234	21
229	TH	29600	10
	PA	29901	14
	U	31215	13
	NP	33850	350
230	TH	30882	10
	PA	32186	10
	U	31625	12
	NP	35201	34
231	AC	35930	100
	TH	33825	10
	PA	33440	10
	U	33800	50
	NP	35650	60
232	TH	35463	11
	PA	35951	22
	U	34606	9
	PU	38360	50
233	TH	38747	11
	PA	37503	10
	U	36933	10
	PU	40056	24
234	TH	40640	12
	PA	40377	12
	U	38164	10
	NP	39972	14
	PU	40359	13
235	PA	42330	100
	U	40930	10
	NP	41053	10
	PU	42180	60
236	PA	45560	200
	U	42456	10
	NP	43434	12
	PU	42897	9
237	PA	47700	50
	U	45403	10
	NP	44885	10
	PU	45109	11
238	PA	51290	300
	U	47330	11
	NP	47476	12
	PU	46182	10
	CM	49415	33
239	U	50599	11
	NP	49321	10
	PU	48598	10
	AM	49402	22
240	U	52737	14
	NP	52230	60
	PU	50136	10
	CM	51717	9
241	NP	54330	100
	PU	52966	10
	AM	52946	10
	CM	53717	11
242	PU	54737	11
	AM	55489	12
	CM	54820	10
243	PU	57772	12
	AM	57184	10
	CM	57190	10
	BK	58698	23
244	PU	59825	14
	AM	59891	10
	CM	58462	10
	CF	61470	9
245	PU	63176	32
	AM	61916	11
	CM	61015	11
	BK	61834	12
	CF	63397	12
246	PU	65310	50
	AM	64940	50
	CM	62636	11
	CF	64114	14
247	CM	65550	13
	BK	65495	12
	ES	68573	38
248	CM	67412	14
	CF	67257	32
	FM	71895	22
249	CM	70770	15
	BK	69860	11
	CF	69735	11
	ES	71140	13
250	BK	72963	14
	CF	71188	11
	FM	74087	33
251	CF	74146	13
	ES	74511	13
252	CF	76051	14
	FM	76835	37
	NO	82866	27
253	CF	79331	16
	ES	79030	12
	FM	79365	11
254	ES	82013	14
	FM	80926	12
	NO	84747	37
255	FM	83813	14
256	FM	85510	15
	NO	87810	40
257	FM	88622	16

General References

Popular Books

(These cover some of the material in this book on a level that should be easily understandable to any college student.)

Adler, I.: "Inside the Nucleus," John Day, New York, 1963.
Cohen, B. L.: "The Heart of the Atom," Doubleday, 1967.
Cook, C. S.: "Structure of Atomic Nuclei," Van Nostrand, Princeton, N.J., 1964.
Gamow, G.: "The Atom and Its Nucleus," Prentice-Hall, Englewood Cliffs, N.J., 1961.
Goldwasser, E. L.: "Optics, Waves, Atoms, and Nuclei," Benjamin, New York, 1965.
Romer, A.: "The Restless Atom," Garden City, New York, 1960.
Stearns, R. L.: "Basic Concepts of Nuclear Physics," Reinhold, New York, 1968.

Elementary Textbooks

(These cover essentially the same material as this book with the exception of Chaps. 2 and 15, at roughly the same level of difficulty.)

Arya, A. P.: "Fundamentals of Nuclear Physics," Allyn and Bacon, Boston, 1966.
Burcham, W. E.: "Nuclear Physics: An Introduction," McGraw-Hill, New York, 1963.
Cork, J. M.: "Radioactivity and Nuclear Physics," Van Nostrand, Princeton, N.J., 1957.
Enge, H. A.: "Introduction to Nuclear Physics," Addison-Wesley, Reading, Mass., 1966.
Evans, R. D.: "The Atomic Nucleus," McGraw-Hill, New York, 1955.
Halliday, D.: "Introductory Nuclear Physics," Wiley, New York, 1955.
Howard, R. A.: "Nuclear Physics," Wadsworth, Belmont, Calif., 1963.

Kaplan, I.: "Nuclear Physics," Addison-Wesley, Cambridge, Mass., 1955.
Meyerhof, W.: "Elements of Nuclear Physics," McGraw-Hill, New York, 1967.

Intermediate Textbooks

(These cover the same material as this book except, generally, Chaps. 2, 9, and 15, and most assume that the student has had a complete course in quantum mechanics.)

Bethe, H. A., and P. Morrison: "Elementary Nuclear Theory," Wiley, New York, 1956.

Elton, L. R. B.: "Introductory Nuclear Theory," Saunders, Philadelphia, 1966.

Green, A. E. S.: "Nuclear Physics," McGraw-Hill, New York, 1955.

Orear, J., A. H. Rosenfeld, and R. A. Schluter: "Nuclear Physics: A Course Given by E. Fermi," The University of Chicago Press, Chicago, 1950.

Roy, R. R., and B. P. Nigam: "Nuclear Physics: Theory and Experiment," Wiley, New York, 1967.

Segre, E.: "Nuclei and Particles," Benjamin, New York, 1965.

Advanced Textbooks

(These give much fuller treatments of the theory, generally do not discuss the material in Chaps. 2, 9, and 15, and require a rather thorough knowledge of quantum mechanics.)

Blatt, J. M., and V. F. Weisskopf: "Theoretical Nuclear Physics," Wiley, New York, 1952.

Bohr, A., and B. R. Mottelson: "Nuclear Structure," Benjamin, New York, 1969.

deBenedetti, S.: "Nuclear Interactions," Wiley, New York, 1964.

McCarthy, I. E.: "Introduction to Nuclear Theory," J. Wiley, New York, 1968.

Preston, M. A.: "Physics of the Nucleus," Addison-Wesley, Reading, Mass., 1962.

Sachs, R. G.: "Nuclear Theory," Addison-Wesley, Cambridge, Mass., 1953.

Review Periodicals

(These carry research-level treatments of essentially every topic discussed in this book.)

Annual Reviews of Nuclear Science, Annual Reviews Inc., Stanford, Calif., annually.

Progress in Nuclear Physics, Pergamon, New York, annually.

Reviews of Modern Physics, American Institute of Physics, New York, quarterly.

Reports on Progress in Physics, The Institute of Physics and the Physical Society, London, annually.

Encyclopedic Treatments

Condon, E. U., and H. Odishaw: "Handbook of Physics," 2d ed., McGraw-Hill, New York, 1967.

Flugge: S.: "Handbuch der Physik," Springer-Verlag, Berlin, 1958 (largely in English in spite of the title).

Index